INTERNATIONAL CENTRE FOR MECHANICAL SCIENCES

COURSES AND LECTURES - No. 314

NONLINEAR FRACTURE MECHANICS

EDITED BY

M.P. WNUK
UNIVERSITY OF WISCONSIN

SPRINGER-VERLAG WIEN GMBH

Le spese di stampa di questo volume sono in parte coperte da
contributi del Consiglio Nazionale delle Ricerche.

This volume contains 226 illustrations.

In order to make this volume available as economically and as rapidly as possible the authors' typescripts have been reproduced in their original forms. This method unfortunately has its typographical limitations but it is hoped that they in no way distract the reader.

ISBN 978-3-211-82246-3 ISBN 978-3-7091-2758-2 (eBook)
DOI 10.1007/978-3-7091-2758-2

PREFACE

Linear elastic fracture mechanics (LEFM), as developed by Griffith, Irwin and Orowan, addresses the problem of fracture in idealised class of materials which remain linearly elastic throughout the loading process up to the point of material separation. At that point a new surface is created at the expense of the work done by the external forces applied to the structure or at the expense of the strain energy contained within the structure (or both). The final act of fracture requires a transformation of one form of energy (elastic) into another form (surface). The elastic strain energy is thus converted into the surface energy associated with the work expanded on severing the inter-atomic bonds in the course of crack propagation.

One important term in the energy balance equation which describes this transformation as proposed by Griffith in 1921, is conspicuously missing. This is the dissipation of energy term. In the non-elastic materials, which abound in nature, a substantial dissipation of energy precedes the initiation of crack growth. It also accompanies a moving crack, either quasi-elastic or dynamic, at any level of the external load. The energy so dissipated provides a source of a very significant contribution to the phenomenon of fracture occurring in any real material capable to sustain an irreversible deformation process. Plasticity in metals, crazing in polymers, microcracking within a large process zone associated with fracture developing in cementitious composites, matrix/fiber debonding and bridging effects encountered in fiber reinforced composite materials are the dominant dissipative mechanisms, all of which contribute to the energy term omitted by Griffith.

New developments in the basic mechanics research, aimed at the establishment of a more realistic mathematical model of fracture, which does incorporate the nonlinear behavior of the materials, are summarized and illuminated from various vantage points in this text. The emerging computational techniques, supplemented by a wealth of experimental data, have a significant impact on practical applications of fracture mechanics. With recent advanced numerical technologies such as adaptive finite element and boundary element methods, implementation of the novel nonlinear concepts into the theory and practice which requires a detailed knowledge of fracture in high performance materials and in high reliability components, became possible.

The text begins with the mathematical preliminaries presented by Marek Matczynski (Poland) and concerning the path independent integrals in two and three dimensions. Included in the discussion is the scope and the limitations of one of the most popular ductile fracture criterion of the past two decades, namely the J-integral concept. This section is followed by a presentation of Vladimir Bolotin, a prestigious researcher from Moscow. Professor Bolotin covers a range of topics related to nonlinear material response, including the Kachanov's and Rabotnov's postulates leading to the formulation of the damage evolution equations. Local and global instabilities due to fracture, defined in the Novozhilov's sense, are further discussed by Michael Wnuk (USA). A model is presented through which both the Griffith and the Dugdale concepts are extended and applied to the elasto-plastic range of material behavior. Stojan Sedmak (Yugoslavia) presents and interprets a spectrum of experimental data collected in the series of fracture tests performed in Belgrade and in Boulder, Colorado, at the laboratories of the National Institute of Standards and Technology. The contribution of John Radon (England) follows the suit and provides a wealth of experimental results. Application of the nonlinear fracture mechanics to metals is explained in the lecture by Dietmar Gross (Germany), and to ceramics and cementitious composites by Victor Li (USA). Various aspects of fracture in concrete are discussed by Alberto Carpinteri (Italy) who puts a particular emphasis on the size-scale effects and the induced brittleness phenomena associated with fracture. Finally, Wojciech Szczepinski (Poland) discusses the results of his own experiments using them to illustrate certain concepts of the continuous damage mechanics.

Values of the specific fracture energy and fracture toughness for a number of materials which fall beyond the LEFM range are used throughout the text.

The sequence of topics described here reflects that used in the CISM Short Course on Nonlinear Fracture Mechanics which took place in Udine in September of 1989. The contents of the book follow the sequence determined by the alphabetical order of authors' names.

Michael P. Wnuk

CONTENTS

Page

MECHANICS OF FATIGUE FRACTURE

V. V. Bolotin
USSR Academy of Sciences, Moscow, USSR

ABSTRACT

A survey of the theory of fatigue cracks initiation and propagation up to the final failure suggested by author is presented. The theory is based on the synthesis of macro- and micromechanics of fracture. The analytical mechanics of fracture is used treating a system "cracked body - loading" as a mechanical one with unilateral non-holonomic constraints. Comparison is performed between the generalized forces of the analytical mechanics of fracture and the conventional concepts of fracture mechanics such as stress intensity factors, energy release rates, crack opening displacements, and path-independent integrals.

The central point of the theory of fatigue is the assumption that generalized forces depend essentially on the level of microdamage accumulated near the crack-tips during all the prehistory of loading and cracks growth. The constitutive equations of the theory include conditions of equilibrium and stability of the system as well as equations governing the microdamage accumulation process. Applications of the theory to high-cycle, low-cycle, and corrosion fatigue are given as well as to the cracks growth in visco-elastic media under long-acting quasistationary loading. References to recent studies of fracture and fatigue of composite materials based on the developed theory are presented.

1. INTRODUCTION

Deformation and fracture of solids develop on various levels characterized with the length scales beginning from the scale of crystal lattice and going up to the dimensions of machines and structures' components. At least two scales are to be considered when we remain in the framework of continuum mechanics. The smaller scale is a characteristic size of grains, inclusions, fibers, microvoids, etc. The larger scale corresponds to the sizes of structural components. Respectively, one says on the two branches of the mechanics of solids, i.e. micro- and macromechanics.

Macromechanics of fracture (in the conventional, narrow interpretation) is in fact the mechanics of solid bodies with macroscopic cracks. The latters are usually treated as mathematical cuts in a body. The main problem is to establish stability conditions for the system "cracked body - - loading" with respect to the cracks growth.

Fundamentals of the macromechanics of fracture are given in a number of textbooks and monographs (see, e.g. [1]). The energy balance approach was suggested in pioneering works by Griffith. Irwin about three decades later introduced stress intensity factors and found the relationship between these factors and the energy release rate in the linear elastic body. Several years later Rice developed an approach based on path-independent integrals accomplishing the framework of the linear fracture mechanics. In the further progress, these concepts were extended upon nonlinear elastic, elasto-plastic, visco-elastic and visco-elasto-plastic media. This a rather broad branch of macromechanics is called now nonlinear fracture mechanics.

The first steps in the theory of fatigue can be traced from pure engineering approachs - the Palmgren - Miner rule of fatigue damage summation. The simpliest phenomenological model of dispersed damage accumulation is in fact a continualistic extention of this approach when a damage measure is introduced in each point of a body [1]. Related stochastic models of fracture originating from the Weibull brittle fracture model also enter into the micromechanics of fracture [2]. Another topic of micromechanics of fracture is the theory of microvoids nucleation and coalescence [3].

Although macro- and micromechanics refer to events on different scale levels, they are in a close relation. Novozhilov was among the first who has shown that the synthesis of macro- and micromechanics of fracture has a far perspective [4]. It is essentially significant for the theory of fatigue. It is a common knowledge that the initiation of fatigue cracks is the result of local damage accumulation near the most stressed or the weakest elements of microstructure. Moreover, the interaction between the microdamage near the crack-tip and the energy release due to the crack growth is the principal mechanism governing the behaviour of fatigue cracks. This idea was realized firstly in paper [5] where the condition of global energy balance was used together with the kinetic equation of microdamage accumulation. This approach was extended further both to classic (high-cycle) and low-cycle fatigue [6, 7, 8] , to cracks propagation in visco-elastic media [9], to corrosion fatigue and corrosion cracking [10]. Stability and growth of delaminations in composite materials was considered in [11, 12]. Kinking and branching of fatigue cracks in non-uniaxial and(or) non-proportional loading, as well as near the boundary of two different materials, was also studied in the frame of the proposed theory [13] .

In principle, the synthesis of macro- and micromechanics of fracture may be realized with the use of one of the concepts of common macromechanics of fracture, say, J-integrals or the crack opening displacements. In [5-17] a generalized approach was used called the analytical fracture mechanics. This generalization is applicable to multiparametric problems, in particular, to bodies containing more then one crack or a crack given with two and more independent parameters. Limiting with quasistatic problems, the principle of virtual work becomes to be the basis for the macromechanics of fracture. Since cracks in convential structural materials are "non-healing", irreversible, the generalized coordinates describing cracks size, shape and position can be chosen in such a way that all the coordinates will be non-decreasing functions of time. Hence, we come to mechanical systems with unilateral nonholonomic constraints.

2. FUNDAMENTALS OF THE ANALYTICAL FRACTURE MECHANICS

Let consider a state Σ_0 of a cracked body under a given loading (Fig. 1). Together with the state Σ_0 called undisturbed, introduce a set of neighbouring disturbed states Σ satisfying to the following conditions: time, given surface and volume force as well as given displacements do not subjected to variation; all equilibrium, compatability, heat-transfer and constitutive equations are satisfied in the volume of the body (except maybe the vicinities of crack-tips). From the viewpoint of continuum mechanics both initial and disturbed states are equivalent. The only parameters subjected to variations are cracks parameters. In addition, we assume that both initial and disturbed states are stable in Lyapunov sense when all the cracks parameters are fixed.

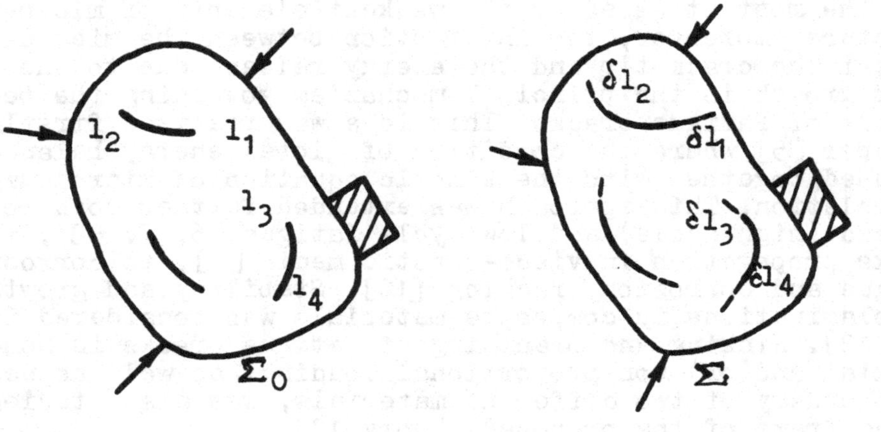

Fig.1. Undisturbed and disturbed states of the system "cracked body-loading"

In paper [5] we called the variations of states described above, in honour of Griffith, the variations in the Griffith sense or, briefly, Griffith variations. Strictly speaking, Griffith had used the energy balance approach and dealt with small actual variations of the crack length. But in fact the concept of virtual work and related variations enter, at least indirectly, in almost all the fundamental studies of conventional fracture mechanics.

When the trajectories of all cracks are known beforehand (for example, from the symmetry considerations), crack sizes take the part of generalized coordinates of the system. Mostly, the number of generalized coordinates is bounded, and small in typical cases. Therefore, we come to finite-degrees-of-freedom systems with unilateral constraints. Equations of continuum mechanics enter into the study on the preparatory stage, when we find the fields of stresses, strains and displacements in the states Σ_0 and Σ. This stage does not belong properly to the mechanics of fracture; the concerned problems are ones of the theories of elasticity, plasticity, etc.

Further on the number of generalized coordinates is assumed limited. Denote the generalized coordinates $l_1, \ldots, l_m$. The set $\tilde{l} = \{l_1, \ldots, l_m\}$ we call the vector of generalized coordinates. The condition of irreversibility of cracks takes the form $\delta l_j \geqslant 0$ where $j = 1, \ldots, m$. Resistance of material against crack propagation we will refer to active forces. Then the constraints on the cracks sizes may be treated as ideal.

Let loading is quasistationary and inertia forces are negligible. Principle of virtual work for systems with ideal unilateral constraints is formulated as follows: a system is in equilibrium then and only then when the summed elementary work of all active forces upon all small arbitrary displacements compatable with the constraints is non-positive:

$$\delta A \leqslant 0. \tag{1}$$

Re-write the virtual work in the form

$$\delta A = \delta A_e + \delta A_i + \delta A_f, \tag{2}$$

where δA_e is virtual work of external forces, δA_i is virtual work of internal forces in the volume of the body except, may be, tip-zones. Virtual work in tip-zones is included into the last term δA_f. This work is connected directly with the cracks propagation, i.e.

$$\delta A_f = -\sum_i \int_{\Lambda_i} \gamma_i | d\underset{\sim}{\sigma}_i \times \delta \underset{\sim}{\lambda}_j | . \tag{3}$$

Here γ_i is specific fracture work, i.e. the amount of work required for formation of the unit of crack surface (the new-born surfaces are counted one time only), $d\underset{\sim}{\sigma}_i$ is a length element of the countour Λ_i of i-th crack, and $\delta\lambda_i$ is an increment of the size of i-th crack. Summation covers all the cracks in the body. Quantities γ_i, generally, vary even in the limits of each crack, as well as depend on crack's sizes and fracture modes.

If the material is dissipative, dissipation of energy is to be included into the work δA_i. Complications occur in the case of hereditary media. As the Griffith variation is isochronic, the initial crack sizes are to be variated in the initial instant $t = 0$. The variations δl_j in following instants $t > 0$ will take into account all the history on the previous time segment $[0, t]$. This approach is illustrated in Fig.2. The initial crack size in the state Σ_0 is denoted l_0, and in instant $t > 0$ is denoted $l(t)$. The initial crack size in the state Σ is taken equal to $l_0 + \delta l_0$ where $\delta l_0 > 0$. Then at $t > 0$ the disturbed crack size is $l(t) + \delta l(t)$, and, generally, $\delta l(t) \neq \delta l_0$. Evidently, these considerations are relative to the two-specimen experimental method for estimation of fracture toughness.

It is expedient to distinguish states for which the virtual work is strictly negative for all $\delta l_j > 0$. We call them sub-equilibrium states. The states for which such variations $\delta l_j > 0$, $j = 1,\ldots,m_1$ exist that $\delta A = 0$, and for remaining variations $\delta A < 0$, are called equilibrium states (with respect to coordinates $l_1,\ldots,l_{m_1}$).

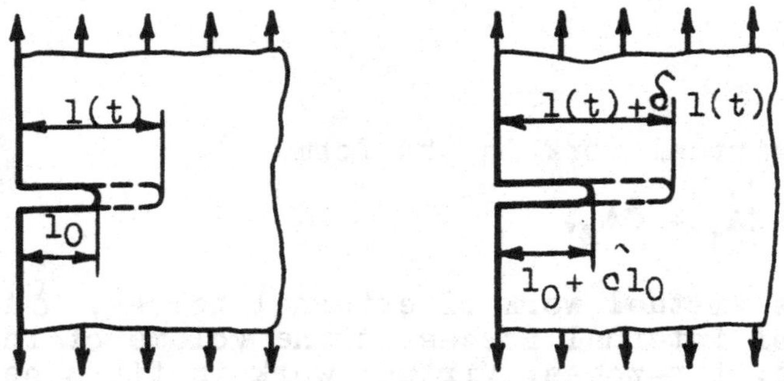

Fig.2. Isochronic variation of a crack in the hereditary material.

Both types of states are equilibriums from the viewpoint of classical mechanics since they satisfy to Eq.(1). If even one variation exists that $\delta A > 0$, i.e. Eq.(1) is violated, we call this state non-equilibrium.

The conditions of stability may be expressed in terms of virtual work also. In particular, the sub-equilibrium states are stable: an additional consumption of energy is needed for transition of the system in any neighbouring state, and there are no such origin in the system. Probably, the term "super-equilibrium" is more suitable for this concept, but we will follow the primary terminology. Non-equilibrium states cannot be realized as equilibrium ones and, therefore, are unstable. Equilibrium states (in Griffith sense) may be both stable and unstable.

To develop the stability conditions for equilibrium states, let introduce the Griffith variation of the virtual work δA, i.e.

$$\delta^2 A \equiv \delta(\delta A) = \sum_{k=1}^{m} \frac{\partial(\delta A)}{\partial l_k}\,\delta l_k . \tag{4}$$

An equilibrium state is stable if the condition $\delta^2 A < 0$ holds for all arbitrary non-vanishing variations δl_j. If such variations exist that $\delta^2 A > 0$, an equilibrium state is unstable. Equilibrium states for which such variations $\delta l_j > 0$ exist that $\delta^2 A = 0$, and for remaining variations $\delta^2 A < 0$, are called neutral states. These states may be critical, e.g. corresponding to a boundary between stability and non-stability, or questionable. In the latter case a study is required of following terms of expansion of δA in power series with respect to δl_j.

This classification of states of cracked bodies was introduced primarily in [5]. It is illustrated in Table, where relationships $\delta A = 0$, $\delta A < 0$, etc., are to be understood in the sense formulated above. The classification has two distinctly different levels: equilibrium and stability. A state enters into one of the classes of the first level (sub-equilibrium, equilibrium and non-equilibrium) depending on the sign of δA. A state is related to one of the classes of the second level (stable, neutral or unstable) depending on signes of δA and $\delta^2 A$.

States of a system "cracked body - loading"				
Sub-equilibrium	Equilibrium $\delta A = 0$			Non-equilibrium
$\delta A < 0$	$\delta^2 A < 0$	$\delta^2 A = 0$	$\delta^2 A > 0$	$\delta A > 0$
Stable		Neutral	Unstable	

By the way, the assumption that all cracks are irreversible is not of primary significance for the theory. We may admit into the consideration reversible, "healing" cracks, too. If for any $\delta l_j < 0$ inequality $\delta A > 0$ holds, it means that the state of the system is unstable with the tendency of corresponding cracks to self-healing. We call such states anti-equilibrium opposite to the non-equilibrium states leading to crack propagation and final failure.

The analogy between the analytical mechanics of fracture and analitical statics was illustrated with a simple model [14]. A heavy body, say, a cylinder is situated on a rigid geared cylindrical surface. The body is restrained against the backword motion by means of a rachet (Fig.3). The condition of restraint is $dl/dt \geq 0$ or, after integration $l(t_2) - l(t_1) \geq 0$ at $t_2 > t_1$. Therefore, the constraint is nonholonomic and unilateral. The state 1 of the system is sub-equilibrium, the state 2 is stable equilibrium. The case 3 corresponds to neutral equilibrium, the state 4 to unstable equilibrium. In case 5 the state of the system is non-equilibrium and, therefore, unstable. The example in Fig 3 is related to the common illustration of the Lagrange - Dirichlet theorem for conservative systems. The main difference is that in the latter case variations δl of both signs are admissible.

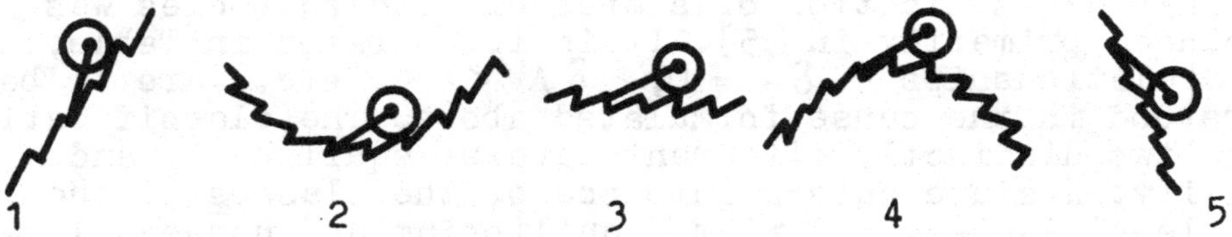

Fig.3. A model illustrating the concepts of the analytical mechanics of fracture. States: 1 - sub-equilibrium, 2 - stable equilibrium, 3 - neutral equilibrium, 4 - unstable equilibrium, 5 - non-equilibrium.

The introduced concepts may be reformulated in terms of generalized forces. Since all the variations are isochronic, the right-hand side of Eq.(3) is a linear form of δl_j. Then

$$\delta A_e + \delta A_i = \sum_{j=1}^{m} G_j \delta l_j, \quad \delta A_f = \sum_{j=1}^{m} \Gamma_j \delta l_j, \tag{5}$$

where multipliers G_j at variations δl_j are the driving (active) generalized forces. Multipliers Γ_j are the resistance (passive) generalized forces. This dividing of forces is rather conditional. For example, dissipative forces in the volume of a body produce a resistance against the crack propagation, but we refer them to active generalized forces. The aim of dividing is to separate the processes localized in tip zones. Together with G_j and Γ_j introduce the resultant generalized forces

$$H_j = G_j - \Gamma_j \quad (j=1,\ldots,m). \tag{6}$$

Let a system "cracked body - loading" is in sub-equilibrium state, i.e. for all δl_j the condition $\delta A < 0$ is fulfilled. With account of Eq.(5) conditions of sub-equilibrium take the form

$$G_j < \Gamma_j \quad (j=1,\ldots,m). \tag{7}$$

The concept of equilibrium requires a special consideration. We say that a system is in the equilibrium state with respect to m_1 generalized coordinates $l_1,\ldots,l_{m_1}$ if the generalized forces satisfy to conditions

$$\begin{aligned} G_j &= \Gamma_j \quad (j=1,\ldots,m_1), \\ G_k &< \Gamma_k \quad (k=m_1+1,\ldots,m). \end{aligned} \tag{8}$$

Equilibrium of multiparametric systems with respect to all generalized coordinates is rather an exception, not counting problems with symmetry of higher order or with the sets of similar cracks. In the most typical situations $m_1 = 1$, i.e. a system is in the equilibrium state with respect to one of the generalized coordinates, and in the sub-equilibrium state with respect to all other coordinates. A system "cracked body - loading" will be in non-equilibrium state, and, therefore, unstable, if only one $\delta l_j > 0$ exists for which $G_j > \Gamma_j$.

The special analysis of stability is to be performed for equilibrium states only. Following the paper [16], the next study is done in terms of generalized coordinates. Let a system is in the equilibrium state with respect to coordinates $l_1,\ldots,l_{m_1}$ and in the sub-equilibrium state with respect to remaining coordinates. Then the latters may be treated as parameters. The second Griffith variation $\delta^2 A$ is a quadratic form of variations of generalized coordinates. Let the generalized forces $H_j = G_j - \Gamma_j$ are differentiable with respect to all $l_1,\ldots, l_{m_1}$. Taking into account Eqs.(5) and (6) we obtain

$$\delta^2 A = \sum_{j=1}^{m_1} \sum_{k=1}^{m_1} \frac{\partial H_j}{\partial l_k} \delta l_j \delta l_k. \tag{9}$$

An equilibrium state is stable if the quadratic form (9) is definitely negative, and unstable if it is definitely positive or sign-indefinite. If the form (9) is non-positive, the study of variations of higher orders becomes necessary.

Let a body is elastic, and external forces are potential. Then the simmetry conditions are satisfied, i.e.

$$\frac{\partial H_j}{\partial l_k} = \frac{\partial H_k}{\partial l_j} \quad (j,k=1,\ldots,m). \tag{10}$$

The form (9) is definitely negative under Cauchy - Sylvester conditions

$$(-1)^n \det\left\| \frac{\partial H_j}{\partial l_k} \right\|_1^n > 0 \quad (n=1,\ldots,m_1). \tag{11}$$

Since cracks are irreversible, it is sufficient to require that $\delta^2 A < 0$ in the first reference ortant only, i.e. at all $\delta l_j > 0$. Therefore the conditions (11) may be relaxed. For example, if $m_1 = 2$ instead of conditions

$$\frac{\partial H_1}{\partial l_1} < 0, \quad \frac{\partial H_1}{\partial l_1}\frac{\partial H_2}{\partial l_1} - \left(\frac{\partial H_1}{\partial l_2}\right)^2 > 0$$

following from Eqs. (11), we arrive to weaker conditions:

$$\frac{\partial H_1}{\partial l_1} < 0, \quad \frac{\partial H_2}{\partial l_2} < 0, \quad \frac{\partial H_1}{\partial l_2} < \left[\frac{\partial H_1}{\partial l_1}\frac{\partial H_2}{\partial l_2}\right]^{1/2}. \tag{12}$$

If external forces are non-potential and/or material is non-elastic, the symmetry conditions may be violated. Then symmetrized terms $1/2\,(\partial H_j/\partial l_k) + 1/2\,(\partial H_k/\partial l_j)$ are to be inserted into Eqs.(11) instead of $\partial H_j/\partial l_k$.

In most typical cases a system is in equilibrium state with respect to a single generalized coordinate. Denoting this coordinate l_1 we will omit the dependence of generalized forces on another coordinates. Let the loading process is given with μ parameters $s_1, \ldots, s_\mu$. The equilibrium condition (8) takes the form $H_1(l_1, s_1, \ldots, s_\mu) = 0$, and the stability condition $\partial H_1(l_1, s_1, \ldots, s_\mu)/\partial l_1 < 0$. Let function $H_1(l_1, s_1, \ldots, s_\mu)$ satisfies to conditions of the theorem on inexplicit function with respect to l_1. Denote $l_1 = L_1(s_1, \ldots, s_\mu)$ the solution of the equilibrium condition. Choosing the loading parameters that $\partial H_1/\partial s_i > 0$ for all i, we arrive to well-known intuitively transparent result:

$$\frac{\partial L_1}{\partial s_i} > 0 \quad (i=1,\ldots,\mu). \tag{13}$$

It means that an equilibrium crack is stable if increasing of loads is needed for the further equilibrium growth of the crack.

3. RELATIONSHIP TO THE CONVENTIONAL FRACTURE MECHANICS

Analytical fracture mechanics is in fact an extension of the conventional (both linear and nonlinear) fracture mechanics upon multiparametric problems. In addition, the area of application of the analytical fracture mechanics is not restricted with the existence of potential energy of a body and (or) loading. Relationship with the concepts of conventional fracture mechanics is illustrated later on special and marginal problems.

Let a crack is one-parametric ($m = 1$), and the system "cracked body - loading" is potential. Then $\delta A_e + \delta A_i = -\delta\Pi$ where Π is the potential energy of the system, and

$$G = -\frac{\partial \Pi}{\partial l} \tag{14}$$

(indices for one-parametric cracks are omited). Evidently, Eq.(14) is the well-known formula for Irwin energy release rate.

If the loading system does not produce work, i.e. $\delta A_e \equiv 0$, and the body is elastic with the potential energy U, then $\delta\Pi = \delta U$. Instead of Eq.(14) we obtain

$$G = -\frac{\partial U}{\partial l}. \qquad (15)$$

In both cases the generalized resistance force Γ is the common fracture toughness measure equal to critical energy release rate G_c. For a plane crack in isotropic linear elastic media and plane strain state we have the Irwin formula

$$G = \frac{1-\nu^2}{E}\left(K_I^2+K_{II}^2\right) + \frac{1+\nu}{E}K_{III}^2. \qquad (16)$$

Here K_I, K_{II}, K_{III} are stress intensity factors, E and ν are Young modulus and Poisson ratio.

The relation to so-called path-invariant integrals is less evident. Virtual work of external forces is

$$\delta A_e = \int_{S_o} p_\alpha \delta u_\alpha dS + \int_{V_o} X_\alpha \delta u_\alpha dV \qquad (17)$$

where p_α is stress vector on the body's surface S_o, X_α is volume force vector in the body's volume V_o (crack-tip zones are excluded), u_α is displacement vector. Virtual work of internal forces is presented, as usually, in the form

$$\delta A_i = -\int_{V_o} \delta W dV, \qquad W = \int_{[o,t]} \sigma_{\alpha\beta} d\varepsilon_{\alpha\beta}. \qquad (18)$$

Here $\sigma_{\alpha\beta}$ and $\varepsilon_{\alpha\beta}$ are stress and strain tensors, and W is equal, with the opposite sign, to the density of internal work produced from the initial instant $t = 0$ up to the considered instant t. Quantities p_α, X_α and $\sigma_{\alpha\beta}$ are to be taken for the undisturbed state, and variations δu_α and δW are to be understood in the

Griffith sense

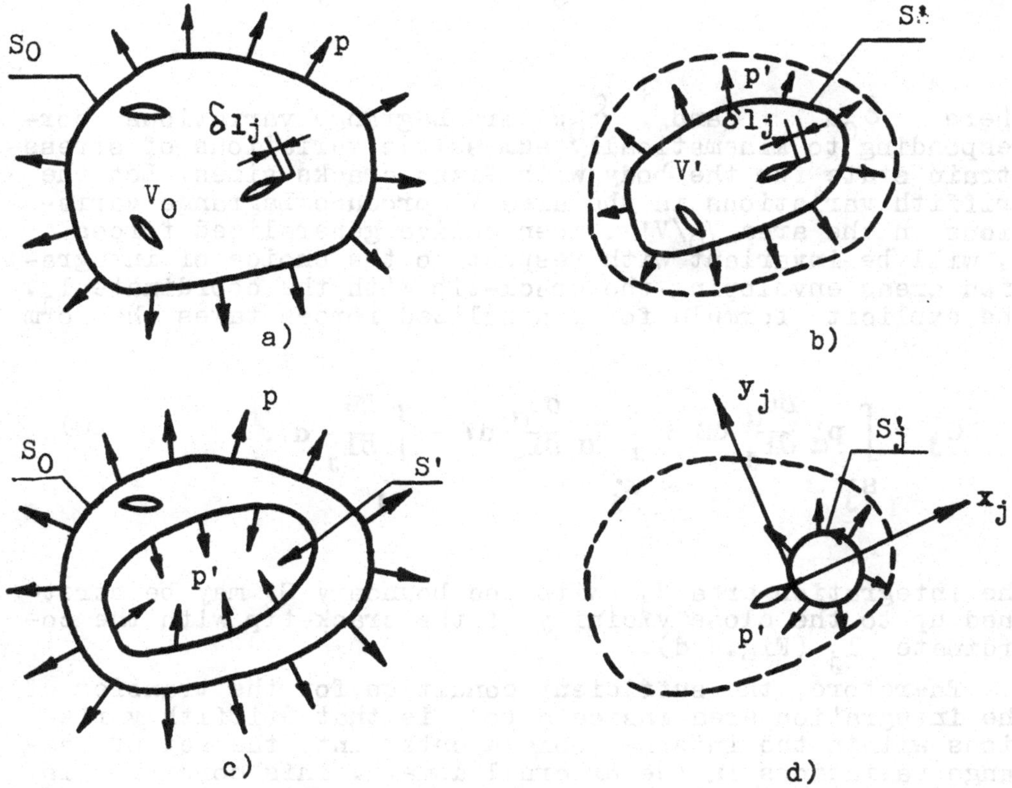

Fig.4. Transfer of integration areas in the calculation of generalized forces: a - given area, b - - transfered area, c - remaining area, d - - relation to J - integral.

Generally, the integration areas from Eqs.(17) and (18) may be deformed continuously up to areas enveloping the crack-tips. Instead of the given system (Fig.4,a) we consider a body with a piece-smooth surface S' situated inside the surface S_0 (Fig.4,b). The embedded volume V' contains the definite crack-tip. Stress vector on S' is denoted $p'_{\alpha\beta}$. The part V_0/V' of the body is in equilibrium. Hence

$$\int_{S_o} p_\alpha \delta_o u_\alpha \, dS + \int_{S'} p'_\alpha \delta_o u_\alpha \, dS + \int_{V_o/V'} X_\alpha \delta_o u_\alpha \, dV - \int_{V_o/V'} \delta_o W \, dV = 0 \quad (19)$$

where $\delta_o u$ and $\delta_o W$ are Lagrange variations corresponding to kinematically admissible variations of stress-strain state for the body with fixed cracks sizes. Let the Griffith variations in the area V' produce Lagrange variations in the area V_o/V'. Then active generalized forces G_j will be invariant with respect to the choice of integration areas enveloping the crack-tip with the coordinate l_j. The explicit formula for generalized forces takes the form

$$G_j = \int_{S'_j} p'_\alpha \frac{\partial u_\alpha}{\partial l_j} dS + \int_{V'_j} X_\alpha \frac{\partial u_\alpha}{\partial l_j} dV - \int_{V'_j} \frac{\partial W}{\partial l_j} dV. \quad (20)$$

The integration area V'_j with the boundary S'_j may be stretched up to the close vicinity of the crack-tip with the coordinate l_j (Fig.4,d).

Therefore, the sufficient condition for the transfer of the integration area inside a body is that Griffith variations within the internal domain enter into the set of Lagrange variations in the external domain. This condition is satisfied for elastic (not necessary linear) bodies and not connected with the potentiality of loading. However, the condition might be violated if material in the external domain is deformed plastically.

The generalized force given with Eq.(20) is relative to J-integral. Let volume forces are absent and the plane stress or strain state is realized approximately near the crack-tip. Denote C_j a cross-section of S_j with a plane orthogonal to the crack front, and Ω_j is the area enclosed with C_j. The local reference axis Ox_j is directed along the external normal to the front (Fig.4,d). Then the force G_j per length unit of the crack front is

$$G_j = \int_{C_j} p'_\alpha \frac{\partial u_\alpha}{\partial l_j} ds - \int_{\Omega_j} \frac{\partial W}{\partial l_j} d\Omega. \tag{21}$$

Let replace the variation with respect to l_j with moving of the countour C_j on the non-variated crack. It means the substitution

$$\frac{\partial}{\partial l_j} = -\frac{\partial}{\partial x_j}. \tag{22}$$

Using Eq.(22) and transforming the second integral in Eq. (21) results in the well-known J-integral:

$$J_j = \int_{C_j} \left[W dy_j - p'_\alpha \frac{\partial u_\alpha}{\partial x_j} ds \right]. \tag{23}$$

Eq.(22) does not mean, generally, an identical transformation. If the size of the tip zone is to be variated together with the size of the crack, Eq.(22) becomes invalid. The simplest example presents the thin plastic zone model by Leonov - Panasyuk - Dugdale (Fig.5). Let 21 is length of a Mode I crack, λ is length of the tip zone, σ_∞ is nominal stress, and σ_0 is ultimate stress assumed constant within the zone $l \leq |x| \leq l + \lambda$. The tip zone length and crack opening are given with formulas [18] :

$$\lambda = l \left[\sec\left(\frac{\pi\sigma_\infty}{2\sigma_o} \right) - 1 \right], \quad \delta = \frac{8\sigma_\infty l}{\pi E} \ln \sec\left(\frac{\pi\sigma_\infty}{2\sigma_o} \right). \tag{24}$$

Applying Eq.(21) to this model, we transfer S' directly to the boundary of the plastic zone assumed as infinitesimally thin. Then

$$G = 2\sigma_o \int_{l}^{l+\lambda} \frac{\partial v}{\partial l}\, dx.$$

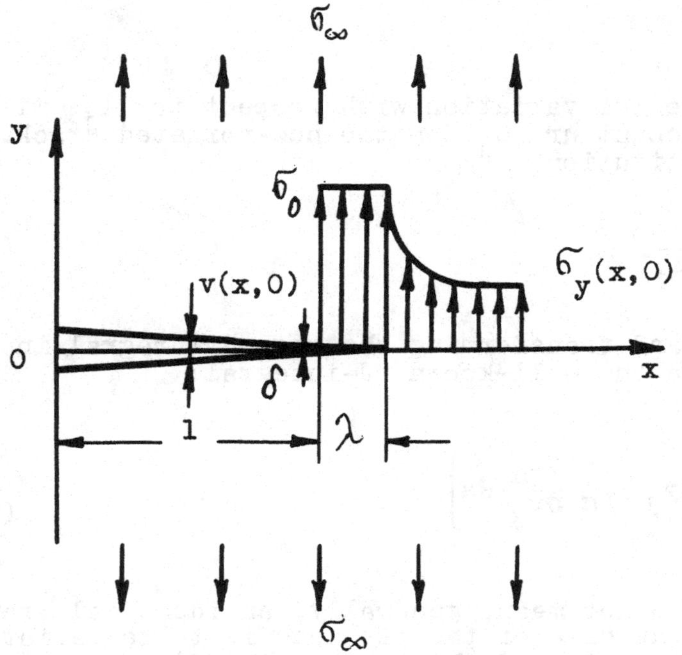

Fig.5. Thin plastic tip zone model

Substitution of Eqs.(24) and the displacement v(x,0) from the solution of Leonov - Panasyuk - Dugdale problem gives:

$$G = \frac{8\sigma_\infty^2 l}{\pi E}\left[\ln \cos\left(\frac{\pi\sigma_\infty}{2\sigma_o}\right) + \frac{\pi\sigma_\infty}{2\sigma_o}\, \mathrm{tg}\left(\frac{\pi\sigma_\infty}{2\sigma_o}\right)\right]. \tag{25}$$

Details of calculations can be found in [18, 19] where they are performed in another context. Under similar assumptions Eq.(23) results in

$$J = -2\sigma_o \int_{l}^{l+\lambda} \frac{\partial v}{\partial x}\, dx = \frac{8\sigma_\infty^2 l}{\pi E} \ln \sec\left(\frac{\pi\sigma_\infty}{2\sigma_o}\right). \tag{26}$$

Comparison of results given with Eqs.(25) and (26) is presented in Fig.6. Here $G_0 = 8\sigma_\infty^2 l/(\pi E)$. The line 1 is plotted according the Irwin formula $G = \pi\sigma_\infty^2 l/E$, the lines 2 and 3 depict generalized forces given with Eqs.(26) and (25), respectively. The difference between G and J-integral is of the same order of magnitude as between J-integral and the approximation of the linear fracture mechanics.

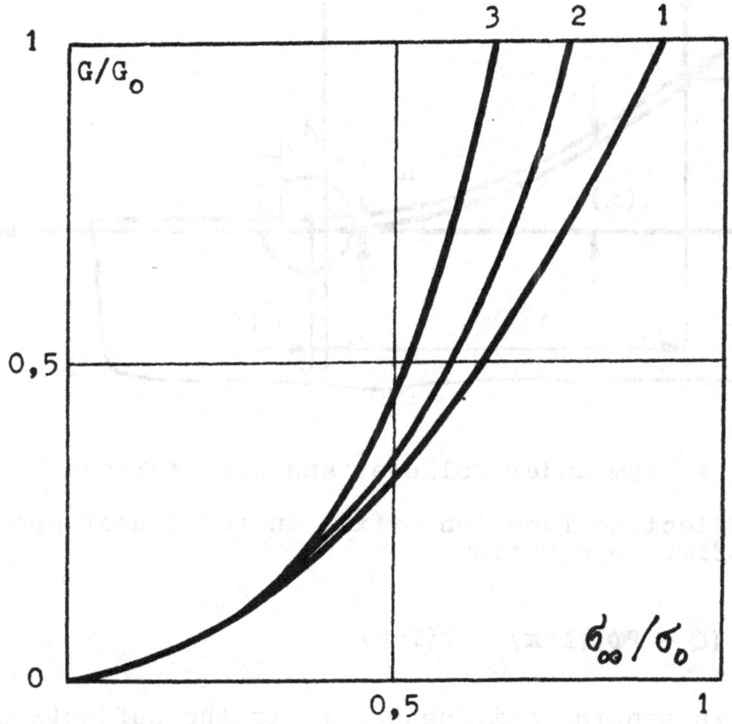

Fig.6. Generalized forces for the thin plastic zone model: 1 - elastic energy release rate, 2 - J-integral, 3 - according to Eq.(25).

It is known [20] that non-existence of potential of external forces effects significantly on stability (instability). Stability of cracks under non-potential loading was studied in paper [15] . Let discuss the simplest example from this paper.

A thin elastic beam (or a plate under cylindrical bending) is clamped at one end. The other end is subjected to longitudinal follower force P and "dead" transversal force Q (Fig.6). These forces as well as other parameters are **refered** to the unity of the beam width. Shear and axial deformations are neglected due to the small thickness ratio h/l. Eventual crack direction is the plane $x > 0$, $y = 0$ dividing the beam and the half-space $y < 0$.

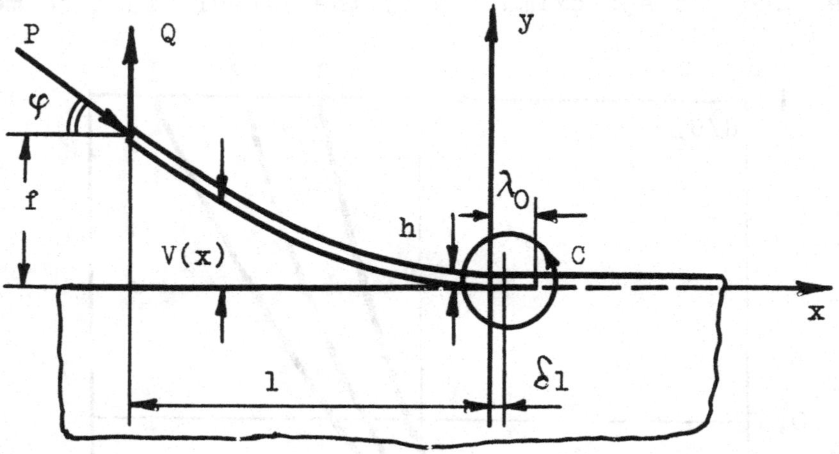

Fig.7. A beam under follower and dead forces.

The deflection function $v(x)$ in the linear approximation satisfies to equation

$$Dv'' = (Q - P\varphi)(l+x) + P(f-v)$$

where D is bending stiffness, f is the deflection on the end of the beam $x = -l$, φ is the rotation angle at the end. Boundary conditions are $v(0) = 0$, $v'(0) = 0$, $v(-l) = f$, $v'(-l) = -\varphi$. Taking $P > 0$ in compression we get $(\beta^2 = P/D)$:

$$v(x) = \frac{Q}{\beta P}\left[-\sin\beta(l+x) + \beta x \cos\beta l + \sin\beta l\right]. \qquad (27)$$

Potential energy of bending is

$$U = \frac{Q^2 l}{4P}\left(1 - \frac{\sin 2\beta l}{2\beta l}\right).$$

Virtual work of external forces $\delta A_e = (Q+P\varphi)(\partial\varphi/\partial l) - P\delta u(-l)$ takes into account the work of the force P on the longitudinal displacement $\delta u(-l)$. The latter is equal to

$$u(-l) = \frac{Q^2 l}{4P}\left[1 + 2\cos^2\beta l - \frac{3\sin 2\beta l}{2\beta l}\right].$$

Substituting these results into relation $\delta A_e + \delta A_i = G\,\delta l$ we obtain the generalized force:

$$G = \frac{Q^2}{2P}\sin^2\beta l. \tag{28}$$

The same formula can be derived using J-integral approach. The integration path C is choosen as shown in Fig.7. In the assumption of this problem $\sigma_x = Dv''(0)\cdot[(h/2) - y]$, $0 \leq y \leq h$; $\varepsilon = \partial u/\partial x = \sigma_x(1 - \nu^2)/E$. Eq.(23) takes the form $J = [v''(-l)]^2/2D$. Substitution of Eq.(27) results into Eq.(28).

Denote with γ the specific tearing work of the beam from the half-space $y < 0$. Then the generalized resistance force $\Gamma = \gamma$. The condition $G = \Gamma$ leads to the following formula for equilibrium sizes of the delaminated beam:

$$\frac{l}{h} = \frac{1}{\sqrt{|p|}}\arcsin\left[\frac{\sqrt{|p|}}{q}\right]. \tag{29}$$

Here notations are introduced $p = Ph^2/D$, $q = Qh(2D\gamma)^{-1/2}$, $p \leq q^2$. By the way, Eq.(29) remains valid if the beam is subjected to tension with the follower force, i.e. $P < 0$. Then

$$\frac{l}{h} = \frac{1}{\sqrt{|p|}}\operatorname{arc\,sh}\left[\frac{\sqrt{|p|}}{q}\right]. \tag{30}$$

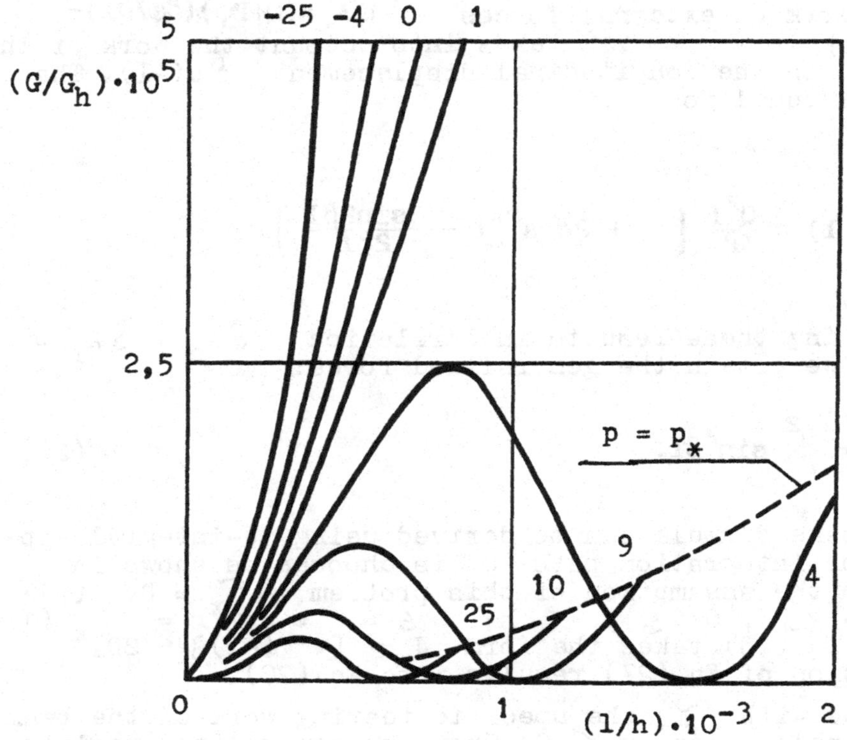

Fig.8. Driving generalized force against the length the delaminations at various load levels. Numbers at **lines** are equal to $p \cdot 10^6$, $p < 0$ for tension.

A diagram for the generalized force G according to Eq.(28) is presented in Fig.8. Here $G_h = Q^2h^2/(2D)$. When the force P is positive, the dependence of G on P and l is non-monotonous. The graphs are terminated in the points corresponding to the critical value of the follower force according to the condition of dynamic stability. If damping proportional to $\partial v/\partial t$ approaches to zero, the critical force $P_* \approx 20.05\ D/l^2$. In notations of Eq.(29) it means that $p_* \approx 20.05\ (h/l)^2$. Due to restriction $p < p_*$ only one of branches of Eq.(29) corresponds to equilibrium state. It is illustrated in Fig.9 where relationship between l/h and p is plotted for fixed values of q.

If l/h is sufficiently small, the system is in a sub-equilibrium state. The curves plotted according to Eqs. (29) and (30) correspond to equilibrium states. Regions of non-equilibrium states are dashed, and that of dynamic instability is double-dashed. Regions of sub-equilibrium

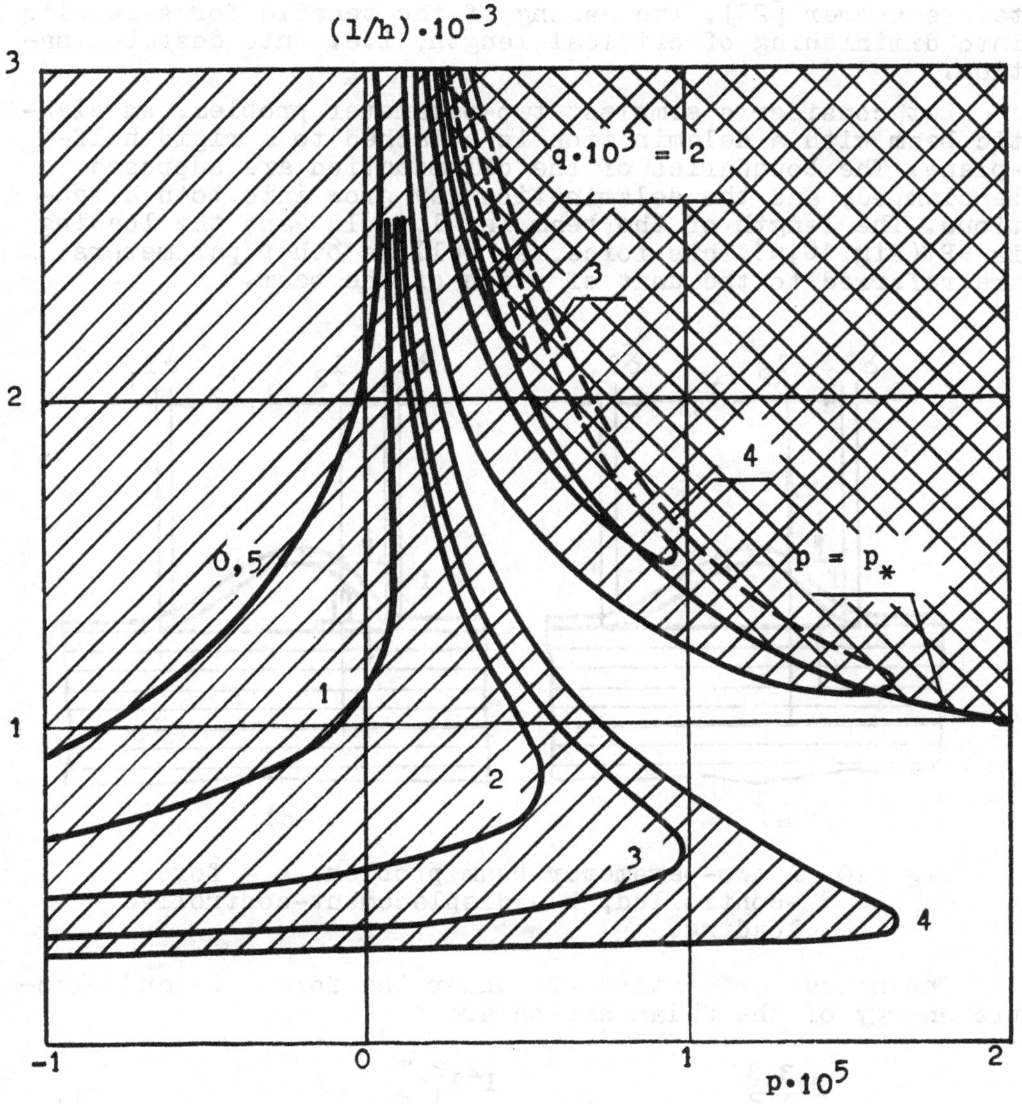

Fig.9. Stability chart for the beam under follower and dead forces. Regions of instability with respect to delamination are dashed, the region of dynamic instability is double-dashed. Follower force is assumed positive in compression.

states remain blank. Transfers from stability to instability and vice versa under compressive force P are rather complicated. Intuition, as it is common in nonconservative problems of stability, does not present here even a qualitative answer [20]. Increasing of the tensile force results into diminishing of critical length, i.e. into destabilisation.

Let consider a simplest two-parameter problem. An elastic beam with a delamination is attached to a rigid half-plane. The boundaries of the delamination are supposed to be clamped, and the delamination may grow into both directions. The length of the beam is $l_1 + l_2$ and the loading is P (Fig. 10,a). This force, as well as other parameters, are referred to the unit of width of the beam.

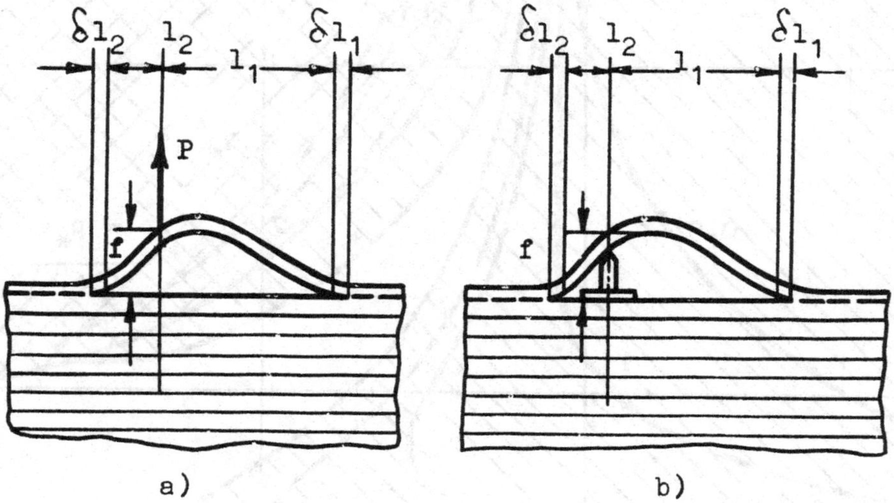

Fig. 10. A two-parameter beam problem: a - force-controlled, b - displacement-controlled loading.

The normal deflection f under the force P and elastic energy of the delamination are

$$f = \frac{P l_1^3 l_2^3}{3D(l_1 + l_2)^3}, \qquad U = \frac{P^2 l_1^3 l_2^3}{6D(l_1 + l_2)^3} \tag{31}$$

where D is bending stiffness of the beam. Virtual work is $\delta A = -\delta U + \delta A_p - \gamma_1 \delta l_1 - \gamma_2 \delta l_2$ where $\delta A_p = P\,\delta f = -2\delta U$, and, generally, $\gamma_1 \neq \gamma_2$. Using (31) we obtain

$$G_1 = \frac{P^2 l_1^2 l_2^4}{2D(l_1 + l_2)^4}, \quad G_2 = \frac{P^2 l_1^4 l_2^2}{2D(l_1 + l_2)^4}. \qquad (32)$$

The resistance generalized forces are $\Gamma_1 = \gamma_1$, $\Gamma_2 = \gamma_2$. If l_1 and l_2 are sufficiently small, $G_1(l_1, l_2, P) < \Gamma_1$. The system is in an equilibrium state with respect to l_1 at $G_1(l_1, l_2, P) = \Gamma_1$, and with respect to l_2 at $G_2(l_1, l_2, P) = \Gamma_2$. Equilibrium dimensions l_1 and l_2 are plotted against P in Fig.11,a. The surface AA_1D corresponds to equilibrium states with respect to l_1, and the surface BB_1D with respect to l_2. Points of the line CD correspond to the states which are in equilibrium with respect to both generalized coordinates. For this line $(l_2/l_1) = (\gamma_1/\gamma_2)^{1/2}$. The surface ACBD separates the regions of sub-equilibrium and non-equilibrium.

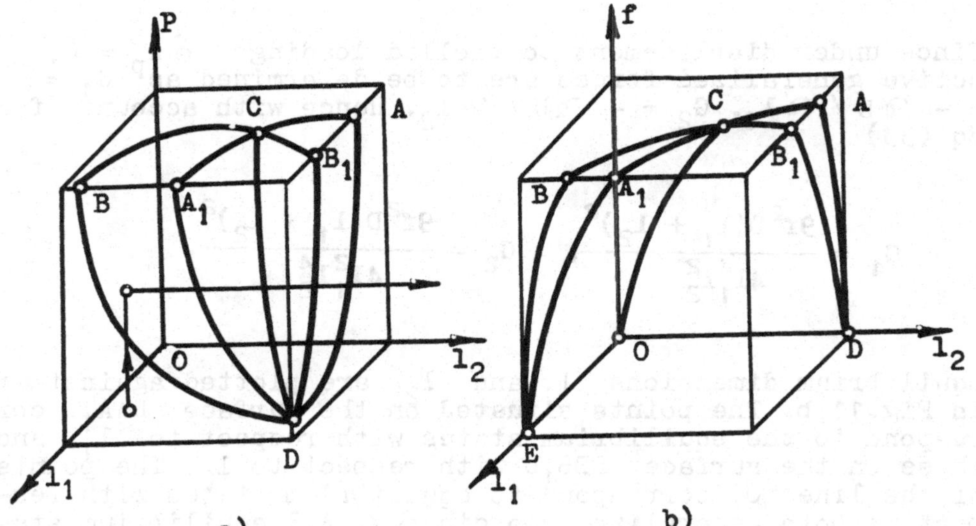

Fig.11. Surfaces of equilibrium states: a - force--controlled, b - dispacement-controlled loading.

To study stability of equilibrium states, Eqs.(12) are to be applied. Substituting into Eqs.(12) the derivatives $\partial H_1/\partial l_1$, $\partial H_2/\partial l_2$ and $\partial H_1/\partial l_2 = \partial H_2/\partial l_1$

with account of Eqs.(32), we see that all equilibrium states are unstable. A case of gradually increasing force P is shown in Fig.11,a. When the equality $G_2(l_1, l_2, P) = \Gamma_2$ primarily becomes attained, the front of the delamination begins to propagate rapidly with respect to l_2.

A related problem is shown in Fig.10,b where the displacement-controlled (rigid) loading is considered. Eqs.(31) for deflection f and elastic energy U remain valid. But now the parameter of loading is not the force P but the deflection f. This effects essentially both on the equilibrium and stability conditions.

Elastic energy of the delamination considered as a function of f takes the form

$$U = \frac{3f^2 D(l_1 + l_2)^3}{2l_1^3 l_2^3}. \qquad (33)$$

Since under displacement-controlled loading $\delta A_p = 0$, active generalized forces are to be determined as $G_1 = -\partial U / \partial l_1$, $G_2 = -\partial U / \partial l_2$. Hence with account of Eq.(33)

$$G_1 = \frac{9f^2 D(l_1 + l_2)^2}{4l_1^4 l_2^2}, \quad G_2 = \frac{9f^2 D(l_1 + l_2)^2}{4l_1^2 l_2^4}.$$

Equilibrium dimensions l_1 and l_2 are plotted against f in Fig.11,b. The points situated on the surface DAA_1E correspond to the equilibrium states with respect to l_1, and those on the surface EBB_1D with respect to l_2. The points of the line OC correspond to equilibrium states with respect to both generalized coordinates. All equilibrium states are stable. Let the initial state with fixed l_1 and l_2 is stable. When under the increasing deflection f an equilibrium state is primarily attained, the further stable growth takes place following to one of the surface DB_ICO or EA_ICO.

4. MECHANICS OF FATIGUE

The theory of fatigue suggested primarily in [5] is based on the synthesis of the analytical macromechanics of fracture and the micromechanics. The main assumption introduced in the paper [5] was that a fatigue crack propagates in a stable way if the conditions of equilibrium and stability in Griffith sense are satisfied taking into account the influence of microdamage on the specific work of fracture. The further development and applications are given in [6 - 17, 28 - 40]. Systematic presentation of the theory can be found in the Chapter 9 of the book [2].

Micromechanics of fracture originates from the known studies on creep fracture, e.g. [1]. A phenomenological scalar measure of damage was used in earlier publications interpreted as specific diminishment of the cross section of a specimen due to formation of microvoids and microcracks. Later on tensor measures were introduced taking into account, in particular, the angular distribution of microcracks. Special models of microdamage for fiber composites were suggested including ruptures of fibers, damage of the fiber -matrix interface as well as combined damage modes of composite materials [2].

Typical equations of microdamage accumulation may be written as follows:

$$\underset{\sim}{\omega}(\underset{\sim}{r},t) = \underset{\tau=o}{\overset{\tau=t}{\Omega}} \{\underset{\sim}{\sigma}(\underset{\sim}{r},t),\underset{\sim}{\varepsilon}(\underset{\sim}{r},t)\},$$

$$\underset{\sim}{\varepsilon}(\underset{\sim}{r},t) = \underset{\tau=o}{\overset{\tau=t}{\Omega}} \{\underset{\sim}{\sigma}(\underset{\sim}{r},t),\underset{\sim}{\omega}(\underset{\sim}{r},t)\}.$$

Here $\underset{\sim}{r}$ is reference vector, t is time, $\underset{\sim}{\sigma}$ and $\underset{\sim}{\varepsilon}$ are stress and strain tensors. The object $\underset{\sim}{\omega}$, generally, is of tensorial nature, e.g. a tensor of second rank. Hereditary functionals $\Omega\{...\}$ and $E\{...\}$ account for the effect of history on stress, strain and microdamage states of the media.

Let the components of the object $\underset{\sim}{\omega}(r, t)$ in vicinities of the crack-tips and on their potential prolongation are characterized with a set of parameters called microdamage vector $\underset{\sim}{\varphi} = \{\varphi_{1},..., \varphi_{\nu}\}$. Level of microdamage close at the crack-tips is given with vector $\underset{\sim}{\psi} = \{\psi_1,..., \psi_{\nu}\}$. Let $l(t)$ is vector process of crack growth, and vector $\underset{\sim}{L}$ relates to the points on the prolongation of cracks. The identity holds

$$\underset{\sim}{\psi} = \underset{\sim}{\varphi}[l(t),t]. \tag{34}$$

Both active and passive generalized forces depend on the level of microdamage accumulated near the crack-tips. A body containing fatigue cracks can be in sub-equilibrium, equilibrium or non-equilibrium states with respect to the generalized coordinate l_j depending on the sign in the relation

$$G_j[\underset{\sim}{l}(t),\underset{\sim}{s}(t),\underset{\sim}{\phi}(t)] \lessgtr \Gamma_j[\underset{\sim}{l}(t),\underset{\sim}{s}(t),\underset{\sim}{\phi}(t)]. \tag{35}$$

Both sides in Eq.(35), generally, depend on the loading vector $\underset{\sim}{s}(t) = \{s_1(t),\ldots, s_\mu(t)\}$. Among the components of $\underset{\sim}{s}(t)$ may be given forces, displacements, heat flows, concentrations of active agents, etc. The first of Eqs.(24) takes the form

$$\underset{\sim}{\varphi}(L,t) = \underset{\tau=o}{\overset{\tau=t}{\Phi}} \{\underset{\sim}{l}(\tau),\underset{\sim}{s}(\tau),\underset{\sim}{L}\}. \tag{36}$$

Here $\Phi\{\ldots\}$ is a functional of the history of loading and cracks growth.

To estimate microdamage at the crack-tips, we have to assume finite stresses, and therefore, either finite curvature radii at the tips or small processing zones ahead of the tips. Denote the set of these new scale parameters $\underset{\sim}{\rho} = \{\rho_1,\ldots, \rho_m\}$. Generally, $\underset{\sim}{\rho}(t)$ depends on the history. Connecting $\underset{\sim}{\rho}(t)$ with the microdamage at the tips, we introduce the equation similar to Eq.(36):

$$\underset{\sim}{\rho}(t) = \underset{\tau=t}{\overset{\tau=t}{R}} \{\underset{\sim}{\phi}(\tau)\}. \tag{37}$$

Discreet argument N is to be introduced in the study of cyclic fatigue equal to the number of a cycle or block of loading. Later on for simplification we use the term "cycle" only. Conditions on δA may be expressed using the upper bounds of differences $G_j - \Gamma_j$ attained during N-th cycle, i.e.

$$H_j(N) = \sup_{t_{N-1} \leqslant t < t_N} \{G_j[\underset{\sim}{l}(t),\underset{\sim}{s}(t),\underset{\sim}{\phi}(t)] - \Gamma_j[\underset{\sim}{l}(t),\underset{\sim}{s}(t),\underset{\sim}{\phi}(t)]\}. \tag{38}$$

Here $[t_{N-1}, t_N)$ is time segment corresponding to N-th cycle. A system "cracked body - loading" is staying in the sub-equilibrium state during N-th cycle if all $H_j(N) < 0$, and becomes unstable if even one of quantities $H_j(N) > 0$. Equality $H_j(N) = 0$ holds for cracks which are in the equilibrium state with respect to l_j.

Let a system is in the equilibrium state with respect to m_1 generalized coordinates. In terms of generalized forces the Griffith variation $\delta(\delta A)$ of virtual work δA is

$$\delta^2 A = \sum_{j=1}^{m_1} \sum_{k=1}^{m_1} \frac{\partial H_j}{\partial l_k} \delta l_j \delta l_k + \sum_{j=1}^{m_1} \sum_{i=1}^{\nu} \frac{\partial H_j}{\partial \Phi_i} \delta l_j \delta \Phi_i . \tag{39}$$

The right-hand side of Eq.(39) contains the isochronic variations $\delta\psi_j$ of the microdamage measures on the moving crack-tips.

The conditions on generalized forces are to be complemented with an equation of microdamage accumulation on the cracks trajectories. Analogously to Eq.(36) we assume

$$\underset{\sim}{\varphi}(\underset{\sim}{L},N) - \underset{\sim}{\varphi}(\underset{\sim}{L},N-1) = \underset{n=1}{\overset{n=N}{\Phi}} \{\underset{\sim}{l}(n), \underset{\sim}{s}(n), \underset{\sim}{L}\}. \tag{40}$$

Let at $t=0$ the system "cracked body - loading" is in the sub-equilibrium state. Then at certain $t > 0$

$$H_j(N) < 0 \quad (j=1,\dots,m). \tag{41}$$

The first violation of these inequalities means the beginning of growth at least of one of the cracks, e.g. the end of the initiation (incubation) stage. Denote with N_* the corresponding cycle number. Usually, only one of $H_j(N)$ attains zero at $N = N_*$ while the other generalized forces satisfy to Eqs.(41). Let it is l_k. The further development of the crack depends on stability of the attained equilibrium state. If $\delta^2 A < 0$, the crack shall be growing continuously with respect to l_k along with the micro-damage accumulation on the moving crack-tip. If $\delta^2 A > 0$, a jump with respect to l_k shall take place. It means either the final failure or a transfer to a new sub-equilibrium state with respect to l_k.

Generally, the condition of jump-like growth with respect to generalized coordinate l_k during N-th cycle is $H_k(N) > 0$. The magnitudes $l_{k*}(N)$ and $l_k^*(N)$ of l_k immediately before and after a jump, respectively, satisfy to the relationship

$$\int_{l_{k*}(N)}^{l_k^*(N)} H_k[\underset{\sim}{L}, \underset{\sim}{s}(t_*), \underset{\sim}{\varphi}(L, t_*)] dL_k \geq 0 \tag{42}$$

which follows from the energy conservation law. Here the crack propagation is considered as an instant process at $t = t_*$, and t_* is time moment when supremum in the right-hand side of Eq.(38) is attained. Neglecting the loss of energy during the jump by heat, acoustic radiation, etc., we obtain the equation for the new value of the generalized coordinate l_k :

$$\int_{l_{k*}(N)}^{l_k^*(N)} G_k[\underset{\sim}{L}, \underset{\sim}{s}(t_*), \underset{\sim}{\varphi}(\underset{\sim}{L}, t_*)] dL_k = \int_{l_{k*}(N)}^{l_k^*(N)} \Gamma_k[\underset{\sim}{L}, \underset{\sim}{s}(t_*), \underset{\sim}{\varphi}(\underset{\sim}{L}, t_*)] dL_k . \tag{43}$$

Therefore, the type of fatigue crack propagation depends on the stability conditions. For illustration, consider an one-parametric crack with the size l in stationary loading, say, under constant amplitudes and average nominal stresses. Let G depend on l and s, and Γ on $\underset{\sim}{\varphi}$ only. Various situations take place depending on the character of functions $G(l,s)$ and $\Gamma(\underset{\sim}{\varphi})$.

One of the situations is shown in Fig.12. The equilibrium firstly attained at N_* is stable. Since Γ increases more rapidly with l than G, the state of the system remains to be stable at $N > N_*$, too. The loss of stability occurs at $N = N_{**}$. The critical size l_{**}, generaly, is less than the critical size l_c determined from the equation $G = \Gamma_0$ and corresponding to short-time monotonic loading.

Fig.12 was plotted for the case when the active generalized force G grows with the crack size l . The typical example is the Mode I crack in an unbounded elastic body. If the body is loaded with two tearing forces applied to the crack surface, the force G will be a decreasing function of l (Fig.13). All equilibrium states of the system are stable.

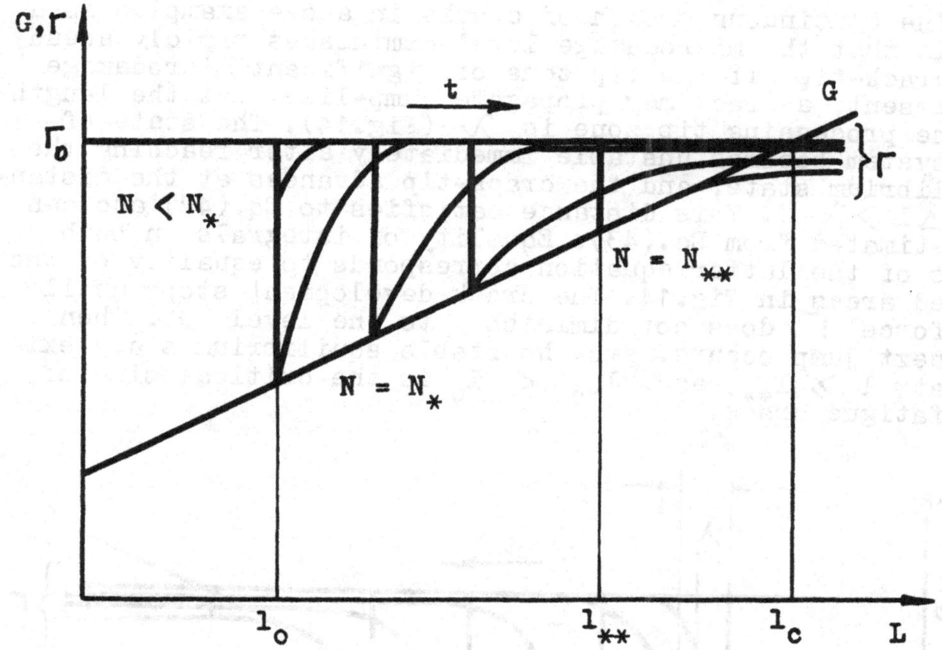

Fig.12. Balance between generalized forces before and during fatigue crack growth

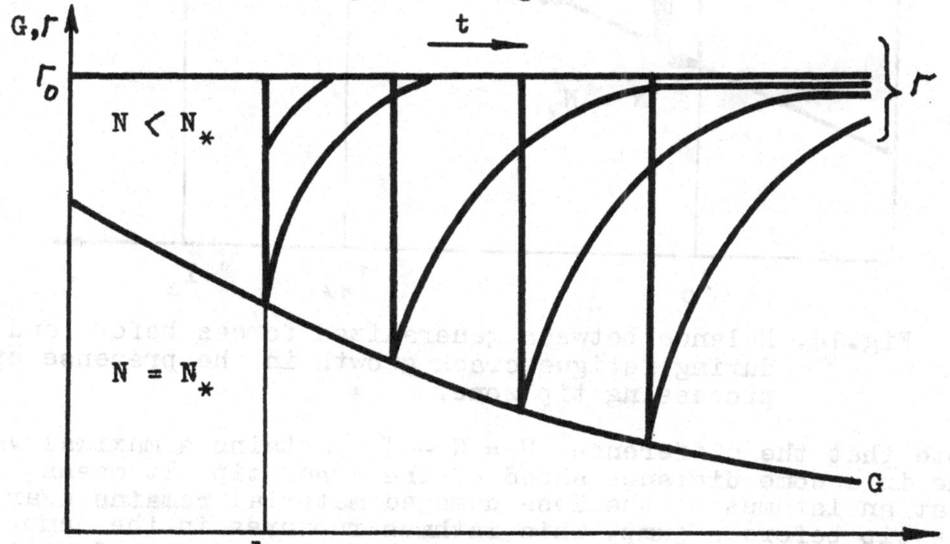

Fig.13. The same as in Fig.12 in the case when G is diminishing with l .

The continuous growth of cracks in above examples is a result that the microdamage level diminishes rapidly ahead the crack-tip. If the tip zone of significant microdamage is present, a crack may propagate jump-like. Let the length of the processing tip zone is λ (Fig.14). The state of the system becomes unstable immediately after reaching the equilibrium state, and the crack-tip advances at the distance $\Delta l > \lambda$. This distance satisfies to Eq.(42) and can be estimated from Eq.(43). Equality of integrals in both parts of the latter equation corresponds to equality of the dashed areas in Fig.14. The crack development stops until the force Γ does not diminish to the level G. Then the next jump occurs, etc. No stable equilibrium state exists at $l \geqslant l_{**}$, and $l_{**} < l_c$ is the critical size of the fatigue crack.

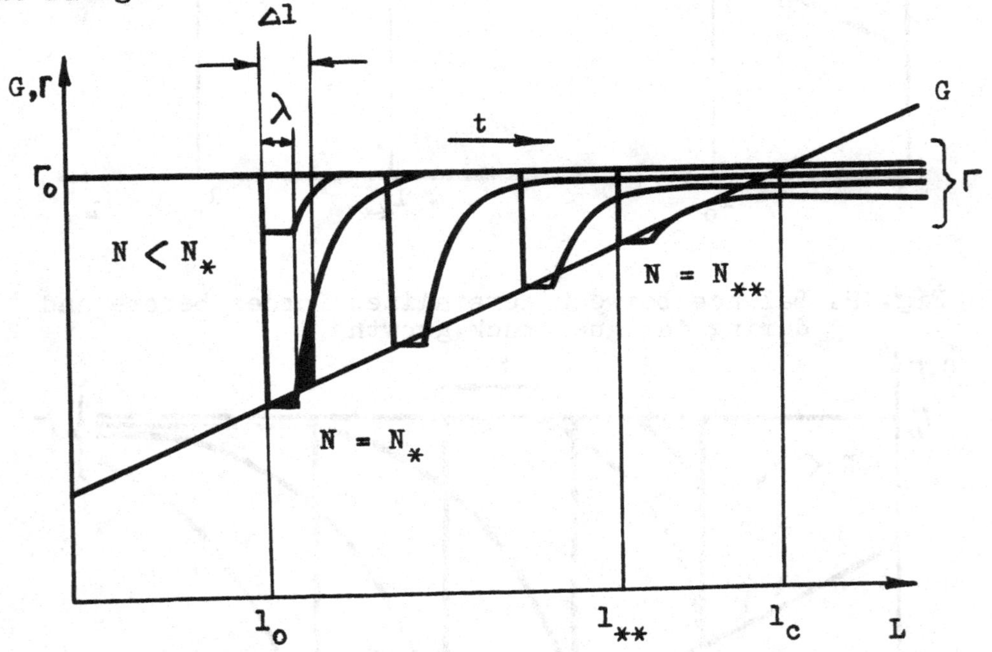

Fig.14. Balance between generalized forces before and during fatigue crack growth in the presense of processing tip zone.

Note that the difference $H = G - \Gamma$ attains a maximal value in a some distance ahead of the crack-tip. It means that an isthmus of the less damaged material remains near the tip before a jump. This isthmus ruptures in the jump instant. If the active generalized force G diminishes rather rapidly with l, but a tip zone exists (Fig.

15), the continuous crack growth takes place, as in cases shown in Fig.12 and 13.

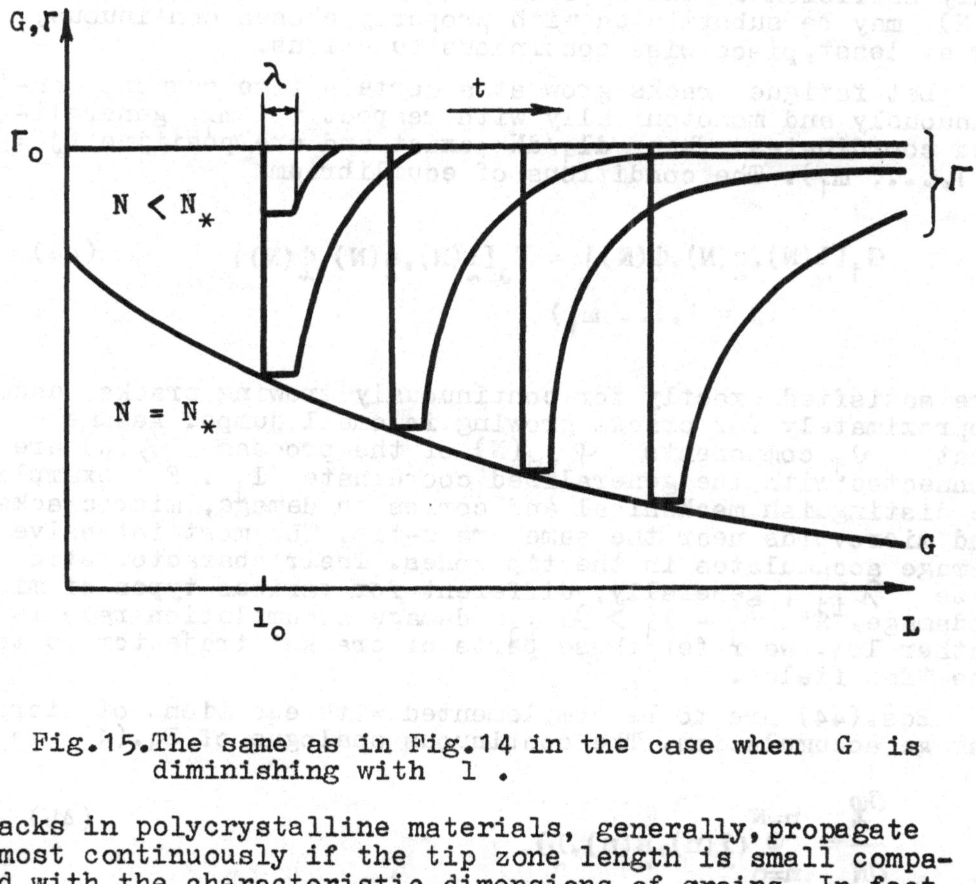

Fig.15.The same as in Fig.14 in the case when G is diminishing with l .

Cracks in polycrystalline materials, generally, propagate almost continuously if the tip zone length is small compared with the characteristic dimensions of grains. In fact, a crack crosses in its development a large number of randomly distributed grains. The front of the crack acts as a kind of "averager" upon the set of grains. But the jump-like propagation becomes more distinctive when the length of the tip zone becomes much larger than the characteristic dimension of grains.

5. QUASISTATIONARY APPROXIMATION

Without account to final failure and large jumps due to the sudden increasing of loads and/or decreasing of material's resistance, fatigue cracks grow rather slowly. Propagation of a crack of conventional (high-cycle) fatigue upon a practically significant length, say, 1 mm requires 10^3 and more cycles of loading. Even for low-cycle fatigue, the cycle number N is usually approximately treated as a continuous argument similar to time t. If loading parameters vary sufficiently slowly from one cycle to another, $l(N)$, $s(N)$ may be substituted with properly chosen continuous, or at least, piece-wise continuous functions.

Let fatigue cracks grow at a certain time segment continuously and monotonically with respect to m_1 generalized coordinates. Then dl_j/dN exist and are positive ($j = 1, \ldots, m_1$). The conditions of equilibrium

$$G_j[\underset{\sim}{l}(N), \underset{\sim}{s}(N), \underset{\sim}{\phi}(N)] = \Gamma_j[\underset{\sim}{l}(N), \underset{\sim}{s}(N), \underset{\sim}{\phi}(N)] \quad (j = 1, \ldots, m_1) \tag{44}$$

are satisfied exactly for continuously growing cracks, and approximately for cracks growing in small jumps. Assume that ν_j components $\varphi_{ij}(N)$ of the process $\underset{\sim}{\varphi}(N)$ are connected with the generalized coordinate l_j. For example, we distinguish mechanical and corrosion damage, microcracks and microvoids near the same crack-tip. The most intensive damage accumulates in the tip zones. Their characteristic size λ_{ij}, generally, different for various types of microdamage. At $L_j - l_j > \lambda_{ij}$ damage accumulation rate is rather low. We refer these parts of cracks' trajectories to the "far field".

Eqs.(44) are to be complemented with equations of microdamage accumulation. The continuous analogue of Eq.(40) is

$$\frac{\partial \underset{\sim}{\varphi}}{\partial N} = \underset{n=o}{\overset{n=N}{\Phi}} \{\underset{\sim}{l}(n), \underset{\sim}{s}(n), \underset{\sim}{L}\}. \tag{45}$$

To build up an approximate solution of Eq.(45), let introduce the characteristic segments of loading process equal to the cycle numbers required for crossing of a tip zone with the corresponding crack front:

$$\Delta N_{ij} = \lambda_{ij} \left(\frac{dl_j}{dN} \right)^{-1} \quad \begin{pmatrix} i=1, \ldots, \nu_j \\ j=1, \ldots, m_1 \end{pmatrix}. \tag{46}$$

We assume that the changes of the loading level during the segments ΔN_{ij} are sufficiently small. In this approximation, stress and damage fields depend much more on the differences $\xi_j = L_j - l_j$ than on the cycle number N. Using the local reference systems moving with the crack-tips, let introduce the new functions $\chi_{ij}(\xi, N) \equiv \varphi_{ij}(L, N)$. Material (substantial) derivatives are to be used in the moving reference systems instead of partial derivatives $\partial \varphi_{ij}/\partial N$. Eq.(45) takes the form (in components)

$$\frac{\partial \chi_{ij}}{\partial N} - \frac{\partial \chi_{ij}}{\partial \xi_j}\frac{dl_j}{dN} = \Phi_{ij}\{\underset{\sim}{l}(n), \underset{\sim}{s}(n), L_j\}_{n=o}^{n=N}. \tag{47}$$

Neglecting in Eq.(47) the derivative with respect to N and integrating the simplified equation under the initial condition $\chi_{ij}(0, N) = \psi_{ij}(N)$, we obtain

$$\chi_{ij}(\xi_j, N) = \phi_{ij}(N) - \left(\frac{dl_j}{dN}\right)^{-1} \int_o^{\xi_j} \Phi_{ij}\{\underset{\sim}{l}(n), \underset{\sim}{s}(n), l_j(N)+\xi_j\}_{n=o}^{n=N} d\xi_j. \tag{48}$$

We identify later on the values of $\chi_{ij}(\xi_{ij}, N)$ at $\xi_j = \lambda_{ij}$ with damage measures $\psi_{ijr}(N)$ in the "far field". Applying to integrals in Eq.(48) the mean value theorem results into

$$\phi_{ij}(N) = \phi_{ij\,r}(N) + \lambda_{ij}\left(\frac{dl_j}{dN}\right)^{-1} \Phi_{ij}\{\underset{\sim}{l}(n), \underset{\sim}{s}(n), l_j(N)+\lambda^o_{ij}\}_{n=N-\Delta N_{ij}}^{n=N} \tag{49}$$

where λ^o_{ij} is of the same order of magnitude as λ_{ij}. But $\psi_{ijr}(N)$ may be estimated approximately with account of nominal stresses and strains. Therefore, $\phi_{ijr}(N)$ are to be considered as known functions.

Substituting Eq.(49) into Eq.(44) we obtain a set of equations with respect to $l_j(N)$, where $j = 1, \ldots, m_1$. If there is no hereditary effects, the resulting equations are differential ones. Under certain additional assumptions [5, 6, 16], the latters can be presented in the form solved explicitly with respect to dl_j/dN.

In the quasistationary approximation isochronic variations of the microdamage measures are

$$\delta\phi_{ij} = -\sum_{k=1}^{m_1}\left(\frac{\partial \varphi_{ij}}{\partial N}\right)_{L_k=l_k}\left(\frac{dl_k}{dN}\right)^{-1}\delta l_k.$$

Hence

$$\delta^2 A = \sum_{j=1}^{m_1}\sum_{k=1}^{m_1}\left\{\frac{\partial H_j}{\partial l_k} - \left(\frac{dl_k}{dN}\right)^{-1}\sum_{i=1}^{\nu_j}\frac{\partial H_j}{\partial \Phi_{ij}}\left(\frac{\partial \varphi_{ij}}{\partial N}\right)_{L_k=l_k}\right\}\delta l_j \delta l_k \quad (50)$$

where $\partial \varphi_{ij}/\partial N$ can be expressed with the right-hand side of Eq.(45).

Let take as an example a Mode I crack in a thick plate. The length of the crack is $2l$, the plate width is b, and the nominal stress is σ_∞. Assume that the crack is symmetricaly placed, and the plane-strain state conditions are fulfilled. We neglect here plastic deformations near the tip, such associated phenomena as crack closure, residual stresses, etc., as well as hardening or softening of the material due to cyclic loading and microdamage accumulation. Some of these effects can be study separatively, see [8]. The reason for this example is, in particular, to demonstrate that the theory results into equations similar to the most used semi-empirical equations of fatigue crack growth.

If $b - 2l \leq 0.1b$, the formulas may be used for the energy release rate G and the stress intensity factor K:

$$G = \frac{K^2(1-\nu^2)}{E}, \qquad K=\sigma_\infty[b\ \mathrm{tg}(\pi l/b)]. \quad (51)$$

Here E is Young modulus, and ν is Poisson ratio. We assume that the resistance force is given as

$$\Gamma=\Gamma_o(1-\phi) \quad (52)$$

where $\varphi(t) = \varphi[l(t), t]$ is a scalar microdamage measure; $= 0$ for the undamaged, and $\varphi = 1$ for completely damaged material. Function $\varphi(x,N)$ satisfies to equation

$$\frac{\partial\varphi}{\partial N} = \begin{cases} 0, & \Delta\sigma < \Delta\sigma_{th} \\ \left[\dfrac{\Delta\sigma - \Delta\sigma_{th}}{\sigma_f}\right]^m, & \Delta\sigma \geq \Delta\sigma_{th} \end{cases} \quad (53)$$

expressing the "threshold-power low" of the damage accumulation 5. In Eq.(53) $\Delta\sigma$ is the range of normal stress $\sigma(x, 0, t)$ during a cycle, σ_f is resistance stress against microdamage, $\Delta\sigma_{th}$ is threshold resistance stress, and m is a positive number relative to exponents entering into S - N curves or the Paris - Erdogan equation [1]. Generally, these material parameters depend on temperature and the stress ratio $R = \sigma_{min}/\sigma_{max}$.

The normal stress distribution at x l, $y = 0$ is assumed $\sigma = \varkappa\sigma_\infty$ at $l \leq |x| \leq l + \lambda$, and $\sigma = \sigma_\infty b/(b - 2l)$ at $|x| > l + \lambda$. Here $\varkappa$ is the stress concentration factor, σ_∞ is nominal stress, and λ is length of the tip zone. If $\varkappa \gg 1$, the approximate relationship [21] exists between $\varkappa$ and K, such as $\varkappa \approx K/(\chi\sigma_\infty\rho)^{1/2}$. Here ρ is interpreted as a characteristic radius of curvature of the crack-tip, and χ is a factor of order of unity. We may consider ρ as a measure of stress concentration at the tip, or, as we want, a characteristic length of a small processing zone. As a special case of Eq.(37), we consider here ρ as a material constant.

The microdamage accumulation within the tip zone is governed with equation

$$\frac{\partial\varphi}{\partial N} = \begin{cases} 0, & \Delta K < \Delta K_{th} \\ \left(\dfrac{\Delta K - \Delta K_{th}}{K_f}\right)^m, & \Delta K \geq \Delta K_{th}. \end{cases} \tag{54}$$

Eq.(54) follows from Eq.(53) under assumptions listed above; ΔK is the range of the stress intensity factor during a cycle; $K_f = \sigma_f(\chi\rho)^{1/2}$ and $\Delta K_{th} = \Delta\sigma_{th}(\chi\rho)^{1/2}$ are material constants (at given stress ratio, temperature and other environmental conditions).

Take, for example, that $\Delta\sigma_\infty = \text{const}$, $\sigma_{\infty,\max} = \text{const}$ during all the loading, and $\Delta K > \Delta K_{th}$. The duration N_* of the initiation stage is the first positive root of the equation

$$G[l(n),\sigma_{\infty,\max}(N)] = \Gamma_o[1-\psi(N)] \tag{55}$$

at $l = l_o$, and l_o is an initial size of the crack. Substituting Eqs.(51) as well as the result of integration of Eq.(54) we obtain

$$N_* = \left(\frac{K_f}{\Delta K - \Delta K_{th}}\right)^m \left(1 - \frac{K^2_{max}}{K^2_{Ic}}\right).$$

Here $K_{IC} = [E\Gamma_0(1-\nu^2)^{-1}]^{1/2}$ is in fact the fracture toughness in the framework of linear fracture mechanics. If at $N > N_*$ Eqs.(46) and (49) are applicable, so

$$\psi(N) = \psi_r(N) + \lambda\left(\frac{\partial l}{\partial N}\right)^{-1}\left(\frac{\Delta K - \Delta K_{th}}{K_f}\right)^m. \qquad (56)$$

The "far-field" damage $\psi_r(N)$ may be found from Eq.(53) at $\Delta\sigma = (\Delta\sigma_\infty b)/(b-2l)$. Substitution of Eq.(56) in Eq. (55) results into the equation of fatigue crack growth

$$\frac{dl}{dN} = \lambda\left(\frac{\Delta K - \Delta K_{th}}{K_f}\right)^m\left(1 - \frac{K^2_{max}}{K^2_{Ic}} - \psi_r(N)\right)^{-1}. \qquad (57)$$

Eq.(57) is of the same structure as the Paris - Erdogan equation transfering into the latter at $\Delta K_{th} = 0$, $K^2_{max} \ll K^2_{IC}$, $\psi_r(N) \ll 1$. All parameters entering into the right-hand side of Eq.(57) can be evaluated from convential fatigue and fracture toughness tests. A preliminary estimate of λ can be found from the Paris - Erdogan equation putting $\Delta K_{th} \approx 0$, and $K_f \approx K_{IC}$. Then $\lambda \approx cK^m_{IC}$ where c is the constant in the equation $dl/dN = c(\Delta K)^m$.

To study stability of the growing fatigue crack, consider the sign of $\delta^2 A$. Eq.(50) transforms in this case as follows:

$$\delta^2 A = \left[\frac{\partial G}{\partial l} + \left(\frac{dl}{dN}\right)^{-1}\left(\frac{\partial\Gamma}{\partial\psi}\right)\left(\frac{\partial\psi}{\partial N}\right)_{x=l}\right]\delta l^2.$$

Derivative dl/dN is taken from Eq.(57), and derivative $\partial\psi/\partial N$ at $x = l$ coincides with the right-hand side from Eq.(54). The condition of stability with account of Eqs.(51) and (52) takes the form

$$K^2_{max} < K^2_{**} = \frac{K^2_{Ic}(1-\psi_r)}{1 + (2\pi l/b)\operatorname{cosec}(2\pi l/b)}. \qquad (58)$$

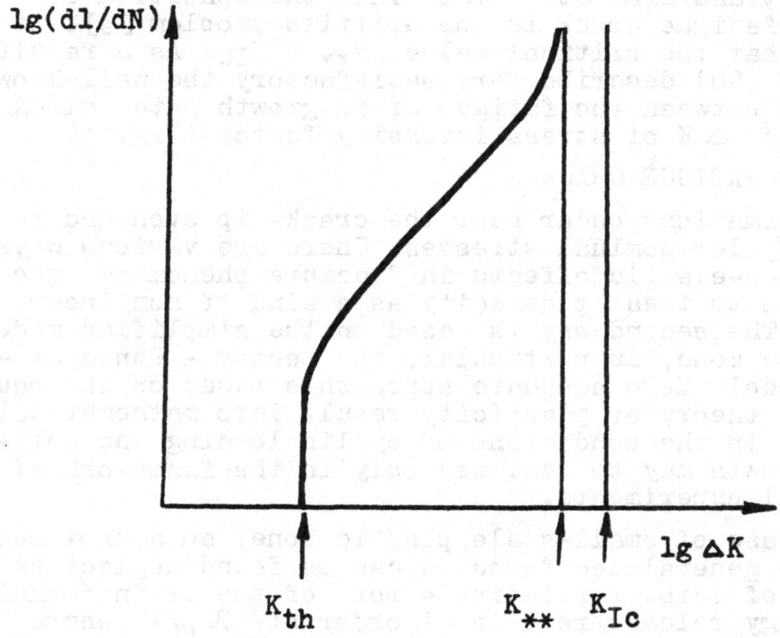

Fig. 16. Crack growth rate versus the range of stress intensity factor according to Eq.(57)

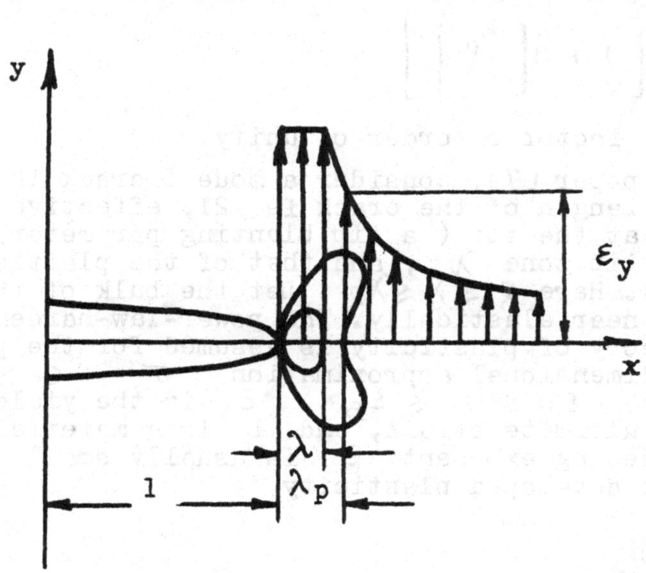

Fig. 17. Plastic zone and processing tip zone in low-cycle fatigue

Eq.(58) transforms at $l \ll b$ into the condition of stability a fatigue crack in the Griffith problem [5]. It is evident that the critical value $K_{**} < K_{IC}$. As a result, Eqs.(57) and (58) describe very satisfactory the well-known relationship between the fatigue crack growth rate dl/dN and the range ΔK of stress intensity factor (Fig.16).

6. LOW-CYCLE FATIGUE CRACK

Plastic deformations occur near the crack-tip even under comparatively low nominal stresses. There are various ways to account non-elastic effects in fracture phenomena. The simpliest one to treat plasticity as a kind of nonlinear elasticity. The second way is based on the simplified models of plastic zone, in particular, the Leonov - Panasyuk - - Dugdale model. More adequate approaches based on the equations of the theory of plasticity result into mathematical models which in the conditions of cyclic loading and fatigue crack growth may be realized only in the framework of computational experiments.

In the case of small-scale plastic zone, an approximate value of the generalized force G can be found neglecting the plastic effects. A relative error of the Irwin formula for the energy release rate is of order of λ_p/l where λ_p is the length of the plastic zone. Thus, for the Mode I crack

$$G = \frac{\pi\sigma_\infty l}{\eta E}\left[1 + O\left(\frac{\lambda_p}{l}\right)\right]. \tag{59}$$

Here η is a factor of order of unity.

Following paper [7], consider a Mode I crack in plane-stress state. Length of the crack is $2l$, effective curvature radius at the tip (a tip blunting parameter) ρ, length of the tip zone λ, and that of the plastic zone λ_p (Fig.17). Here $\rho \leq \lambda \leq \lambda_p$. Let the bulk of the body is deformed linear elastically. The power-law-hardening model of the theory of plasticity is assumed for the plastic zone. In one-dimensional approximation $\sigma = E\varepsilon + B(\varepsilon - \varepsilon_Y)^\beta$ where $\varepsilon_Y \leq \varepsilon \leq \varepsilon_B$, ε_Y is the yield strain, ε_B is the ultimate strain, and B is a material constant. The hardening exponent β is usually small compared with unity. At developed plasticity

$$\sigma \approx B\varepsilon^\beta. \tag{60}$$

Let effects of crack closure, residual stresses and secondary plastic deformation are negligible, and the loading process is pulsating, i.e. $\Delta\sigma_\infty = \sigma_{\infty,max}$. To estimate the stress and strain concentration near the crack-tip, we use the Neuber equation $\varkappa_\sigma \varkappa_\varepsilon = \varkappa^2$ relating the stress and strain concentration factor $\varkappa_\sigma$ and $\varkappa_\varepsilon$ with the stress concentration factor for Hook material. In the case of sufficiently sharp crack $\varkappa \approx (1/\chi\rho)^{1/2}$. With account of Eq. (60)

$$\varkappa_\sigma = (1/\chi\rho)^{\beta/(1+\beta)}, \quad \varkappa_\varepsilon = (1/\chi\rho)^{1/(1+\beta)}. \tag{61}$$

The strain range at $y = 0$, $|x| \geqslant 1$ is

$$\Delta\varepsilon = \begin{cases} \Delta\varepsilon_\infty (1/\chi\rho)^{1/(1+\beta)}, & 0 \leqslant |x-1| \leqslant \lambda, \\ \Delta\varepsilon_\infty [1\lambda/(\chi\rho|x-1|)]^{1/(1+\beta)}, & \lambda < |x-1| \leqslant \lambda_p. \end{cases} \tag{62}$$

Here $\Delta\varepsilon_\infty$ is the nominal strain range. The length λ_p of the plastic zone may be found from condition $\Delta\varepsilon(1+\lambda_p) = \varepsilon_Y$. Hence

$$\lambda_p = \frac{1\lambda}{\chi\rho}\left[\frac{\Delta\varepsilon_\infty}{\varepsilon_Y}\right]^{1+\beta}. \tag{63}$$

Eqs.(62) and (63) are in agreement with results of the more accurate analysis of the stress-strain state near a crack in a power-hardening material [21, 22].

The active generalized force may be determined using the Irwin formula with the error of order $\lambda_p/1$. The resistance generalized force we assume in the form $\Gamma = \gamma_0 g(\psi)$ with, say, $g = 1 - \psi^\alpha$ where is microdamage measure near the crack-tip. In general case, the cyclic hardening can be incorporated into the function $g(\psi)$.

The accumulation of microdamage is governed with strains at the prolongation of the crack path. Assume that the increment of the microdamage measure during one cycle of loading is $\Delta\psi = (\Delta\varepsilon/\varepsilon_f)^m$. Here ε_f is a characteristics of the material, and exponent m is related to the exponent entering in the Coffin equation for the low-cycle fatigue. For processes with slowly varying parameters it results into equation

$$\frac{\partial\varphi}{\partial N} = \left(\frac{\Delta\varepsilon}{\varepsilon_f}\right)^m. \tag{64}$$

Solution of Eq.(64) in quasistationary approximation with account of Eqs.(62) and (63) takes the form

$$\phi(N) = \phi_r(N)+\lambda\left(\frac{dl}{dN}\right)^{-1}\left(\frac{\Delta\varepsilon_\infty}{\varepsilon_f}\right)^m\left(\frac{1}{\chi\rho}\right)^{m/(1+\beta)}\left\{1+ \frac{1+\beta}{m-(1+\beta)}\left[1-\left(\frac{\lambda}{\lambda_p}\right)^{[m-(1+\beta)]/(1+\beta)}\right]\right\}. \tag{65}$$

Here is assumed $m \neq 1 + \beta$. If $m = 1 + \beta$ we obtain instead of Eq.(65)

$$\phi(N) = \phi_r(N) + \lambda\left(\frac{dl}{dN}\right)^{-1}\left(\frac{\Delta\varepsilon_\infty}{\varepsilon_f}\right)^m\left(\frac{1}{\chi\rho}\right)\left[1 + \ln\left(\frac{\lambda_p}{\lambda}\right)\right]. \tag{66}$$

Substitution Eq.(65) into the equilibrium condition at $g(\psi) = 1 - \psi^\alpha$ after solution with respect to dl/dN results into the differential equation of low-cycle fatigue crack growth [7]

$$\frac{dl}{dN} = \lambda\left(\frac{\Delta e}{\varepsilon_f}\right)^m\left(\frac{1}{\lambda\rho}\right)^{m/(1+\beta)}\left\{1 + \frac{1+\beta}{m-(1+\beta)}\left[1 + \left(\frac{\lambda}{\lambda_p}\right)^{[m-(1+\beta)]/(1+\beta)}\right]\right\}\left[\left(1-\frac{K^2_{max}}{K^2_{Ic}}\right)^{1/\alpha} - \phi_r\right]^{-1}. \tag{67}$$

Here a pseudo-stress intensity factor is introduced as $K = Ee(\pi l)^{1/2}$, and $K_c = (\eta\gamma_0 E)^{1/2}$.

The structure of Eq.(67) is similar to that of Eq.(57). The equation can be simplified putting $\lambda = \lambda_p$, i.e. identifying the tip and plastic zones. Then

$$\frac{dl}{dN} = \lambda_p\left(\frac{\Delta e}{\varepsilon_f}\right)^m\left(\frac{1}{\chi\rho}\right)^{m/(1+\beta)}\left[\left(1-\frac{K^2_{max}}{K^2_{Ic}}\right)^{1/\alpha} - \phi_r\right]^{-1}. \tag{68}$$

If $\lambda \ll \lambda_p$, $m \neq 1 + \beta$, Eq.(68) takes the form

$$\frac{dl}{dN} = \frac{\lambda m}{m-(1+\beta)} \left(\frac{\Delta e}{\varepsilon_f}\right)^m \left(\frac{1}{\chi\rho}\right)^{m/(1+\beta)} \left[\left(1 - \frac{K_{max}^2}{K_{Ic}^2}\right)^{1/\alpha} - \phi_r\right]^{-1}. \quad (69)$$

Let consider Eqs.(67) - (69) at a moderate loading level. For carbon steels the exponent in Eq.(64) is usually taken $m = 2$. It results into the same exponent at $\Delta e/\varepsilon_f$ in Eqs.(67) - (69). But Δe enters into Eq.(68) with the size of plastic zone. Therefore, the exponent at Δe in Eq.(68) is in fact $m + 1 + \beta$. If $m = 2$ and $0 \leq \beta \leq 1$ we obtain $3 \leq m + 1 + \beta \leq 4$. The upper bound corresponds to the traditional exponent in the Paris - Erdogan equation. On the other side, if is assumed in Eq.(69) a material constant, the exponent remains $m = 2$. These considerations at some extent clarify the range of variation of exponents in semi-empirical equations of fatigue crack growth. For example, if $m = 4$, the similar discussion results into the enequalities $5 \leq m + 1 + \beta \leq 6$.

Up to now we have assumed that the material remains linear elastic in all the bulk of the body except vicinities of the crack-tip. A related problem of fatigue crack growth in a material which follows to the power-law in all the bulk was considered in [7]. The driving force is [21]

$$G = \sigma_\infty^{(1+\beta)/\beta} I(\beta) B^{-1/\beta} \quad (70)$$

where $I(\beta)$ is the factor tabulated in 22 . The further calculations are similar to the former ones. Details can be found in [7].

7. APPLICATION OF THE MODEL OF THIN PLASTIC ZONE

The model of thin plastic zone by Leonov - Panasyuk - Dugdale appears to be a successful simplest approximation to nonlinear fracture mechanics. This model estimates rather correctly the size of the plastic zone as well as the opening displacement on the crack-tip. Together with the concept of dispersed damage, the model of thin plastic zone was applied to fatigue cracks in [23]. It was assumed that a crack propagates when the damage measure at the tip attains a certain ultimate level. The approach suggested in papers [5, 6] takes into account the global energy balance, and a crack propagates when the damage measure at the tip is, as a rule, much less than the ultimate level [8].

The scheme of the crack in a thin elasto-plastic plate

was depicted in Fig.5. The length λ_p of the plastic tip zone and the crack opening displacement at the tip are given with Eqs.(24). Condition of equilibrium in the Leonov - Panasyuk - Dugdale model are usually stated in terms of crack opening displacements, i.e.

$$\delta = \delta_c. \tag{71}$$

Here δ_c is critical displacement considered as a material constant depending on temperature only. It is easy to rewrite the equilibrium condition in terms of generalized forces, i.e. in the form $G = \Gamma$. The driving force G is given with Eq.(25). The resistance force $\Gamma = G_c$ is equal to G with substitution of δ_c instead of δ. The result is rather cumbersome [19].

Evidently that the condition $G = G_c$ is equivalent to Eq.(71). We consider later a pulsating loading, i.e. $\Delta\sigma_\infty = \sigma_{\infty,\,max}$ neclecting the influence of residual stresses and cyclic hardening (softening). Assume that

$$\frac{\partial\varphi}{\partial N} = \left[\frac{\sigma_o - \Delta\sigma_{th}}{\sigma_f}\right]^m, \quad \delta_c = \delta_o(1 - \varphi^\alpha) \tag{72}$$

where σ_o is limit opening stress, and the other notations are to be understood as earlier. The stress range in the tip zone is taken equal to $\sigma_o > \Delta\sigma_{th}$. Substitution of Eq.(71) results into the equation of quasistationaly approximation

$$\frac{\partial l}{\partial N} = \lambda_p\left[\frac{\sigma_o - \Delta\sigma_{th}}{\sigma_f}\right]^m\left[\left[1 - \frac{\delta}{\delta_o}\right]^{1/\alpha} - \phi_r\right]^{-1}. \tag{73}$$

At $\lambda_p \ll l$ instead of Eq.(73) we obtain

$$\frac{\partial l}{\partial N} = \frac{\pi^2\sigma_\infty^2 l}{8\sigma_o^2}\left[\frac{\sigma_o - \Delta\sigma_{th}}{\sigma_f}\right]^m\left[\left[1 - \frac{l}{l_c}\right]^{1/\alpha} - \phi_r\right]^{-1}.$$

where l_c is the critical size of the crack in the virgin material, i.e. $l_c = E\sigma_o\delta_o/(\pi\sigma_\infty^2)$.

In the case of low-cycle fatigue, it is natural to relate the rate of accumulation of microdamage with the range of plastic deformations. Neglecting the residual deformations, let postulate that

$$\frac{\partial\varphi}{\partial N} = \left[\frac{2v - \delta_{th}}{\delta_f}\right]^m . \tag{74}$$

Here $2v$ is the crack opening displacement within the plastic zone, i.e. at $l \leqslant |x| \leqslant l + \lambda_p$. The mechanical meaning of material constants δ_f, δ_{th} and m is evident. At $2v < \delta_{th}$ the right-hand side in Eq.(74) vanishes.

The analytical expression for $v(x)$ is rather cumbersome. Assume the simplified, linear distribution of the opening displacement along the plastic zone: $2v = (l + \lambda_p - |x|)(\delta/\lambda_p)$. Solving Eq.(75) in quasistationary approximation, we arrive to

$$\psi = \psi_r + \frac{\lambda_p}{m+1}\left[\frac{dl}{dN}\right]^{-1}\left[\frac{\delta}{\delta_f}\right]^m\left[1 - \frac{\delta_{th}}{\delta}\right]^{m+1},$$

and the equation of the stable crack propagation takes the form

$$\frac{dl}{dN} = \frac{\lambda_p}{m+1}\left[\frac{\delta}{\delta_f}\right]^m\left[1 - \frac{\delta_{th}}{\delta}\right]^{m+1}\left[\left[1 - \frac{\delta}{\delta_0}\right]^{1/\alpha} - \psi_r\right]^{-1}. \tag{75}$$

Eq.(75) is valid at $\delta_{th} < \delta < \delta_0(1 - \psi_r^\alpha)$. If $\sigma_\infty^2 \ll \sigma_c^2$, and δ is not nearly to the mentioned boundaries, Eq.(75) shows that dl/dN is proportional to $(\Delta K)^{m+2}$. When $m = 2$, we obtain the resulting exponent at a moderate loading level between two and four, which is in agreement with the conclusion of Section 6, where the quite different model was used.

8. CRACKS OF STATIC FATIGUE IN LINEAR VISCOELASTIC MEDIA

Cracks in viscoelastic media, say, in polymers and polymer-based material, may grow both under cyclic and slowly varying loading. We will say on cyclic and static fatigue cracks, respectively. A survey of earlier studies in the mechanics of fracture of polymers is given in [24]. The survey in the book [25] is extended till 1987. A theory of static fatigue cracks growth in viscoelastic microdamaging media was suggested in [9].

Two problems arize when we deal with the cracks propagation in visco-elastic media. Firstly, it is the determination of the driving forces similar to the energy release rate G . The energy transfer and transformation in such media is complicated due to the strong hereditary effects, dissipation and production of both thermal and non-thermal components of entropy. The arguments around the thermodynamic aspects of these problems are rather hot and not at all very decisive. Hence, the pure mechanical approach to determine G becomes useful. The idea of Griffith variation in the presence of hereditary effects was introduced in Fig.2. Virtual work corresponding to the isochronic variation δl is in fact the result of the initial variation δl_c of the crack size. In the absense of volume forces

$$G(t)\delta l = \int_S p_\alpha(t)\left[u_\alpha^{II}(t)-u_\alpha^{I}(t)\right]dS + \int_V\left[W^{II}(t)- W^{I}(t)\right]dV \qquad (76)$$

where the same notations as in Eqs.(17) - (23) are used. Indices I and II correspond to the two similar bodies in neighbouring states (Fig.2).

The second problem is connected with the modelling of microdamage accumulation. Contrary to the microcracks in metallic and ceramic solids at moderate temperatures, the microcracks in polymers may be self-healing even at the room temperature. Hence, instead of Eqs.(53), (64), etc. for non-healing microdamage, hereditary models are to be used, such as

$$\varphi(x,t) = \int_0^t \Phi(t-\tau)f[\sigma(x,t)]d\tau \qquad (77)$$

with the function $f[\cdot]$ of the stresses on the crack path, and the hereditary kernel $\Phi(t-\tau)$. Eq.(77) may take into account both the irreversible microdamage, and the relaxation of microdamage due to the molecular motion in polymers.

There are a few attempts to simplify the determination of the driving force G using euristical considerations relating to the Volterra analogy principle. In [19, 26] it was assumed that

$$G(t) = D(\lambda/c)K^2(t) \qquad (78)$$

where D(t) is visco-elastic compliance for the considered crack mode and stress-strain constitutive equations, and K(t) is the transient stress intensity factor. The value of the compliance is taken at characteristic time $\theta_c=\lambda/c$, with a scale parameter λ and the crack growth rate $c \equiv dl/dt$. The question arises, what are the values of λ in real polymers. Knauss [26] estimated that λ for a polyurethan rubber is of order of 10^{-10}m, i.e. extremely small making the continuum interpretation questionable. An alternative presentation was proposed by Shapery [27]:

$$G(t) = \int_0^t D(t-\tau)\left[dK^2(\tau)\right]. \tag{79}$$

Another proposal was given in [14]

$$G(t) = K(t)\int_0^t D(t-\tau)dK(\tau). \tag{80}$$

reflecting the fact, that the stress field around a moving crack is elastic, and the strain field is the hereditary transform with the kernel $D(t-\tau)$. But Eq.(80) is an euristic relationship, too.

In two limiting cases, i.e. for very rapidly and very slowly growing cracks, Eqs.(78) - (80) yield the same results. For example, for the Mode I in the plane strain-state

$$\frac{K^2(t)(1-\nu_0^2)}{E_0} \leqslant G(t) \leqslant \frac{K^2(t)(1-\nu_\infty^2)}{E_\infty} \tag{81}$$

where E_0 and ν_0 are instant values of Young modulus and Poisson ratio, and E_∞ and ν_∞ are corresponding long time values.

Comparison of numerical results following from Eqs.(76) (79) and (80) is given in Fig. 18. Computations where done for the case c = const. Referring to Eq.(76) and Fig.2 $l_I(t) = l_0 + ct$, $l_{II}(t) = l_0 + \delta l_0 + ct$, i.e. $\delta l = \delta l_0$ = const. The compliance function for the Mode I is taken as

$$D(t) = \frac{1-\nu^2}{E_\infty}\left[1 - \frac{E_0 - E_\infty}{E_0}\exp(-t/\theta_r)\right] \tag{82}$$

where θ_r is retardation time. It is assumed that Poisson ratio is constant, namely $\nu = 0,3$, and $E_0 = 10E_\infty$. At $t = 0$ nominal stresses σ_∞ are applied, and later on being kept constant. The ratio G/G_0 is plotted in Fig. 18 versus t/θ_r where $G_0 = \pi l_0 \sigma_\infty / E_0$. Nondimensional parameter $\eta = c\theta_r/l_0$ characterizes the relation between

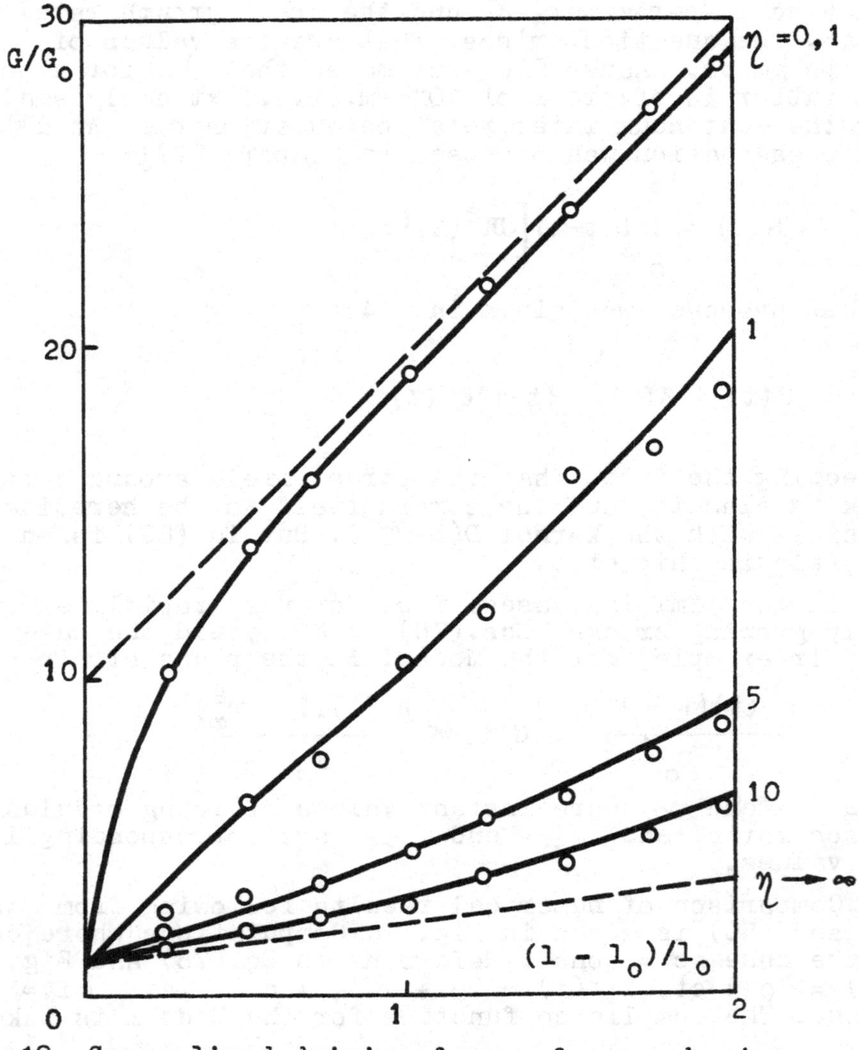

Fig.18. Generalized driving forces for cracks in a linear visco-elastic material at various crack growth rates.

times of the hereditary deformation and the crack growth. Broken lines in Fig.18 correspond to limiting cases $\eta \to \infty$ and $\eta \to 0$ when the generalized force may be determined from the Irwin formula putting $E = E_0$ and $E = E_\infty$, respectively. Solid lines are plotted according to Eq.(79). Special signs denote results computed with Eq.(76); the strain field $\varepsilon_{\alpha\beta}(t)$ and displacement fields $u_\alpha(t)$ are determined applying numerically the hereditary transform with the kernel $D(t-\tau)$ to the corresponding elastic fields.

Similar results were obtained in other examples, in particular, for an exponentional-power kernel with the weak singularity, as well as for the case of displacement-controlled loading.

If the crack growth is treated as a result of the interaction between the microdamage accumulation and the global energy balance in the system, the way of approximation $G(t)$ loses its significance. In fact, the form of $G(t)$ becomes important when the crack size approaches to the critical one. But then the $G(t)$ may be estimated more or less accurately using the instant modulus approximation. In other cases the crack growth is governed predominantly with the microdamage accumulation.

Some numerical results taken from the paper 9 are presented in Fig.19-21. Computation was performed using Eqs.(78) and the quasistationary approximation. The numerical dates are as follows: $E_0 = 10E_\infty$, $\nu = 0,3$, $\lambda = \rho =$ = const. The power-threshold model of microdamage similar to Eq.(54) was assumed with $K_f = K_0$, $K_{th} = 0$, $m = 4$. The level of nominal stresses σ_∞ = const is characterized with the nondimensional parameter $q=(\sigma_\infty/K_0)(\pi\rho)^{1/2}$, and the initial crack size $l_0 = 100\rho$. The vertical strokes in Fig.19 indicate the termination of the initiation stage. The points corresponding to the time $t = t_{**}$ of final failures are located in the broken line.

The times t_* and t_{**} are plotted against the loading level q in Fig.20. When q increases from 10^{-2} up to $5 \cdot 10^{-2}$, the times t_* and t_{**} vary in three - four orders of magnitude. The type of curves changes rather evidently in vicinity of $q = 3,13 \cdot 10^{-2}$. This value corresponds to the relatioship $K = K_\infty = (\gamma_0 E_0)^{1/2}$. A similar change can be found in Fig.19, too.

It is interesting to follow the variation of the microdamage measure ψ when a crack grows. Until the crack-tip is fixed, ψ increases monotonously (Fig.21). When the crack growth accelerates, ψ decreases. At more higher nominal stress level, the microdamage on the tip becomes very small. For example, $\psi < 2,5 \cdot 10^{-3}$ at $q = 5 \cdot 10^{-2}$, and the hereditary deformation is the decisive factor on this stage. On the contrary, at low nominal stress levels the microdamage measure is rather close to unity. As it is seen from Fig.21, a crack begins to propagate at $q = 10^{-2}$ when $\psi \approx$ ≈ 0.9. The initiation time t_* is much more than the retardation time θ_r , and the crack growth is governed almost entirely with the microdamage accumulation near the crack-tip.

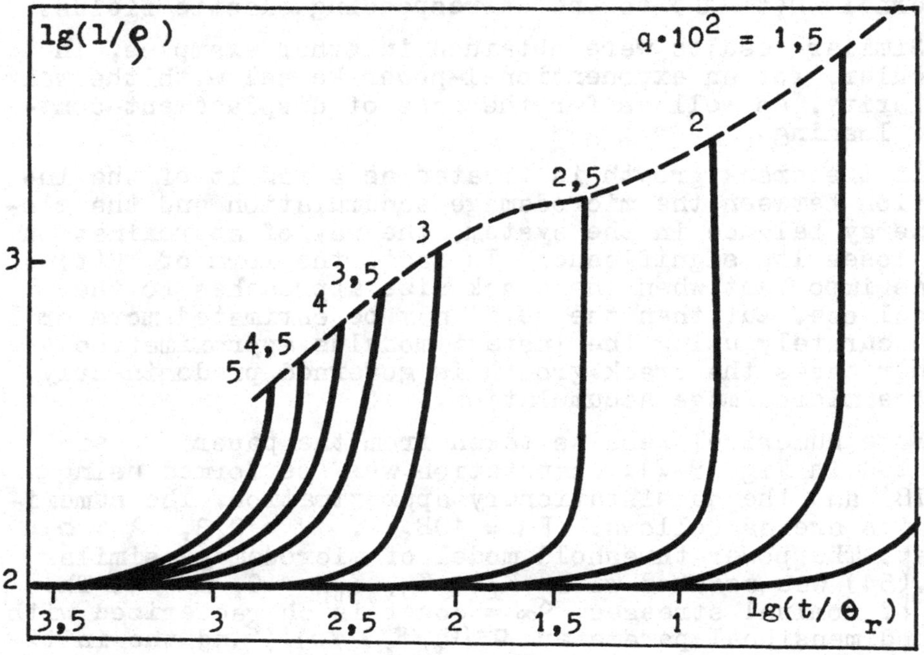

Fig. 19. Crack propagation in a linear visco-elastic material at various nominal stress levels.

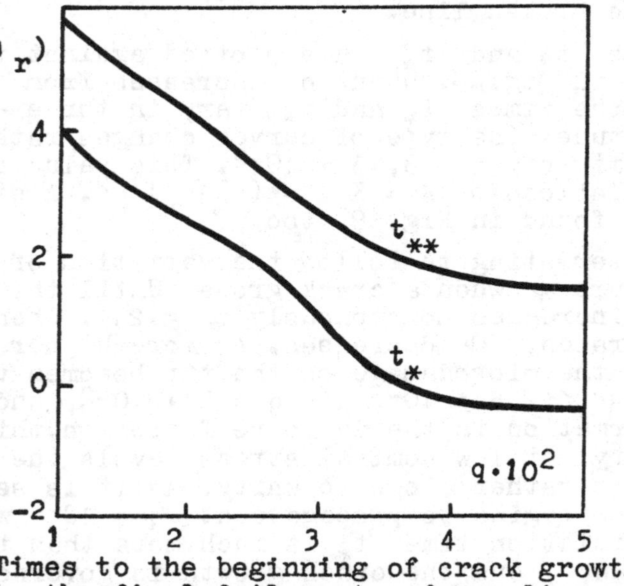

Fig. 20. Times to the beginning of crack growth t_*, and to the final failure t_{**} for a linear visco-elastic material versus the non-dimensional nominal stress.

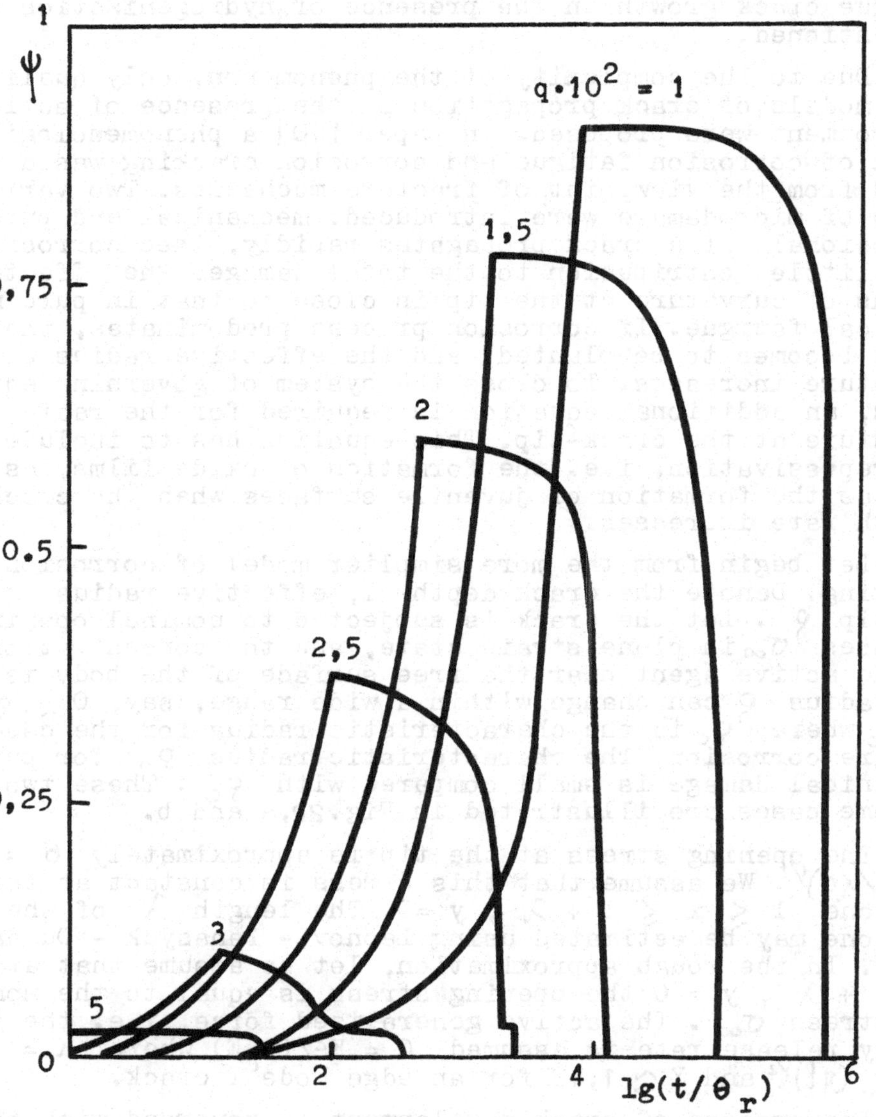

Fig. 21 . Microdamage measure at the tips of cracks in a linear visco-elastic material at various nominal stress levels.

9. CORROSION FATIGUE AND STRESS CORROSION CRACKING

Corrosion as a physico-chemical phenomenon of interaction of metals with the environment was investigated by a number of physicists, electrochemists and metallurgists. However, electrochemical and physical phenomena are closely connected with mechanical processes. As an example, the accelerated fatigue crack growth in the presence of hydrogenization may be mentioned.

Due to the complexity of the phenomenon, only qualitative models of crack propagation in the presence of active environment were proposed. In paper [10] a phenomenological model of corrosion fatigue and corrosion cracking was developed from the viewpoint of fracture mechanics. Two various types of microdamage were introduced: mechanical and pure corrosional. If a crack propagates rapidly, then corrosion does little contribution to the total damage. The effective radius of curvature at the tip is close to that in pure mechanical fatigue. If corrosion process predominates, the crack becomes to be blunted, and the effective radius of curvature increases. To close the system of governing equations, an additional equation is required for the radius of curvature at the crack-tip. This equation has to include the repassivation, i.e. the formation of oxide films, as well as the formation of juvenile surfaces when the crack growth rate increases.

Let begin from the more simplier model of corrosion cracking. Denote the crack depth l, effective radius at the tip ρ. Let the crack is subjected to nominal opening stresses σ_∞ in plane strain state, and the concentration of the active agent near the free surface of the body is c. The radius ρ can change within a wide range, say, $0 \leq \rho \leq \rho_c$ where ρ_c is the characteristic radius for the case of pure corrosion. The characteristic radius ρ_s for pure mechanical damage is small compared with ρ_c. These two extreme cases are illustrated in Fig.22,a and b.

The opening stress at the tip is approximately $\sigma = (K/\pi\rho)^{1/2}$. We assume that this stress is constant at the tip zone $l \leq x \leq l + \lambda$, $y = 0$. The length λ of the tip zone may be estimated using Leonov - Panasyuk - Dugdale model. In the rough approximation, let us assume that at $x > l + \lambda$, $y = 0$ the opening stress is equal to the nominal stress σ_∞. The active generalized force, i.e. the energy release rate is assumed $G = K^2/(\eta E)$ where $K = Y\sigma_\infty(\pi l)^{1/2}$, and $Y \approx 1,12$ for an edge Mode I crack.

The process of crack development is governed with the microdamage accumulation close to the crack-tip. For corrosion cracking, at least two phenomenological measures of microdamage are to be introduced, $0 \leq \psi_s \leq 1$ for pure mechanical microdamage, and $0 \leq \psi_c \leq 1$ for pure corrosion one. In general case both forces depend on $\psi_s(t)$ and $\psi_c(t)$. For simplication only, we assume later that only $\Gamma(t)$ depends on microdamage, say, $\Gamma = \gamma_0[1-(\psi_s+\psi_c)^\alpha]$ where γ_0 is specific fracture work for nondamaged material, and $\alpha > 0$.

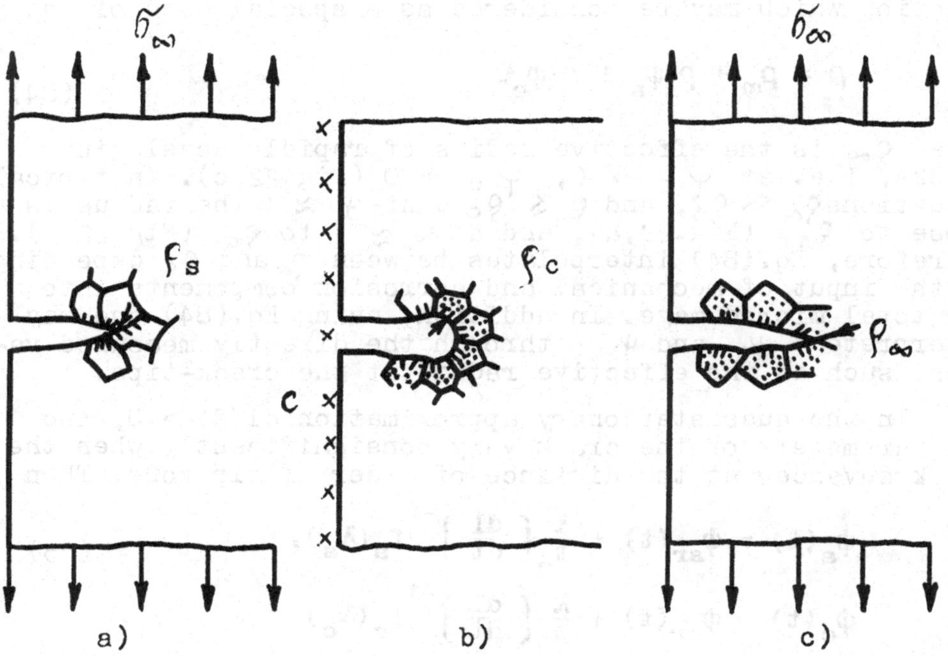

Fig.22. Interaction of mechanical and corrosion factors: a - pure mechanical loading, b - blunting of the crack due to corrosion, c - sharpening of the crack due to mechanical loading.

To evaluate the microdamage measures at the crack-tip, the prehistory of loading and crack growth is to be taken into account. Denote by $\varphi_s(x,t)$ and $\varphi_c(x,t)$ the microdamage measures at the prolongation of the crack, i.e. at $x > l(t)$. Let these measures satisfy to equations

$$\frac{\partial\varphi_s}{\partial t} = \frac{1}{t_c} f_s(\sigma_\infty, c, l, \rho, \varphi_s, \varphi_c, x);$$
$$\frac{\partial\varphi_c}{\partial t} = \frac{1}{t_c} f_c(\sigma_\infty, c, l, \varphi_s, \varphi_c, x). \qquad (83)$$

Here t_c is a time constant, f_s and f_c are functions of the given processes $\sigma_\infty(t)$ and $c(t)$, as well as the functions of $l(t)$, $\rho(t)$, $\varphi_s(x,t)$, and $\varphi_c(x,t)$ which are to be found in the process of solution of the problem.

To close the set of equations, an equation for the effective radius $\rho(t)$ is required. The simplest one relates to the measures $\psi_s(t)$ and $\psi_c(t)$ by means of the finite equation which may be considered as a special case of Eq. (37):

$$\rho = \rho_\infty + \rho_s\psi_s + \rho_c\psi_c. \tag{84}$$

Here ρ_∞ is the effective radius of rapidly developing cracks, i.e. at $\psi_s \to 0$, $\psi_c \to 0$ (Fig.22,c). In typical situations $\rho_\infty \ll \rho_s$, and $\rho_s \lesssim \rho_c$. At $\psi_s \approx 1$ the radius is close to ρ_s (Fig.22,a), and at $\psi_c \approx 1$ to ρ_c (Fig.22,b). Therefore, Eq.(84) interpolates between ρ_∞ and ρ_c depending on the input of mechanical and corrosion components into the total microdamage. In addition, using Eq.(84) one may interprete ψ_s and ψ_c through the directly measured values, such as the effective radii at the crack-tips.

In the quasistationary approximation $dl/dt > 0$, and all parameters of the crack vary nonsignificantly when the crack advances at the distance of order of tip zone. Then

$$\psi_s(t) = \psi_{sr}(t) + \frac{\lambda}{t_c}\left(\frac{dl}{dt}\right)^{-1} f_s(\lambda_s), \tag{85}$$

$$\psi_c(t) = \psi_{cr}(t) + \frac{\lambda}{t_c}\left(\frac{dl}{dt}\right)^{-1} f_c(\lambda_c)$$

where $\psi_{sr}(t)$ and $\psi_{cr}(t)$ are microdamage measures in the "far field", i.e. accumulated before particles of the materials had attained with the tip zone. The sizes λ_s and λ_c satisfy to conditions $0 < \lambda_s \leq \lambda$, $0 < \lambda_c \leq \lambda$.

Substituting Eqs.(85) into the equilibrium condition $G(t) = \gamma_o[1-(\psi_s+\psi_c)^\alpha]$, we come to the equation of crack growth

$$\frac{dl}{dt} = \frac{\lambda}{t_c} \frac{f_s(\lambda_s) + f_c(\lambda_c)}{[1 - (K/K_c)^2]^{1/\alpha} - (\psi_{sr} + \psi_{cr})}. \tag{86}$$

Qualitative features of corrosion cracking are shown in Fig.23 and 24. Lines 1 relates to the case $c_1 = 0$, lines 2 and 3 to the cases $c_1 > 0$ and $c_1 > c_2$, respectively. The shape of the lines $dl/dt = f(K)$ varies depending on the relation between the mechanical and corrosional damage. If corrosion is intensive, the second term in the numerator of the right-hand side of Eq.(86) dominates. It results into a "plateau" in Fig.24 and that agrees with expemental data . Generally, it follows from Eq.(86) that no

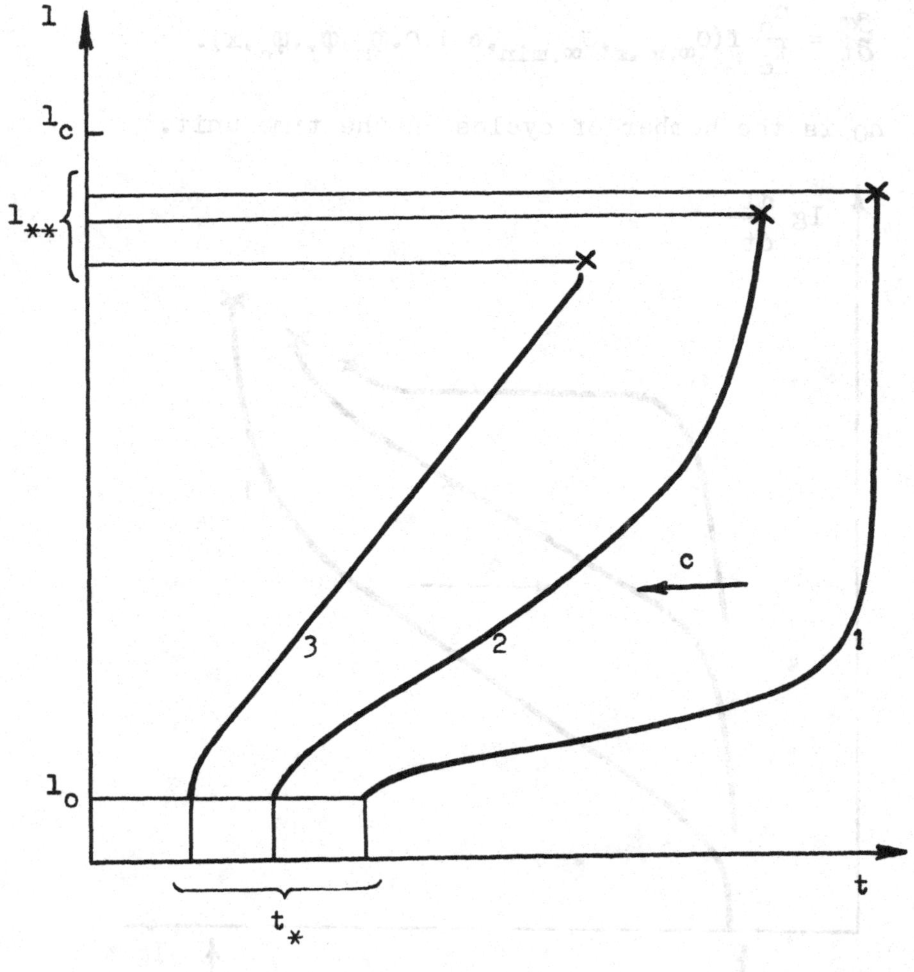

Fig.23. Stress corrosion crack growth at various concentrations of the active agent.

universal connection exists between K and l, as well as between lg K and lg(dl/dt). For example, the initial size of a crack is of significance leading to the scatter of experimental data. Eq.(86) takes into account all the history of loading and environment entering both with $\psi_{sr}(t)$ and $\psi_{cr}(t)$ and with the radius $\rho(t)$.

The extension of the model on cracks of corrosion fatigue is quite natural. In addition to φ_s and φ_c let introduce the measure of microdamage due to the cyclic fatigue. Eqs.(85) are to be complemented with the similar equation with respect to the measure $\varphi_f(x,t)$, say, such one:

$$\frac{\partial \varphi}{\partial t} = \frac{n_o}{t_c} f(\sigma_{\infty,max}, \sigma_{\infty,min}, c, l, \rho, \varphi_s, \varphi_f, \varphi_c, x).$$

Here n_0 is the number of cycles in the time unit.

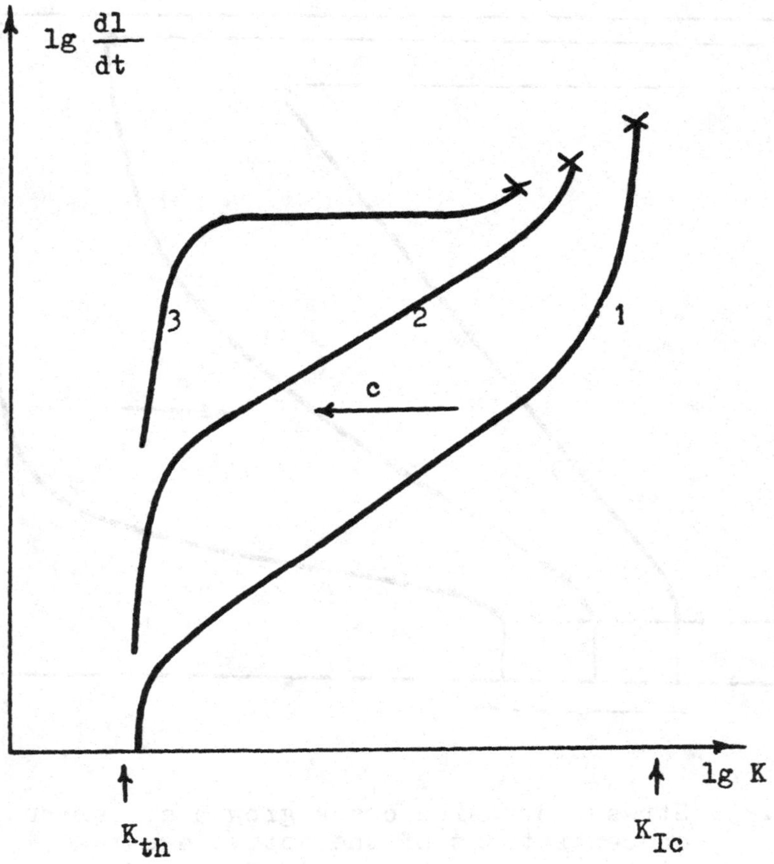

Fig.24. Stress corrosion crack growth rate versus the stress intensity factor at various concentrations of the active agent.

Extremal values of nominal stresses are assumed varying slowly with cycles, and the cycle number $N(t)$ is a slowly varying time function. Keeping the former assumptions, the equation of crack growth in the quasistationary approximation is

$$\frac{dl}{dt} = \frac{\lambda}{t_0} \frac{n_0 f(\lambda_f) + f_s(\lambda_s) + f_c(\lambda_c)}{[1 - (K/K_c)^2]^{1/\alpha} - (\phi_{fr} + \phi_{sr} + \phi_{cr})}.$$

For effective radius of curvature we follow Eq.(84) with the evident generalization, i.e. $\rho = \rho_\infty + \rho_f \psi_f + \rho_s \psi_s + \rho_c \psi_c$.

Some numerical results are shown in Fig.25. The measures ψ_f, ψ_c and $\psi = \psi_f + \psi_c$ are plotted against the summed cycle number $N = \int n_0(t)dt$. Microdamage due to the quasistatic loading was not taken into account.

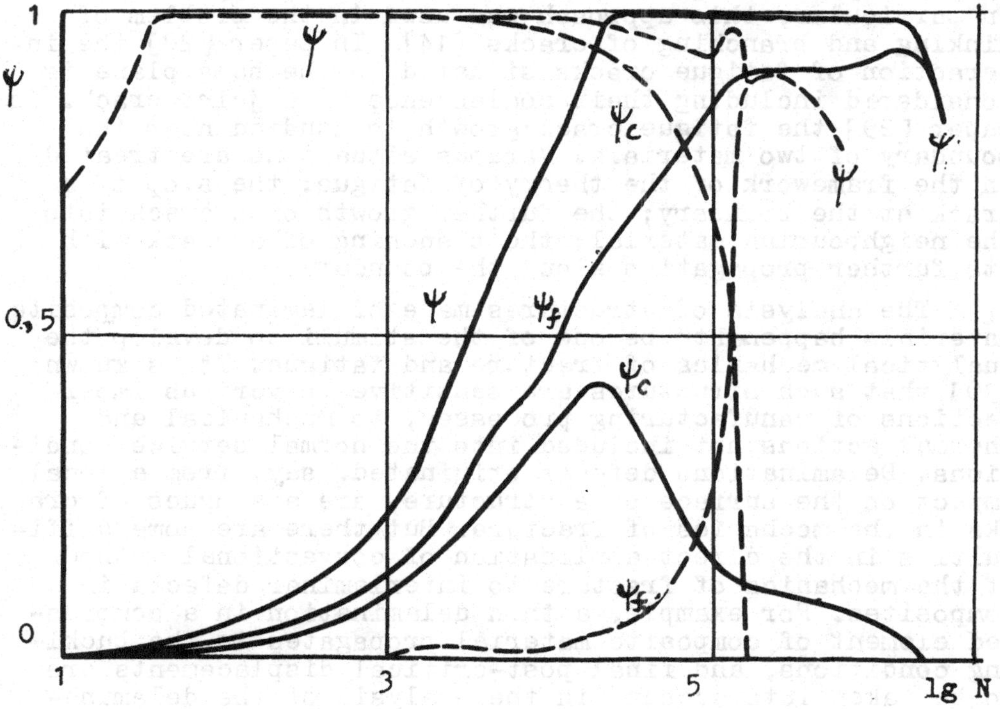

Fig.25. Microdamage accumulation due to fatigue ψ_f, corrosion ψ_c and summed microdamage ψ.

The solid lines correspond to an approximately equal input of cyclic loading and corrosion into the damage near the crack-tip. All three measures grow monotonously during the initiation stage, i.e. up to $N_* \approx 10^4$. After the start of the crack, the corrosion damage at the moving tip begins to decrease, diminishing almost up to zero at the instant of final failure. The mechanical damage continues to increase, and the most part of the crack growth occurs at values $\psi = \psi_f + \psi_c$ close to unity. Near the final failure both ψ_f and ψ_c begin to decrease.

The brocken lines in Fig.25 relate to the case when the concentration of the active agent is six times higher than in the former case. Corrosion damage near the tip prevails almost up to the final failure, and $\psi \approx \psi_c \approx 1$. The part of pure mechanical damage increases only at the final stage of the crack growth.

10. CONCLUSION

The unified approach [5,6] to the fracture mechanics and the theory of fatigue covers a wide area of applications. In particular, this approach was used in the problem of kinking and branching of cracks [14]. In paper [28] the interaction of fatigue cracks situated in the same plane is considered including their coalescence in a joint crack. In paper [29] the fatigue crack growth is studied near the boundary of two materials. Various situations are treated in the framework of the theory of fatigue: the stop of a crack at the boundary; the further growth of a crack into the neighbouring material; the branching of a crack with its further propagation along the boundary.

The analysis of structures made of laminated composite materials happened to be one of the stimuli to develop the analytical mechanics of fracture and fatigue. It is known [30] that such structures are sensitive to various imperfections of manufacturing processes, to mechanical and thermal actions not included into the normal service conditions. Delaminations defects originated, say, from a local impact on the surface of a structure, are analogues of cracks in the mechanics of fracture. But there are some difficulties in the direct application of conventional methods of the mechanics of fracture to interlaminar defects in composites. For example, a thin delamination in a compressed element of composite material propagates in the buckling conditions, and final post-critical displacements are to be taken into account in the analysis of the delamination's stability and growth.

Theory of stability and growth of delamination defects in composites is developed in [11, 12, 31]. Applications to cylindrical and spherical shells with interlaminar defects can be found in [32, 33]. Defects under combination of transversal tenson and interlaminar shear are studied in [34 - 37]. In [38] the secondary cracks occuring on delaminated parts of composite structures are considered. Delaminations subjected to thermal action are studied in [39, 40].

REFERENCES

1. Rabotnov, Yu.N.: Mechanics of deformable solids, Nauka, Moscow 1979 (in Russian).
2. **Bolotin, V.V.**: Prediction of service life for mashines and structures, ASME Press, New York, 1989 (translated from the Russian edition of 1984).
3. Hutchinson, J.W.: Micro-mechanics of damage in deformation and fracture, Technical University of Denmark, Lyngby, 1987.
4. Novozhilov, V.V.: On prospects of phenomenological approach to fracture problems, in: Mechanics of deformable bodies and structures, Mashinostroyenie, Moscow, 1975, 349-359 (in Russian).
5. Bolotin, V.V.: Equations of fatigue cracks growth, Sov. Mech. Solids, 4 (1983), 153-160 (in Russian).
6. Bolotin, V.V.: Unified models in fracture mechanics,Sov. Mech.Solids, 3 (1984), 127-137 (in Russian).
7. Bolotin, V.V.: Growth of low-cycle fatigue cracks, in: Problems of Shipbuilding. Ships Design, Rumb, Leningrad, 1985, 19-25 (in Russian).
8. Bolotin, V.V.: A model of fatigue cracks with the tip zone, Sov. Appl. Mech., 23 (1987), 67-73 (in Russian).
9. Bolotin, V.V.: Griffith's cracks in viscoelastic media with microdamage, in: Strength Design, Mashinostroyenie, Moscow, 25 (1987), 12-33 (in Russian).
10. Bolotin, V.V.: Mechanical model of corrosion cracking, Sov. Mech. Engng., 4 (1987), 20-26 (in Russian).
11. Bolotin, V.V.: Delamination defects in structures of composite materials, Sov. Mech. Compos. Mater., 2 (1984), 239-255 (in Russian).
12. Bolotin, V.V.: Fracture and fatigue of composite plate and shells, in: Inelastic behaviour of plates and shells Proc. of IUTAM Symposium, Rio de Janeiro, 1985, (Eds. J.Bevilacqua, R.Feijoo, R.Valid), Springer, Heidelberg, 1986, 131-161.
13. Bolotin, V.V.: A unified approach to damage accumulation and fatigue crack growth, Engng. Fract. Mech, 22 (1985), 387-398.
14. Bolotin, V.V.: A unified model of fatigue and fracture with applications of structural reliability, in: Random Vibration. Status and Recent Developments (S.H.Crandall Festschrift), Elsevier, Amsterdam, 1986, 47-58.
15. Bolotin, V.V.: Stability and growth of cracks under nonconservative loading, Sov. Mech. Solids, 5 (1987), 148--157 (in Russian).
16. Bolotin, V.V.: Multiparameter fracture mechanics, in: Strength Design, 25 (1984), Mashinostroyenie, Moscow, 12-33 (in Russian).

17. Bolotin, V.V.: Stability and growth of fatigue cracks, Sov. Mech. Solids, 4 (1988), 133-140 (in Russian).
18. Parton, V.Z. and Morozov, E.M.: Mechanics of elasto-plastic fracture, Nauka, Moscow, 1985 (in Russian).
19. Wnuk, M.P.: Subcritical growth of fracture (inelastic fatigue), Intern.J. Fracture, 7 (1971), 383-408.
20. Bolotin, V.V.: Nonconcervative problems of elastic stability, Pergamon Press, Oxford, 1963 (translated from the Russian edition of 1961).
21. Rice, J.R.: Mathematical analysis in the mechanics of fracture, in: Fracture. An advanced treatise (Ed. H.Liebowitz), v.2, Academic Press, New York, 1968, 191-311.
22. Hutchinson, J.W.: Singular behaviour at the end of a tensile crack in a hardening material, J. Mech. Phys. Solids, 16 (1968), 13-31.
23. Janson, J.: Dugdale-crack in a material with continuous damage formation, Engng.Fract.Mech., 9(1977), 891-899.
24. Knauss, W.G.: The mechanics of polymer fracture, Appl. Mech. Rev., 26 (1973), 1-17.
25. Kaminskiy, A.A. and Gavrilov, D.A.: Mechanics of fracture of polymers, Naukova Dumka, Kiev, 1988 (in Russian)
26. Knauss, W.G.: Delayed fracture - the Griffith problem in lineary viscoelastic materials, Intern.J. Fract., 6 (1970), 7-20.
27. Shapery, R.A.: Correspondence principles and a generalized J - integral for large deformation and fracture analysis of viscoelastic media, Intern.J. Fract.,25, (1985), 195-223.
28. Kovekh, V.M.: Interaction of fatigue crack, Sov. Mech. Solids, 2 (1988), 155-158 (in Russian).
29. Kovekh, V.M.: Behaviour of a fatigue crack near the boundary of two media, Sov. Mech. Solids, 4 (1988), 141-145 (in Russian).
30. Blagonadezhin, V.L., Vorontsov, A.N. and Murzakhanov, G.Kh.: Mechanical problems in manufacturing of composite structures, Sov. Mech. Compos. Mater., 5 (1987), 859-877 (in Russian).
31. Bolotin, V.V.: Damage and fracture of composites in delamination modes, Sov. Mech. Compos. Mater., 3 (1987), 424-432 (in Russian).
32. Kislyakov, S.A.: Stability and growth of delamination in a cylindrical composite shell in compression, Sov. Mech. Compos. Mater, 4 (1985), 653-657 (in Russian).
33. Nefedov, S.V.: Study of growth of elliptical delaminations in composites under long-acting quasistatic, Sov. Mech. Compos. Mater., 5 (1988), 827-833 (in Russian).
34. Bolotin, V.V.: Interlaminar fracture of composites under combined loading, Sov.Mech.Compos.Mater., 3 (1988),

410-418 (in Russian).
35. Shchugorev, V.N.: Delamination defects under combined tearing and interlaminar shear, Sov. Mech. Compos. Mater., 3 (1987), 539-542 (in Russian).
36. Shchugorev, V.N.: Kinetics of delamination defects under combined tearing and interlaminar shear, Sov. Mech. Compos. Mater., 2 (1988), 227-237 (in Russian).
37. Nesin, D.N.: Low-cycle fatigue of composites with interlaminar defects, Sov. Mech. Compos. Mater., 1 (1985), 227-237 (in Russian).
38. Murzakhanov, G.Kh. and Shchugorev, V.N.: Effect of secondary cracks on stability and growth of delaminations in composite structures, Sov. Mech. Compos. Mater. 6 (1988), 1120-1124 (in Russian).
39. Bolotin, V.V., Yefimov, A.E., Mesentsev, N.S., Shebunin, I.V. and Shchugorev, V.N.: Fracture toughness of composites with polymer matrix at elevated temperatures, Sov. Mech. Compos. Mater., 5 (1988), 839-844 (in Russian).
40. Shebunin, I.V.: Behaviour of delaminations in composites under thermal action, Sov. Mech. Compos. Mater., 4 (1988), 644-650 (in Russian).

NON-LINEAR PHENOMENA ASSOCIATED WITH FRACTURE IN STRAIN-SOFTENING MATERIALS

A. Carpinteri
Politecnico di Torino, Torino, Italy

ABSTRACT

- Progressive cracking in structural elements of concrete is considered. Two simple models are applied, which, even though different, lead to similar predictions for the fracture behaviour. Both Virtual Crack Propagation Model and Cohesive Limit Analysis (Section 2), show a trend towards brittle behaviour and catastrophical events for large structural sizes. Such a trend is fully confirmed by more refined finite element models, as well as by experimental testing on plain and reinforced concrete members.
- A numerical Cohesive Crack Model is proposed (Section 3) to describe strain softening and strain localization in concrete. Such a model is able to predict the size effects of fracture mechanics accurately. A general explanation to the well-known decrease in tensile strength and increase in fracture toughness by increasing the member sizes, is given in terms of Dimensional Analysis. The true values of such two intrinsecal material properties may be found exactly only with comparatively low values of the brittleness numbers

$$s = \frac{K_{IC}}{\sigma_u b^{1/2}} \quad \text{and} \quad s_E = \frac{G_F}{\sigma_u b}.$$

- Whereas for Mode I, only untieing of the finite element nodes is applied to simulate crack growth, for Mixed Mode a topological variation is required at each step (Section 4). In the case of the four point shear specimen, the load vs. deflection diagrams reveal snap-back instability for large sizes. By increasing the specimen sizes, such instability tends to reproduce the classical LEFM instability. Experimentally, the fracture toughness of concrete appears to be unique and represented by the Mode I fracture energy even for Mixed Mode problems.

- Remarkable size effects are theoretically predicted and experimentally confirmed also for reinforced concrete (Section 5). The brittleness of the flexural members increases by increasing size and/or decreasing steel content. On the other hand, a physically similar behaviour is revealed in the cases where the brittleness number

$$N_P = \frac{f_y\, b^{1/2}}{K_{IC}} \frac{A_s}{A}$$

is the same. On the basis of these results, the empirical code rules regarding the minimum amount of reinforcement could be considerably revised.

1. INTRODUCTION

In the last few years it has been pointed out that very considerable scale effects appear when the size of the structure (the shape being constant) is varied. The mechanical behaviour is brittler and the crack propagation faster for larger structure size [1-4].

Concrete-like materials present a tensile softening behaviour after the peak-load, i.e., the load sustained by the material decreases when the deformation is increased [5-12]. If the loading process is deformation-controlled, the material behaves in a stable manner and the descending load vs. deformation law can be obtained experimentally.

When strain-softening is involved in the mechanical behaviour of a structural member not homogeneous or not homogeneously loaded, it is accompanied by strain-localization [5,8,13,14]. The suitable constitutive law to describe the mechanical behaviour in the localized damage zone appears to be "stress vs. displacement", so that energy is dissipated on a crack area rather than in a material volume. Fracture energy is defined as the energy dissipated on a unit crack area [8,15-18].

Whereas in classical plasticity and damage theory [19,20] geometrically similar structures behave in the same way as only energy dissipation per unit volume is allowed, when also energy dissipation per unit area is contemplated (strain or curvature localization) the global brittleness becomes scale-dependent.

Size-scale and slenderness are demonstrated to have fundamental influence on the global structural behaviour, which can range from

ductile to brittle when strain softening and strain localization are taken into account. The brittle behaviour coincides with a snap-back instability [21-22] in the load vs. deflection path, which shows a positive slope in the softening branch.
The snap-back load-deflection branch may be captured experimentally only if the loading process is controlled by a monotonically increasing function of time, e.g. the crack mouth opening or sliding displacement. On the other hand, the snap-back load-deflection branch may be captured numerically only if the loading process is controlled by a monotonically increasing function of the crack length. An example of such function is provided by the "indirect displacement control scheme" [23]. This technique uses a displacement norm as controlling parameter. On the other hand, as a monotonically increasing function of the crack length, it is possible to use the crack length itself, in Mode I [7,17] as well as in Mixed Mode [24,25]. Such technique, called "crack length control scheme", will be presented in this report. FEM-crack propagation requires a continuous modification of the mesh. Whereas for Mode I, only node untieing is applied to simulate crack growth, for Mixed Mode interelement crack propagation a topological variation must be performed at each step automatically [9].

The non-linear and dissipative phenomena occurring at the crack tip of concrete-like materials are different from those occurring in metals, and theory of plasticity is not able to describe them in a consistent manner. Microcracking of mortar and debonding between mortar and aggregates are the principal damage mechanisms in the vicinity of a crack tip. When a sufficiently high number of microcracks and debondings coalesce, a macroscopical growth of the macrocrack occurs and this is often only a stable stage of the crack propagation process. The firstly partial and then total stress relaxation at the crach tip is the result of the strain-softening constitutive law of concrete-like materials.

The amount of energy dissipated in the localized Mixed Mode fracture zone results to be experimentally equal to the product of Mode I fracture energy and total fracture area. Therefore, the introduction of an additional fracture toughness parameter for Mixed Mode problems, appears unnecessary. The assumption of the "Maximum Circumferential Stress Criterion" [26], for which any crack growth step is produced by an elementary Mode I (or opening) mechanism along the curvilinear trajectory, seems to be confirmed by the experimental results.

A general explanation to the well-known decrease in apparent strength and increase in fracture toughness by increasing the member sizes, is given in terms of Dimensional Analysis. Due to the different physical dimensions of tensile strength $[FL^{-2}]$ and fracture toughness $[FL^{-1}]$, the true values of such two intrinsecal material properties may be found exactly only with comparatively large members, when the brittleness numbers [1-4,6]

$$s = \frac{K_{IC}}{\sigma_u\, b^{1/2}} \qquad \text{and} \qquad s_E = \frac{G_F}{\sigma_u\, b}$$

are sufficiently low, K_{IC} being the fracture toughness, G_F the fracture energy and σ_u the tensile strength of concrete, and b a characteristic specimen size.

When a small percentage of steel is required to reinforce high strength concrete, the crushing failure of the beam edge in compression is usually avoided. On the other hand, one or more cracks originate at the beam edge in tension and the material is so brittle in this case that the size of the crack tip process zone is very small if compared with the size of the zone where the stress singularity field is dominant. For these reasons, it will be shown that a Linear Elastic Fracture Mechanics model [27-29] is able to capture the most relevant aspects and trends in the mechanical and failure behaviour of low reinforced high strength concrete beams in flexure.

The present theoretical and experimental investigation aims at evaluating the size-scale effects on reinforced concrete members, by means of three point bending tests. Thirty (30) R.C. beams are tested [29], with thickness t = 150 mm and depth b = 100, 200, 400 mm, respectively. The span is assumed to be six (6) times the beam depth b.

Five different values of the non-dimensional number

$$N_P = \frac{f_y b^{1/2}}{K_{IC}} \frac{A_s}{A}$$

are considered (about equal to 0.00, 0.10, 0.30, 0.75, 1.20), f_y being the steel yield strength, K_{IC} the concrete fracture toughness and A_s/A the steel percentage. Both size-scale b and steel percentage A_s/A are varied. The ratio of the distance of the bars from the lower edge of the beam, to the beam depth is assumed to be constant and equal to 0.1.

The loading process is carried out on initially uncracked R.C. beams, by controlling the tensile strain on the lower edge of the beam or, after cracking, the crack mouth opening displacement. Even the compressive strain on the upper edge of the beam and the beam deflection are recorded.

Remarkable size-scale effects are theoretically predicted and experimentally confirmed. The brittleness of the system increases by increasing size-scale and/or decreasing steel area. On the other hand, a physically similar behaviour is revealed in the cases where the non-dimensional number N_P is the same.

Based on the concepts of Fracture Mechanics, fracture criteria for, respectively, plain and reinforced concrete structures will be proposed, analyzed, numerically implemented and experimentally verified.

2. ELEMENTARY FRACTURE MECHANICS MODELS

2.1 Application of dimensional analysis

Due to the different physical dimensions of ultimate tensile strength, σ_u, and fracture toughness , K_{IC}, scale effects are always present in the usual fracture testing of common engineering materials. This means that, for the usual size-scale of the laboratory specimens, the ultimate strength collapse or the plastic collapse at the ligament tends to anticipate and obscure the brittle crack propagation. Such a competition between collapses of a different nature can easily be shown by considering the ASTM formula [30] for the three point bending test evaluation of fracture toughness (Fig.1):

$$K_I = \frac{P\,l}{t\,b^{3/2}} f(a/b), \tag{2.1}$$

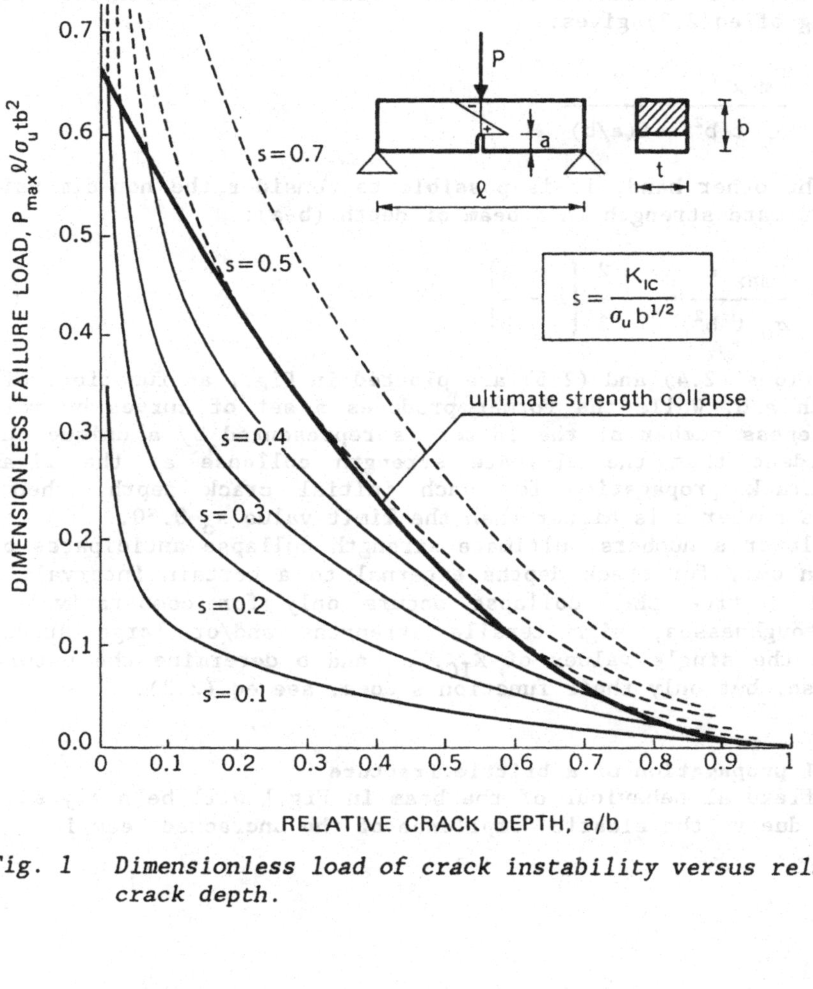

Fig. 1 Dimensionless load of crack instability versus relative crack depth.

with

$$f(a/b)=2.9(a/b)^{1/2}-4.6(a/b)^{3/2}+21.8(a/b)^{5/2}-37.6(a/b)^{7/2}+38.7(a/b)^{9/2}.$$

At the crack propagation eq.(2.1) becomes:

$$K_{IC} = \frac{P_{max}\, l}{t\, b^{3/2}} f(a/b), \tag{2.2}$$

where P_{max} is the external load of brittle fracture.
If both members of eq.(2.2) are divided by $\sigma_u\, b^{1/2}$ we obtain:

$$\frac{K_{IC}}{\sigma_u\, b^{1/2}} = s = \frac{P_{max}\, l}{\sigma_u\, t\, b^2} f(a/b), \tag{2.3}$$

where s is the *brittleness number* defined by Carpinteri [1-4]. Rearranging of eq(2.3) gives:

$$\frac{P_{max}\, l}{\sigma_u\, t\, b^2} = \frac{s}{f(a/b)}. \tag{2.4}$$

On the other hand, it is possible to consider the non-dimensional load of ultimate strength in a beam of depth (b-a):

$$\frac{P_{max}\, l}{\sigma_u\, t\, b^2} = \frac{2}{3}\left(1 - \frac{a}{b}\right)^2. \tag{2.5}$$

Equations (2.4) and (2.5) are plotted in Fig.1 as functions of the crack depth a/b. While the former produces a set of curves by varying the brittleness number s, the latter is represented by a unique curve. It is evident that the ultimate strength collapse at the ligament precedes crack propagation for each initial crack depth, when the brittleness number s is higher than the limit-value $s_o=0.50$.

For lower s numbers, ultimate strength collapse anticipates crack propagation only for crack depths external to a certain interval. This means that a true LEFM collapse occurs only for comparatively low fracture toughnesses, high tensile strengths and/or large structure sizes. Not the single values of K_{IC}, σ_u and b determine the nature of the collapse, but only their function s does, see eq.(2.3).

2.2 Virtual propagation of a brittle fracture

The flexural behaviour of the beam in Fig.1 will be analyzed. The deflection due to the elastic compliance of the uncracked beam is:

$$\delta_e = \frac{P\, l^3}{48\ E\ I}\ , \tag{2.6}$$

where I is the inertial moment of the cross-section. On the other hand, the deflection due to the local crack compliance is [31]:

$$\delta_c = \frac{3}{2}\ \frac{P\, l^2}{t\ b^2\ E}\ g(a/b)\ , \tag{2.7}$$

with:

$$g(a/b) = \left(\frac{a/b}{1-a/b}\right)^2 \times$$

$$\times \quad \left\{5.58-19.57(a/b)+36.82(a/b)^2-34.94(a/b)^3+12.77(a/b)^4\right\}\ . \tag{2.8}$$

The superposition principle provides:

$$\delta = \delta_e + \delta_c\ ,$$

and, in non-dimensional form:

$$\frac{\delta\, l}{\epsilon_u\ b^2} = \frac{P\, l}{\sigma_u\ t\ b^2}\ \left[\frac{1}{4}\ (l/b)^3 + \frac{3}{2}\ (l/b)^2\ g(a/b)\right]\ , \tag{2.9}$$

where $\epsilon_u = \sigma_u/E$. The term within square brackets is the dimensionless compliance, which is a function of the beam slenderness, l/b, as well as of the crack depth, a/b. Some linear load-deflection diagrams are represented in Fig.2, by varying the crack depth a/b and for the fixed ratio $l/b=4$.

Through eqs.(2.4) and (2.5), it is possible to determine the point of crack propagation as well as the point of ultimate strength on each linear plot in Fig.2. Whereas the former depends on the brittleness number s, the latter is unique. The set of the crack propagation points with s=constant and by varying the crack depth, represents a virtual load-deflection path, where point by point the load is always that producing crack instability.

When the crack grows, the load of instability decreases and the compliance increases, so that the product at the right member of eq.(2.9) may result to be either decreasing or increasing. The diagrams in Fig.2 show the deflection decreasing (with the load) up to the crack depth $a/b \simeq 0.3$ and then increasing (in discordance with the load).

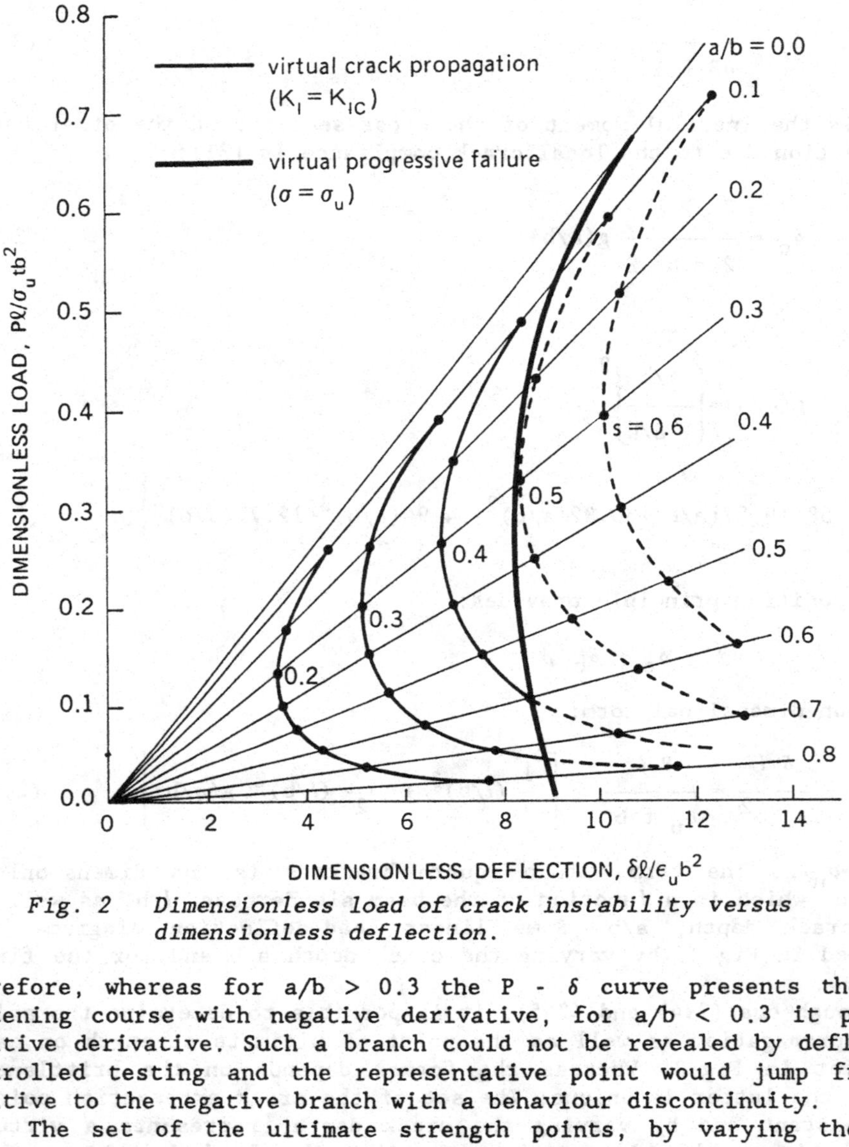

Fig. 2 Dimensionless load of crack instability versus dimensionless deflection.

Therefore, whereas for $a/b > 0.3$ the $P - \delta$ curve presents the usual softening course with negative derivative, for $a/b < 0.3$ it presents positive derivative. Such a branch could not be revealed by deflection-controlled testing and the representative point would jump from the positive to the negative branch with a behaviour discontinuity.

The set of the ultimate strength points, by varying the crack depth, is represented by the thick line in Fig.2. Such a line intersects the virtual crack propagation curves with $s \leq s_o = 0.50$, analogous to what is shown in Fig.1, and presents a slight indentation with $dP/d\delta > 0$.

The crack mouth opening displacement w_1 is a function of the specimen geometry and of the elastic modulus [31]:

$$w_1 = \frac{6 P l a}{t b^2 E} h(a/b) , \tag{2.10}$$

with:

$$h(a/b)=0.76-2.28(a/b)+3.87(a/b)^2-2.04(a/b)^3+\frac{0.66}{(1-a/b)^2}. \qquad (2.11)$$

In non-dimensional form eq.(2.10) becomes:

$$\frac{w_1\, l}{\epsilon_u\, b^2}=\frac{P\, l}{\sigma_u\, t\, b^2}\left[6\ (l/b)\quad (a/b)\quad h(a/b)\right]. \qquad (2.12)$$

The term within square brackets is the dimensionless compliance which, also in this case, depends on beam slenderness and crack depth. Some linear load-crack mouth opening dislacement diagrams are reported in Fig.3, by varying the crack depth a/b and for $l/b = 4$.

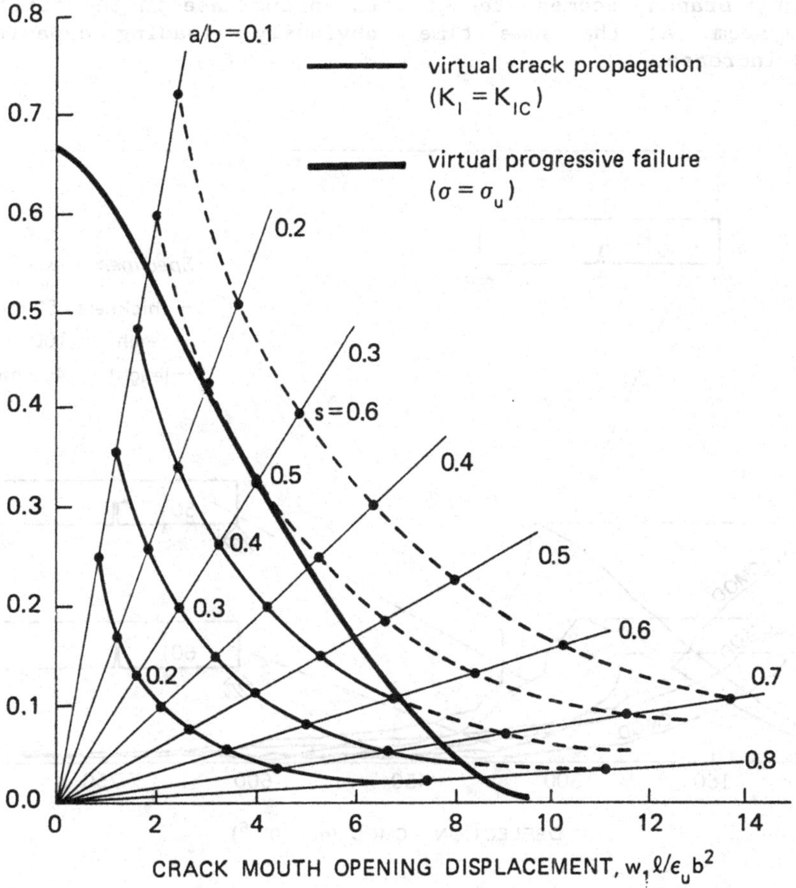

Fig. 3 Dimensionless load of crack instability versus dimensionless crack mouth opening displacement.

The set of the crack propagation points with s = constant and by varying the crack depth, represents a virtual process even in this case. When the crack grows, the product at the right member of eq.(2.12) always increases, the compliance increase prevailing over the critical load decrease for each value of a/b. The P - w_1 curve presents always negative derivative and the crack mouth opening displacement w_1 increases even when load and deflection both decrease ($dP/d\delta > 0$) in the catastrophic P - δ branch. If the crack mouth opening displacement is controlled, i.e., if w_1 increases monotonically without jumping, it would be possible to go along the virtual P - δ path with positive slope.

Such a theoretical statement was confirmed through an experimental investigation on high strength concrete beams [32]. The mechanical response of the specimens with deep cracks appeared stable (Fig.4-a). Both load-deflection and load-CMOD curves showed the same shape with a softening branch of negative slope. By decreasing the relative crack depth such a branch becomes steeper with an increase in the brittleness of the system. At the same time, obviously, loading capacity and stiffness increase.

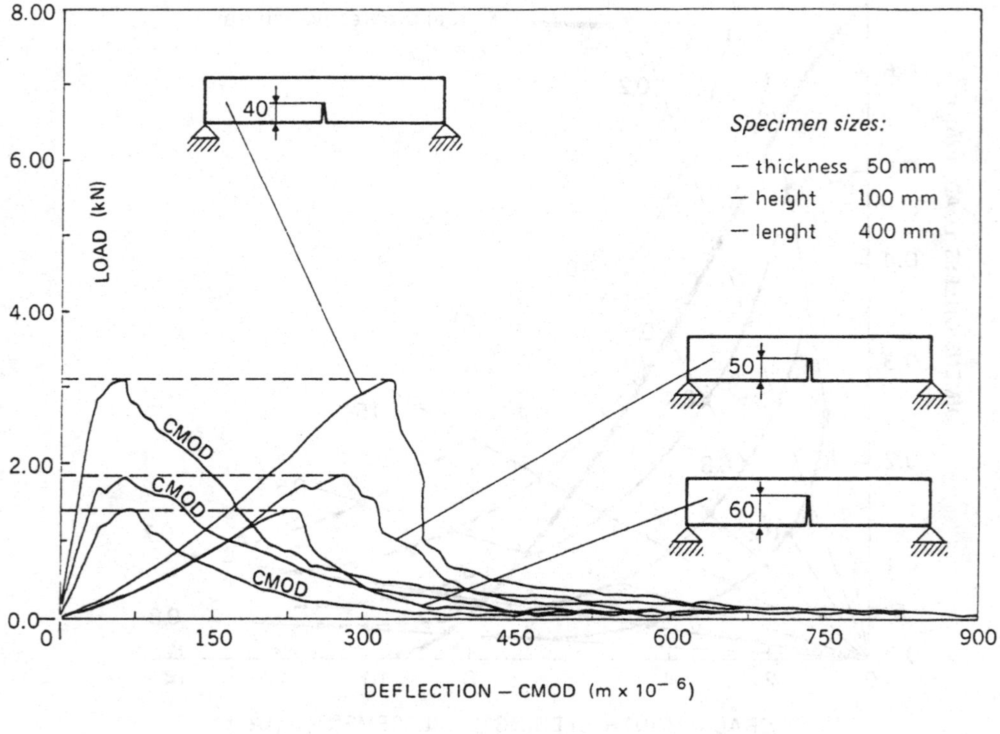

Fig. 4-a Experimental load-deflection and load-CMOD diagrams (a_o/b = 0.4, 0.5, 0.6).

The specimens with shallow cracks (Fig.4-b) on the contrary, presented a very unstable behaviour. Whereas the load-CMOD curves have a softening tail with negative slope, the load-deflection curves are characterized by a snap-back softening instability with a softening branch of partial positive slope. More precisely, the case a_o = 30 mm shows an almost vertical drop in the loading capacity when the maximum load is achieved. This experimental finding confirms the theoretical result in Fig.2. In fact, the relative crack depth a_o/b = 0.3 is the critical condition between stability and instability for deflection-controlled loading processes. If the loading process had been deflection-controlled, then, once reached the bifurcation point of the loading path, the load would have presented a negative jump down to to the lower softening branch with a negative slope.

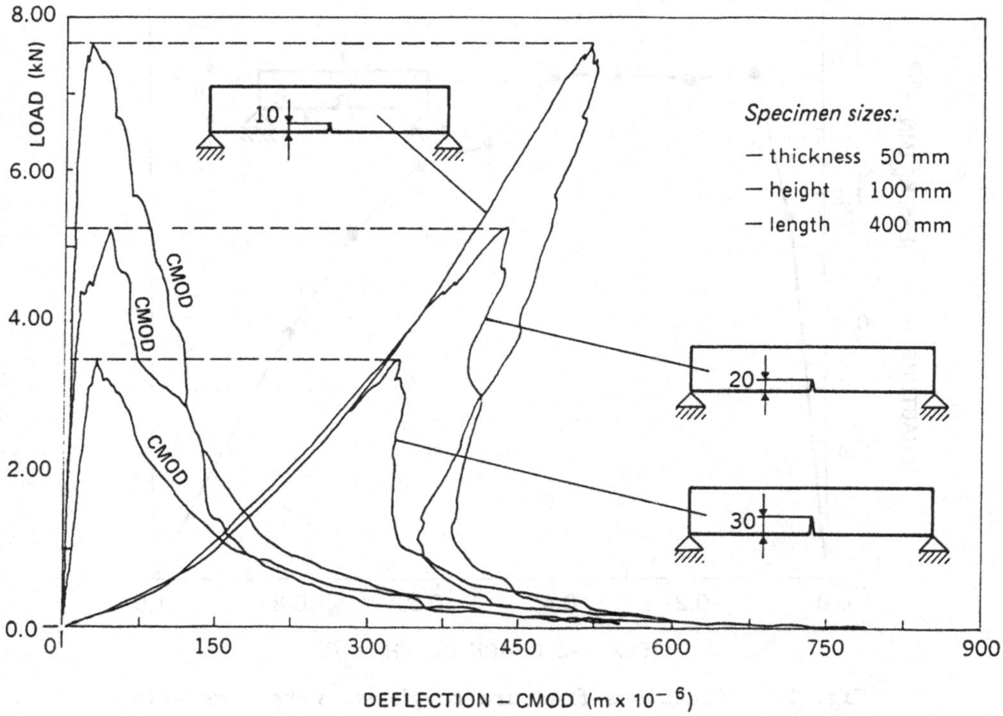

Fig. 4-b Experimental load-deflection and load-CMOD diagrams (a_o/b = 0.1,0.2, 0.3).

Therefore, it is evident that, although the process is unstable in nature, it can develop in a stable manner if CMOD-controlled.

All the diagrams in Fig.4 converge towards the same asymptotic tail, the limit situation being independent of the initial crack length.

The previous theoretical and experimental analysis emphasize that the (brittle or ductile) structural behaviour is connected with a geometrical feature, as is the case of the crack depth. More generally, all the geometrical features of the specimen influence the global brittleness (or ductility), and particularly slenderness and size-scale. The values of K_{IC}, according to equation (2.1), are reported in Fig.5 against the relative crack depth. They appear nearly constant for $0.1 < a_o/b < 0.4$, and then decreasing with the crack depth. Even the fictitious K_{IC} related to the crack depths $a_o/b = 0.05$ and $a_o/b = 0.7$ and 0.8 are reported.

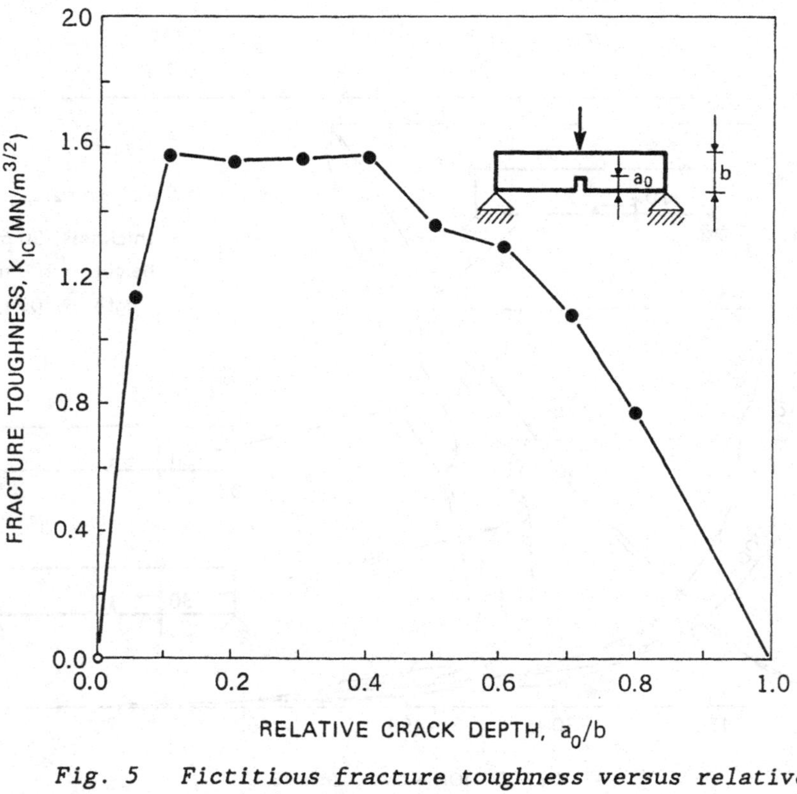

Fig. 5 Fictitious fracture toughness versus relative crack depth.

They present values much lower than the others since in these cases the ultimate tensile strength collapse at the ligament precedes decidedly the LEFM instability (Figs. 1 and 2, s = 0.52). This fracture toughness decrease for extreme relative crack depths (tending to zero or to unity) was discussed in detail by Carpinteri [1-4].

2.3 Uniaxial tensile loading of slabs

Let us consider an elastic-softening material with a double constitutive law: (a) tension σ vs. dilatation ϵ, and (b) tension σ vs. crack opening displacement w, after reaching the ultimate tensile strength σ_u or strain $\epsilon_u = \sigma_u/E$ (Fig.6):

$$\sigma = E\,\epsilon\ , \qquad \text{for } \epsilon \le \epsilon_u\ , \tag{2.13-a}$$

$$\sigma = \sigma_u(1-w/w_c)\ , \qquad \text{for } w \le w_c\ , \tag{2.13-b}$$

$$\sigma = 0\ , \qquad \text{for } w > w_c\ . \tag{2.13-c}$$

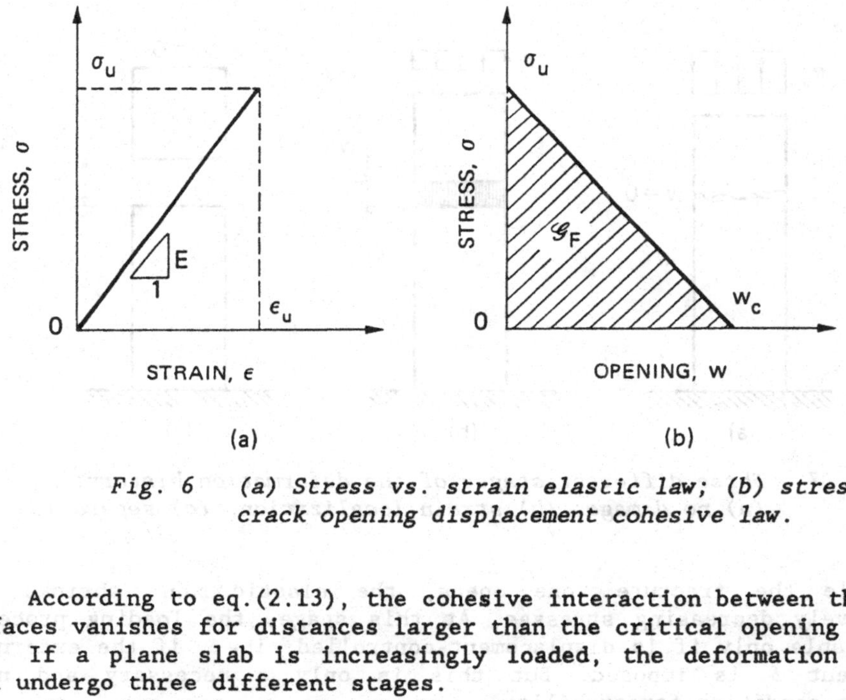

Fig. 6 (a) Stress vs. strain elastic law; (b) stress vs. crack opening displacement cohesive law.

According to eq.(2.13), the cohesive interaction between the crack surfaces vanishes for distances larger than the critical opening w_c.

If a plane slab is increasingly loaded, the deformation history will undergo three different stages.

(A) - The slab behaves elastically without damage or fracture zones (Fig. 7-a). The displacement of the upper edge is:

$$\delta = \frac{\sigma}{E}\, l\ , \qquad \text{for } \epsilon \le \epsilon_u\ . \tag{2.14}$$

(B) - After reaching the ultimate tensile strength σ_u, a fracture cohesive zone develops in the weakest section of the slab. Observe that, as the stress field is homogeneous, another cause of inhomogeneity must be assumed for strain-localization. The slab behaves elastically only outside the fracture zone (Fig. 7-b). The displacement of the upper edge is:

$$\delta = \frac{\sigma}{E} l + w \,, \qquad \text{for } w \le w_c \,. \tag{2.15}$$

Recalling eq.(2.13-b), eq.(2.15) gives:

$$\delta = \frac{\sigma}{E} l + w_c \left(1 - \frac{\sigma}{\sigma_u}\right) , \qquad \text{for } w \le w_c \,. \tag{2.16}$$

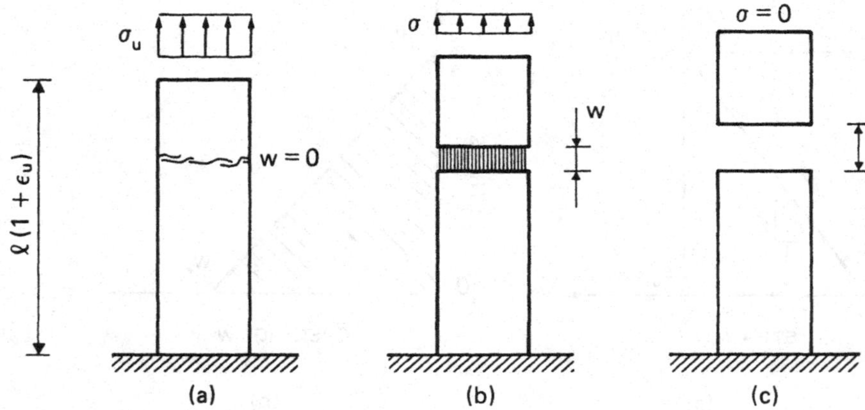

Fig. 7 Three different stages of the deformation history: (a) no damage; (b) strain localization; (c) separation.

While the fracture zone opens, the elastic zone shrinks at progressively decreasing stresses. At this stage, the loading process may be stable only if is displacement-controlled, i.e., if the external displacement δ is imposed. But this is only a necessary and not sufficient condition for stability.

(C) - When $\delta \ge w_c$ the reacting stress σ vanishes, the cohesive forces disappear and the slab is completely separated into two pieces (Fig. 7-c).

Rearranging of eq.(2.14) gives:

$$\sigma = E \frac{\delta}{l} \,, \qquad \text{for } \delta \le \epsilon_u l \,, \tag{2.17}$$

while the condition of complete separation (stage C) reads:

$$\sigma = 0 \ , \qquad \text{for } \delta \geq w_c \ . \qquad (2.18)$$

When $w_c > \epsilon_u l$, the softening process is stable only if displacement-controlled, since the slope $d\sigma/d\delta$ at stage (B) is negative (Fig. 8-a).

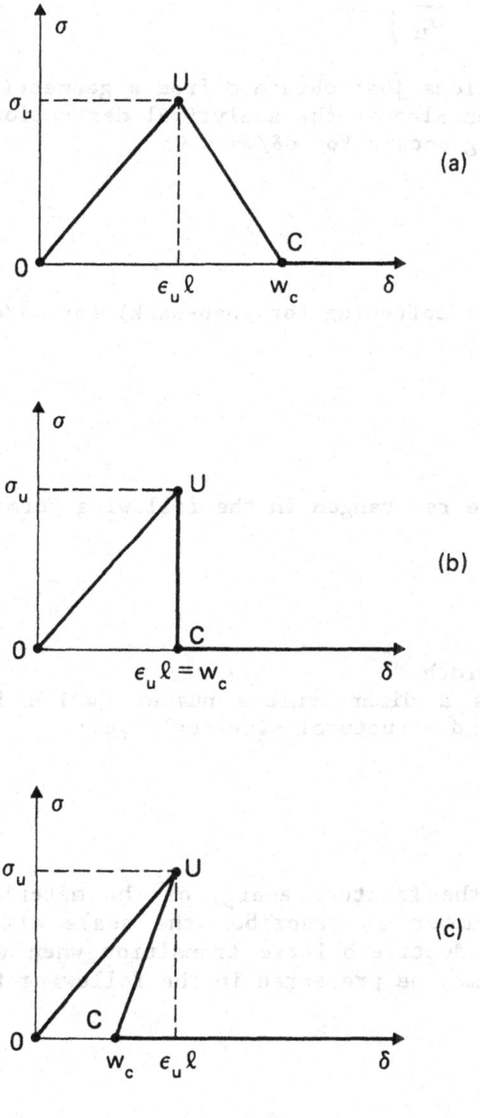

Fig. 8 Stress-displacement response: (a) externally unstable; (b) and (c) internally unstable.

When $w_c = \epsilon_u l$, the slope $d\sigma/d\delta$ is infinite and a drop in the loading capacity occurs, even if the loading is displacement-controlled (Fig. 8-b). Eventually, when $w_c < \epsilon_u l$, the slope $d\sigma/d\delta$ becomes positive (Fig. 8-c) and the same negative jump occurs like that shown in Fig. 8-b.

Rearranging of eq.(2.16) provides:

$$\delta = w_c + \sigma \left(\frac{l}{E} - \frac{w_c}{\sigma_u} \right) . \tag{2.19}$$

The same conditions just obtained from a geometrical point of view (Fig. 8), may be given also by the analytical derivation of eq.(2.19).

Normal softening occurs for $d\delta/d\sigma < 0$:

$$\left(\frac{l}{E} - \frac{w_c}{\sigma_u} \right) < 0 , \tag{2.20}$$

whereas catastrophical softening (or snap-back) for $d\delta/d\sigma \geq 0$:

$$\left(\frac{l}{E} - \frac{w_c}{\sigma_u} \right) \geq 0 . \tag{2.21}$$

Eq.(2.21) may be rearranged in the following form:

$$\frac{(w_c/2b)}{\epsilon_u(l/b)} \leq \frac{1}{2} , \tag{2.22}$$

where b is the slab width.

The ratio $(w_c/2b)$ is a dimensionless number, which is a function of material properties and structural size-scale [6]:

$$s_E = \frac{w_c}{2b} = \frac{G_F}{\sigma_u b} , \tag{2.23}$$

$G_F = \frac{1}{2} \sigma_u w_c$ being the fracture energy of the material (Fig. 6). The energy brittleness number s_E describes the scale effects of fracture mechanics, i.e., the ductile-brittle transition when the size-scale is increased. Eq.(2.22) may be presented in the following final form:

$$\frac{s_E}{\epsilon_u \lambda} \leq \frac{1}{2} , \tag{2.24}$$

with λ = slenderness = l/b.

When the size-scale and slab slenderness are relatively large and the fracture energy relatively low, the global structural behaviour is brittle. Not the single values of parameters s_E, ϵ_u and λ, but only their combination $B = s_E/\epsilon_u \lambda$ is responsible for the global brittleness or ductility of the structure considered.

When $B \leq 1/2$ the plane rectangular slab of Fig. 7 shows a mechanical behaviour which can be defined *brittle* or *catastrophic*. A snap-back instability of the global equilibrium occurs since, if point U in Fig. 8-c is reached and then the imposed external displacement δ is decreased by a very small amount $d\delta$, the global unloading may occur along two alternative paths: the elastic UO or the virtual softening UC.

The global brittleness of the slab can be defined as the ratio of the ultimate elastic energy contained in the body to the energy dissipated by fracture:

$$\text{Brittleness} = \frac{\frac{1}{2}\frac{\sigma_u^2}{E} \times (\text{Area}) \times l}{G_F \times (\text{Area})} = \frac{1}{2B} . \tag{2.25}$$

Such a ratio is higher than unity when eq.(2.21) is verified and a catastrophical softening instability occurs.
Recalling the relationship between energy brittleness number s_E (see eq.2.23) and stress brittleness number:

$$s = \frac{K_{IC}}{\sigma_u b^{1/2}} , \tag{2.26}$$

which reads as follows:

$$s_E = \epsilon_u s^2 , \tag{2.27}$$

it is possible to relate the brittleness ratio in eq.(2.25) to the stress brittleness number s, through the slab slenderness λ:

$$\text{Brittleness} = \frac{\lambda}{2s^2} . \tag{2.28}$$

2.4 Three point bending of beams

The linear elastic behaviour of a three point bending initially uncracked beam may be represented by the following dimensionless equation:

$$\tilde{P} = \frac{4}{\lambda^3} \tilde{\delta} , \tag{2.29}$$

where the dimensionless load and central deflection are respectively given by:

$$\tilde{P} = \frac{P\,l}{\sigma_u t b^2}, \tag{2.30}$$

$$\tilde{\delta} = \frac{\delta\,l}{\epsilon_u\, b^2}, \tag{2.31}$$

with l = beam span, b = beam depth, t = beam thickness.

Once the ultimate tensile strength σ_u is achieved at the lower beam edge, a fracturing process in the central cross-section is supposed to start. Such a process admits a limit-situation like that in Fig.9. The limit stage of the fracturing and deformation process may be considered as that of two rigid parts connected by the hinge A in the upper beam edge. The equilibrium of each part is ensured by the external load, the support reaction and the closing cohesive forces. The latter depend on the distance between the two interacting surfaces: increasing the distance w, the cohesive forces decrease till they vanish for w ≥ w_c.

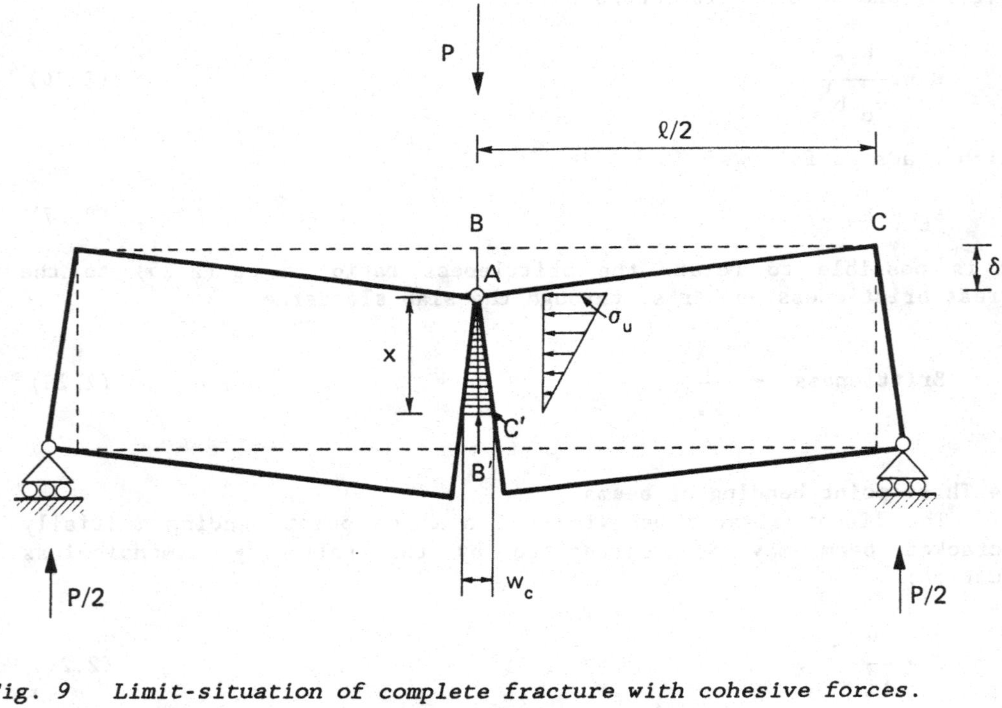

Fig. 9 Limit-situation of complete fracture with cohesive forces.

The geometrical similitude of the triangles ABC and AB'C' in Fig.9 provides:

$$\frac{\delta}{l/2} = \frac{w_c/2}{x} , \tag{2.32}$$

where x is the extension of the triangular distribution of cohesive forces. Eq.(2.32) can be rearranged as:

$$x = \frac{w_c \, l}{4\delta} . \tag{2.33}$$

The rotational equilibrium round point A is possible for each beam part only if the moments of support reaction and cohesive forces respectively are equal:

$$\frac{P}{2}\frac{l}{2} = \frac{\sigma_u x t}{2}\frac{x}{3} . \tag{2.34}$$

Recalling eq.(2.33), the relation between load and deflection may be obtained:

$$P = \frac{\sigma_u t l w_c^2}{24} \frac{1}{\delta^2} . \tag{2.35}$$

Eq.(2.35) can be put into dimensionless form:

$$\tilde{P} = \frac{1}{6}\left(\frac{s_E \, \lambda^2}{\epsilon_u \, \tilde{\delta}}\right)^2 . \tag{2.36}$$

While the linear eq.(2.29) describes the elastic behaviour of the beam, initially uncracked, the hyperbolic eq.(2.36) represents the asymptotical behaviour of the same beam, totally cracked. Eq.(2.29) is valid only for load values lower than that producing the ultimate tensile strength σ_u at the lower beam edge:

$$\tilde{P} \leq \frac{2}{3} . \tag{2.37}$$

On the other hand, eq.(2.36) is valid only for deflection values higher than that producing a cohesive zone of extension x equal to the beam depth b:

$$x \leq b . \tag{2.38}$$

From eqs.(2.33) and (2.38) it follows:

$$\tilde{\delta} \geq \frac{s_E \lambda^2}{2 \epsilon_u} . \tag{2.39}$$

The bounds (2.37) and (2.39), upper for load and lower for deflection respectively, can be transformed into two equivalent bounds, both upper for deflection and load. Eqs.(2.29) and (2.37) provide:

$$\tilde{\delta} \leq \frac{\lambda^3}{6} , \tag{2.40}$$

whereas eqs.(2.36) and (2.39):

$$\tilde{P} \leq \frac{2}{3} . \tag{2.41}$$

Conditions (2.37) and (2.41) are identical. Therefore, a stability criterion for elastic-softening beams may be obtained comparing eqs.(2.39) and (2.40). When the two domains are separated, it is presumable that the two P - δ branches - linear and hyperbolic - are connected by a regular curve (Fig. 10-a). On the other hand, when the two domains are partially overlapped, it is well-founded to suppose them as connected by a curve with higly negative or even positive slope (Fig. 10-b).

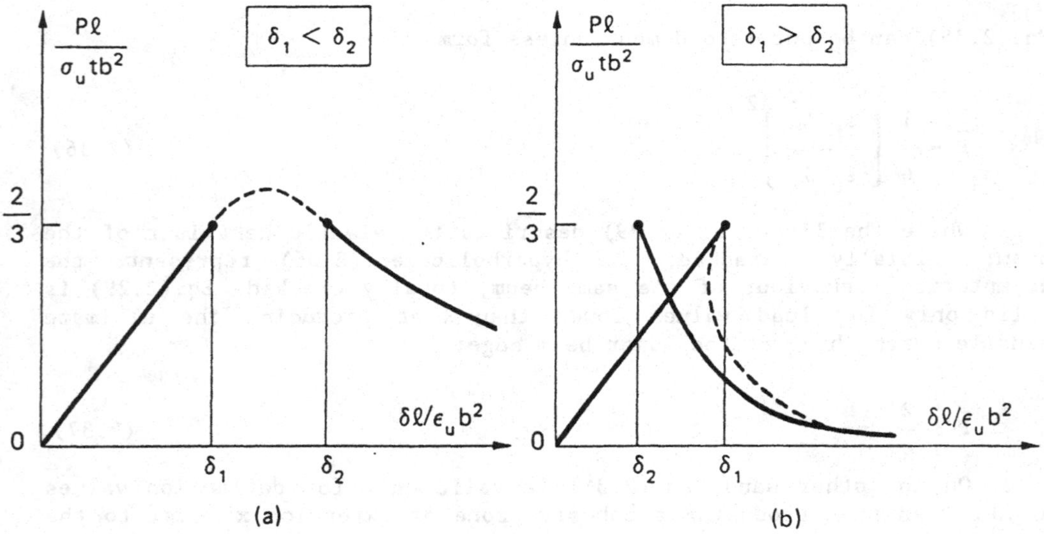

Fig. 10 Load-deflection diagrams: (a) ductile and (b) brittle condition. $\delta_1 = \lambda^3/6$; $\delta_2 = s_E \lambda^2/2\epsilon_u$.

Unstable behaviour and catastrophical events (snap-back) may be possible for:

$$\frac{s_E \lambda^2}{2 \epsilon_u} \leq \frac{\lambda^3}{6} , \tag{2.42}$$

and the brittleness condition for the three point bending geometry becomes:

$$\frac{s_E}{\epsilon_u \lambda} \leq \frac{1}{3} . \tag{2.43}$$

Even in this case, the system is brittle for low brittleness numbers s_E, high ultimate strains ϵ_u and large slendernesses λ. Observe that the same dimensionless number $B = s_E/\epsilon_u\lambda$ appears in eq.(2.24) also, where the upper bound for brittleness is equal to 1/2.

It is therefore evident that the relative brittleness for a structure is dependent on loading condition and external constraints, in addition to material properties, size-scale and slenderness. For instance, uniaxial tension is more unstable than three point bending (Fig. 11).

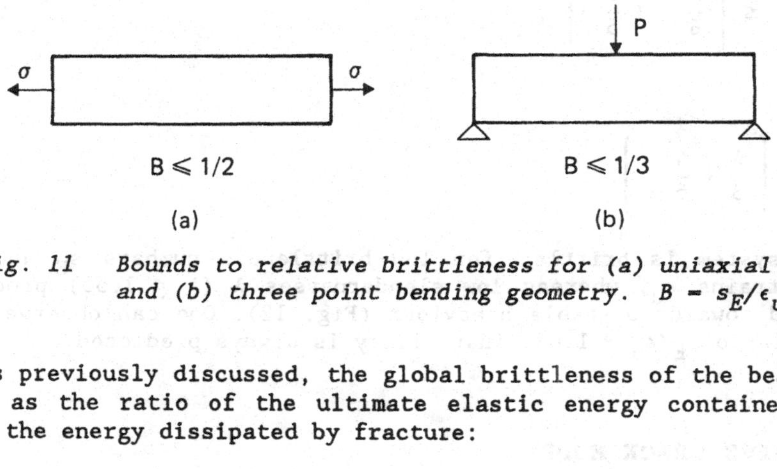

Fig. 11 Bounds to relative brittleness for (a) uniaxial tension and (b) three point bending geometry. $B = s_E/\epsilon_u\lambda$.

As previously discussed, the global brittleness of the beam can be defined as the ratio of the ultimate elastic energy contained in the body to the energy dissipated by fracture:

$$\text{Brittleness} = \frac{\frac{1}{2} P_u \delta_u}{G_F \times (\text{Area})} = \frac{\frac{1}{18} \sigma_u \epsilon_u b t l}{G_F b t} = \frac{1}{18\, B} . \tag{2.44}$$

Such a ratio is higher than unity when:

$$\frac{s_E}{\epsilon_u \lambda} \leq \frac{1}{18} . \tag{2.45}$$

Eq.(2.45) represents a stricter condition for global structural brittleness compared with eq.(2.43).

2.5 Three point bending of slabs

When the shear forces can not be neglected (deep beams) and the Poisson ratio is negligible, eq.(2.29) is replaced by [33]:

$$\tilde{\delta} = \tilde{P} \left(\frac{1}{4} \lambda^3 + \frac{3}{5} \lambda \right) , \tag{2.46}$$

whereas eq.(2.40) becomes:

$$\tilde{\delta} \leq \left(\frac{\lambda^3}{6} + \frac{2}{5} \lambda \right) . \tag{2.47}$$

Snap-back is then expected for:

$$\frac{s_E \lambda^2}{2 \epsilon_u} \leq \left(\frac{\lambda^3}{6} + \frac{2}{5} \lambda \right) , \tag{2.48}$$

or:

$$\frac{s_E}{\epsilon_u} \leq \left(\frac{\lambda}{3} + \frac{4}{5} \frac{1}{\lambda} \right) . \tag{2.49}$$

The system is brittle for low brittleness numbers s_E and high ultimate strains ϵ_u, whereas low slendernesses λ ($\lambda \lesssim 1.55$) produce a clear trend towards unstable behaviour (Fig. 12). One can observe that, below the ratio $s_E/\epsilon_u \simeq 1.03$, instability is always predicted.

3. COHESIVE CRACK MODEL

3.1 Numerical implementation

The cohesive crack model is based on the following assumptions [6-8,18]

(1) The cohesive fracture zone (plastic or process zone) begins to develop when the maximum principal stress achieves the ultimate tensile strength σ_u (Fig. 6-a).

(2) The material in the process zone is partially damaged but still able to transfer stress. Such a stress is dependent on the crack opening displacement w (Fig. 6-b).

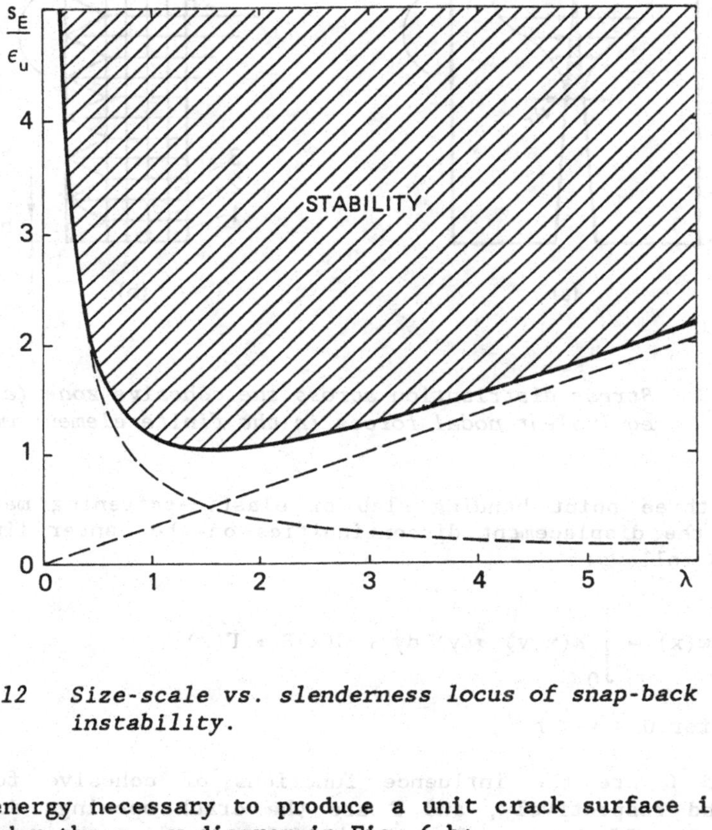

Fig. 12 Size-scale vs. slenderness locus of snap-back instability.

The energy necessary to produce a unit crack surface is given by the area under the σ - w diagram in Fig. 6-b:

$$G_F = \int_0^{w_c} \sigma \, dw = \frac{1}{2} \sigma_u w_c \quad .$$

The *real crack tip* is defined as the point where the distance between the crack surfaces is equal to the critical value of crack opening displacement w_c and the normal stress vanishes (Fig. 13-a). On the other hand, the *fictitious crack tip* is defined as the point where the normal stress attains the maximum value σ_u and the crack opening vanishes (Fig. 13-a).

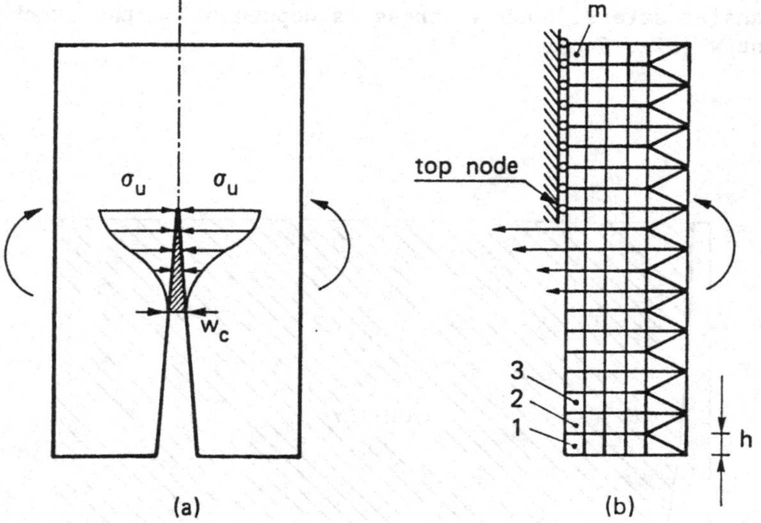

Fig. 13 *Stress distribution across the cohesive zone (a) and equivalent nodal forces in the finite element mesh (b).*

If a three point bending slab of elastic-softening material is considered, the displacement discontinuities on the center line may be expressed as follows:

$$w(x) = \int_0^b K(x,y)\ \sigma(y)\ dy + C(x)P + \Gamma(x)\ , \qquad (3.1)$$

for $0 \leq x < b$,

where K and C are the influence functions of cohesive forces and external load respectively, and Γ are the crack openings due to the specimen weight. If a stress-free crack of length a has developed with a cohesive zone of length Δa, the following additional conditions are to be taken into account:

$$\sigma(y) = 0\ , \qquad \text{for } 0 \leq y \leq a\ , \qquad (3.2\text{-a})$$

$$\sigma(y) = \sigma_u \left(1 - \frac{w(y)}{w_c} \right)\ , \qquad \text{for } a \leq y \leq a+\Delta a\ , \qquad (3.2\text{-b})$$

$$w(x) = 0\ , \qquad \text{for } (a+\Delta a) \leq x < b\ . \qquad (3.2\text{-c})$$

Eqs.(3.1) and (3.2) can be rearranged as follows:

$$w(x) = \int_{a}^{a+\Delta a} K(x,y) \left(1 - \frac{w(y)}{w_c}\right) \sigma_u \, dy + \int_{a+\Delta a}^{b} K(x,y)\, \sigma(y)\, dy + C(x)P + \Gamma(x) ,$$

$$\text{for} \quad 0 \le x \le (a+\Delta a) , \qquad (3.3\text{-a})$$

$$w(x) = 0 , \qquad \text{for} \quad (a+\Delta a) \le x < b . \qquad (3.3\text{-b})$$

The function $\sigma(y)$ depends on the distribution $w(x)$ and on the external load P. Therefore, for each value of P, eq.(3.3-a) represents an integral equation for the unknown function w. On the other hand, the beam deflection is given by:

$$\delta = \int_{0}^{b} C(y)\, \sigma(y)\, dy + D_p P + D_\gamma , \qquad (3.4)$$

where D_p is the deflection for P = 1 and D_γ is the deflection due to the specimen weight.

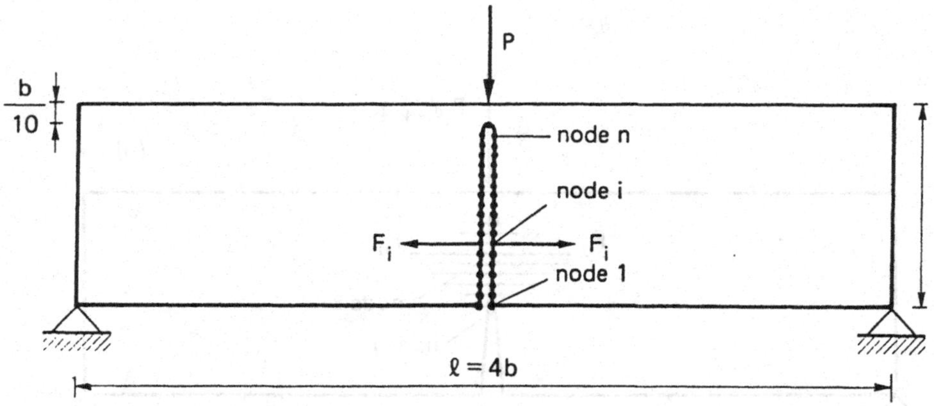

Fig. 14 Finite element nodes along the potential fracture line.

A numerical procedure is implemented to simulate a loading process where the parameter incremented step by step is the fictitious crack depth. Real (or stress-free) crack depth, external load and deflection are obtained at each step after an iterative computation. The closing stresses acting on the crack surfaces (Fig. 13-a) are replaced by nodal forces (Fig. 13-b). The intensity of these forces depends on the opening of the fictitious crack, w, according to the σ - w constitutive law of the material (Fig. 6-b). When the tensile strength σ_u is achieved at the fictitious crack tip (Fig. 13-b), the top node is opened and a cohesive force starts acting across the crack, while the fictitious crack tip moves to the next node.

With reference to the three point bending test (TPBT) geometry in Fig. 14, the nodes are distributed along the potential fracture line. The coefficients of influence in terms of node openings and deflection are computed by a finite element analysis where the fictitious structure in Fig. 14 is subjected to (n + 1) different loading conditions.

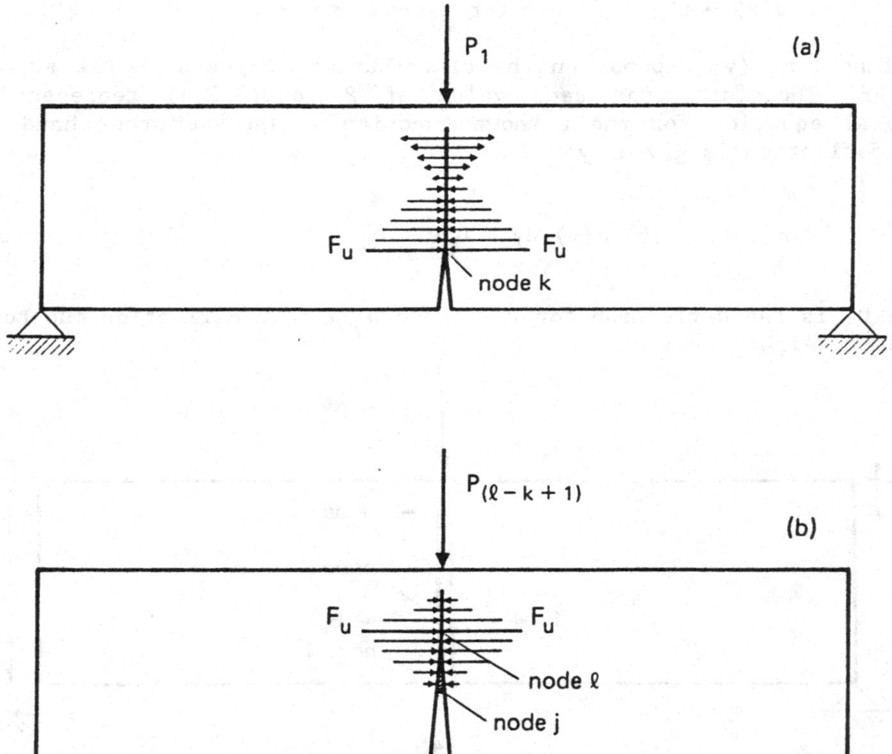

Fig. 15 Cohesive crack configurations at the first (a) and (l-k+1)-th (b) crack growth increment.

Consider the TPBT in Fig. 15-a with the initial crack tip in the node k. The crack opening displacements at the n fracture nodes may be expressed as follows:

$$\underline{w} = \underline{K}\ \underline{F} + \underline{C}\ P + \underline{\Gamma}\ , \tag{3.5}$$

being:

$\underline{w}$ = vector of the crack opening displacements,
$\underline{K}$ = matrix of the coefficients of influence (nodal forces),
$\underline{F}$ = vector of the nodal forces,
$\underline{C}$ = vector of the coefficients of influence (external load),
P = external load,
$\underline{\Gamma}$ = vector of the crack opening displacements due to to the specimen weight.

On the other hand, the initial crack is stress-free and therefore:

$$F_i = 0, \qquad \text{for } i = 1,2,\ldots,(k-1)\ , \tag{3.6-a}$$

while at the ligament there is no displacement discontinuity:

$$w_i = 0, \qquad \text{for } i = k,(k+1),\ldots,n. \tag{3.6-b}$$

Eqs(3.5) and (3.6) constitute a linear algebraical system of 2n equations and 2n unknowns, i.e., the elements of vectors $\underline{w}$ and $\underline{F}$. If load P and vector $\underline{F}$ are known, it is possible to compute the beam deflection, δ:

$$\delta = \underline{C}^T\ \underline{F} + D_P\ P + D_\gamma\ . \tag{3.7}$$

After the first step, a cohesive zone forms in front of the real crack tip (Fig. 15-b), say between nodes j and *l*. Then eqs.(3.6) are replaced by:

$$F_i = 0\ , \qquad \text{for } i = 1,2,\ldots,(j-1)\ , \tag{3.8-a}$$

$$F_i = F_u \left(1 - \frac{w_i}{w_c}\right), \qquad \text{for } i = j,(j+1),\ldots,l\ , \tag{3.8-b}$$

$$w_i = 0\ , \qquad \text{for } i = l,(l+1),\ldots,n\ , \tag{3.8-c}$$

where F_u is the ultimate strength nodal force (Fig. 13-b):

$$F_u = b\ \sigma_u/m\ . \tag{3.9}$$

Eqs.(3.5) and (3.8) constitute a linear algebraical system of (2n+1) equations and (2n+1) unknowns, i.e., the elements of vectors $\underline{w}$ and $\underline{F}$ and the external load P.

At the first step, the cohesive zone is missing (l=j=k) and the load P_1 producing the ultimate strength nodal force F_u at the initial crack tip (node k) is computed. Such a value P_1, together with the related deflection δ_1 computed through eq.(3.7), gives the first point of the P - δ curve.

At the second step, the cohesive zone is between the nodes k and (k+1), and the load P_2 producing the force F_u at the second fictitious crack tip (node k+1) is computed. Eq.(3.7) then provides the deflection δ_2. At

the third step, the fictitious crack tip is in the node (k+2), and so on. The present numerical program simulates a loading process where the controlling parameter is the fictitious crack depth.
On the other hand, real (or stress-free) crack depth, external load and deflection are obtained at each step after an iterative procedure.

The program stops with the untieing of the node n and, consequently, with the determination of the last couple of values F_n and δ_n. In this way, the complete load-deflection curve is automatically plotted by the computer.

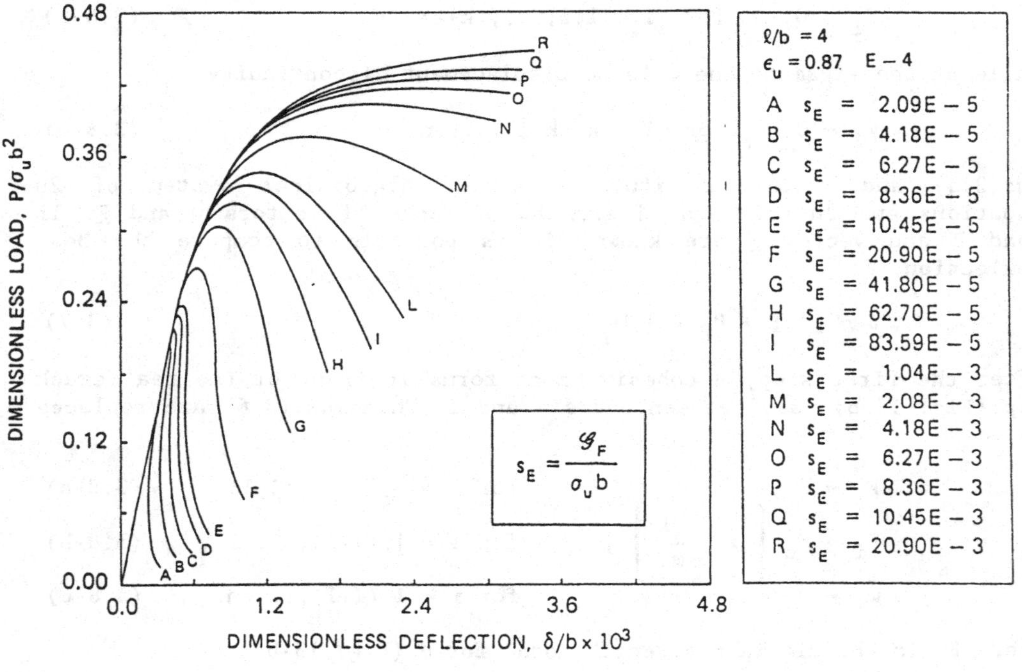

Fig. 16 Dimensionless load vs. deflection diagrams by varying the brittleness number $s_E = G_F/\sigma_u b$ ($\lambda = 4$, $a_o/b = 0.0$, $\epsilon_u = 0.87 \times 10^{-4}$).

3.2 Strain localization and apparent strength of initially uncracked slabs

Some dimensionless load-deflection diagrams for a concrete-like material are plotted in Fig. 16, with $a_o/b = 0.0$, $\epsilon_u = 0.87 \times 10^{-4}$, $\nu = 0.1$, $t = b$, $l = 4b$, and by varying the non-dimensional number s_E.
The specimen behaviour is brittle (snap-back) for low s_E numbers, i.e., for low fracture toughnesses G_F, high tensile strengths, σ_u, and/or

large sizes, b. For $s_E \lesssim 10.45 \times 10^{-5}$, the P - δ curve presents positive slope in the softening branch and a catastophical event occurs if the loading process is deflection-controlled. Such indenting branch is not virtual only if the loading process is controlled by a monotonically increasing function of time [34,35], like, for example, the displacement discontinuity across the crack [32]. On the other hand, eq.(1.43) gives: $s_E \leq 11.60 \times 10^{-5}$. Such a condition reproduces that shown in Fig. 16 very accurately, whereas eq.(1.45) appears too severe. When the post-peak behaviour is kept under control up to the complete structure separation, the area delimited by load-deflection curve and deflection-axis represents the product of fracture energy, G_F, and initial cross-section area, bt.

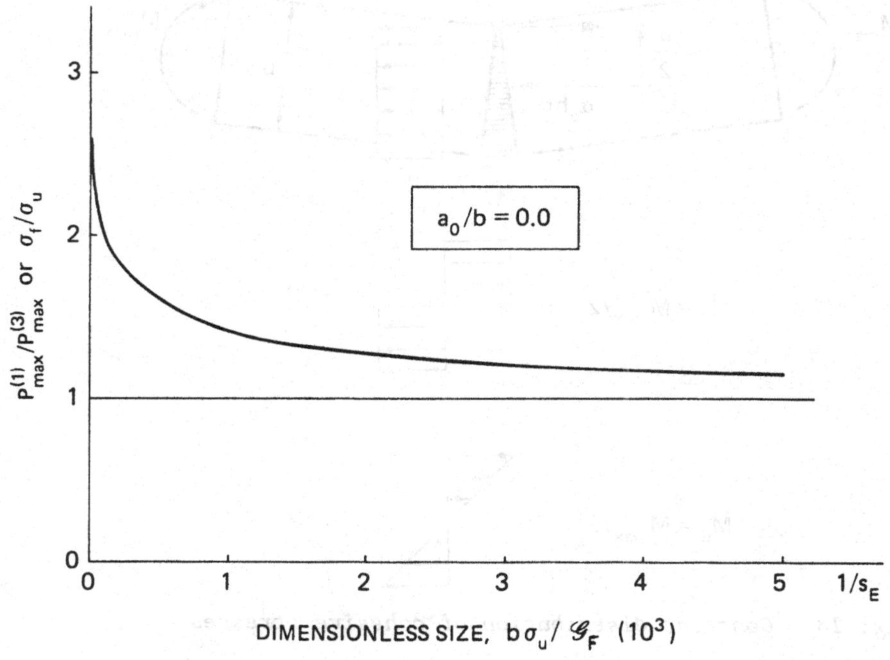

Fig. 17 Decrease in apparent strength by increasing the specimen size ($\lambda = 4$, $a_o/b = 0.0$, $\epsilon_u = 0.87 \times 10^{-4}$).

The maximum loading capacity $P^{(1)}_{max}$ of initially uncracked specimens with $l = 4b$ is obtained from Fig. 16. On the other hand, the maximum load $P^{(3)}_{max}$ of ultimate strength is given by:

$$P^{(3)}_{max} = \frac{2}{3} \frac{\sigma_u t b^2}{l} \quad . \qquad (3.10)$$

The values of the ratio $P^{(1)}_{max}/P^{(3)}_{max}$ may also be regarded as the ratio of the apparent tensile strength σ_f (given by the maximum load $P^{(1)}_{max}$ and applying eq.(3.10)) to the true tensile strength σ_u (considered as a material constant). It is evident from Fig. 17 that the results of the cohesive crack model tend to those of the ultimate strength analysis for low s_E values:

$$\lim_{s_E \to 0} P^{(1)}_{max} = P^{(3)}_{max} . \tag{3.11}$$

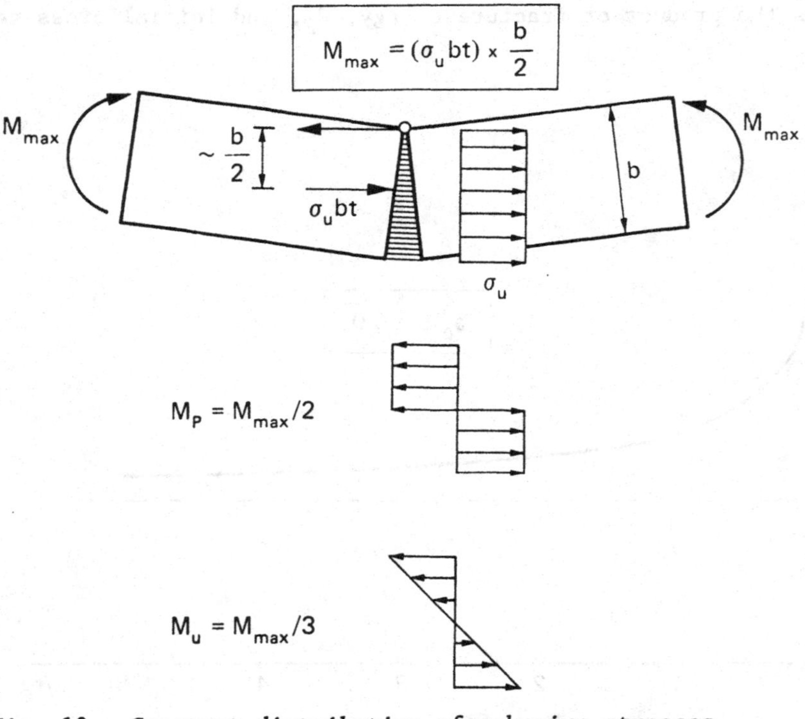

Fig. 18 Constant distribution of cohesive stresses.

Therefore, only for comparatively large specimen sizes the tensile strength σ_u can be obtained as $\sigma_u = \sigma_f$. With the usual laboratory specimens, an apparent strength higher than the true one is always found.

As a limit case, for the size $b \to 0$ or fracture energy $G_F \to \infty$ (elastic-perfectly plastic material in tension), i.e., for $s_E \to \infty$, the apparent strength $\sigma_f \to 3\sigma_u$. In fact, in the centre of the beam, the uniform stress distribution (Fig. 18) produces a plastic hinge with a resistant moment M_{max} which is twice the classical moment of the bi-rectangular limit stress distribution (elastic-perfectly plastic material in tension and compression).

The fictitious crack depth at the maximum load is plotted as a function of $1/s_E$ in Fig. 19. The brittleness increase for $s_E \to 0$ is evident also from this diagram, the process zone at $dP/d\delta = 0$ tending to disappear (brittle collapse), whereas it tends to cover the whole ligament for $s_E \to \infty$ (ductile collapse). On the other hand, the real (or stress-free) crack depth at the maximum load is always zero for each value of s_E. This means that the slow crack growth does not start before the softening stage. Therefore, neither the slow crack growth occurs nor the cohesive zone develops before the peak, when $s_E \to 0$.

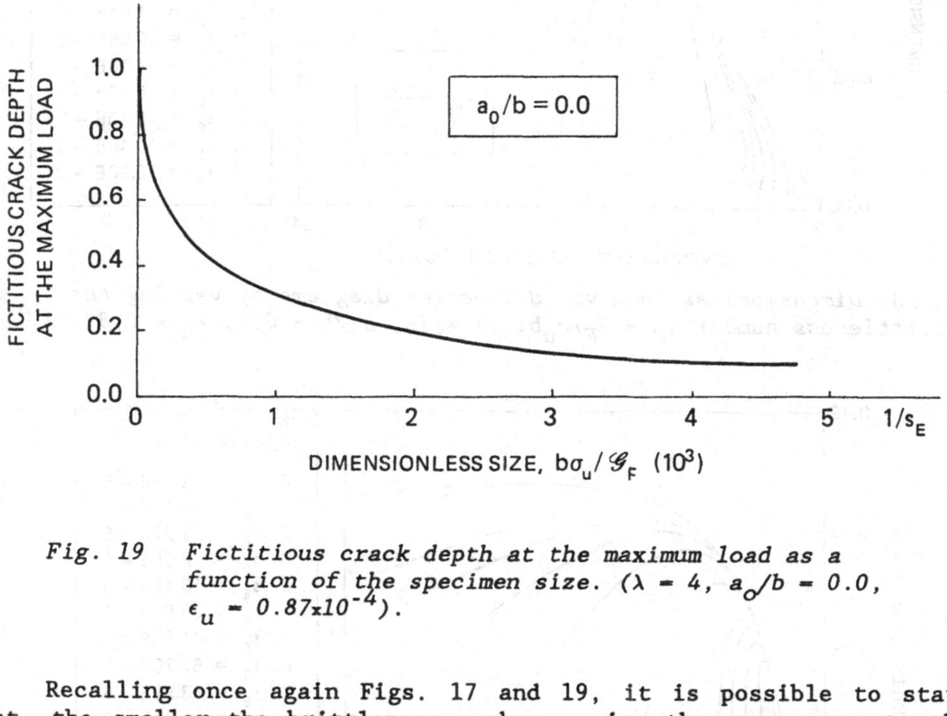

Fig. 19 *Fictitious crack depth at the maximum load as a function of the specimen size. ($\lambda = 4$, $a_o/b = 0.0$, $\epsilon_u = 0.87 \times 10^{-4}$).*

Recalling once again Figs. 17 and 19, it is possible to state that, the smaller the brittleness number s_E is, the more accurate the snap-back is in reproducing the perfectly-brittle ultimate strength instability ($a_o/b = 0$).

The diagrams in Fig. 20 are related to an higher beam slenderness, $\lambda = 16$. The same brittleness increase by decreasing s_E is obtained as previously, but in this case it is easier to achieve the snap-back instability of the beam, when $s_E \lesssim 62.70 \times 10^{-5}$. On the other hand, eq.(2.43) provides: $s_E \leq 46.40 \times 10^{-5}$, which is a good approximation.

The diagrams in Fig. 21 are related to an higher ultimate strain which is four times the preceding one, $\epsilon_u = 3.48 \times 10^{-4}$ ($\lambda = 4$). Also in this case, the snap-back appears for $s_E \lesssim 62.70 \times 10^{-5}$. Obviously eq.(2.43) provides again: $s_E \leq 46.40 \times 10^{-5}$, whereas eq.(2.49) gives $s_E \leq 53.36 \times 10^{-5}$, which is a better approximation.

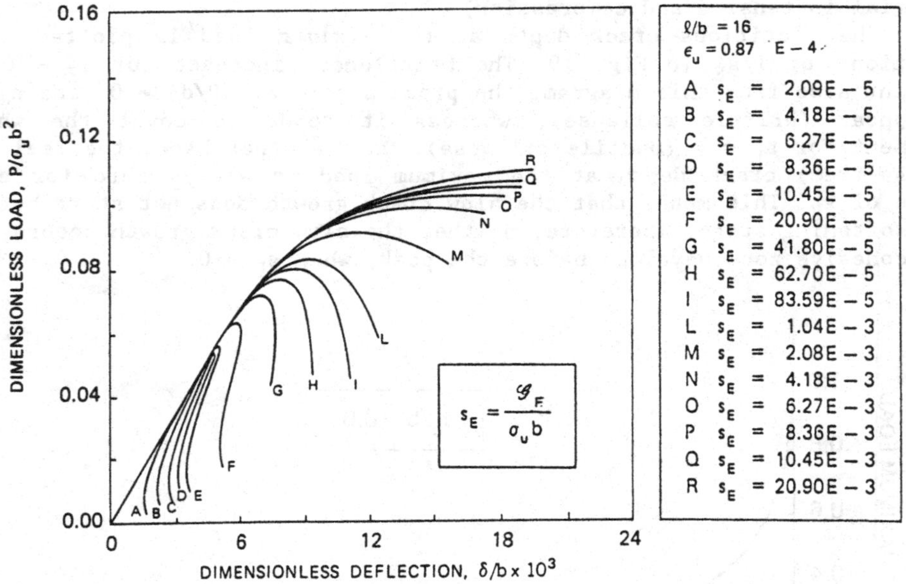

Fig. 20 Dimensionless load vs. deflection diagrams by varying the brittleness number $s_E = G_F/\sigma_u b$. ($\lambda = 16$, $a_o/b = 0.0$, $\epsilon_u = 0.87 \times 10^{-4}$).

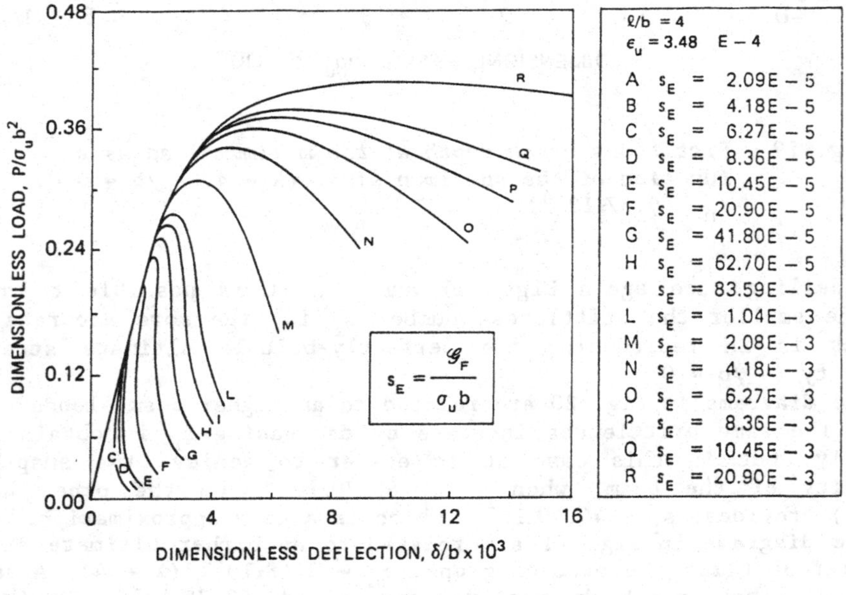

Fig. 21 Dimensionless load vs. deflection diagrams by varying the brittleness number $s_E = G_F/\sigma_u b$. ($\lambda = 4$, $a_o/b = 0.0$, $\epsilon_u = 3.48 \times 10^{-4}$).

3.3 Cohesive crack propagation and fictitious fracture toughness of initially cracked slabs

The mechanical behaviour of three point bending slabs with initial cracks is investigated on the basis of the same cohesive numerical model presented in the preceding section. Some dimensionless load-deflection diagrams are represented in Fig. 22, for $a_o/b = 0.5$, $\epsilon_u = 0.87 \times 10^{-4}$, $\nu = 0.1$, $t = b$, $l = 4b$, and by varying the brittleness number s_E. The initial crack makes the specimen behaviour more ductile than in the case of initially uncracked specimen. For the set of s_E numbers considered in Fig. 22, the snap-back does not occur.

The area delimited by load-deflection curve and deflection-axis represents the product of fracture energy, G_F, and initial ligament area, $(b-a_o)t$. The areas under the $P - \delta$ curves are thus proportional to the respective s_E numbers, in Fig. 22 as well as in Figs. 16, 20 and 21. This simple result is due to the assumption that energy dissipation occurs only on the fracture surface, whereas in reality energy is also dissipated in a damage volume around the crack tip, as assumed by Carpinteri and Sih [36] and proved by Cedolin, Dei Poli and Iori [37].

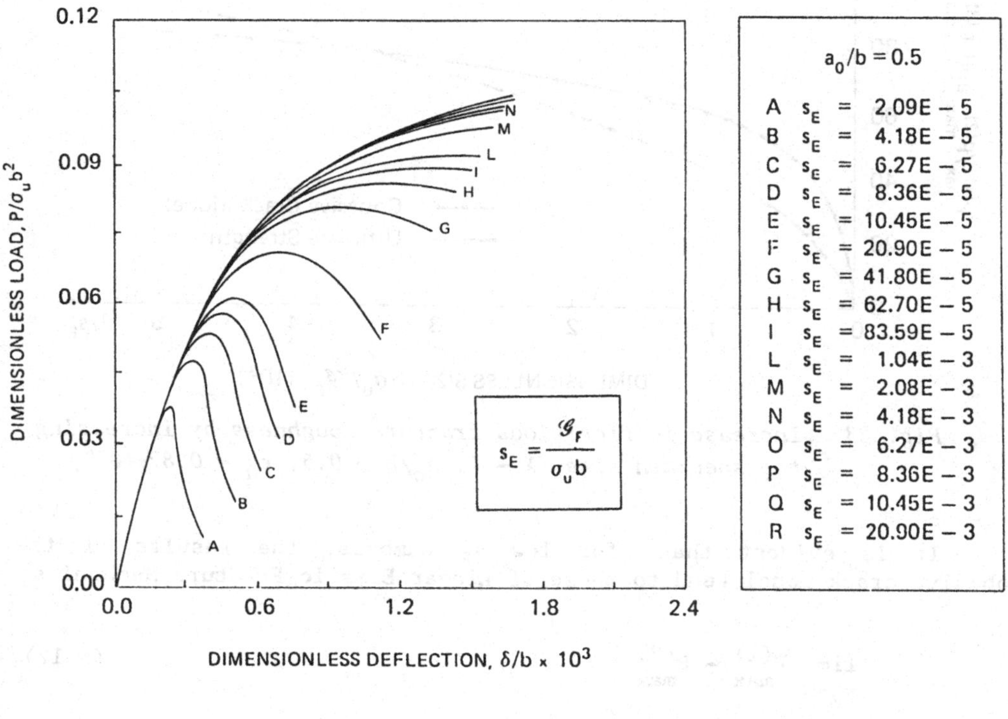

Fig. 22 Dimensionless load vs. deflection diagrams by varying the brittleness number $s_E = G_F/\sigma_u b$. ($\lambda = 4$, $a_o/b = 0.5$, $\epsilon_u = 0.87 \times 10^{-4}$).

The maximum loading capacity $P_{max}^{(1)}$ according to the cohesive crack model, is obtained from Fig. 22. On the other hand, the maximum load $P_{max}^{(2)}$ of brittle fracture can be obtained from the Linear Elastic Fracture Mechanics equation (1.1), with $K_{IC} = [G_F\ E]^{1/2}$ (plane stress condition).

The values of the ratio $P_{max}^{(1)}/P_{max}^{(2)}$ are reported as functions of the inverse of the brittleness number s_E in Fig. 23. Such a ratio may also be regarded as the ratio of the fictitious fracture toughness (given by the maximum load $P_{max}^{(1)}$) to the true fracture toughness (considered as a material constant).

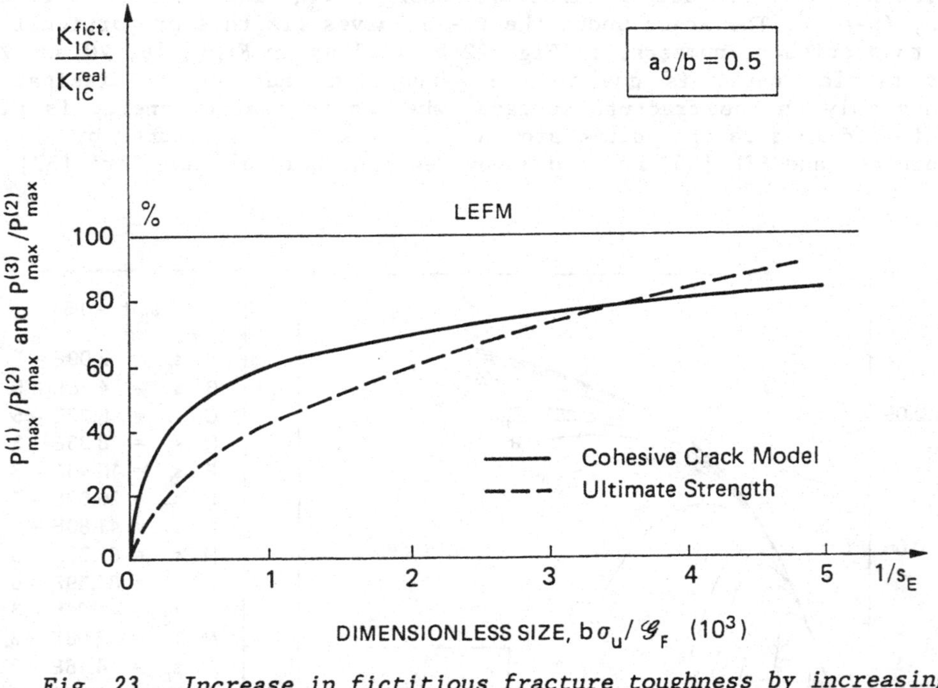

Fig. 23 Increase in fictitious fracture toughness by increasing the specimen size ($\lambda = 4$, $a_o/b = 0.5$, $\epsilon_u = 0.87 \times 10^{-4}$).

It is evident that, for low s_E numbers, the results of the cohesive crack model tend to those of Linear Elastic Fracture Mechanics:

$$\lim_{s_E \to 0} P_{max}^{(1)} = P_{max}^{(2)} , \qquad (3.12)$$

and therefore, the maximum loading capacity can be predicted applying the simple condition $K_I = K_{IC}$. It appears that the true fracture toughness K_{IC} of the material can be obtained only with very large specimens. In fact, with the laboratory specimens, a fictitious fracture toughness lower than the true one is always measured [1,2,6,38-46].

4. MIXED MODE CRACK PROPAGATION

4.1 Fundamentals

The Principle of Virtual Work can be used as the integral statement to formulate the elastic-softening problem in terms of finite element approximation:

$$\int_V d\underline{\epsilon}^T \underline{\sigma}\, dV = \int_V d\underline{u}^T \underline{F}\, dV + \int_S d\underline{u}^T \underline{p}\, dS \,, \tag{4.1}$$

where $\underline{\sigma}^T = [\sigma_x, \sigma_y, \sigma_z, \tau_{xy}, \tau_{yz}, \tau_{xz}]$ is the stress vector, $d\underline{\epsilon}^T$ is the vector of incremental virtual strain, $\underline{F}^T = [F_x, F_y, F_z]$ is the vector of body forces acting per unit volume, $d\underline{u}^T = [du, dv, dw]$ is the vector of incremental virtual displacement and $\underline{p}^T = [p_x, p_y, p_z]$ is the vector of tractions acting per unit area of external surface S.

Eq.(4.1) is the weak form of the equilibrium equations and is valid for linear as well as for non-linear stress-strain constitutive laws.

According to the cohesive crack model, the process zone near the crack tip can be represented by means of closing tractions $\underline{p}_c$ acting on both the crack faces. Therefore, the last term in eq.(4.1) can be decomposed as follows (Fig. 24):

$$\int_S d\underline{u}^T \underline{p}\, dS = \int_{S_c} d\underline{u}^T \underline{p}_c\, dS + \int_{S-S_c} d\underline{u}^T \underline{p}\, dS \,, \tag{4.2}$$

where S_c is the process zone, i.e., the crack surface where the cohesive forces are active.

Assuming a linear softening constitutive law, the traction versus displacement relationship can be written (Fig.24):

$$\underline{p}_c = \underline{p}_u + \underline{N}^T \underline{L}\, \underline{N}\, (\underline{u}^+ - \underline{u}^-) \tag{4.3}$$

where $\underline{p}_u$ is the ultimate tensile strength in vectorial form, $\underline{N}$ is the transformation matrix from the global to the local reference system, varying point by point on the crack surface, $\underline{L}$ is the cohesive constitutive matrix in a local cartesian system, the index + refers to the positive side of the crack, while the index - refers to the negative one.

From equilibrium considerations across the crack surface, it is possible to write:

$$\underline{p}_c^+ = -\underline{p}_c^- , \quad \underline{p}_u^+ = -\underline{p}_u^- , \quad s_c^+ = s_c^- = s_c/2 . \tag{4.4}$$

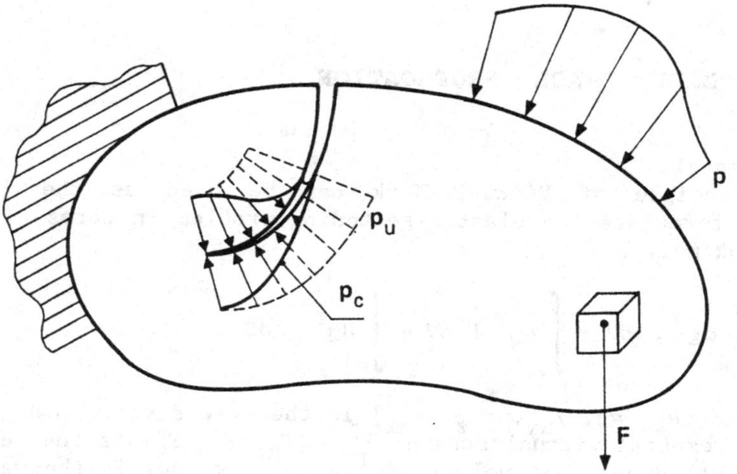

Fig 24. Mixed Mode cohesive crack propagation.

The first term in the right-hand side of eq.(4.2) can be written:

$$\int_{S_c} d\underline{u}^T \underline{p}_c \, dS = \int_{S_c^+} d\underline{u}^{+T} \underline{p}_u^+ \, dS + \int_{S_c^-} d\underline{u}^{-T} \underline{p}_u^- \, dS +$$

$$\int_{S_c^+} d\underline{u}^{+T} \underline{N}^T \underline{LN}(\underline{u}^+ - \underline{u}^-) \, dS - \int_{S_c^-} d\underline{u}^{-T} \underline{N}^T \underline{LN}(\underline{u}^+ - \underline{u}^-) \, dS . \tag{4.5}$$

The last two terms in eq.(4.5) can be represented as follows:

$$\int_{\frac{1}{2}S_c} \begin{Bmatrix} d\underline{u}^+ \\ d\underline{u}^- \end{Bmatrix}^T \begin{bmatrix} \underline{N}^T & \underline{0} \\ \underline{0} & \underline{N}^T \end{bmatrix} \begin{bmatrix} \underline{L} & -\underline{L} \\ -\underline{L} & \underline{L} \end{bmatrix} \begin{bmatrix} \underline{N} & \underline{0} \\ \underline{0} & \underline{N} \end{bmatrix} \begin{Bmatrix} \underline{u}^+ \\ \underline{u}^- \end{Bmatrix} dS . \tag{4.6}$$

The Principle of Virtual Work, eq.(4.1), can be developed according to eqs.(4.2), (4.5) and (4.6):

$$\int_V d\underline{\epsilon}^T \underline{\sigma} \, dV = \int_V d\underline{u}^T \underline{F} \, dV + \int_{S-S_c} d\underline{u}^T \underline{p} \, dS + \int_{S_c^+} d\underline{u}^{+T} \underline{p}_u^+ \, dS + \int_{S_c^-} d\underline{u}^{-T} \underline{p}_u^- \, dS +$$

$$+ \int_{\frac{1}{2}S_c} \begin{Bmatrix} d\underline{u}^+ \\ d\underline{u}^- \end{Bmatrix}^T \begin{bmatrix} \underline{N}^T & \underline{0} \\ \underline{0} & \underline{N}^T \end{bmatrix} \begin{bmatrix} \underline{L} & -\underline{L} \\ -\underline{L} & \underline{L} \end{bmatrix} \begin{bmatrix} \underline{N} & \underline{0} \\ \underline{0} & \underline{N} \end{bmatrix} \begin{Bmatrix} \underline{u}^+ \\ \underline{u}^- \end{Bmatrix} dS. \quad (4.7)$$

4.2 Finite element discretization

Subdividing the domain in a finite number of elements and expressing the internal displacements by means of locally based shape functions $\underline{H}$, it is possible to write:

$$\underline{u}(x,y,z) = \underline{H}(x,y,z)\, \underline{u} \, . \quad (4.8)$$

From derivation of eq.(4.8), the strain versus displacement relationship can be obtained:

$$\underline{\epsilon} = \underline{B}\, \underline{u} \, . \quad (4.9)$$

Selecting an appropriate constitutive law for the uncracked zone, the stress versus strain relationship appears as follows:

$$\underline{\sigma} = \underline{D}\, (\underline{\epsilon} - \underline{\epsilon}_o) + \underline{\epsilon}_o \, . \quad (4.10)$$

Substituting eqs.(4.8), (4.9) and (4.10) in eq.(4.7), and indicating by the letter e the generic element, it is possible to write:

$$d\underline{u}^T \left[\sum_e \int_V \underline{B}^T \underline{D}\, \underline{B}\, dV \right] \underline{u} - \begin{Bmatrix} d\underline{u}^+ \\ d\underline{u}^- \end{Bmatrix}^T \left[\sum_e \int_{\frac{1}{2}S_c} \underline{H}^T [\underline{N}^T][\underline{L}][\underline{N}]\underline{H}\, dS \right] \begin{Bmatrix} \underline{u}^+ \\ \underline{u}^- \end{Bmatrix} =$$

$$= d\underline{u}^T \sum_e \int_V (\underline{H}^T \underline{F} - \underline{B}^T \underline{\sigma}_o + \underline{B}^T \underline{D}\, \underline{\epsilon}_o)\, dV + d\underline{u}^T \left[\sum_e \int_{S-S_c} \underline{H}^T \underline{p}\, dS \right] +$$

$$+ d\underline{u}^{+T} \left[\sum_e \int_{S^+} \underline{H}^T \underline{p}_u^+\, dS \right] + d\underline{u}^{-T} \left[\sum_e \int_{S^-} \underline{H}^T \underline{p}_u^-\, dS \right] . \quad (4.11)$$

Since:

$$\{\underline{u}^+\} \subset \{\underline{u}\} \, , \quad \{\underline{u}^-\} \subset \{\underline{u}\} \, , \quad \{d\underline{u}^+\} \subset \{d\underline{u}\} \, , \quad \{d\underline{u}^-\} \subset \{d\underline{u}\} \, , \quad (4.12)$$

an assemblage procedure can be carried out [47]:

$$(\underline{K} - \underline{C})\, \underline{u} = \underline{F}_V + \underline{F}_S + \underline{F}_u^+ + \underline{F}_u^- \, , \quad (4.13)$$

where:

$\underline{K}$ = stiffness matrix,
$\underline{C}$ = softening matrix,
$\underline{F}_V$, $\underline{F}_S$, $\underline{F}_u^+$, $\underline{F}_u^-$ = loading vectors,
$(\underline{K} - \underline{C})$ = effective stiffness matrix.

Neglecting the tangential cohesive tractions, the constitutive matrix $[\underline{L}]$ becomes:

$$[\underline{L}] = \begin{bmatrix} 0 & 0 & 0 \\ 0 & 0 & 0 \\ 0 & 0 & l_{33} \end{bmatrix} . \qquad (4.14)$$

Only the component of the mutual displacement normal to the crack surface, w (crack opening displacement), is taken into account. The remaining components are disregarded.
The scalar quantity l_{33} is assumed as follows:

$$l_{33} = \frac{\sigma_u}{w_c} , \quad \text{for } 0 < w < w_c \qquad (4.15\text{-a})$$

$$l_{33} = 0 , \quad \text{for } w \geq w_c \qquad (4.15\text{-b})$$

where σ_u is the ultimate tensile strength of the material and w_c is the critical value of the crack opening displacement w. For crack opening displacements greater than the critical value w_c, the interaction forces disappear, and both the crack surfaces are stress-free. During the irreversible fracture process, the crack opening displacement w rusults to be a monotonic increasing function of time.

At the first step the cohesive zone is absent, matrix $\underline{C}$ vanishes and matrix $\underline{K}$ is positive definite. A linear elastic solution can be found, giving position and orientation of the growing crack. The crack surface S_c starts propagating by a pre-defined length ΔS_c. Such an incremental length is chosen so small that matrix $(\underline{K} - \underline{C})$ remains positive definite, and the maximum cohesive crack opening displacement is less than w_c. Eq.(4.13) can be solved for two right-hand side vectors:

$$(\underline{K} - \underline{C})\, \underline{u}_1 = \underline{F}_v + \underline{F}_s , \qquad (4.16\text{-a})$$

$$(\underline{K} - \underline{C})\, \underline{u}_2 = - \underline{F}_u^+ - \underline{F}_u^- . \qquad (4.16\text{-b})$$

If σ_θ is the circumferential stress at the fictitious crack tip, for each value of the angle θ it is possible to write:

$$\lambda(\sigma_\theta)_1 - (\sigma_\theta)_2 = \sigma_u , \qquad (4.17)$$

and then, solving with respect to the loading multiplier λ:

$$\lambda = \frac{\sigma_u + (\sigma_\theta)_2}{(\sigma_\theta)_1} . \qquad (4.18)$$

Eq.(4.18) is interpretable as a function $\lambda = \lambda(\theta)$. The minimum of $\lambda = \lambda(\theta)$, and the related displacement vector:

$$\underline{u} = \lambda\,(\min)\, \underline{u}_1 - \underline{u}_2 , \qquad (4.19)$$

describe the second step of the cracking process, providing the orientation of the subsequent crack branch.

At the following steps the same procedure is repeated, without moving the real crack tip, until one of the following conditions is verified.

(1) The crack opening displacement at the real crack tip reaches its critical value w_c.

(2) Matrix ($\underline{K}$-$\underline{C}$) in eq.(4.16) becomes positive semidefinite.

In both cases, the real crack tip moves and the cohesive crack surface S_c shrinks, until the crack opening displacement at the real crack tip is less than w_c, or, respectively, matrix ($\underline{K}$-$\underline{C}$) is positive definite.

4.3 Snap-back softening instability and brittle mixed mode fractures

For Mixed Mode fracture, a topological variation is required at each step of the interelement crack propagation. The numerical response of the four point shear specimen (Fig.25) is analyzed according to the cohesive crack model. The geometrical features of the specimen are the following: l = 4 b, t = b, a_o = 0.2 b, c = 0.8 b and 0.4 b, whereas the material is assumed to present the ultimate strain $\epsilon_u = \sigma_u/E = 0.741 \times 10^{-4}$ and $\nu = 0.1$.

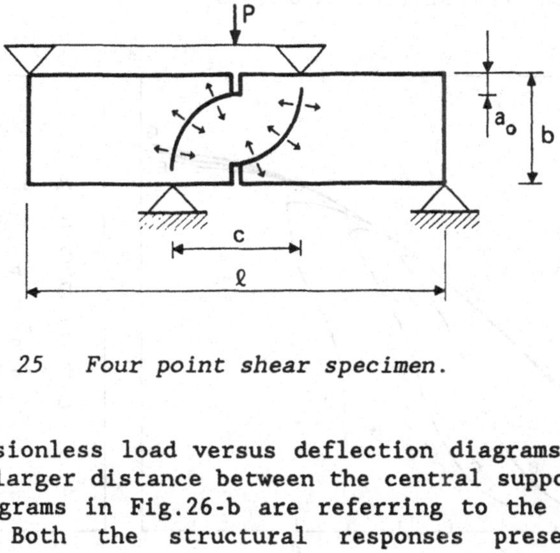

Fig. 25 Four point shear specimen.

The dimensionless load versus deflection diagrams in Fig.26-a are related to the larger distance between the central supports (c/b = 0.8), whereas the diagrams in Fig.26-b are referring to the smaller distance (c/b = 0.4). Both the structural responses present a snap-back instability for $s_E \lesssim 0.001$.

On the other hand, an evident difference in the P - δ shape transpires. Whereas for c/b = 0.8, the snap-back branch ($dP/d\delta > 0$) is followed by a normal softening tail ($dP/d\delta < 0$), passing through the stationary condition $d\delta/dP = 0$, for c/b = 0.4 the normal softening tail does not appear after the snap-back behaviour, and the snap-back branch tends to go back to the origin. The latter kind of equilibrium path reveals a potentially brittler behaviour. In fact, if the loading process is deflection-controlled, the loading capacity would show a vertical drop down to zero, without any possibility for the crack to arrest.

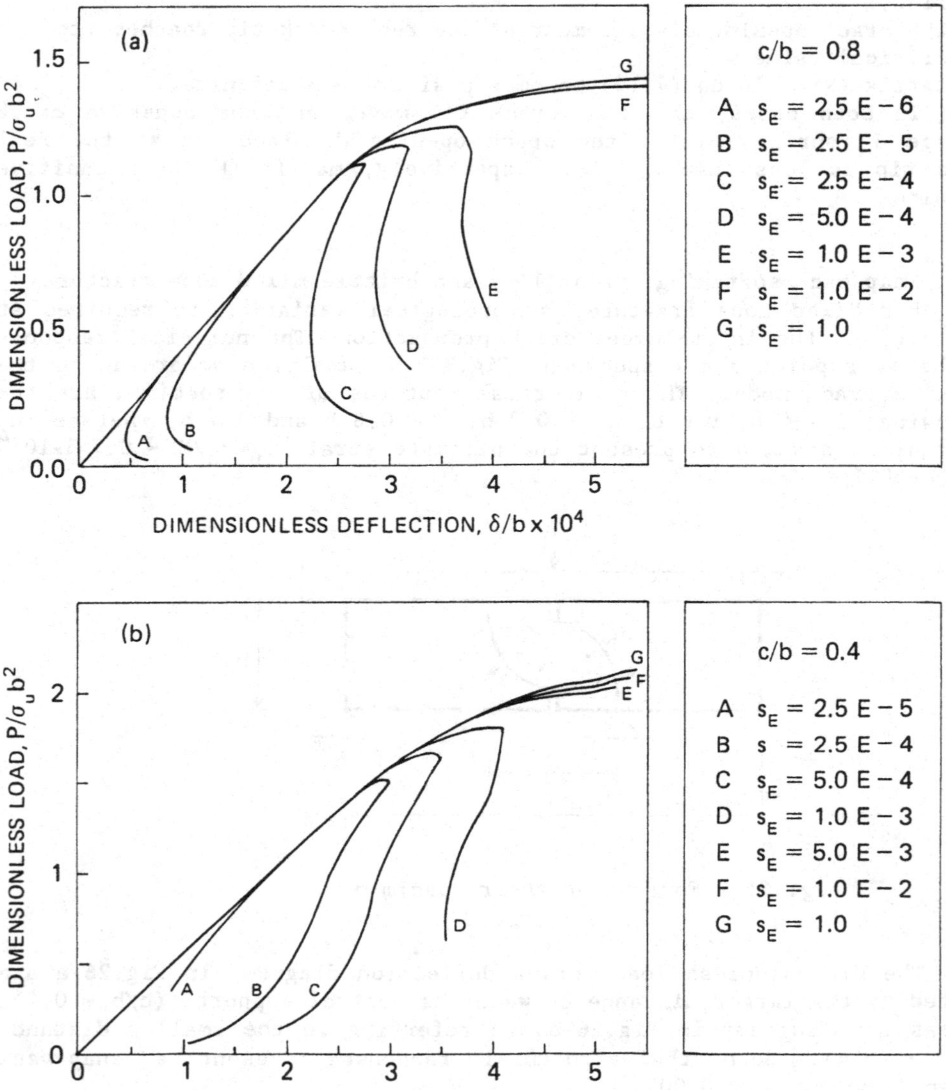

Fig. 26 Ductile-brittle transition by varying the brittleness number $s_E = G_F/\sigma_u b$ (four point shear specimen). (a) c/b = 0.8; (b) c/b = 0.4.

The maximum loading capacity P_{COHES} according to the cohesive crack model, is provided by the diagrams in Fig.26. On the other hand, the maximum load P_{LEFM} of brittle fracture can be derived from the application of the Maximum Circumferential Stress Criterion [26]:

$$\frac{d\sigma_\theta}{d\theta} = 0 \quad , \quad \sigma_\theta (2\pi r)^{1/2} = K_{IC} = (G_F E)^{1/2} \ . \qquad (4.20)$$

Stress intensification is produced in both the crack tip regions and the stress-intensity factors for Mode I and Mode II can be expressed respectively as:

$$K_I = \frac{P}{tb^{1/2}} f_I \left(\frac{l}{b}, \frac{a}{b}, \frac{c}{b} \right) \ , \qquad (4.21\text{-a})$$

$$K_{II} = \frac{P}{tb^{1/2}} f_{II} \left(\frac{l}{b}, \frac{a}{b}, \frac{c}{b} \right) \ , \qquad (4.21\text{-b})$$

f_I and f_{II} being the shape functions.

The angle θ_o of crack branching is provided by the following equation:

$$f_I \sin\theta_o + f_{II} (3 \cos\theta_o - 1) = 0 \ , \qquad (4.22)$$

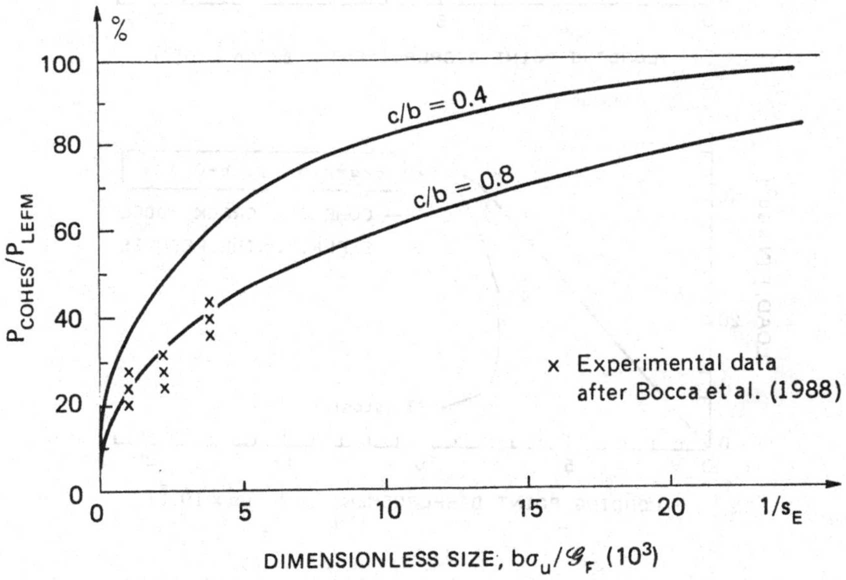

Fig. 27 Size-scale transition towards Mixed Mode-LEFM-instability.

whereas the Mixed Mode crack instability is predicted by the condition:

$$P_{LEFM} \cos \frac{\theta_o}{2} \left(f_I \cos^2 \frac{\theta_o}{2} - \frac{3}{2} f_{II} \sin\theta_o \right) = t b^{\frac{1}{2}} K_{IC} \quad . \tag{4.23}$$

The values of the ratio P_{COHES}/P_{LEFM} are represented in Fig.27 against the dimensionless size $1/s_E$. A transition is evident towards LEFM by increasing the size-scale of the structure. For the brittler geometry, $c/b = 0.4$, the transition appears to be faster, and already for $b\sigma_u/G_F = 2 \times 10^4$ or $s_E = 5 \times 10^{-5}$, the asymptotical LEFM condition is achieved.

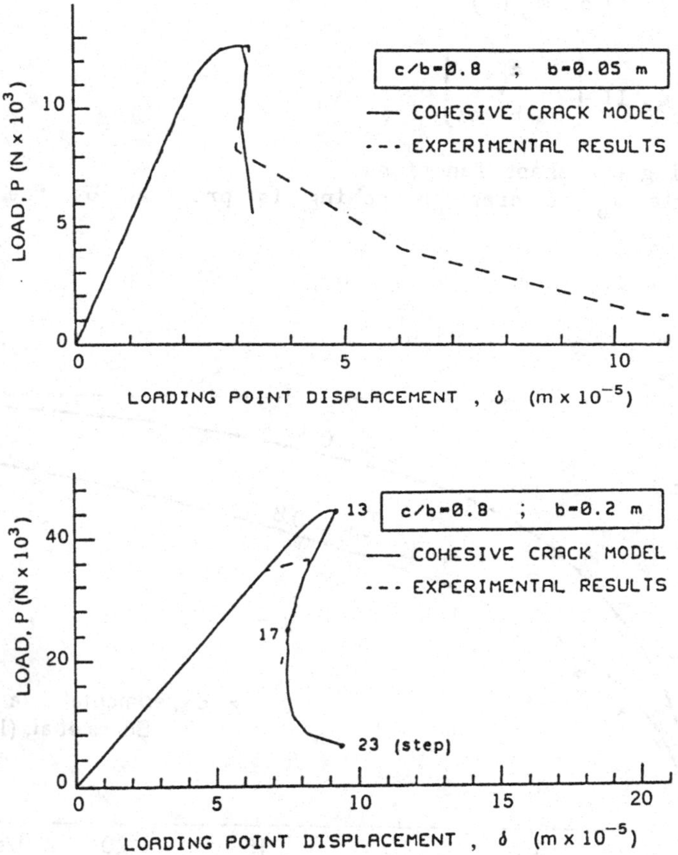

Fig. 28 Experimental load vs. deflection curves and numerical cohesive crack simulation, for c/b = 0.8.- (a) b = 5 cm; - (b) b = 20 cm.

In this case, the size of the cohesive zone is negligible with respect to the size of the zone where the $r^{-1/2}$ LEFM-stress-singularity is dominant.

For c/b = 0.8, the total load versus loading point deflection diagrams are plotted in Figs.28-a and b, in the cases b = 5 and 20 cm respectively. The Mixed Mode cohesive crack model describes both the experimental curves [25] satisfactorily. The size b = 20 cm (Fig.28-b) produces snap-back instability in the experimental as well as in the numerical curve. The mechanical properties utilized in the numerical analysis are: Young's modulus E = 27000 MPa, ultimate tensile strength σ_u = 2 MPa, fracture energy G_F = 100 N/m. The area enclosed between numerical curve and deflection axis is approximately equal to the product of the Mode I fracture energy G_F and the total fracture area, and represents the amount of energy dissipated in the localized fracture zone. The amount of energy dissipated by punching at the supports, was deliberately neglected, assuming ascending elastic branches consistent with the elastic modulus of the material (Figs.28).

It is remarkable that the application of the usual Mode I fracture energy G_F only, was able to provide consistent results. It was unnecessary to introduce additional fracture toughness parameters, like, for example, the Mode II fracture energy G_F^{II} [23,48]. As a matter of fact, the Mixed Mode fracture energy results approximately equal to the Mode I fracture energy G_F, each elementary crack growth step being produced by a Mode I (or opening) mechanism along the curvilinear trajectory.

The sequence of the finite element meshes utilized for the case b = 20 cm, c/b = 0.8, is reported in Fig.29. The trajectory of the finite element rosette reproduces the experimental fracture trajectory accurately. It is remarkable that the real crack (complete disconnection) starts propagating only at the 13th step, when the fictitious crack (cohesive interaction) is beyond one half of the beam depth. On the other hand, at the 22nd step, both fictitious and real crack tip are close to the upper beam edge. The single steps are indicated also in the diagram of Fig.28-b.

The sequence of the finite element meshes utilized for the case b = 20 cm, c/b = 0.4, is reported in Fig.30. Also in this case, the numerical simulation describes the experimental fracture trajectory very accurately, included the deviations at the beam edges [25].

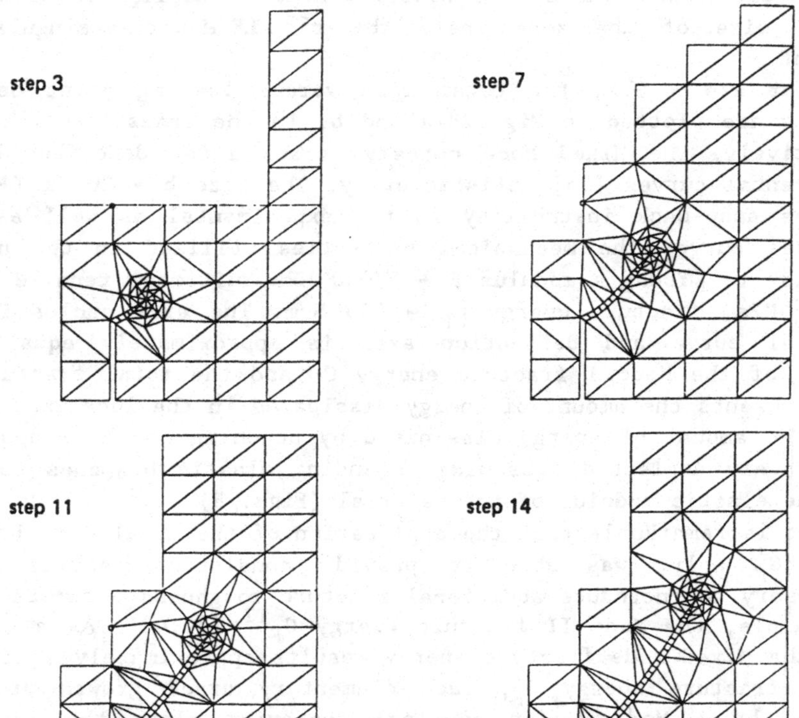

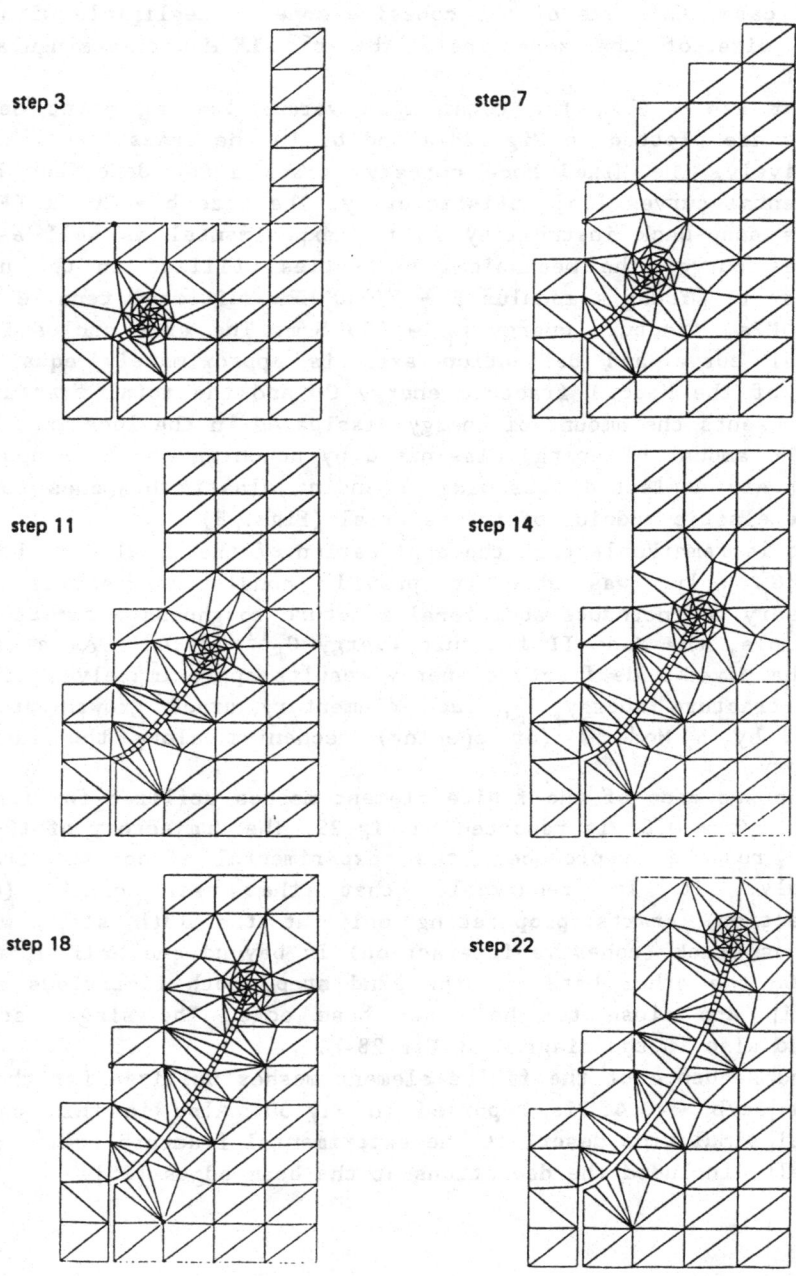

Fig. 29 Finite element remeshing. c/b = 0.8; b = 20 cm.

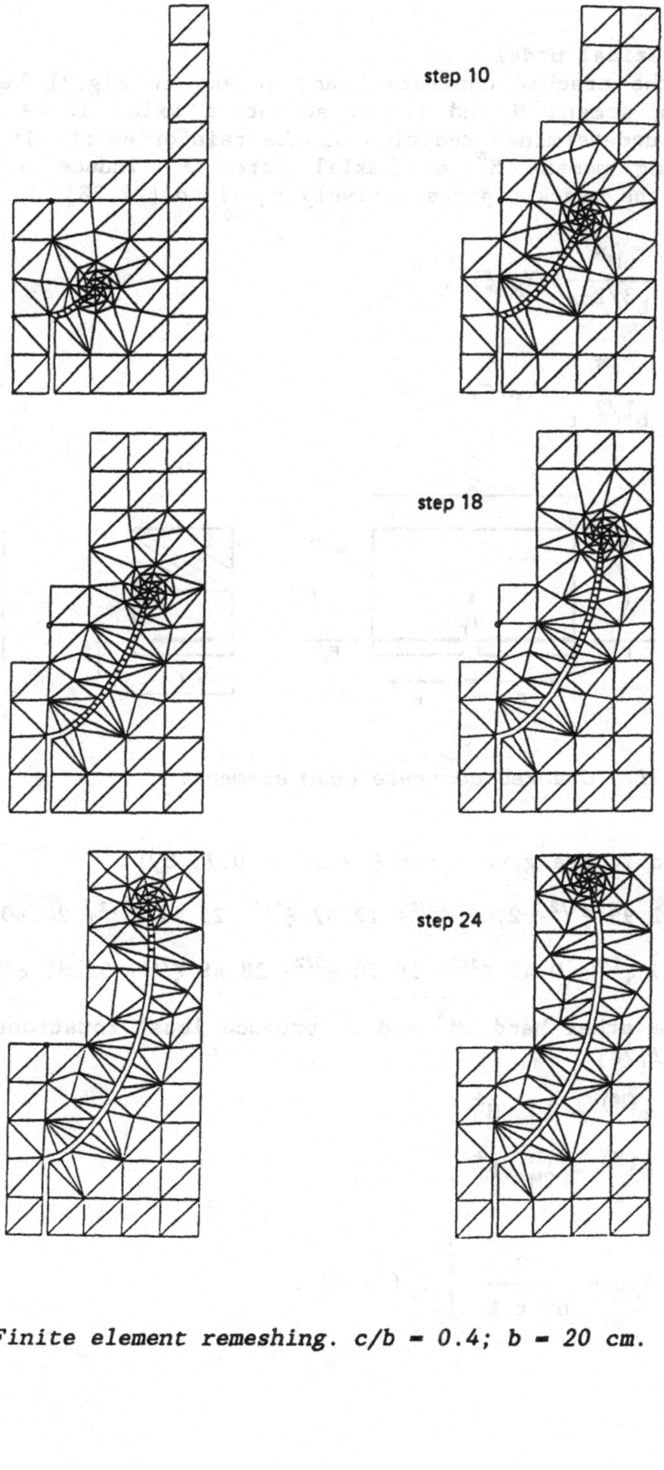

Fig. 30 Finite element remeshing. c/b = 0.4; b = 20 cm.

5. SIZE EFFECT IN THE FAILURE PROCESS OF REINFORCED CONCRETE

5.1 Theoretical model

Let the cracked concrete beam element in Fig.31 be subjected to the bending moment M and to an eccentric axial force F due to the statically undetermined reaction of the reinforcement. It is well-known that bending moment M^* and axial force F^* induce stress-intensity factors at the crack tip respectively equal to [27,28]:

$$K_I^{(M)} = \frac{M^*}{b^{3/2}\, t} Y_M(\xi) , \tag{5.1-a}$$

$$K_I^{(F)} = \frac{F^*}{b^{1/2}\, t} Y_F(\xi) , \tag{5.1-b}$$

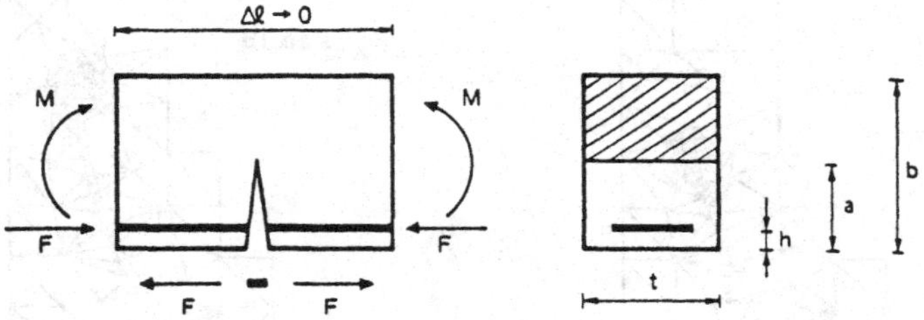

Fig. 31 Cracked concrete beam element.

where Y_M and Y_F are given, for $\xi = a/b \leq 0.7$, by:

$$Y_M(\xi) = 6\,(1.99\,\xi^{1/2} - 2.47\,\xi^{3/2} + 12.97\,\xi^{5/2} + 23.17\,\xi^{7/2} + 24.80\,\xi^{9/2}), \tag{5.1-c}$$

$$Y_F(\xi) = 1.99\,\xi^{1/2} - 0.41\,\xi^{3/2} + 18.70\,\xi^{5/2} - 38.48\,\xi^{7/2} + 53.85\,\xi^{9/2}. \tag{5.1-d}$$

On the other hand, M^* and F^* produce local rotations respectively equal to [27,28]:

$$\varphi^{(M)} = \lambda_{MM}\, M^* , \tag{5.2-a}$$

$$\varphi^{(F)} = \lambda_{MF}\, F^* , \tag{5.2-b}$$

where:

$$\lambda_{MM} = \frac{2}{b^2\, t\, E} \int_0^{\xi} Y_M^2(\xi)\, d\xi , \tag{5.3-a}$$

$$\lambda_{MF} = \frac{2}{b\ t\ E} \int_0^{\xi} Y_M(\xi)\ Y_F(\xi)\ d\xi\ . \tag{5.3-b}$$

Up to the moment of steel yielding or slippage, the local rotation in the cracked cross-section is equal to zero:

$$\varphi = \varphi^{(M)} + \varphi^{(F)} = 0\ . \tag{5.4}$$

Eq.(5.4) is the congruence condition giving the unknown force F. Recalling that (Fig.31):

$$M^* = M - F\ (b/2 - h)\ , \tag{5.5-a}$$

$$F^* = -\ F\ , \tag{5.5-b}$$

eqs.(5.2) and (5.4) provide:

$$\frac{F\ b}{M} = \frac{1}{(0.5 - h/b) + r(\xi)}\ , \tag{5.6}$$

where:

$$r(\xi) = \frac{\int_0^{\xi} Y_M(\xi)\ Y_F(\xi)\ d\xi}{\int_0^{\xi} Y_M^2(\xi)\ d\xi}\ . \tag{5.7}$$

If a perfectly plastic behaviour of the reinforcement is considered (yielding or slippage), from eq.(5.6) the moment of plastic flow for the reinforcement results:

$$M_P = F_P\ b \left[(0.5 - h/b) + r(\xi) \right]\ . \tag{5.8}$$

However, it should be observed that, if concrete presents a low crushing strength and steel a high yield strength, crushing of concrete can precede plastic flow of reinforcement.

The mechanical behaviour of the cracked reinforced concrete beam section is rigid until the bending moment M_P is exceeded, i.e., $\varphi = 0$ for $M \leq M_P$. On the other hand, for $M > M_P$ the $M - \varphi$ diagram becomes linear hardening:

$$\varphi = \lambda_{MM}\ (M - M_P)\ . \tag{5.9}$$

After the plastic flow of reinforcement, the stress-intensity factor at the crack tip is given by the superposition principle:

$$K_I = K_I^{(M)} + K_I^{(F)} . \tag{5.10}$$

Recalling eqs.(5.1) and considering the loadings:

$$M^* = M - F_P \,(b/2 - h) \;, \tag{5.11-a}$$

$$F^* = - F_P \;, \tag{5.11-b}$$

the global stress-intensity factor results:

$$K_I = \frac{Y_M(\xi)}{b^{3/2}\, t} \left[M - F_P \left(\frac{b}{2} - h \right) \right] - \frac{F_P}{b^{1/2}\, t} Y_F(\xi) \; . \tag{5.12}$$

The moment of crack propagation is then:

$$\frac{M_F}{K_{IC}\, b^{3/2}\, t} = \frac{1}{Y_M(\xi)} + N_P \left[\frac{Y_F(\xi)}{Y_M(\xi)} + \frac{1}{2} - \frac{h}{b} \right] , \tag{5.13}$$

with:

$$N_P = \frac{f_y\, b^{1/2}}{K_{IC}} \, \frac{A_S}{A} \;, \tag{5.14}$$

while the rotation at crack propagation is:

$$\varphi_F = \lambda_{MM} \,(M_F - M_P) \; . \tag{5.15}$$

The crack propagation moment is plotted in Fig.32 as a function of the crack depth ξ and varying the brittleness number N_P. For low N_P values, i.e., for low reinforced beams or for small cross-sections, the fracture moment decreases while the crack extends, and a typical phenomenon of unstable fracture occurs. For $N_P \gtrsim 0.7$, a stable branch follows the unstable one, while for $N_P \gtrsim 8.5$ only the stable branch remains. The locus of the minima is represented by a dashed line in Fig.32. In the upper zone the fracture process is stable whereas it is unstable in the lower one.

Rigid behaviour ($0 \leq M \leq M_P$) is followed by linear hardening ($M_P < M < M_F$). The latter stops when crack propagation occurs. If the fracture process is unstable, diagram $M - \varphi$ presents a discontinuity and drops from M_F to $F_P b$ with a negative jump. In fact, in this case, a complete and instantaneous disconnection of concrete occurs. The new moment $F_P b$ can be estimated according to the scheme in Fig.33.

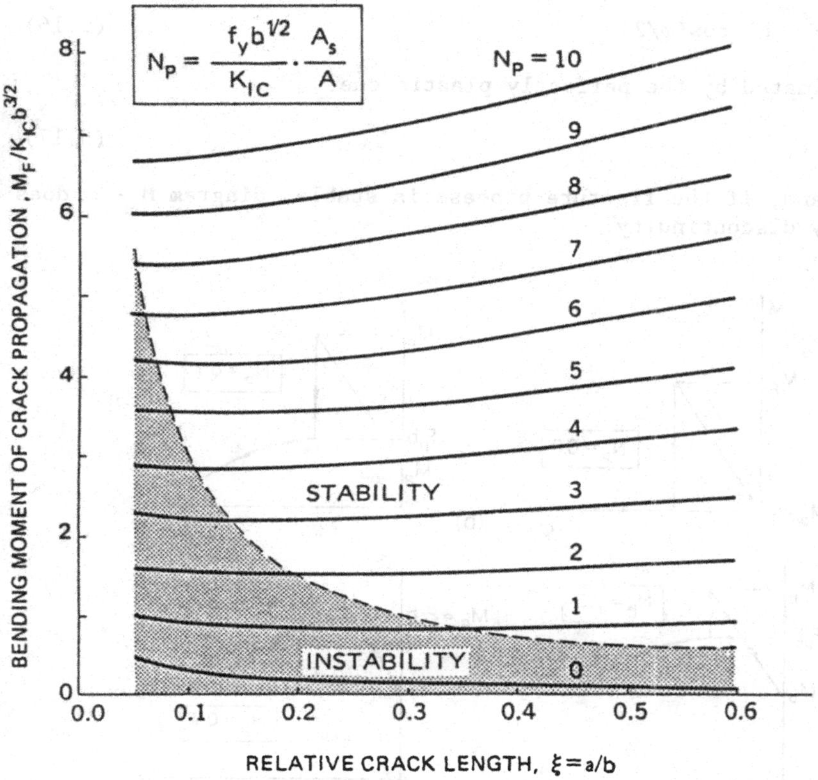

Fig. 32 Bending moment of crack propagation against relative crack length.

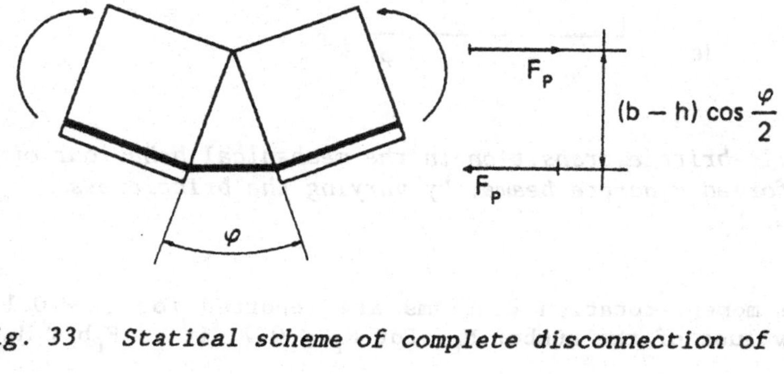

Fig. 33 Statical scheme of complete disconnection of concrete.

The non-linear descending law:

$$M = F_P\ (b - h)\ \cos(\varphi/2)\ , \tag{5.16}$$

is thus approximated by the perfectly plastic one:

$$M = F_P\ b\ . \tag{5.17}$$

On the other hand, if the fracture process is stable, diagram M - φ does not present any discontinuity.

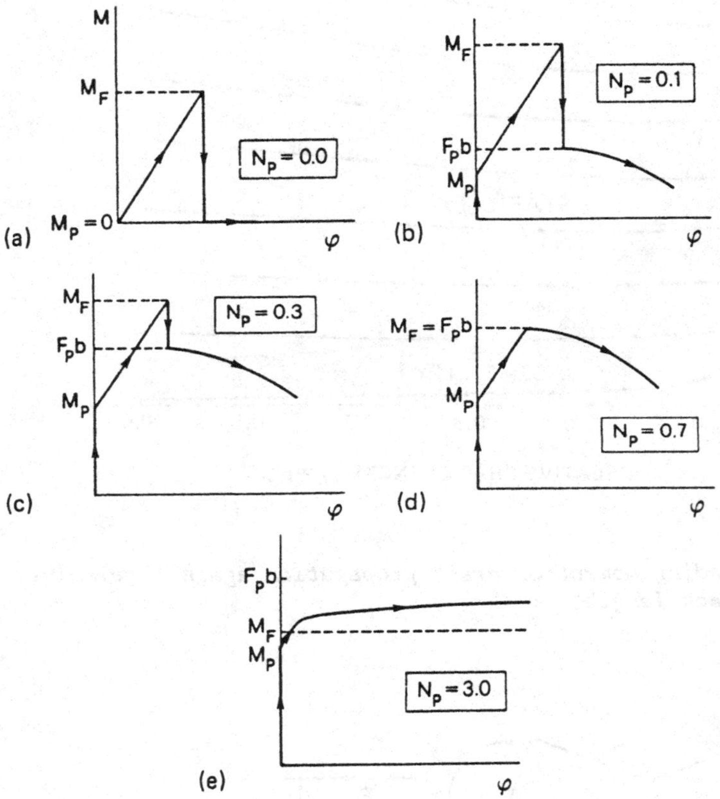

Fig. 34 Ductile-brittle transition in the mechanical behaviour of reinforced concrete beams, by varying the brittleness number N_P.

In Fig.34 the moment-rotation diagrams are reported for $\xi = 0.1$ and five different values of the number N_P. For $N_P \lesssim 0.7$, it is $F_Pb < M_F$

and therefore a discontinuity appears in the M - φ diagram (Figs.34-a to c). On the other hand, for $N_P \lesssim 0.7$, the curves in Fig.32 lie in the unstable zone completely. In conclusion, the above presented theoretical model predicts unstable behaviour for low content of steel and/or for large beam depth.

It is worth noting that the initial crack of the theoretical model was not present in the experimental beams, although a crack formed during the loading process. In addition, the reinforcement plastic flow of the theoretical model, in practice was due to the steel bar slippage. Nevertheless, the theoretical ductile-brittle transition was confirmed by the experiments, as will be shown in the following sections.

5.2 Material properties

The beams of the present experimental investigation [29] were made of concrete with crushed aggregate of maximum size D_{max} = 12.7 mm. The amount of cement (type 525) is 480 kg/m^3, and the water/cement ratio is equal to 0.27. Considerable attention was spent to avoid cracking due to hydration and shrinkage.

The compressive strength (after 28 days) was obtained with twenty (20) cubic specimens of side 160 mm. The average value resulted to be R_{cm} = 91.2 N/mm^2, with a standard deviation of 8.8 N/mm^2.

The curing time of the beams was 3 days at 30 °C and then a second period followed at 20 °C. As an average, the tests were carried out after 20 days from moulding.

The tests of elastic modulus were performed on three (3) specimens of size 150x150x450 mm and provided an average value of the secant modulus E (between zero and 1/3 of the ultimate load) equal to 34,300 N/mm^2.

The fracture energy G_F was determined by three point bending testing on three (3) specimens of size b = 100 mm, t = 150 mm, l = 750 mm. The span was equal to L = 720 mm and the beams were pre-notched on the center-line, the notch depth being equal to one half of the beam depth (a/b = 0.5) and the notch width to 5 mm. The average value of the fracture energy results to be G_F = 0.0484 N/mm, so that the critical value of the stress-intensity factor can be evaluated:

$$K_{IC} = (G_F\, E)^{1/2} = 40.75 \text{ N/mm}^{3/2}.$$

The utilized steel bars present a nominal diameter of 4, 5, 8, 10 mm, respectively. The bars of 4 and 5 mm do not present yielding and the conventional limit, with 0.2% permanent deformation, is equal to 637 N/mm^2 and 569 N/mm^2 respectively. The yield strength for the bars of 8 and 10 mm, on the other hand, is equal to 441 N/mm^2 and 456 N/mm^2 respectively.

5.3 Description of the R.C. beam specimens

Thirty (30) reinforced concrete (R.C.) beams were tested, with the cross-section of thickness t = 150 mm and depth b = 100, 200, 400 mm. The span between the supports was assumed to be equal to six (6) times the beam depth b and, therefore, to 600, 1200, 2400 mm for the specimens

A, B, C respectively.

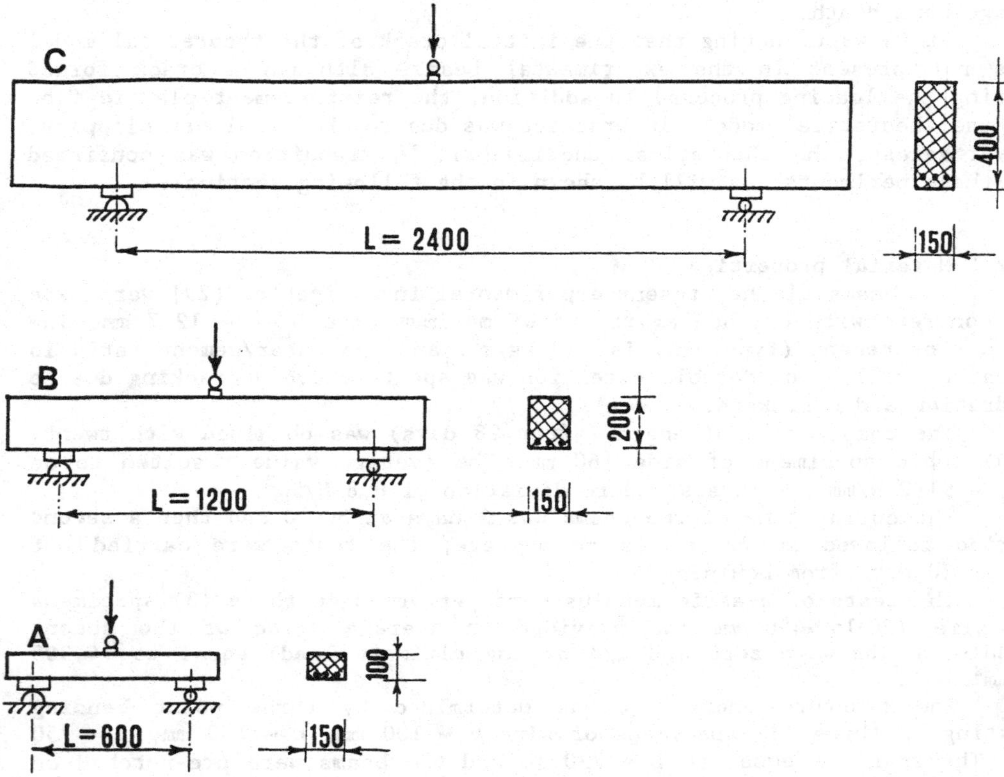

Fig. 35 Different sizes of geometrically similar reinforced concrete beam specimens.

The specimens were marked in the following way.

- By varying the beam size (Fig.35):
  - (A) beam depth b = 100 mm (t = 150 mm; L = 600 mm);
  - (B) beam depth b = 200 mm (t = 150 mm; L = 1200 mm);
  - (C) beam depth b = 400 mm (t = 150 mm; L = 2400 mm).
- By varying the brittleness class:
  - (0) brittleness number $N_P = 0$ (no reinforcement);
  - (1) brittleness number $N_P \simeq 0.13$ (on the average);
  - (2) brittleness number $N_P \simeq 0.36$ (on the average);
  - (3) brittleness number $N_P \simeq 0.72$ (on the average);
  - (4) brittleness number $N_P \simeq 1.18$ (on the average).

The content of steel depends on the beam size and on the brittleness number, see eq.(5.14). It is reported for each beam in Table 1.

The distance of the bars from the lower beam edge is, in each case, equal to 1/10 of the beam depth ($h/b = 0.1$). For each beam size (A,B,C) and for each brittleness class (0, 1, 2, 3, 4) two (2) R.C. beams were realized, with a total number of 30 specimens. All the beams were initially unnotched and uncracked. The following results, when not otherwise specified, are related to the average value of each case contemplated by the experimental investigation.

5.4 Testing apparatus and procedure

The experimental investigation was carried out at the Department of Structural Engineering of the Politecnico di Torino. The three point bending tests on R.C. beams were realized by a M.T.S. machine. The beams were supported by a cylindrical roller and a spherical connection respectively at the two extremities. The load was applied through a hydraulic actuator and the loading process was controlled by a strain gage DD1, placed on the lower beam edge, parallel to the beam axis and symmetrical with respect to the force. Its length was equal to the beam depth, i.e. 100, 200 or 400 mm, respectively for the beam sizes A, B and C. The sensitivity of the strain gage utilized is 1 mV/1V/1 mm, for a feed-voltage of 5 V. The strain rate was imposed at a constant and very low value. On the average, the crack formation in the middle of the beam, was achieved after about 7' (minutes) and the steel yielding after about 45' (minutes).

Transducers with a sensitivity of 1 mV/0.01 mm were used to measure the central deflection. The latter was referred to a bar, connected with the concrete beam at the middle of the depth and in correspondence of the two supports.

The load of first cracking was detected by means of a brittle enamel, applied in the zone where the first crack formation is expected.

The loads of first crack formation, of steel yielding and of final collapse, are summarized in Table 1.

Beam Size	Brittleness Class	Sizes t x b (mm)	Nominal content of steel	Actual percentage of steel	Yield limit of steel (N/mm^2)	Actual value of N_P	Cracking load (kN)	Yielding load (kN)	Ultimate load (kN)
A	0	150 x 100	0	0.00	0	0	11.77	0.00	0.00
	1	"	1 ϕ 4	$0.85\ 10^{-3}$	637	0.134	11.77	7.02	5.78
	2	"	2 ϕ 5	$2.56\ 10^{-3}$	569	0.360	12.50	15.20	11.28
	3	"	2 ϕ 8	$6.53\ 10^{-3}$	441	0.710	13.53	27.94	22.06
	4	"	2 ϕ 10	$10.03\ 10^{-3}$	456	1.170	14.90	34.51	47.81
B	0	150 x 200	0	0.00	0	0	22.55	0.00	0.00
	1	"	1 ϕ 5	$6.40\ 10^{-4}$	569	0.128	19.52	10.29	5.80
	2	"	3 ϕ 5	$1.90\ 10^{-3}$	569	0.380	20.84	23.10	17.14
	3	"	3 ϕ 8	$4.90\ 10^{-3}$	441	0.760	22.36	41.43	56.72
	4	"	3 ϕ 10	$7.75\ 10^{-3}$	456	1.240	26.67	64.95	76.56
C	0	150 x 400	0	0.00	0	0	40.20	0.00	0.00
	1	"	2 ϕ 4	$4.27\ 10^{-4}$	637	0.135	36.67	15.69	8.40
	2	"	4 ϕ 5	$1.28\ 10^{-3}$	569	0.360	38.73	32.36	24.40
	3	"	4 ϕ 8	$3.27\ 10^{-3}$	441	0.720	43.14	53.93	65.00
	4	"	4 ϕ 10	$5.17\ 10^{-3}$	456	1.170	48.93	84.42	97.65

TABLE 1 *Description of the reinforced concrete specimens and related loads of first cracking, steel yielding and final collapse.*

5.5 Experimental results and discussion

The load-deflection diagrams are plotted in the Figs.36-a, b and c, for each beam size and by varying the brittleness class (each curve

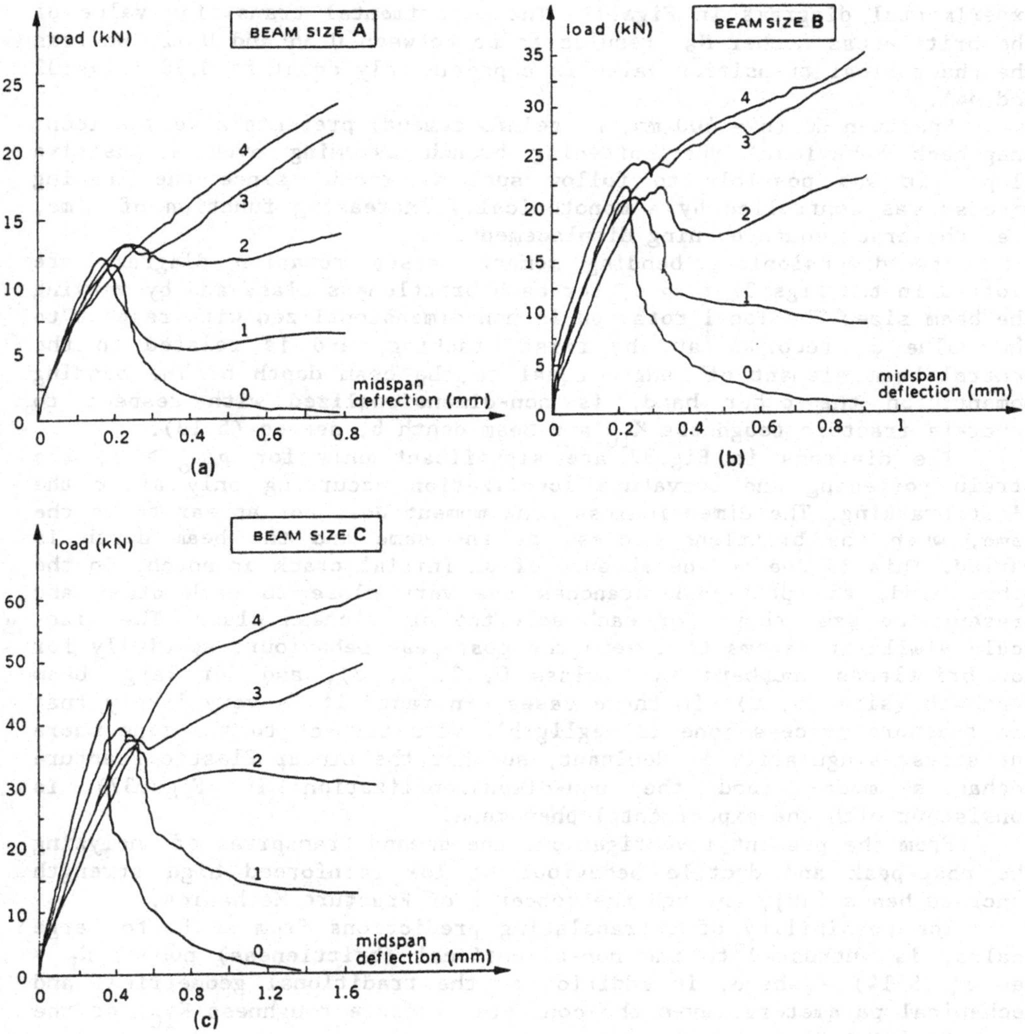

Fig. 36 Load vs. deflection diagrams of R.C. beams.
(a) beam depth b - 100 mm; (b) beam depth b - 200 mm;
(c) beam depth b - 400 mm.

is related to a single specimen of the two considered). As is possible to verify in Table 1, the peak or first cracking load is decidedly lower than the steel yielding load only in the cases 3 and 4, i.e. for high brittleness numbers N_P. In the cases 0 and 1, the opposite result is clearly obtained. On the other hand, case 2 demonstrates to represent a transition condition between hyperstrength and plastic collapse, the two critical loads being very close. Therefore, the same brittleness transition theoretically predicted in Fig.34, is reproposed by the experimental diagrams in Figs.36. The experimental transition value of the brittleness number N_P, results to be between 0.36 and 0.72, whereas the theoretical transition value is approximately equal to 0.70 (Figs.32 and 34).

Specimen CO (b = 400 mm, no reinforcement) presents a very evident snap-back behaviour, the softening branch assuming even a positive slope. It was possible to follow such a branch, since the loading process was controlled by a monotonically increasing function of time, i.e. the crack mouth opening displacement.

The dimensionless bending moment versus rotation diagrams are plotted in the Figs.37-a to e, for each brittleness class and by varying the beam size. The local rotation is non-dimensionalized with respect to the value φ_o recorded at the first cracking, and is related to the central beam element of length equal to the beam depth b. The bending moment, on the other hand, is non-dimensionalized with respect to concrete fracture toughness K_{IC} and beam depth b, see eq.(5.13).

The diagrams in Fig.37 are significant only for $\varphi/\varphi_o > 1$, the strain softening and curvature localization occurring only after the first cracking. The dimensionless peak moment does not appear to be the same, when the brittleness class is the same and the beam depth is varied. This is due to the absence of an initial crack or notch. On the other hand, the post-peak branches are very close to each other and present the same shape for each selected brittleness class. The size-scale similarity seems to govern the post-peak behaviour, specially for low brittleness numbers N_P (class 0, 1, 2, 3), and for large beam depths b (sizes B, C). In these cases, in fact, it is very likely that the fracture process zone is negligible with respect to the zone where the stress-singularity is dominant, so that the Linear Elastic Fracture Mechanics model (and the non-dimensionalization in Figs.37) is consistent with the experimental phenomena.

From the present investigation, the demand transpires of analyzing the post-peak and ductile behaviour of low reinforced high strength concrete beams [49], through the concepts of Fracture Mechanics.

The possibility of extrapolating predictions from small to large scales, is entrusted to the non-dimensional (brittleness) number N_P - see eq.(5.14) - where, in addition to the traditional geometrical and mechanical parameters, even the concrete fracture toughness K_{IC}, or the concrete fracture energy G_F, appears.

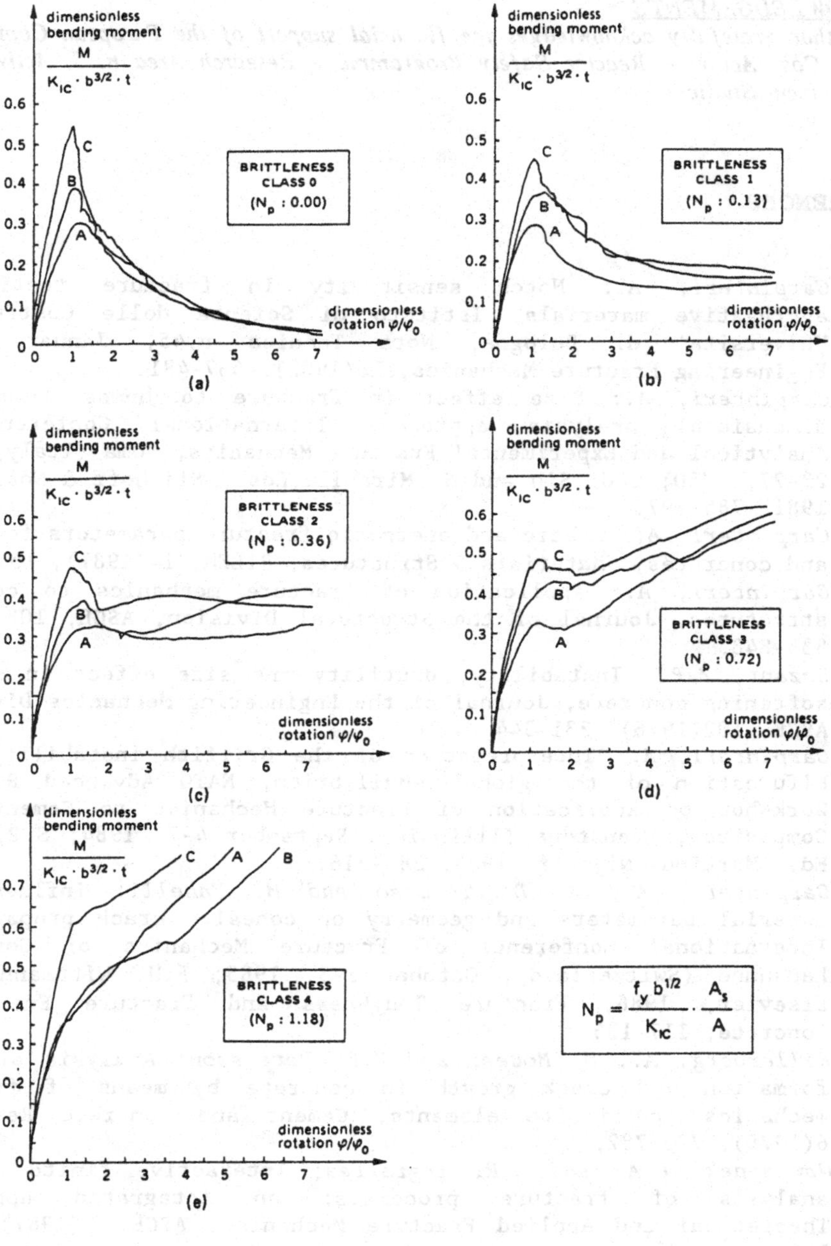

Fig. 37 Dimensionless bending moment vs. rotation diagrams. (A) beam depth b = 100 mm; (B) beam depth b = 200 mm; (C) beam depth b = 400 mm.

ACKNOWLEDGEMENTS
The author gratefully acknowledges the financial support of the European Communities Shared Cost Action - Reactor Safety Programme - Research Area n. 5, Activity n. 4 (Containment Studies).

REFERENCES

1. *Carpinteri, A.:* Notch sensitivity in fracture testing of aggregative materials, Istituto di Scienza delle Costruzioni, Universita' di Bologna, Nota Tecnica n.45, January 1980; Engineering Fracture Mechanics, 16(1982), 467-481.
2. *Carpinteri, A.:* Size effect in fracture toughness testing: a dimensional analysis approach, International Conference on Analytical and Experimental Fracture Mechanics, Roma (Italy), June 23-27, 1980; G.C. Sih and M. Mirabile,Eds., Sijthoff & Noordhoff, 1981, 785-797.
3. *Carpinteri, A.:* Static and energetic fracture parameters for rocks and concretes, Materials & Structures, RILEM, 14(1981), 151-162.
4. *Carpinteri, A.:* Application of fracture mechanics to concrete structures, Journal of the Structural Division, ASCE, 108(1982), 833-848.
5. *Bazant, Z.P.:* Instability, ductility and size effect in strain-softening concrete, Journal of the Engineering Mechanics Division, ASCE, 102(1976), 331-344.
6. *Carpinteri, A.:* Interpretation of the Griffith instability as a bifurcation of the global equilibrium, NATO Advanced Research Workshop on Application of Fracture Mechanics to Cementitious Composites, Evanstone (Illinois), September 4-7, 1984; S.P. Shah, Ed., Martinus Nijhoff, 1985, 287-316.
7. *Carpinteri, A., A. Di Tommaso and M. Fanelli*: Influence of material parameters and geometry on cohesive crack propagation, International Conference of Fracture Mechanics of Concrete, Lausanne (Switzerland), October 1-3, 1985; F.H. Wittmann, Ed., Elsevier, 1986, Fracture Toughness and Fracture Energy of Concrete, 117-135.
8. *Hillerborg, A., M. Modeer and P.E. Petersson*: Analysis of crack formation and crack growth in concrete by means of fracture mechanics and finite elements, Cement and Concrete Research, 6(1976), 773-782.
9. *Wawrzynek, P.A. and A.R. Ingraffea*: Interactive finite element analysis of fracture processes: an integrated approach, Theoretical and Applied Fracture Mechanics, ASCE, 8(1987), 137-150.
10. *Jenq, Y.S. and S.P. Shah*: Two parameter fracture model for concrete, Journal of Engineering Mechanics, ASCE, 111(1985), 1227-1241.

11. *Roelfstra, P.E. and F.H. Wittmann*: Numerical method to link strain softening with failure of concrete, International Conference of Fracture Mechanics of Concrete, Lausanne (Switzerland), October 1-3, 1985; F.H. Wittmann, Ed., Elsevier, 1986, Fracture Toughness and Fracture Energy of Concrete, 163-175.
12. *Rots, J.G., D.A. Hordijk and R. de Borst*: Numerical simulation of concrete fracture in direct tension, Fourth International Conference on Numerical Methods in Fracture Mechanics, San Antonio (Texas), March 23-27, 1987, Pineridge Press, 457-471.
13. *Bazant, Z.P.*: Size effect in blunt fracture: concrete, rock, metal, Journal of Engineering Mechanics, ASCE, 110(1984), 518-535.
14. *Maier, G.*: On the unstable behaviour in elastic-plastic beams in flexure, (in Italian), Istituto Lombardo, Accademia di Scienze e Lettere, Rendiconti, Classe di Scienze (A), 102(1968), 648-677, .
15. *Carpinteri, A.*: Size effect on the brittleness of concrete structures, (in Italian), A.I.T.E.C. Conference, Parma (Italy), October 17-18, 1985, 109-123.
16. *Carpinteri, A.*: Limit analysis for elastic-softening structures: scale and slenderness influence on global brittleness, Euromech Colloquium 204, Structure and Crack Propagation in Brittle Matrix Composite Materials, Jablonna (Poland), November 12-15, 1985; A.M. Brandt and I.H. Marshall, Eds., Elsevier Applied Science, 1986, Brittle Matrix Composites 1, 497-508.
17. *Carpinteri, A. and M. Fanelli*: Numerical analysis of the catastrophical softening behaviour in brittle structures, Fourth International Conference on Numerical Methods in Fracture Mechanics, San Antonio (Texas), March 23-27, 1987, Pineridge Press, 369-386.
18. *Petersson, P.E.*: Crack growth and development of fracture zones in plain concrete and similar materials, Report TVBM 1006, Lund Institute of Technology, 1981.
19. *Dougill, J.W.*: On stable progressively fracturing solids, Zeitschrift für Angewandte Mathematik und Physik (ZAMP), 27(1976), 423-437.
20. *Mazars, J. and J. Lemaitre*: Application of continuous damage mechanics to strain and fracture behaviour of concrete, NATO Advanced Research Workshop on Application of Fracture Mechanics to Cementitious Composites, Evanston (Illinois), September 4-7, 1984; S.P. Shah, Ed., Martinus Nijhoff, 1985, 507-520.
21. *Carpinteri, A., G. Colombo and G. Giuseppetti*: Accuracy of the numerical description of cohesive crack propagation, International Conference of Fracture Mechanics of Concrete, Lausanne (Switzerland), October 1-3, 1985; F.H. Wittmann, Ed., Elsevier, 1986, Fracture Toughness and Fracture Energy of Concrete, 189-195.
22. *Carpinteri, A., G. Colombo, G. Ferrara and G. Giuseppetti*: Numerical simulation of concrete fracture through a bilinear softening stress-crack opening displacement law, SEM-RILEM International Conference on Fracture of Concrete and Rock, Houston (Texas), June 17-19, 1987, S.P. Shah and S.E. Swartz, Eds., Proceedings, 178-191.

23. *Rots, J.G. and R. de Borst*: Analysis of mixed mode fracture in concrete, Journal of Engineering Mechanics, ASCE, 113(1987), 1739-1758.
24. *Carpinteri, A. and S. Valente*: Numerical modelling of mixed mode cohesive crack propagation, International Conference on Computational Engineering Science, Atlanta (Georgia), April 10-14, 1988, S.N. Atluri and G. Yagawa, Eds., Springer-Verlag, 12-VI.
25. *Bocca, P., A. Carpinteri and S. Valente*: Size effects in the mixed mode crack propagation: softening and snap-back analysis, International Conference on Fracture and Damage of Concrete and Rock, Vienna (Austria), July 4-6, 1988; to appear on Engineering Fracture Mechanics.
26. *Erdogan, F. and G.C. Sih*: On the crack extension in plates under plane loading and transverse shear, Journal of Basic Engineering, 85(1963), 519-527.
27. *Carpinteri, A.*: A fracture mechanics model for reinforced concrete collapse, IABSE Colloquium on Advanced Mechanics of Reinforced Concrete, Delft, 1981, 17-30.
28. *Carpinteri, A.*: Stability of fracturing process in RC beams, Journal of Structural Engineering, ASCE, 110(1984), 544-558.
29. *Bosco, C., A. Carpinteri and P.G. Debernardi*: Fracture of reinforced concrete: scale effects and snap-back instability, International Conference on Fracture and Damage of Concrete and Rock, Vienna (Austria), July 4-6, 1988; to appear on Engineering Fracture Mechanics.
30. Standard Method of Test for Plane Strain Fracture Toughness of Metallic Materials, E 399-74, ASTM.
31. *Tada, H., P. Paris and G. Irwin*: The stress analysis of cracks handbook, Del Research Corporation, St. Louis (Missouri), 1963, 2.16-17.
32. *Biolzi, L., S. Cangiano, G.P. Tognon and A. Carpinteri*: Snap-back softening instability in high strength concrete beams, SEM-RILEM International Conference on Fracture of Concrete and Rock, Houston (Texas), June 17-19, 1987, S.P. Shah and S.E. Swartz, Eds., (in press on Materials & Structures).
33. *Biolzi, L.*: private communication.
34. *Fairhurst, C., J.A. Hudson and E.T. Brown*: Optimizing the control of rock failure in servo-controlled laboratory test, Rock Mechanics, 3(1971), 217-224.
35. *Rokugo, K., K. Ohno and W. Koyanagi*: Automatical measuring system of load-displacement curves including post-failure region of concrete specimens, International Conference of Fracture Mechanics of Concrete, Lausanne (Switzerland), October 1-3, 1985; F.H. Wittmann, Ed., Elsevier, 1986, Fracture Toughness and Fracture Energy of Concrete, 403-411.
36. *Carpinteri, A. and G.C. Sih*: Damage accumulation and crack growth in bilinear materials with softening, Theoretical and Applied Fracture Mechanics, 1(1984), 145-159.

37. *Cedolin, L., S. Dei Poli and I. Iori*: Tensile behaviour of concrete, Journal of Engineering Mechanics, ASCE, 113(1987), 431-449.
38. *Barr, B. and T.J. Bear*: Fracture toughness tests for concrete, International Journal of Fracture, 13(1977), 92-96.
39. *Ingraffea, A.R. and V. Saouma*: Numerical modeling of discrete crack propagation in reinforced and plain concrete, Fracture Mechanics of Concrete: Structural Application and Numerical Calculation, G.C. Sih and A. Di Tommaso, Eds., Martinus Nijhoff Publishers, 1985, 171-225.
40. *Kasperkiewicz, J., D. Dalhuisen and P. Stroeven*: Structural effects in the fracture of concrete, Euromech Colloquium 204, Structure and Crack Propagation in Brittle Matrix Composite Materials, Jablonna (Poland), November 12-15, 1985; A.M. Brandt and I.H. Marshall, Eds., Elsevier Applied Science, 1986, Brittle Matrix Composites 1, 537-548.
41. *Li, V.C. and E. Liang*: Fracture processes in concrete and fiber reinforced cementitious composites, Journal of Engineering Mechanics, ASCE, 112(1986), 566-586.
42. *Li, V.C., C.M. Chan and K.Y. Leung*: Experimental determination of the tension-softening relations for cementitious composites, Cement and Concrete Research, 17(1987), 441-452.
43. *Shah, S.P.:* Dependence of concrete fracture toughness on specimen geometry and on composition, Fracture Mechanics of Concrete: Material Characterization and Testing, A. Carpinteri and A.R. Ingraffea, Eds., Martinus Nijhoff Publishers, 1984, 111-135.
44. *Walsh, P.F.:* Fracture of plain concrete, Indian Concrete Journal, 44(1972), 469-470.
45. *Zaitsev, Y.V. and K.L. Kovler*: Effect of specimen geometry, stress state and structure heterogeneity of cementitious composite materials on KIC, Euromech Colloquium 204, Structure and Crack Propagation in Brittle Matrix Composite Materials, Jáblonna (Poland), November 12-15, 1985; A.M. Brandt and I.H. Marshall, Eds., Elsevier Applied Science, 1986, Brittle Matrix Composites 1, 559-570.
46. *Ziegeldorf, S., H.S. Müller and H.K. Hilsdorf*; A model law for the notch sensitivity of brittle materials, Cement and Concrete Research, 10(1980), 589-599.
47. *Carpinteri, A. and S. Valente*: Size-scale transition from ductile to brittle failure: a dimensional analysis approach, CNRS-NSF Workshop on Strain Localization and Size Effect due to Cracking and Damage, E.N.S. Cachan (France), Sept. 6-9, 1988.
48. *Bazant, Z.P. and P.A. Pfeiffer*: Shear fracture tests of concrete, Materials and Structures, 19(1986), 111-121.
49. *Jaccoud, J.P. and H. Charif*: Armature minimal pour le contrôle de la fissuration, Rapport final des essais série C, Publication IBAP n.114, Ecole Polytechnique Fédérale de Lausanne, Juillet, (1986).

PATH INDEPENDENT INTEGRALS AND CRACK GROWTH PARAMETERS IN NONLINEAR FRACTURE MECHANICS

D. Gross
T. H. Darmstadt, Darmstadt, FRG

1 Introduction

In this paper some important aspects of nonlinear fracture mechanics are discussed; we will concentrate to fracture initiation and quasistatic crack growth of a macroscopic crack in an inelastic material, where plasticity or creep play a considerable role. From the engineering point of view these items mainly refer to ductile structural metals. Not considered are many other topics of modern nonlinear fracture mechanics like distributed microcracking, damage mechanics aspects, composite materials, fatigue or dynamic problems. Mainly the basic theoretical facts will be discussed, no attention is payd to specific problems of engineering application. In addition, this analysis in most cases is restricted to plane problems and monotonic unidirectional loading; the material is assumed to be isotropic and generally homogeneous.

Basis of fracture mechanics philosophy is the existence of singular dominant fields in the surrounding of a crack tip, which are characterized by only one parameter. These parameters control and determine (in the range of their validity) indirectly the state in the fracture process zone and by this the onset and the growth of a crack. In general they are related to path independent integrals.

2 Elements of linear fracture mechanics

2.1 Crack tip field

Considered is a plane structure with a crack under mode I condition. A linear elastic material is assumed with E = Young's modulus, ν = Poisson's ratio, $G = E/2(1+\nu)$ = shear modulus, $\kappa = 3 - 4\nu$ in plane strain and $\kappa = (3-\nu)/(1+\nu)$ in plane stress.

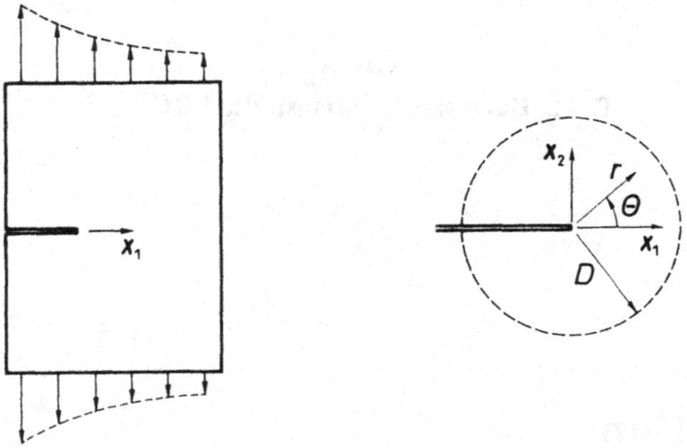

Figure 1: Crack in a plane structure

Independent of the specific loading and geometry of the structure the dominant singular field at the crack tip ($r \to 0$) is

$$\sigma_{ij} = \frac{K}{\sqrt{2\pi r}} \tilde{\sigma}_{ij}(\theta) \quad , \tag{1}$$

$$u_i - u_i^0 = \frac{K}{2G} \sqrt{\frac{r}{2\pi}} \tilde{u}_i(\theta) \quad . \tag{2}$$

$\tilde{\sigma}_{ij}(\theta)$ and $\tilde{u}_i(\theta)$ are well known eigenfunctions, fulfilling the boundary conditions at the crack tip; special values are $\tilde{\sigma}_{22}(0) = \tilde{\sigma}_{11}(0) = 1$, $\tilde{u}_2(\pm\pi) = \pm(\kappa+1)$. K is the mode I ***stress intensity factor***; it is determined by the solution of the complete boundary value problem for the structure. By K the crack tip field is fully characterized. The size of the region (its characteristic length is denoted by D), where the field really dominates, depends on the geometry (e.g. the crack length) and the loading of the structure.

2.2 Small scale yielding

On account of the high stresses there is yielding at the crack tip even if the material is brittle. In linear fracture mechanics it is assumed, that small scale yielding (ssy) holds.

That is to say, the plastic zone size at the crack tip has to be small compared with all relevant geometric length quantities (e.g. thickness), especially with the size (D) of the dominant K-field. Under these circumstances K provides an unique measure of the intensity of the strain field at the tip, which is independent on the details of geometry and loading.

Assuming a constant yield stress σ_0 along the plastic zone ligament and a K-field stress distribution outside one can estimate the size r_p of the plastic zone by use of (1):

$$r_p = \begin{cases} \frac{1}{3\pi}\left(\frac{K}{\sigma_0}\right)^2 & \text{plane strain} \ , \\ \frac{1}{\pi}\left(\frac{K}{\sigma_0}\right)^2 & \text{plane stress} \ . \end{cases} \tag{3}$$

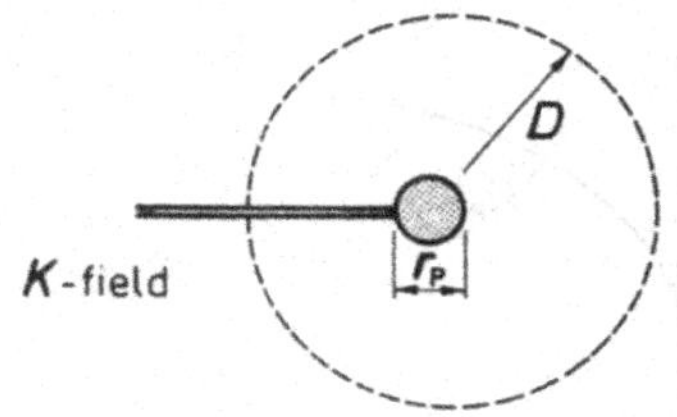

Figure 2: Small scale yielding

2.3 K–concept

At the crack tip there exists a region (D) where the K-field dominates. Under ssy-conditions a plastic zone of size r_p is embedded. The fracture process takes place in a zone of characteristic length $\rho < r_p$. If $D \gg r_p, \rho$ holds, the state of the process zone is controlled by only the K-field. This hypothesis is the basis of the K-concept: K is used as an unique measure, determining indirectly the state of the process zone at the

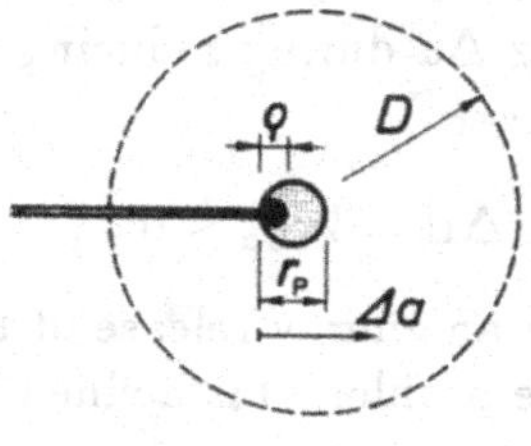

Figure 3: K-concept

crack tip. If K reaches a material specific critical value K_c, the state at the crack tip becomes critical: the crack starts to propagate. The criterion for fracture initiation reads

$$K = K_c \quad . \tag{4}$$

The material parameter K_c is called *fracture toughness*.

For further crack growth one could expect that K_c remains constant for an ideal brittle material. Real materials (e.g. intermediate and low strength metals) show a behavior sketched in Figure 4. The critical K is not longer a constant but a material dependent function of the crack advance Δa. This function is called *resistance curve* and denoted by $K_R(\Delta a)$. For continued crack advance the condition

$$K = K_R(\Delta a) \tag{5}$$

must be fulfilled.

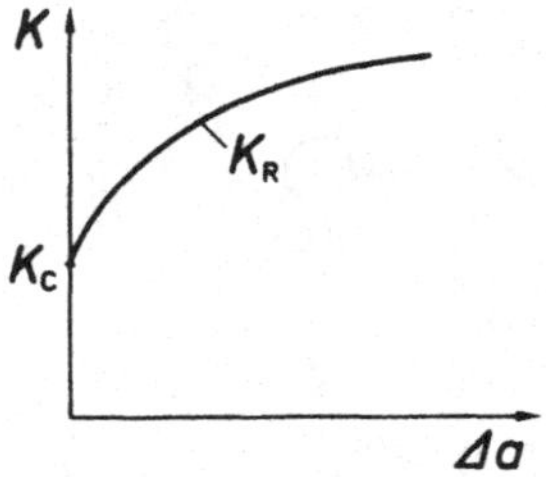

Figure 4: Resistance curve

2.4 Energy release rate

We now consider the energy release during crack growth in an ideal elastic body. If the external forces can be derived from a potential (e.g. dead loads, elastic springs) the total energy Π of the system consists of the inner potential (strain energy of the body) and the external potential (potential of the external forces). On account of the crack advance Δa the system moves from one (1) to another (2) equilibrium state. The change of total energy $\Delta\Pi = \Pi_2 - \Pi_1$ of the system can be expressed by the work $W_{\Delta a}$ done by the tractions along Δa during reducing them (quasistatically) to zero. Because this work is negativ (or zero)

$$\Delta\Pi = W_{\Delta a} \leq 0 \quad . \tag{6}$$

Crack growth is accompanied by an energy release of the system. The *energy release rate* (per unit thickness for plane problems) is defined by

$$\mathcal{G} = -\frac{\mathrm{d}\Pi}{\mathrm{d}a} \quad . \tag{7}$$

For linear elastic bodies the energy release rate is directly related to the K-factor. It is calculated as mentioned above. In state (1) the traction acting along the ligament ahead of the tip is $\sigma_y = K(a)/\sqrt{2\pi x}$ (see (1)). For sufficient small Δa the displacement of the upper (+) and the lower (−) crack face along the same line is $v^{\pm} = \pm K(a + \Delta a)\frac{\kappa+1}{2G}\sqrt{(\Delta a - x)/2\pi}$. Thus

$$W_{\Delta a} = \Delta\Pi = -\frac{1+\kappa}{4\pi G}K(a)K(a+\Delta a)\int_0^{\Delta a}\sqrt{\frac{\Delta a - x}{x}}\,dx = -\frac{1+\kappa}{8G}K(a)K(a+\Delta a)\Delta a \quad .$$

From this with $\Delta a \to 0$ we find

$$\mathcal{G} = \frac{1+\kappa}{8G}\,K^2 = \begin{cases} \dfrac{1-\nu^2}{E}K^2 & \text{plane strain} \\ \dfrac{K^2}{E} & \text{plane stress} \end{cases} \tag{8}$$

As a consequence we can rewrite the criterion for fracture initiation (4) and for continued quasistatic crack propagation as

$$\mathcal{G} = \mathcal{G}_c \quad . \tag{9}$$

For a perfectly elastic (brittle) material, where the process zone degenerates to a point at the crack tip, $\mathcal{G}_c$ is twice the surface energy per unit area. For a real material under ssy condition $\mathcal{G}_c$ is interpreted as *fracture energy*: that is energy per unit area necessary for the formation of new surfaces, the accompanied processes in the process zone and the plastic deformation. Equation (9) can also be seen as a consequence of the first law of thermodynamics; it is a necessary condition for crack propagation. Although (9) and (4) are mathematically equivalent (equation (9) is historically older) the interpretation of the fracture criterion today is totaly shifted to (4) because its basic idea can be extended to large scale yielding and creep (nonlinear fracture mechanics).

3 Path independent integrals

3.1 Conservation integrals of J-type

Considered is a homogeneous elastic (linear or nonlinear, isotropic or anisotropic) solid with a strain energy density $W(\varepsilon_{ij})$ such that

$$\sigma_{ij} = \frac{\partial W}{\partial \varepsilon_{ij}} \quad . \tag{10}$$

First we show the path independence of the integral (J-vector)

$$J_k = \int_\Gamma b_{kj}n_j\,ds = \int_\Gamma (W\delta_{jk} - \sigma_{ij}u_{i,k})n_j\,ds \quad , \tag{11}$$

where Γ is some closed surface (or contour in 2D), ds is an area element (or arc length in 2D) and $\boldsymbol{n}$ is the outer normal to Γ. The tensor $b_{kj} = W\delta_{jk} - \sigma_{ij}u_{i,k}$ is called the *energy momentum tensor* of elastostatics. If there are no body forces ($\sigma_{ij,j} = 0$) and all fields are sufficiently continuous and differentiable, then

$$\begin{aligned} \text{Div } \boldsymbol{b} = b_{kj,j} &= \frac{\partial W}{\partial \varepsilon_{mn}} \frac{\partial \varepsilon_{mn}}{\partial x_j} \delta_{jk} - \sigma_{ij,j}u_{i,k} - \sigma_{ij}u_{i,kj} \\ &= \sigma_{mn}u_{m,nk} - \sigma_{ij}u_{i,kj} = 0 \quad . \end{aligned}$$

Thus, by use of the divergence theorem, we get

$$J_k = 0 \tag{12}$$

for any closed surface Γ surrounding a homogeneous defect free material with no singularitiers or discontinuities. For an inhomogeneous material (12) generally does not hold. However for a material inhomogeneity of the special type $W = W(\varepsilon_{ij}, x_2, x_3)$ (translationally homogeneous material = properties do not change with x_1) it is easy to verify that $b_{1j,j} = 0$ and as a consequence

$$J_1 = 0 \quad . \tag{13}$$

Beside (11) there exist two other integrals of the J-type:

$$L_k = \int_\Gamma \epsilon_{klm}(x_l b_{mj} + u_l \sigma_{mj}) n_j ds \quad , \tag{14}$$

$$M = \int_\Gamma \left[b_{ij}x_i + \frac{1}{2}\sigma_{ij}(2-\alpha)u_i \right] n_j ds \quad . \tag{15}$$

Herein L_k is a vector quantity and M a scalar; ϵ_{klm} is the permutation tensor and $\alpha = 3$ in 3D respectively $\alpha = 2$ in 2D. Similar as (12)

$$L_k = 0 \tag{16}$$

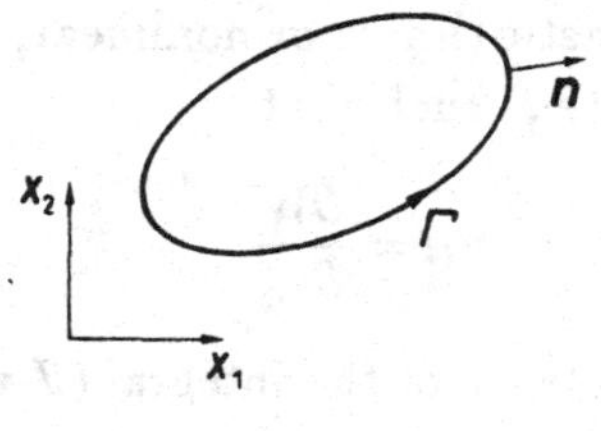

Figure 5: J-integral contour

for any closed surface Γ surrounding a defect free material, if the material is assumed to be isotropic and homogeneous. It can be shown, that the validity of $J_k = 0$ and $L_k = 0$ is not restricted to small (infinitesimal) deformations. Contrary to this

$$M = 0 \tag{17}$$

only holds in the special case of linear, infinitesimal elasticity.

There are different possibilities to derive (12), (16) and (17) from the basic equations of elastostatics. One is to apply Noether's theorem to the variational principle of elastostatics and to introduce several types of transformations (translation for J_k, rotation for L_k and similarity for M). Another direct derivation follows from the translational and the rotational invariance of the energy balance equation.

3.2 Energetic forces

For simplicity we consider a contour Γ surrounding a body with the boundary Γ_t. This surface Γ_t can be seen as a discontinuity surface, dividing the "continuum" in a part with no material and a part with elastic material; Γ_t may be loaded by the external traction $t_i = \sigma_{ij}n_j$. Now assume that by removing material, without change of the external load, the surface Γ_t is displaced by a constant increment $\mathrm{d}s_k$ (= translation of Γ_t). The so induced change $\mathrm{d}\Pi$ of total energy of the system consists of the strain energy

$$\int_{\Gamma_t} W n_k \mathrm{d}s_k \mathrm{d}s$$

stored in the removed layer and the difference

$$-\int_{\Gamma_t} t_i u_i(x_k + \mathrm{d}s_k)\mathrm{d}s + \int_{\Gamma_t} t_i u_i(x_k)\mathrm{d}s = -\int_{\Gamma_t} \sigma_{ij} u_{i,k} n_j \mathrm{d}s_k \mathrm{d}s$$

of the external potential. This leads to

$$\mathrm{d}\Pi = \mathrm{d}s_k \int_{\Gamma_t} (W\delta_{jk} - \sigma_{ij}u_{i,k})n_j \mathrm{d}s \quad .$$

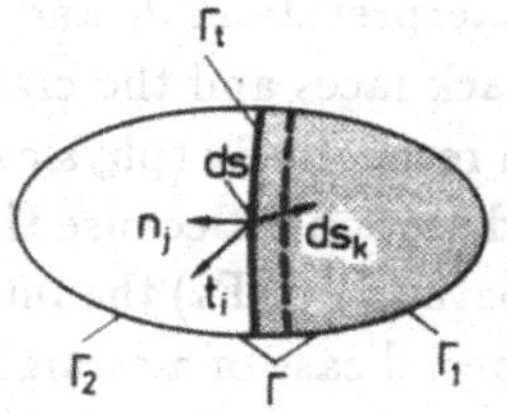

Figure 6: Displacement of a discontinuity surface

Because $\int_{\Gamma_1} + \int_{\Gamma_t} = 0$ (closed surface, see (12)) we can rewrite this as

$$d\Pi = -ds_k \int_{\Gamma_1} (W\delta_{jk} - \sigma_{ij}u_{i,k})n_j ds \tag{18}$$

or with $\int_{\Gamma_2} = 0$ and $\Gamma = \Gamma_1 + \Gamma_2$ as

$$d\Pi = -ds_k J_k \ . \tag{19}$$

This result is valid not only for this example, but it can be generalized to arbitrary discontinuity surfaces (area defects) and point defects (e.g. singularities on account of dislocations or of centers of dilatation). In other words: the energy change of an elastic system due to the translational displacement of a point or an area defect can be described by the path independent integral J_k taken arround the defect. In this energetic sense J_k may be interpreted as the *material force* acting on the defect. Similarly L_k is a path independent integral describing a *material moment* acting on the defect, wich leads to an energy change of the system, if the defect rotates. M describes the energy change on account of a self similar growth of the defect.

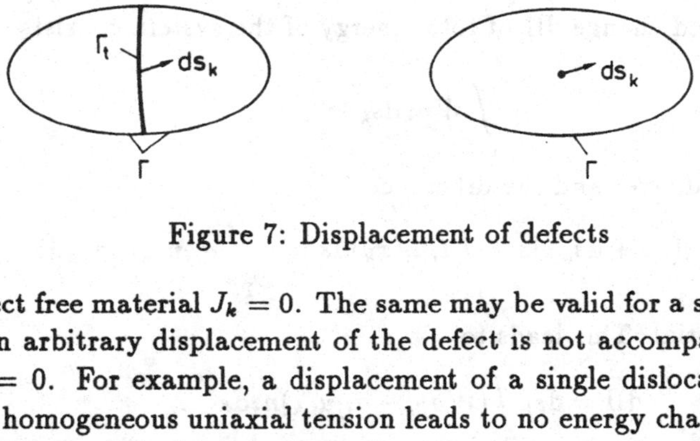

Figure 7: Displacement of defects

For a defect free material $J_k = 0$. The same may be valid for a surface surrounding a defect, if an arbitrary displacement of the defect is not accompanied by an energy change: $d\Pi = 0$. For example, a displacement of a single dislocation in an infinite region under homogeneous uniaxial tension leads to no energy change.

Next we apply (18), (19) to a crack in 2D with traction free faces, assuming contours Γ_1, Γ_2 surrounding the crack tip (starting end ending at the opposite crack faces, see figure 8). According to the force interpretation, J_1 and J_2 describe the energy release of the system, if the part of the crack faces and the crack tip surrounded by Γ_1 or Γ_2 is displaced in x_1- and x_2-direction respectively (physically senseful and kinematically possible is only a movement in x_1-direction). Because the length and the shape of this discontinuity line depends on the path (Γ_1 or Γ_2) the integrals J_1 and J_2 in general are path dependent. However in the special case of a straight crack J_1 is *path independent* on account of $\int_{\Gamma_1} + \int_{\Gamma^+} - \int_{\Gamma_2} + \int_{\Gamma^-} = 0$ (see (12)) and $\int_{\Gamma^+} = \int_{\Gamma^-} = 0$, from which $\int_{\Gamma_1} = \int_{\Gamma_2}$. This is not valid for J_2 applied to a straight crack and not valid for J_1 and J_2 applied to a curved crack.

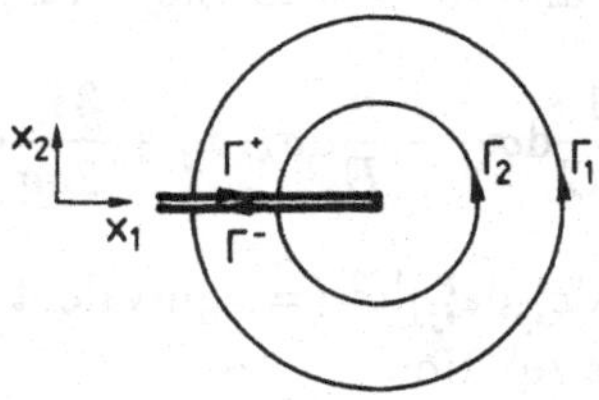

Figure 8: Path independence of J

The component J_1 in fracture mechanics is called the *J-integral* (subscript 1 is ommited)

$$J = \int_\Gamma (W\delta_{j1} - \sigma_{ij}u_{i,1})n_j \mathrm{d}s \quad ; \tag{20}$$

in its energetic interpretation it describes the energy release rate during quasistatic crack propagation in an elastic solid:

$$J = \mathcal{G} \quad . \tag{21}$$

For ssy under mode I conditions we get from (8)

$$J = \mathcal{G} = \frac{1+\kappa}{8G}K^2 \quad .$$

3.3 Inelastic materials

In conventional plasticity it is assumed, that the strains consist of an elastic and a plastic part:

$$\varepsilon_{ij} = \varepsilon^e_{ij} + \varepsilon^p_{ij} \quad . \tag{22}$$

The stress–strain relationship for the elastic part is usually linear. For the plastic part formulations in strain increments (incremental plasticity) or in total strains (deformation plasticity) are used. For both a yield condition is necessary; widely used is the von Mises condition

$$f = J_2 - k^2 = 0 \tag{23}$$

with $J_2 = s_{ij}s_{ij}/2$ (= second invariant of the stress deviator) and $k = \sigma/\sqrt{3}$ (σ = uniaxial yield stress).

Associated to the yield condition is in incremental plasticity the flow rule

$$\mathrm{d}\varepsilon^p_{ij} = \mathrm{d}\lambda \frac{\partial f}{\partial \sigma_{ij}} \quad ,$$

which leads with (23) for $f = 0$ and $d\sigma > 0$ to the Prandtl–Reuss law

$$d\varepsilon_{ij} = \frac{1}{2G}d\sigma_{ij} - \frac{\nu}{E}d\sigma_{kk}\delta_{ij} + \frac{3s_{ij}}{2g\sigma}d\sigma \quad , \tag{24}$$

where $g = d\sigma/d\varepsilon^p$ and $d\varepsilon^p = [\frac{2}{3}d\varepsilon^p_{ij}d\varepsilon^p_{ij}]^{1/2}$ (= equivalent uniaxial plastic increment).

In deformation plasticity the relation

$$\varepsilon^p_{ij} = \lambda s_{ij}$$

is assumed. By use of (23) λ is determined as $\lambda = 3\varepsilon^p/2\sigma$, which leads for $f = 0$ to the Hencky–Ilyushin law

$$\varepsilon_{ij} = \frac{1}{2G}\sigma_{ij} - \frac{\nu}{E}\sigma_{kk}\delta_{ij} + \frac{3\varepsilon^p}{2\sigma}s_{ij} \quad . \tag{25}$$

In deformation plasticity the inelastic material behavior is described like a nonlinear elastic behavior. Physically senseful can this description be used for monotonic loading, when no unloading occurs. Especially if proportional loading ($s_{ij} = \alpha s^0_{ij}$, α = load parameter) is assumed, incremental flow theory (24) and finite deformation theory (25) coincide.

Because deformation plasticity is equivalent to nonlinear elasticity, the path independent integrals of section 3.1 and 3.2 can be applied. The only difference is, that W in deformation plasticity cannot be considered as an elastic (recoverable) strain energy but should be interpreted as stress working density

$$W = \int_0^t \sigma_{ij}(x_k, t)\frac{\partial\varepsilon_{ij}(x_k, t)}{\partial t}dt \quad . \tag{26}$$

Consequently the (energetic) force interpretation of the path independent integrals cannot be applied in the elastic–plastic case. Especially the J-integral (20) in plasticity has nothing to do with an energy release rate. As we will see J is within the frame of deformation plasticity an unique measure of the stress and the deformation state at the crack tip.

Now we consider a linear viscoelastic solid under quasistatic conditions. It is well known, that the Laplace transformed field equations of linear viscoelasticity have the same structure as the equations of linear elasticity. This property is called elastic viscoelastic analogy or correspondence principle. So, if a functional $\overline{W}(\overline{\varepsilon}_{ij})$ is introduced, such that

$$\overline{\sigma}_{ij} = \frac{\partial\overline{W}}{\partial\overline{\varepsilon}_{ij}} \quad ,$$

then path independent integrals analogous to those of sections 3.1 and 3.2 exist. Herein the bar indicates a quantity as $\mathcal{L}$-transformed. Note that the inverse of $\overline{W}$ is not equal to the strain energy density, but has a complicated physical interpretation. In fracture mechanics an integral

$$\overline{J} = \int_\Gamma (\overline{W}\delta_{1j} - \overline{\sigma}_{ij}\overline{u}_{i,1})n_j \mathrm{d}s \tag{27}$$

can be introduced, which in analogy to the J-integral is path independent for a straight crack. It is obvious, that the inverse of $\overline{J}$ is not an energy release rate. As result of the correspondence principle it can be shown, that in mode I the $\overline{J}$-integral is proportional to $\overline{K}^2$, where $\overline{K}$ is the $\mathcal{L}$-transformed $K(t)$. It shall be mentioned, that the integral (27) has attained no great importance in fracture mechanics.

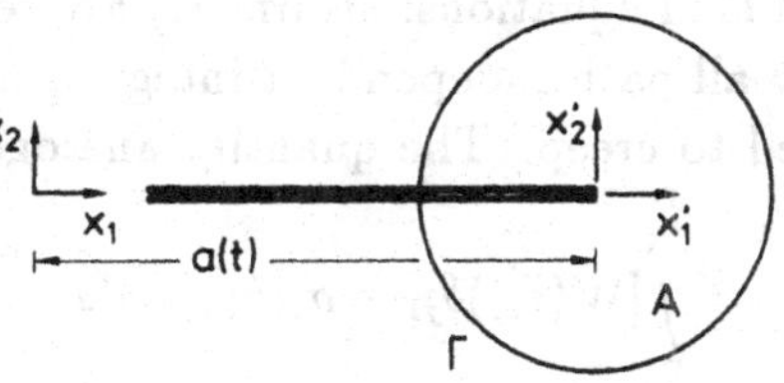

Figure 9: Growing crack

Finally let us consider a growing straight crack of length $a(t)$ in an arbitrary (elastic or inelastic) material. Relativ to the fixed system x_k moves the system x'_k (fixed at the moving crack tip) and the arbitrary contour Γ. Then the time rate $\mathcal{G}\dot{a}$ of energy flow to the tip is the difference of the work rate of tractions acting on Γ and the rate of stress work (26) in the material wich coincides instantanously with the region A (Figure 9):

$$\mathcal{G}\dot{a} = \int_\Gamma \sigma_{ij}n_j\dot{u}_i \mathrm{d}s - \frac{\mathrm{d}}{\mathrm{d}t}\int_A W \mathrm{d}A + \dot{a}\int_\Gamma W n_1 \mathrm{d}s$$

(the last term is the convective term). With $\dot{u}_i(x_k,t) = \partial u_i(x'_k,t)/\partial t - \dot{a}\partial u_i(x_k,t)/\partial x_1$ and (20) this expression becomes

$$\mathcal{G}\dot{a} = J\dot{a} + \left[\int_\Gamma \sigma_{ij}n_j \frac{\partial u_i(x'_i,t)}{\partial t}\mathrm{d}s - \frac{\mathrm{d}}{\mathrm{d}t}\int_A W(x'_k,t)\mathrm{d}A\right] ,$$

from which

$$\mathcal{G} = J \quad , \tag{28}$$

if the bracketed quantity vanishes. This case occurs for a homogeneous elastic material (as already discussed) and for any inelastic (in x_1-direction homogeneous) material under steady state conditions (= time derivatives in the brackets vanish).

3.4 Creep

Secondary (steady state) creep can be described as nonlinear viscous flow. Its constitutive relation is analogeous to nonlinear elastic behavior:

$$\sigma_{ij} = \frac{\partial W(\dot{\varepsilon}_{ij})}{\partial \dot{\varepsilon}_{ij}} \quad , \tag{29}$$

with the stress working rate density

$$W(\dot{\varepsilon}_{ij}) = \int_0^{\dot{\varepsilon}_{ij}} \sigma_{ij} \mathrm{d}\dot{\varepsilon}_{ij} \quad .$$

This analogy is valid for all field equations: strains ε_{ij} are replaced by strain rates $\dot{\varepsilon}_{ij}$ and u_i by $\dot{u}_i$. Consequently all path independent integrals and arguments of sections 3.1 and 3.2 can be extended to creep. The quantity analogeous to J is the so called C^*-integral

$$C^* = \int_\Gamma [W(\dot{\varepsilon}_{ij})\delta_{j1} - \sigma_{ij}\dot{u}_{i,1}]n_j \mathrm{d}s \quad . \tag{30}$$

To make the analogy complete C^* can be expressed as

$$C^* = -\frac{\mathrm{d}\dot{\Pi}}{\mathrm{d}a} \quad ,$$

where $\dot{\Pi}$ is the total work rate.

4 Elastic-plastic fracture mechanics

4.1 Dugdale model

In thin sheets often strip shaped plastic zones ahead of a crack tip are observed. They are primarely formed by cross slip on planes at 45° to the sheet plane; the zone height is of the same order as the thickness of the sheet. Assuming an elastic perfect plastic material and the length d of the plastic zone is long compared with t (Figure 10), Dugdale modelled the plastic zone as a yield line along which the yield stress σ_y acts. The elastic-plastic problem of a crack of length $2a$ in an infinite region under uniform tension σ then can be described as an elastic problem. The unknown length d of the plastic zone is determined from the condition that no stress singularity occurs.

The solution can be obtained by superposition of the two sketched problems with the K-factors

$$\begin{aligned} K^{(1)} &= \sigma\sqrt{\pi b} \quad , \\ K^{(2)} &= -\frac{2}{\pi}\sigma_y\sqrt{\pi b}\arccos\frac{a}{b} \end{aligned}$$

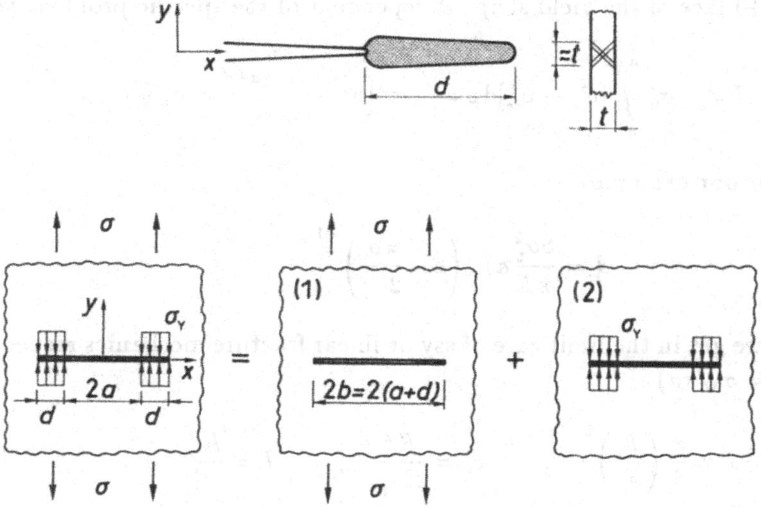

Figure 10: Dugdale model

and the displacements in y-direction (plane stress)

$$v^{(1)}(a) = \frac{2\sigma}{E}\sqrt{b^2 - a^2} \quad ,$$

$$v^{(2)}(a) = \frac{4\sigma_y}{\pi E}\left[-\sqrt{b^2 - a^2}\arccos\frac{a}{b} + a\ln\frac{b}{a}\right] \quad .$$

From $K^{(1)} + K^{(2)} = 0$ we get

$$d = b - a = a\left[\left(\cos\frac{\pi\sigma}{2\sigma_y}\right)^{-1} - 1\right] \tag{31}$$

what is in good agreement with experimental datas for $\sigma < 0.9\sigma_y$. With (31) the *crack tip opening displacement* δ_t (CTOD) is

$$\delta_t = 2[v^{(1)}(a) + v^{(2)}(a)] = \frac{8\sigma_y}{\pi E} a \ln\left(\cos\frac{\pi\sigma}{2\sigma_y}\right)^{-1} \quad . \tag{32}$$

It can be interpreted as a result of the plastic deformation and a possible measure of stretch at the crack tip.

Now we determine the J-integral by using a contour running along the lower $(-)$ and the upper $(+)$ face of the yield strip. Independent of the specific problem we find

$$J = -\sigma_y \int_a^{a+d} [v_{,x}^- - v_{,x}^+]\mathrm{d}x = -\sigma_y \left[v^+ - v^-\right]_a^{a+d} = \sigma_y \delta_t \tag{33}$$

and with (32) for our example

$$J = \frac{8\sigma_y^2}{\pi E} a \ln\left(\cos\frac{\pi\sigma}{2\sigma_y}\right)^{-1} \quad . \tag{34}$$

From (31)–(34) we get in the limit case of ssy or linear fracture mechanics respectively $(\sigma/\sigma_y \ll 1, \;\; K = \sigma\sqrt{\pi a})$

$$d = \frac{\pi}{8}\left(\frac{K}{\sigma_y}\right)^2 \quad , \qquad \delta_t = \frac{K^2}{E\sigma_y} \quad , \qquad J = \frac{K^2}{E} \quad . \tag{35}$$

These results on hand it is possible to formulate (within the framework of the Dugdale model) a fracture criterion for arbitrary plastic zone sizes on the basis of δ_t:

$$\delta_t = \delta_{tc} \quad . \tag{36}$$

The crack starts to grow, if δ_t reaches a material specific critical value δ_{tc}. By (33) an equivalent criterion is

$$J = J_c \quad , \tag{37}$$

where $J_c = \sigma_y \delta_{tc}$. Note that (36) and (37) in ssy lead to the K-concept (see (35)).

4.2 HRR-field and J-concept

There exists no solution in analytic form for the plastic crack tip field for an arbitrary (incremental) elastic-plastic constitutive law. The most important approximation to this problem was found by Hutchinson, Rice and Rosengreen and is called *HRR-field.* It bases on deformation plasticity (see section 3.3) and the description of strain hardening by the Ramberg-Osgood power law. Neglecting the elastic strains $(\varepsilon_{ij} = \varepsilon_{ij}^p)$ and assuming the von Mises yield condition it reads

$$\frac{\varepsilon_{ij}}{\varepsilon_y} = \frac{3}{2}\alpha\left(\frac{\sigma}{\sigma_y}\right)^{n-1}\frac{s_{ij}}{\sigma_y} \tag{38}$$

with the equivalent uniaxial stress $\sigma^2 = 3s_{ij}s_{ij}/2$ and the material constants α, n (n is the "hardening exponent"). As limits it involves for $n = 1$ the linear elastic case and for $n \to \infty$ the perfect plastic case. By (38) proportional loading is anticipated.

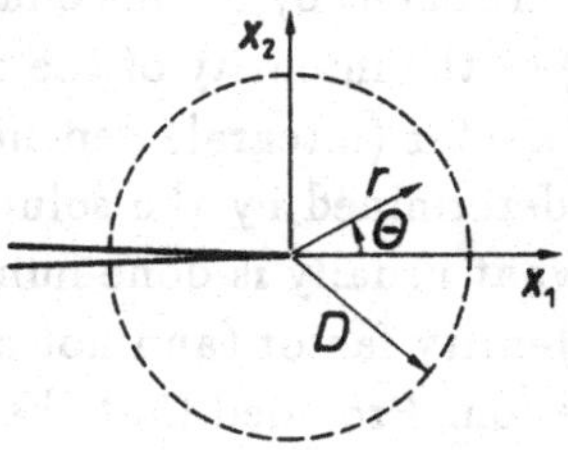

Figure 11: Crack geometry and J-controlled region D

Since the J-integral is path independent for a material of type (38), we chose Γ to be a circle of arbitrary radius r around the crack tip (figure 11):

$$J = \int_{-\pi}^{+\pi} (W\delta_{ij} - \sigma_{ij}u_{i,1})n_j r\mathrm{d}\theta \quad .$$

For J to be nonzero for $r \to \infty$ it is necessary that

$$\sigma_{ij}\varepsilon_{ij} \sim \frac{f(\theta)}{r} \quad , \qquad W \sim \frac{\widetilde{W}(\theta)}{r} \quad .$$

As a consequence from (10) and (38) it follows

$$\sigma_{ij} \sim \left(\frac{J}{r}\right)^{\frac{1}{1+n}} \tilde{\sigma}_{ij}(\theta) \quad , \qquad \varepsilon_{ij} \sim \left(\frac{J}{r}\right)^{\frac{n}{1+n}} \tilde{\varepsilon}_{ij}(\theta) \quad .$$

The angular functions $\tilde{\sigma}_{ij}(\theta)$, $\tilde{\varepsilon}_{ij}(\theta)$ can only be found by solution of the crack tip boundary value problem. It leads to a nonlinear eigenvalue problem, what usually is solved numerically; the eigenfunctions are tabulated in handbooks. Putting all together the dominant singular field at the crack tip ($r \to 0$) can be written as

$$\begin{aligned}
\sigma_{ij} &= \sigma_y \left[\frac{J}{\alpha\sigma_y\varepsilon_y I_n r}\right]^{\frac{1}{1+n}} \tilde{\sigma}_{ij}(\theta) \quad , \\
\varepsilon_{ij} &= \alpha\varepsilon_y \left[\frac{J}{\alpha\sigma_y\varepsilon_y I_n r}\right]^{\frac{n}{1+n}} \tilde{\varepsilon}_{ij}(\theta) \quad , \qquad (39) \\
u_i - u_i^0 &= \alpha\varepsilon_y \left[\frac{J}{\alpha\sigma_y\varepsilon_y I_n}\right]^{\frac{n}{1+n}} r^{\frac{1}{1+n}} \tilde{u}_i(\theta) \quad ,
\end{aligned}$$

where I_n is a known scaling factor. Analogous to the K-field in the linear elastic case the character of the field (39) is independent of the specific geometry and loading of a structure. It is fully determined (controlled) by J; this quantity now can be considered as an unique measure (for $n \neq \infty$) of the intensity of the crack tip field even for large scale yielding. Similar as K the J-factor (integral) depends on the geometry and the loading of a structure. It can be determined by the solution of the complete elastic-plastic boundary value problem, what usually is done numerically.

In this interpretation, as an intensity factor (and not as an energy release rate), J is used in the plastic fracture criterion. Provided that the characteristic size D of the region, where the J-controlled field (39) dominates, is large compared with the size of the process zone, then the state within the process zone is indirectly determined by J. The state becomes critical, that is the crack starts to grow, if J reaches a material specific value J_c:

$$J = J_c \quad . \tag{40}$$

On account of its direct relation with J (see (39)) any field quantity at a certain position of the near field can be used to formulate an equivalent fracture criterion. For example a δ_t often is introduced; it may be defined by the crack opening at the intercepts of the crack tip profile and two 45° lines. From (39) then we get

$$\delta_t = d_n \frac{J}{\sigma_y} \quad , \tag{41}$$

where d_n is a factor determined by (39).

It must be emphasized, that the J-concept bases on the existence of a J-dominant region at the crack tip. This requires a sufficient high strain hardening. For a nonhardening material ($n \to \infty$) the solution (39) partly degenerates and is not unique.

4.3 Crack growth and stability

Crack growth under large scale yielding conditions is accompanied by elastic unloading and nonproportional loading. This cannot be properly modelled by deformation plasticity, and as a consequence J in general cannot be applied for such problems. However for small amounts of crack growth J may be used under certain conditions. In such a case for continued crack growth

$$J = J_R(\Delta a) \tag{42}$$

is valid, where J_R is the J-resistance curve of a material.

A J-controlled crack growth only is possible, if the deviations in the near field (39) on account of the crack advance are small. Obviously this requires $\Delta a \ll D$. A more detailed result can be found by considering the change of the near field strains (see (39))

$$\varepsilon_{ij}(J, r, \theta) = k_n \left[\frac{J}{r}\right]^{\frac{n}{1+n}} \tilde{\varepsilon}_{ij}(\theta)$$

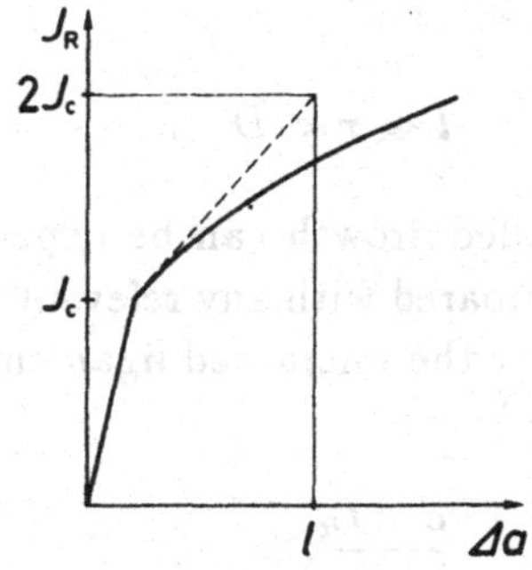

Figure 12: Resistance curve

due to increments dJ and da:

$$\mathrm{d}\varepsilon_{ij} = \frac{\partial \varepsilon_{ij}}{\partial J}\mathrm{d}J - \frac{\partial \varepsilon_{ij}}{\partial x_1}\mathrm{d}a \quad .$$

With

$$\frac{\partial}{\partial x_1} = \frac{\partial}{\partial r}\cos\theta - \frac{\partial}{r\partial\theta}\sin\theta$$

we get

$$\mathrm{d}\varepsilon_{ij} = k_n \left[\frac{J}{r}\right]^{\frac{n}{1+n}} \left[\frac{n}{1+n}\frac{\mathrm{d}J}{J}\tilde{\varepsilon}_{ij} + \frac{\mathrm{d}a}{r}\tilde{\beta}_{ij}\right] \tag{43}$$

where

$$\tilde{\beta}_{ij} = \frac{n}{1+n}\cos\theta\tilde{\varepsilon}_{ij} + \sin\theta\frac{\partial \varepsilon_{ij}}{\partial\theta} \quad .$$

While the first term in (43) describes proportional loading, the second term is non-

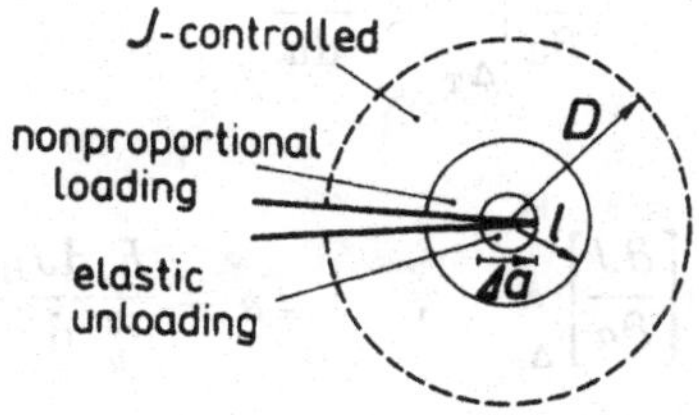

Figure 13: Crack tip region during crack growth

proportional. Thus nearly proportional loading occurs if

$$\frac{\mathrm{d}a}{r} \ll \frac{\mathrm{d}J}{J} \quad .$$

Introducing a characteristic (material specific) length $l \ll D$ for crack growth by $J/l = \mathrm{d}J/\mathrm{d}a$, this can be rewritten as

$$l \ll r < D \quad . \tag{44}$$

Under this condition a J-controlled growth can be expected. Because the size D of the dominant J-field is small compared with any relevant geometry parameter c of the structure (e.g. the crack length or the uncracked ligament) this condition can also be written as $l \ll c$ or (with $J = J_R$)

$$\frac{c}{J_R}\frac{\mathrm{d}J_R}{\mathrm{d}a} \gg 1 \quad . \tag{45}$$

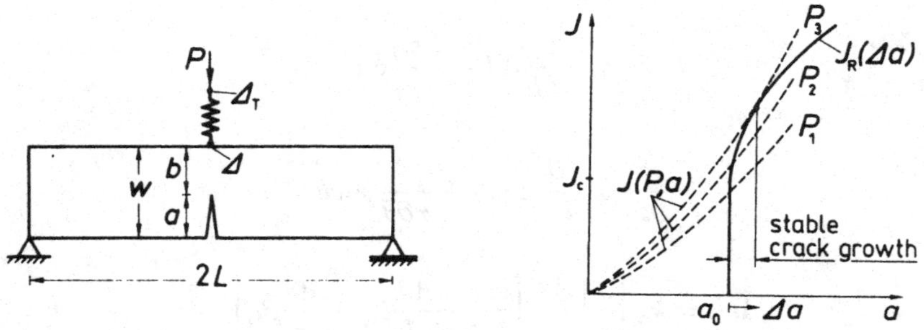

Figure 14: Stability of crack growth

We now consider crack growth ($J = J_R$) in a structure under compliant loading, where the displacement Δ_T is prescribed. The condition of stability of an equilibrium state is

$$\left.\frac{\partial J}{\partial a}\right|_{\Delta_T} < \frac{\mathrm{d}J_R}{\mathrm{d}a} \quad . \tag{46}$$

Introducing the quantities

$$T = \frac{E}{\sigma_y^2}\left[\frac{\partial J}{\partial a}\right]_{\Delta_T} \quad , \qquad T_R = \frac{E}{\sigma_y^2}\frac{\mathrm{d}J_R}{\mathrm{d}a} \quad , \tag{47}$$

where T_R is denoted as *tearing modulus*, the stability condition (46) becomes

$$T < T_R \quad . \tag{48}$$

The stability limit is given by

$$T = T_R \quad . \tag{49}$$

As depicted in figure 14, from the two equations (42) and (49) the load (respectively the corresponding J) and the crack length at the stability limit can be determined.

References

1. Rice, J.R.: Fracture mechanics, Appl. Mech. Rev., 38 (1985), 1271-1275
2. Rice, J.R.: The mechanics of quasi-static crack growth, Proc. 8th US Natl. Congr. of Appl. Mech., 1978, 191-216
3. Hutchinson, J.W.: Nonlinear fracture mechanics, Dept. of solid mechanics, The Technical University of Denmark, 1979
4. Rice, J.R.: Conserved integrals and energetic forces, in: Fundamentals of Deformation and Fracture, Cambridge University Press 1985, 33-56
5. Budiansky, B. and J.R. Rice: Conservation Laws and Energy-Release rates, J. of Appl. Mech., 40 (1973), 201-203
6. Buggisch, H., Gross, D. and K.-H. Krüger: Einige Erhaltungssätze der Kontinuumsmechanik vom J-Integral-Typ, Ing. Archiv, 50 (1981), 103-111
7. Nilsson, F.: A path independent integral for transient crack problems, Int. J. Solids Structures, 9 (1973), 1107-1115
8. Kanninen, M.F. and C.H. Popelar: Advanced Fracture Mechanics, Oxford University Press, 1985
9. Riedel, H.: Recent Advances in Modelling Creep Crack Growth, ICF 7, vol.2, 1495-1523, Pergamon Press 1989
10. Gao, J.C., Zhang, X.T. and K.C. Hwang: The asymptotic near tip solution for mode III crack in steady growth in power law hardening media, Int. J. of Fract., 21 (1983), 301-317

NON-LINEAR FRACTURE MECHANICS OF INHOMOGENEOUS QUASI-BRITTLE MATERIALS

V. C. Li
University of Michigan, Ann Arbor, Michigan, USA

ABSTRACT

This chapter presents a survey of non-linear fracture mechanics of inhomogeneous quasi-brittle material. The large scale bridging in the fracture process zone is explicitly modelled as cohesive spring-like tractions and its implications on crack formation and limitations of linear elastic fracture mechanics are discussed. A model of tension-softening is described to illustrate the possibility of relating the micromechanical mechanisms and the material structure to composite non-linear fracture property. Finally, recent research on experimental determination of tension-softening behavior in inhomogeneous quasi-brittle materials is reviewed.

1. INTRODUCTION

A large class of engineering material, including brittle matrix fiber and particulate reinforced composites, single phase polycrystalline ceramics, and rocks, display quasi-brittle behavior as a result of a toughening mechanism known as bridging. Bridging occurs when, for example, fibers crossing a matrix crack in a fibrous composite provide a closing pressure on the crack flanks. Debonding and frictional pulling out of the fibers drains part of the available elastic strain energy release into the crack tip zone, and is reflected as requiring a higher crack driving force. That is, the material appears to be tougher than when no bridging action is present.

In fiber composites, bridging actions have been observed in fiber reinforced mortar [e.g. 1,2], in fiber reinforced thermoplastics [e.g. 3,4] and in whisker reinforced ceramics [e.g. 5,6]. In particulate composites such as ceramics reinforced with ductile particles [e.g. 7], or in concrete -- cement reinforced with aggregates [e.g. 8], the bridging action is contributed by the particles and aggregates which may elastically rupture, plastically yield, or frictionally pulled out. In materials like polycrystalline ceramics and rocks, bridges may be provided by partially broken grains or grain ligaments. Such bridging actions have been observed in alumina, graphite and glass ceramics [9, 10, 11, 12]. In polymers, bridging by crazing at a crack tip [e.g. 13] and bridging in certain rubber toughened polymers under specific conditions have been observed.

The bridging forces, the extent of the bridging zone and the resulting toughening effect can differ significantly from material system to material system. However, it is possible to formulate the fracture problem in such materials in a general way based on the notion of a smeared-out stress versus displacement relationship which could be regarded as a kind of 'cohesive' traction acting across the crack tip process zone. This averaged stress-displacement relationship would include all effects of ligament volume fraction, stiffness, fracture or yield strength, interface bond properties and other relevant factors. Mathematically, it is convenient to think of this cohesive traction as if provided by (possibly non-linear) springs since the cohesive traction generally depends on the displacement across the crack flanks.

The analysis of fracture propagation in the presence of large scale bridging zone is motivated by the need of having an analytic tool for predicting structural behavior, particularly the ultimate strength, when the material fails by fracture advance, and yet the LEFM theory suitable for small scale yielding and elastic plastic fracture theory suitable for ductile material, are inapplicable. This occurs in the quasi-brittle material systems mentioned above, especially when the part size is small in comparison to the bridging zone size. In addition, there is increasing interest in understanding the formation process of a macroscopic crack, and the prediction of the level of toughening that could be achieved by various bridging mechanisms in inhomogeneous composite bodies. In this chapter, we first discuss in Section 2 the modelling of non-linear fracture propagation and the development of the bridging process zone, based on a given relationship between the traction and crack flank opening (the spring law) in the bridging zone. A specific spring law is then developed in Section 3 for fiber reinforced concrete to illustrate the dependence of composite fracture behavior, including R-curve behavior, on the material structure and on micromechanisms of deformation. Finally, in Section 4, experimental techniques for determination of the spring laws of any quasi-brittle material are presented.

2. FRACTURE PROPAGATION AND PROCESS ZONE DEVELOPMENT

In this section, we shall first examine how the spring laws relate to the uniaxial tensile behavior of a material system. The contribution of bridging to toughness will be presented through a generalized spring law description. Subsequently the relationship between a special type of spring law, a tension-softening curve, and the development of the bridging process zone is described. The limitation of the theory of linear elastic fracture mechanics (LEFM), in the presence of a large scale bridging zone will be discussed.

2.1 The Concept of Bridging Forces and its Contributions to Toughening

For convenience, it may be useful to envision a uniaxial tensile response of a brittle matrix composite reinforced with short frictionally bonded fibers aligned in the direction of applied loading. Prior to matrix failure, the composite response may be approximated by a linear elastic stress-strain curve with stiffness given by the law of mixture. After the matrix cracks, the maximum post-cracking stress may or may not be higher than the cracking stress, depending on the fiber volume, fiber aspect ratio and the bond strength (Figure 2.1). Prior to the maximum post-cracking stress, only the fibers along this fracture surface are carrying the load, and elastic debonding at the fiber/matrix interface may occur, if the fibers are strong enough not to rupture. When the maximum post-cracking strength is reached the fibers are fully debonded and slide out with frictional resistance, leading to a softening behavior with continued crack opening.

If the uniaxial specimen is unloaded at the matrix cracking strength, and then reloaded from zero crack opening, the tensile load σ_b versus crack opening displacement δ would give the required spring law (Figure 2.2). It should be noted that the spring law is given by a relationship between stress and displacement, not stress versus strain. This distinction is

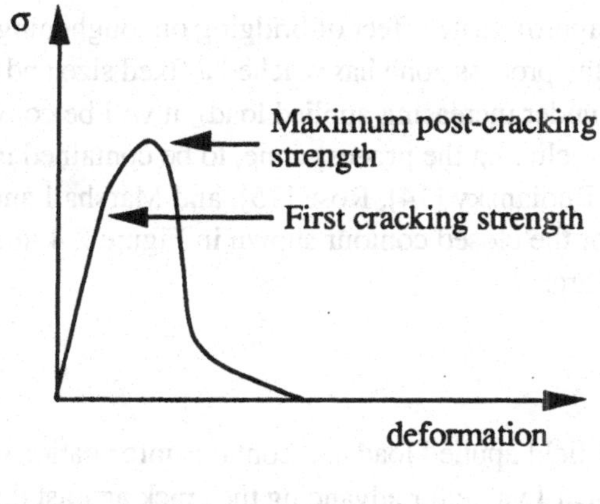

Fig. 2.1: Schematics of a tensile stress-deformation relationship of a composite

made necessary by the fact that in most quasi-brittle material systems, macroscopic scale fracture occurs on essentially a single plane such that classical strain measures would loose their physical meaning.

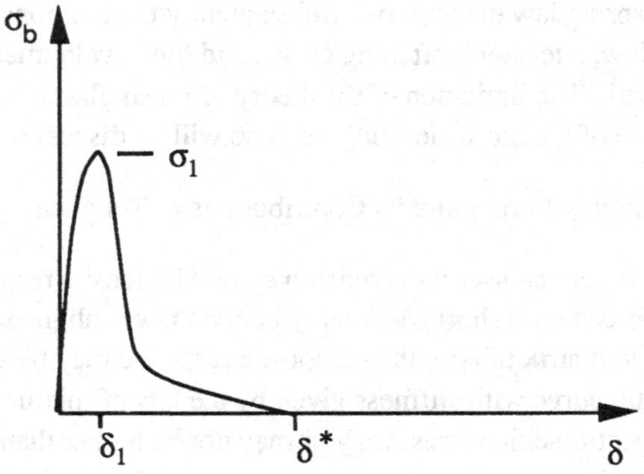

Fig. 2.2: Schematics of a stress-displacement spring law

If we now apply this spring law as a constitutive relationship for the crack tip zone of a structure containing a large steady state crack, the stress as a function of location along the crack line would look something like that shown in Figure 2.3a. The stress ahead of the crack tip can be described by the classical linear elastic crack tip stress field with intensity given by K_m corresponding to the toughness of the matrix. In the process zone, the stress rises with the crack opening following the spring law. At some point in the process zone behind the crack tip, the opening will be large enough such that the softening behavior of the spring will dominate.

In order to consider the approximate effect of bridging on toughening at steady state, (steady state here means that the process zone has reached a fixed size and simply translates as the physical crack extends under increasing applied load), it will be convenient to consider the crack tip region, including the process zone, to be contained in a far field K-dominant region. Following Budiansky [14], Rose [15], and Marshall and Evans [16], applying the J-integral [17] for the closed contour shown in Figure 2.4, the three contributions of the J-integral are:

$$J_\infty + J_b + J_{tip} = 0 \qquad (2.1)$$

The J_∞ term represents the far field applied load and contains information on the structural geometry. The J_{tip} is the critical J value for advancing the crack against the matrix toughness. It is generally more rigorous to describe this 'matrix toughness' as one that

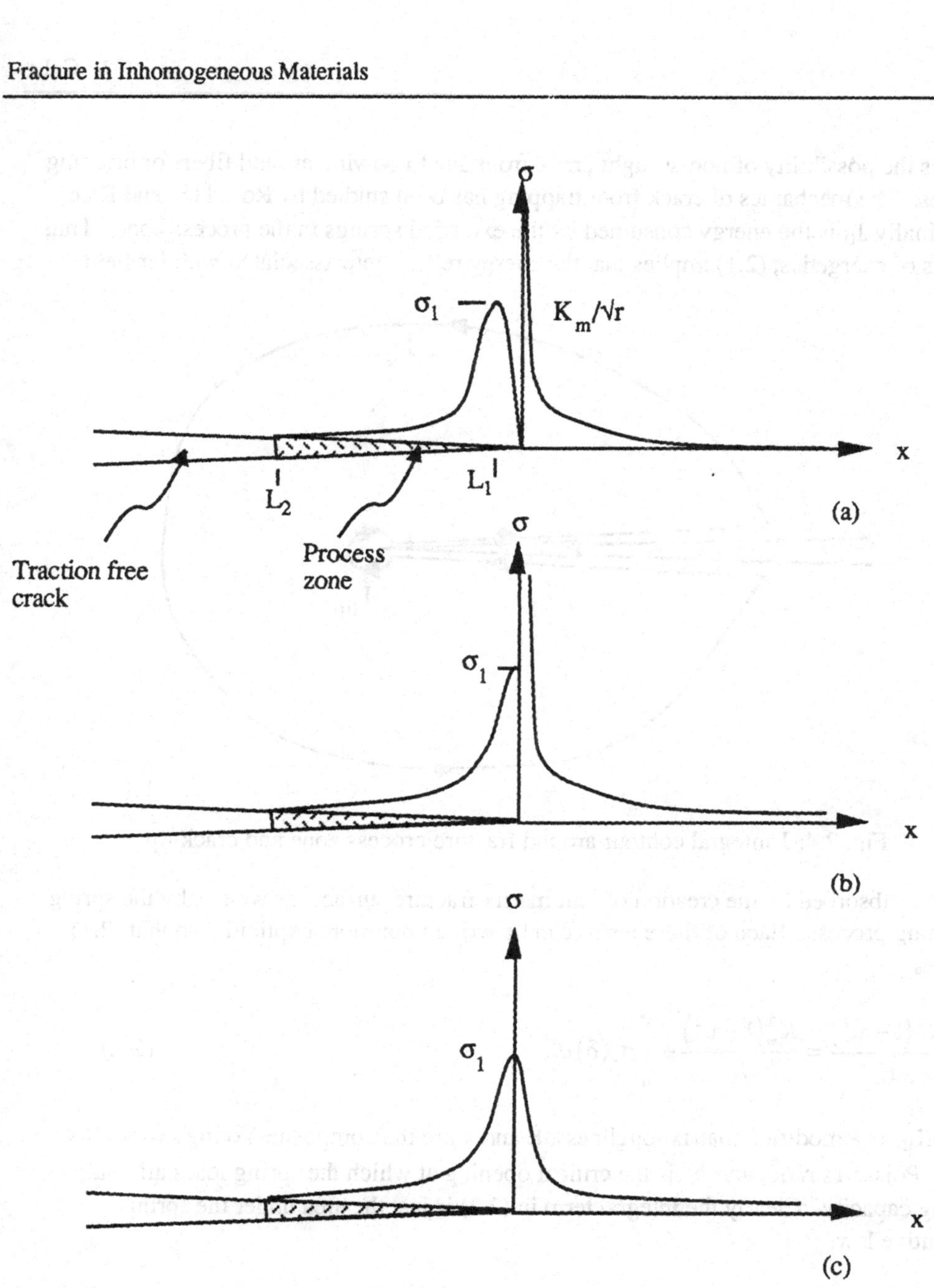

Fig. 2.3: Stress distribution along the crack line for (a) the most general situation, (b) the case when rising part of the spring law is unimportant or lumped as part of the matrix toughness, and (c) the case when the softening part of the spring law dominates the fracture propagation behavior

includes the possibility of non-straight crack front due to bowing around fibers or bridging particles. The mechanics of crack front trapping has been studied by Rose [18] and Rice [19]. Finally J_b is the energy consumed by the extended springs in the process zone. Thus in terms of energetics, (2.1) implies that the energy release rate associated with far field

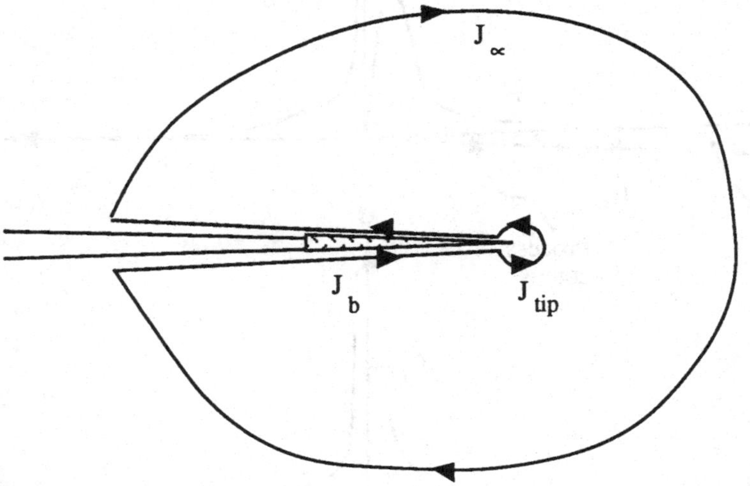

Fig. 2.4: J-integral contour around fracture process zone and crack tip

loading is absorbed by the creation of unit matrix fracture surface, as well as by the spring deforming process. Each of these terms can be written out more explicitly, so that (2.1) becomes

$$\frac{K^2(1-\upsilon^2)}{E}=\frac{K_m^2(1-\upsilon^2)}{E}+\int_0^{\delta^*}\sigma_b(\delta)\,d\delta \qquad (2.2)$$

where K_m is a modified matrix toughness, E and ν are the composite Young's Modulus and the Poisson's ratio, and δ^* is the critical opening at which the spring loses all load carrying capacity. Clearly the integral term in (2.2) is just the area under the spring constitutive law.

Equation (2.2) may be regarded as a generalization of the now familiar fracture criteria of Irwin [20], Barenblatt [21] and Dugdale [22]. Equation (2.2) reduces to Irwin's fracture criterion if the bridging zone does not exists, so that fracture propagation is resisted by the matrix toughness K_m only. If, on the other hand, K_m is negligible compared to the bridging term, as in the case of certain fiber reinforced composites with low toughness matrix, then equation (2.2) reduces to a form similar to the fracture criterion expressed by

Barenblatt. Finally, if again, K_m is negligible, and that σ_b is a constant independent of δ, then Dugdale's fracture criterion will result.

In some material systems, such as those with particle bridges which deform elastically and then fractures (as may be the case for certain particulate reinforced ceramics and high strength concrete), the softening tail part of the $\sigma_b(\delta)$ curve does not exist, and the bridging process zone goes out to L_1 only (Figure 2.3a). The amount of toughening for such a system can be written in terms of a positive spring stiffness (meaning spring force increases with increasing displacement) determined by the volume fraction, fracture strength and elastic stiffness of the particles (see, e.g. [14]). In some cases (probably in normal strength concrete and in alumina), the area under the $\sigma_b(\delta)$ curve up to δ_1 is likely to be small compared to the area up to δ^* so that the energy in opening the springs up to δ_1 could be lumped into an effective K_m, and the process zone contains only a tension-softening part (Figure 2.3b). In materials like short fiber reinforced concrete, the integral term in (2.2) contributed mainly by a softening spring reflecting the large amount of energy absorbed by fiber friction pull-out overwhelms the effective K_m term, and the stress distribution along the crack line will be well approximated by that shown in Figure 2.3c.

In certain engineering systems where crack opening must be restricted, it may be appropriate to neglect the contribution of the softening part of the $\sigma_b(\delta)$ curve to the overall toughness if stringent crack opening limitation is required. In other engineering systems such as in concrete structures, certain amount of crack opening is routinely tolerated, and the part or all of the softening $\sigma_b(\delta)$ curve may be counted to provide composite toughness. In all cases, this softening part of the curve is important to withstand accidental overloads to provide structural integrity.

2.2 Relation Between Tension-Softening And Process Zone Development

We now turn our attention to the case for a composite with a bridging zone in which the cohesive traction is well represented by a softening $\sigma_b(\delta)$ curve, as shown in Figure 2.3b or 2.3c, and when the steady state has not been reached. Other cases where the rising part of the curve is important could be similarly analyzed (see, e.g. [23]). It may be useful to consider the crack as being in the formation process prior to steady state.

For simplicity, consider a crack growing in an elastic body loaded remotely by σ_∞. The governing equation for equilibrium gives the tensile stress on the crack line as:

$$\sigma = \sigma_\infty + \int_{crack} G(x - x')\frac{\partial\delta(x')}{\partial x'}dx' \tag{2.3}$$

where the integral is to be carried over the crack line. A supplementary equation providing the condition for crack extension is needed and can be written as:

$$K_\infty + \int_{bridge} f(c,x')\sigma_b(x')dx' = K_m \qquad (2.4)$$

The influence functions G and f in (2.3) and (2.4) embody information on the structural geometry and the loading configurations. For the numerical illustrations shown below, we have used those corresponding to a crack of total (including the length of the process zone) length c in an infinite body (see, e.g. [24]). In (2.3), the stress σ is identified as the bridging traction σ_b and forms a functional relationship with δ in the process zone. Thus (2.3) and (2.4) form a set of integral equations in δ which in general must be solved numerically. Various simplifications have been made such as by assuming a linear $\sigma_b(x)$ traction distribution in the process zone [25], or by assuming a linear or parabolic shape of the crack face [26] in the solution of these equations. A complete numerical solution of these equations have been obtained by Li and Liang [24] for the case where K_m is small compared to the bridging integral term in (2.4). This assumption is appropriate for certain fiber reinforced composite systems but would not be correct for single phase ceramics, where the K_m term and the bridging contribution to toughness are probably on the same order of magnitude [27]. The non-linear fracture model represented by (2.3) and (2.4) have been employed in various forms by a number of researchers. These include, e.g. Hillerborg et al [28] in concrete, Ingraffea and coworkers in concrete [29] and rock [30], Mai and Lawn [31] in alumina ceramics, and Wnuk and co-workers (see, e.g. references in the chapter by Wnuk in this text) in polymer.

Following [24], the complete solution of (2.3) and (2.4) allows the crack shape profile and the stress distribution profiles to be determined. Figure 2.5 illustrates a quarter of the crack shape profile symmetrical about both axes. The characteristics of this profile is that the crack tip zone closes smoothly as u(x) -> $x^{3/2}$ whereas the classical LEFM crack profile u(x) is quadratic in x. This is the case if the presence of the fracture process zone eliminates the stress singularity at the crack tip.

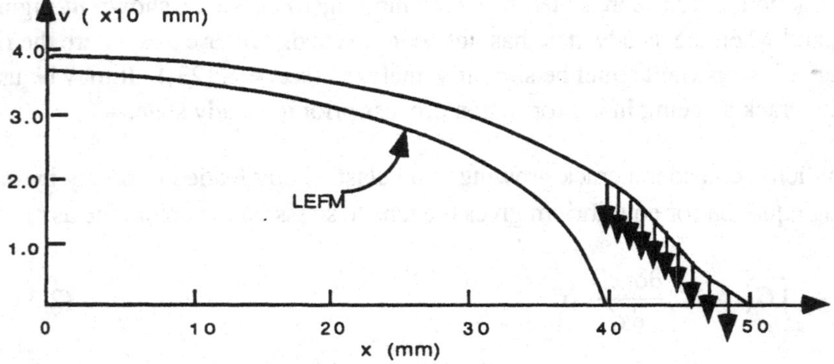

Fig. 2.5: The calculated crack opening profile symmetric about the x and y-axis

The resulting stress profile is shown in Figure 2.6. In approaching the crack tip, the stress rises to the composite tensile strength and decays to zero as the crack opens in the process zone. The LEFM stress profile is also shown for reference. These two curves are expected to overlap should the process zone size l_p reduces to zero. The tension-softening curve used in this calculation is shown in the insert of Figure 2.6.

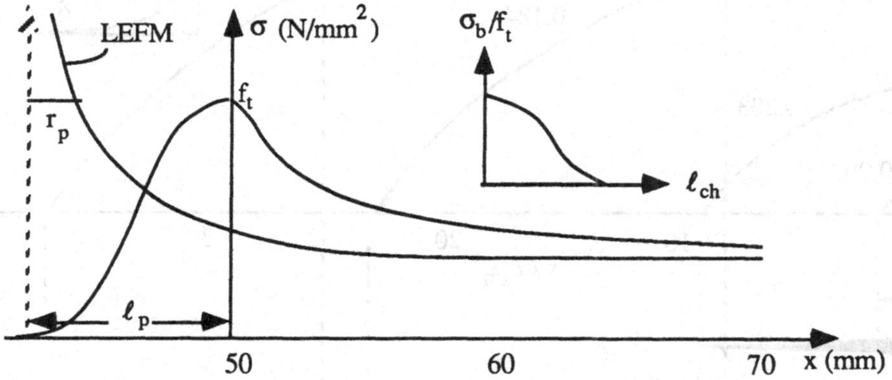

Fig. 2.6: The calculated stress distribution along the crack line

An interesting result of this analysis is the evolution of the process zone (Figure 2.7). The horizontal dashed line represents the critical opening value δ^*. When the opening exceeds this value, the traction is reduced to zero. Four stages of increasing crack opening displacements are schematically shown on the lower half of Figure 2.7. The first two of these stages illustrate the growth of this process zone without growth of the physical crack. In the third stage, the physical crack grows with a further extension of the process zone length. In the fourth stage, the physical crack grows further with a small decrease in the process zone length. After this, steady state is reached, whereby the process zone length remains constant and translates to the right as the physical crack grows.

Figure 2.8 schematically illustrates the same process observed by ultrasonics and in-situ microscopy technique in a crack wedged open in a Westerly granite [32]. The light shaded zone denoted as "partially separated with geometrical or frictional resistance to crack opening" can probably be associated with the inelastic processes induced by the K_m singularity or the increasing branch of the σ_b-δ curve and the part of the process zone labelled L_1 in Figure 2.3a. This part is not modelled in the present analysis but could be included as described earlier. The dark shaded zone denoted as "frictional resistance to crack opening" is likely associated with the softening branch of the σ_b-δ curve. The experimental observations illustrated in Figure 2.8 suggest that only after some growth of this process zone does the physical crack begin to extend, just as predicted by the theoretical analysis.

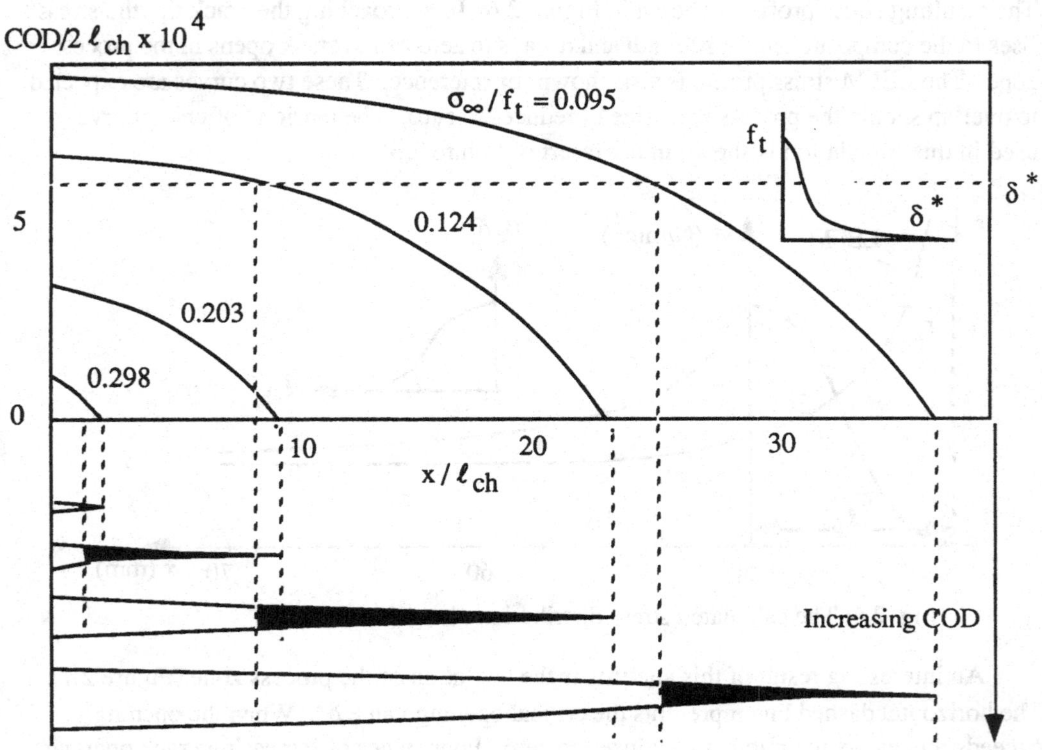

Fig. 2.7: The calculated stages of fracture process zone development

The R-curve of quasi-brittle material, denoting the increasing amount of energy absorbed in the process zone, has been measured by a number of investigators (e.g. [31], [33]) in the form of increasing critical energy release rate or apparent fracture toughness with a certain measure of increase in crack length. It should be clear that in quasi-brittle material, this increased energy absorption is due to the growth of the process zone as described above, and that once the process zone is fully developed, the plateau of the R-curve would be reached. Examples of R-curve calculations are given in section 3.4.

The details of the process zone evolution and translation is obviously related to the shape of the σ_b-δ curve, but less obvious is that they are also influenced by the stress field surrounding the process zone. This implies that with different loading configurations and specimen geometry, it may be expected that the details of the process zone growth observed

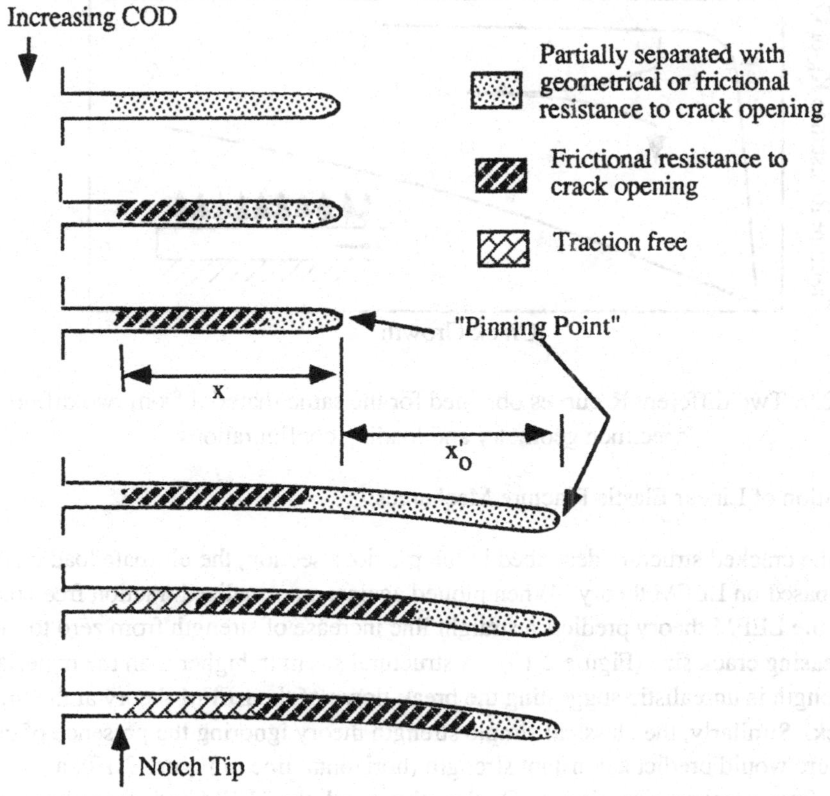

Fig. 2.8: Schematics showing the development of the fracture process zone in Granite by ultrasonic and in-situ microscopy techniques (adapted from [32])

would also be different [24]. Because the R-curve is associated with the process zone development, its shape may be expected to be influenced by these same factors. Hence the R-curve should not be regarded as a material property. As an example, Figure 2.9 illustrates a double cantilevered beam with different loading configurations and the corresponding R-curves. This point is particularly important as the slope of the R-curve in association with the point of specimen instability, is sometimes used to extract information regarded as a material property (e.g. [34]). In some cases where the specimen geometry and loading configuration is such that the crack tip stress field is similar, such as in a three point bend specimen and in a compact tension specimen, the resulting R-curve may look quite similar.

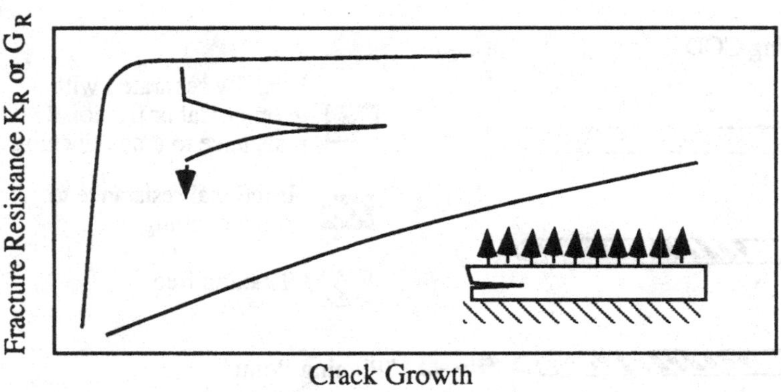

Fig. 2.9: Two different R-curves obtained for the same material from two different specimen geometry and loading configurations

2.3 Limitation of Linear Elastic Fracture Mechanics

For the cracked structure described in the previous section, the ultimate load can be predicted based on LEFM theory. When plotted against a normalized traction free crack length a_0, the LEFM theory predicts a straight line increase of strength from zero to infinity with decreasing crack size (Figure 2.10). A structural strength higher than the material tensile strength is unrealistic suggesting the break-down of the LEFM theory at the limit of small crack. Similarly, the classical simple strength theory ignoring the presence of cracks in a structure would predict a constant strength (horizontal line in Figure 2.10), again unrealistic from practical experience. On the other hand, the LEFM and strength theories have been proven adequate for ideally brittle Griffith type materials and for ductile yielding materials. These theories are therefore very useful ones in the limiting cases of suitably (will be explained later) large and small cracks respectively. Between these two limiting cases, we need a theory more suitable for the quasi-brittle material. The formalism presented in section 2.2 above can form the basis for such a non-linear fracture theory. This analysis results in the curve bridging the two limiting extremes, shown in Figure 2.10. The merging of this curve with those predicted by LEFM and material strength theories suggests the range of crack sizes when these simplified theories perform adequately.

The analysis presented above results in a natural "material characteristic length" l_{ch} parameter, defined as G_cE/f_t^2 or $(K_{IC}/f_t)^2$ where G_c is the fracture energy, K_{IC} is the fracture toughness, E the Young's Modulus and f_t is the tensile strength. The l_{ch} parameter was first introduced to concrete by Hillerborg [28] (with a slightly different definition of fracture energy). l_{ch} can be conveniently interpreted as the ratio between the slope of the stress-strain curve (the elastic Young's Modulus) and the slope of the tension-softening curve of a material if the tension-softening curve is approximated by a linear fit.

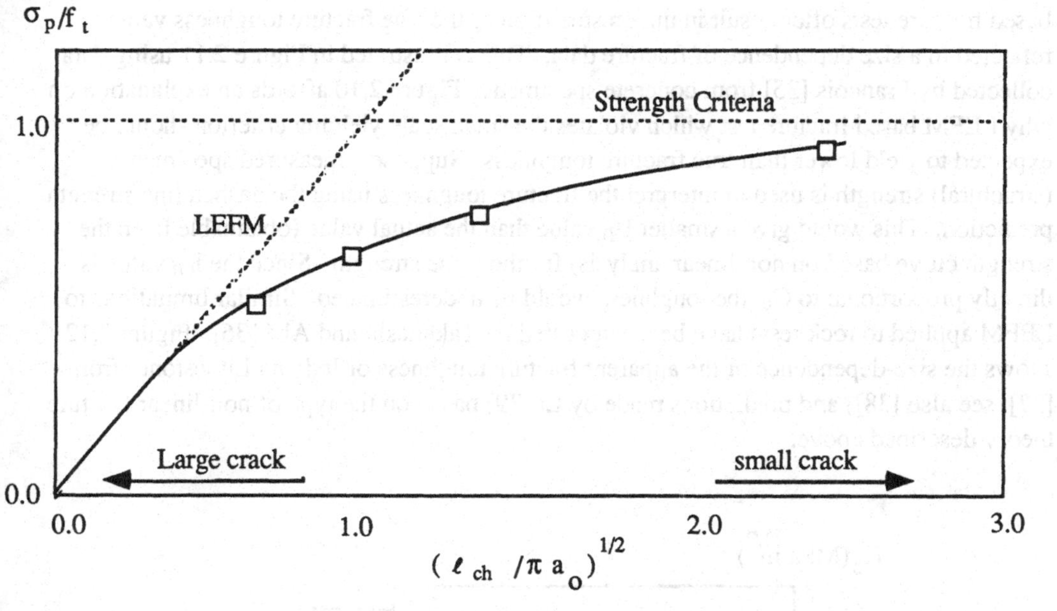

Fig. 2.10: Strength dependence on the relative crack length based on strength theory, linear elastic fracture mechanics theory and the present non-linear elastic fracture mechanics theory

The l_{ch} value of a quasi-brittle material is an extremely important material characterization parameter because it is a stronger indicator of brittleness than say the fracture energy or the fracture toughness itself. For example, the fracture toughness of high strength concrete and most ceramics would be higher than that for normal strength concrete, and yet these materials are generally perceived as more brittle than normal strength concrete. This is because the l_{ch} value for them is very small due to the disproportionate rise in tensile strength over fracture toughness. Hence the notch sensitivity increases from concrete, to high strength concrete to ceramics, due to their decreasing l_{ch} values. The contribution of fiber reinforcement or particle reinforcement in many such material systems is in the resulting disproportionate increase in toughness over tensile strength by providing an energy absorption process zone. [Note: the tensile strength and flexural strength in reinforced composites are not proportional. Toughness increase helps in the flexural strength but does not necessarily relate directly to the tensile strength.] For example, toughness increase in FRC is typically in the range of 10 to 50 times while strength increase is typically in the range of 1.1 to 2 times [2]. For ceramics composites, the toughness increase due to whisker and particulate reinforcements is typically in the range of 2 to 5 times while strength increase is typically not as significant [7].

In the concrete and rock community, there is increasing awareness that classical LEFM based fracture tests often result in underestimation of the true fracture toughness value, reflected in a size dependence of fracture data. This is illustrated in Figure 2.11 using data collected by Francois [35] from concrete specimens. Figure 2.10 affords an explanation on why LEFM based fracture test which violates the small scale yielding criterion should be expected to yield lower than true fracture toughness. Suppose a measured specimen (structural) strength is used to interpret the fracture toughness using the dashed line strength prediction. This would give a smaller l_{ch} value than the actual value (obtainable from the strength curve based on non-linear analysis) for the same strength. Since the l_{ch} value is directly proportional to G_c, the toughness would be underestimated. Similar limitations to LEFM applied to rock tests have been suggested by Takahashi and Abé [36]. Figure 2.12 shows the size-dependence of the apparent fracture toughness of Indiana Limestone (from [37], see also [38]) and predictions made by Li [39] based on the type of non-linear fracture theory described above.

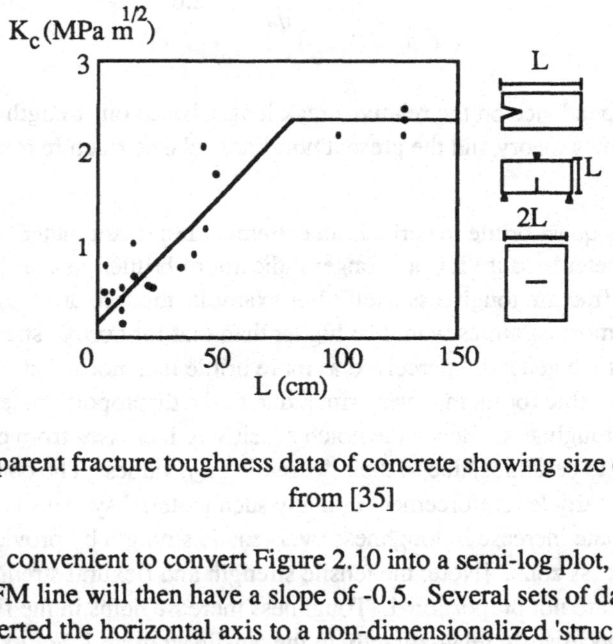

Fig. 2.11: Apparent fracture toughness data of concrete showing size dependence, adapted from [35]

It is often convenient to convert Figure 2.10 into a semi-log plot, as shown in Figure 2.13. The LEFM line will then have a slope of -0.5. Several sets of data (e.g. Bazant et al [40] who converted the horizontal axis to a non-dimensionalized 'structural size') have demonstrated that concrete as a quasi-brittle material follows the type of behavior predictable by a non-linear fracture theory which explicitly account for the presence of a bridging process zone.

The material characteristic lengths of some structural materials are given in Table 2.1. As an order of magnitude estimate, the process zone size may be regarded as roughly one to

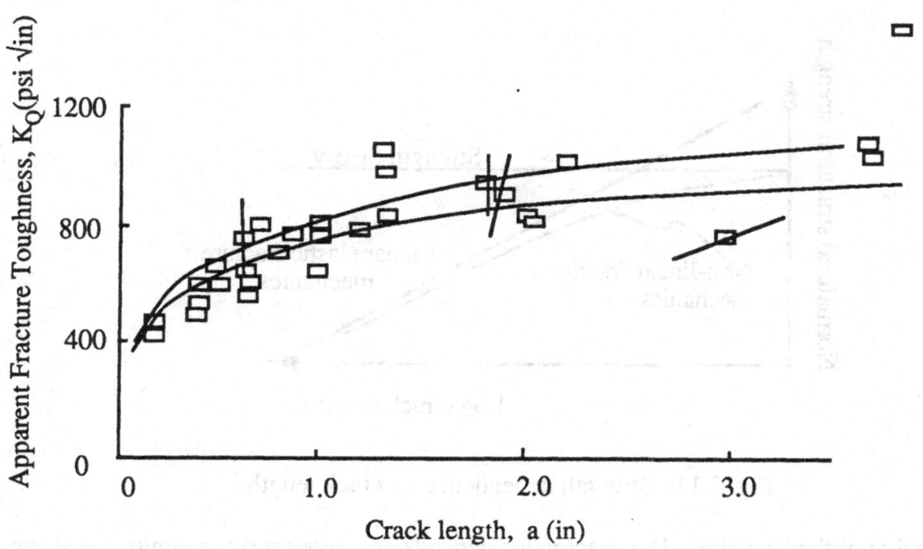

Fig. 2.12: Apparent fracture toughness of limestone showing size dependence, adapted from [37]. Lines are predicted size dependence from non-linear fracture mechanics theory

Table 2.1: Characteristic Length (l_{ch}) of some structural materials

Material	Material Characteristic Length (l_{ch}) in mm
Ceramics/Glass	
Monolithic Ceramics	0.02 - 0.4
SiC Whisker/Alumina (V_f=0.2)	0.27
SiC Fiber/LAS (V_f=0.45)	0.42
Carbon Fiber/Pyrex (V_f=0.4)	1.43
Cementitious Materials	
Concrete	60 - 160
High Strength Concrete	2 - 30
Fiber Reinforced Concrete	200 - 1000
Polymers	
Thermoplastics	0.3 - 20
Fiber Reinforced Thermoplastics	1.3 - 3
Thermosets	0.035 - 0.63
Fiber Reinforced Thermosets	1 - 25

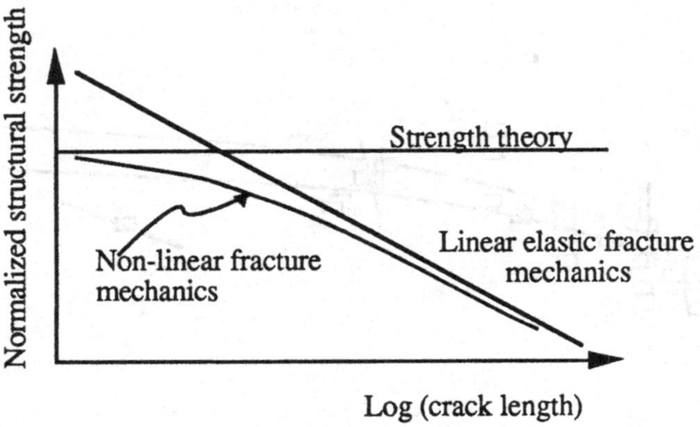

Fig. 2.13: Strength dependence on crack length

ten times that of the l_{ch} value. The exact value depends on, amongst other things, the shape of the tension-softening curve. This rough estimate may be useful in estimating the required minimum dimensions for LEFM based fracture tests, and in the determination of validity of LEFM based fracture analysis by comparisons to the minimum physical dimensions in a boundary value problem.

2.4 Concluding Remarks on Fracture Modelling in Quasi-Brittle Materials

In many inhomogeneous materials, formation of a crack tip process zone enhances the toughness of the material. Fracture analysis of structures of such materials by classical LEFM techniques may be hampered by the presence of such a process zone, if the zone size causes the small scale yielding condition to be violated. A non-linear formulation of fracture formation and propagation is reviewed for the case where the softening part of a force-displacement relation describing the process zone spring-type behavior is important. Results of a sample analysis suggest that the tension-softening curve for such a material is a fundamental property governing structural strength and R-curve behavior. For such materials, the material characteristic length l_{ch} is a useful indicator in determining the range of validity of the simpler LEFM theory.

The general σ_b-δ curve is fundamental in another sense. It directly reflects the material structure and processing conditions, and could be used as a measure of the effect or the effectiveness of microstructure tailoring in engineered advanced composite systems. Unfortunately techniques for measurement of the σ_b-δ curve is at present still not well developed although increasing amount of research in this area is inevitable. The following sections will look at theoretical models relating the material structures of fiber reinforced concrete to the composite tension-softening curve. Presentation of a new technique for measuring the σ_b-δ curve based on the J-integral of Rice [17] applied to a quasi-brittle material concludes this chapter.

3. Tension-Softening and R-Curve Behavior in Fiber Reinforced Concrete

The introduction of fibers can significantly improve the fracture energy of concrete. Figure 3.1 [2] shows that with only one percent volume fraction of high modulus polyethylene fibers, the fracture energy improve by 600 times that without reinforcement. This toughness increase is mainly due to the alteration of the post-cracking behavior, enhancing the load bearing capacity even as the matrix crack opens. The resulting tension-softening curve of fiber reinforced concrete (FRC) is strongly influenced by the micromechanical failure mechanisms related to the specific type of fiber reinforcement.

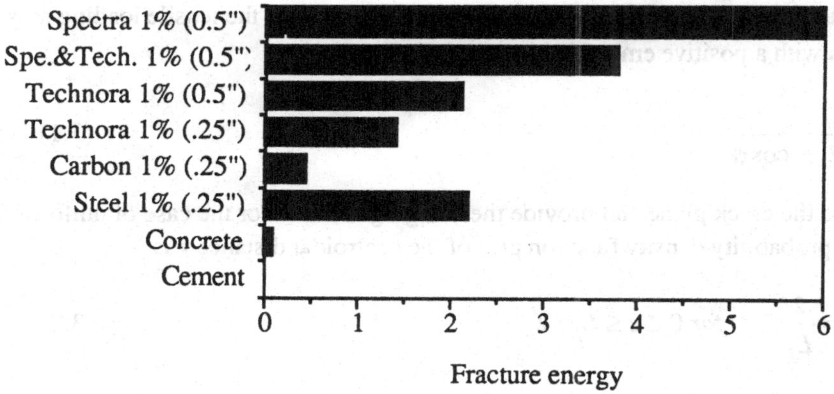

Fig. 3.1: Fracture energy of some synthetic FRC and other cementitious materials

The enhanced tension-softening process implies an enhanced bridging zone at a macroscopic crack tip in which energy is absorbed by fiber actions. These actions could include fiber pulled out against frictional work, fiber pulled out at an angle to the matrix crack as if going over a friction pulley (a snubbing process), and fiber elastically or inelastically elongated and/or ruptured. Some of these actions are most conspicuous in a high modulus polyethylene FRC [41] and may also occur in other composite systems. In this section we review a theoretical model [42] which combines the statistics of fiber distribution and the mechanics of fiber action to predict the tension-softening curve of such FRCs. It should be noted that the assumptions used in this model may not be applicable to other material systems. However, the model does serve to illustrate how a certain non-linear fracture resistance property of an inhomogeneous material could be predicted based on the material structure and failure mechanisms. This section ends with a discussion of the effects of fiber pre-treatment on composite properties as indicated through an R-curve behavior analysis in relation to fiber bridging properties.

3.1 Composite Model

The following assumptions are adopted: (1) the fibers have 3-D random distribution in location and orientation; (2) the fibers are straight with cylindrical geometry; (3) the fibers behave linear elastically; (4) the fibers rupture when its axial stress reaches the fiber strength σ_f^u; (5) the Poisson's effect of the fiber on pull-out is neglected; (6) the fiber/matrix bond is frictional and the elastic bond strength is neglected. The frictional bond strength may exhibit slip hardening or weakening behavior. In addition, the model assumes that the effect of fiber pull-out from matrix at an oblique angle can be characterized by a snubbing friction coefficient, f.

Figure 3.2 shows a fiber of length L_f arbitrarily located with its centroid at a distance z from the matrix crack plane, and with an orientation angle ϕ to the tensile loading axis. Only fibers with a positive embedded length ℓ, defined by

$$\ell = \frac{L_f}{2} - \frac{z}{\cos\phi} \tag{3.1}$$

would cross the crack plane and provide the bridging action. For the case of uniform fiber length, the probability density function p(z) of the centroidal distance z is

$$p(z) = \frac{2}{L_f} \qquad for\ 0 \le z \le L_f / 2 \tag{3.2}$$

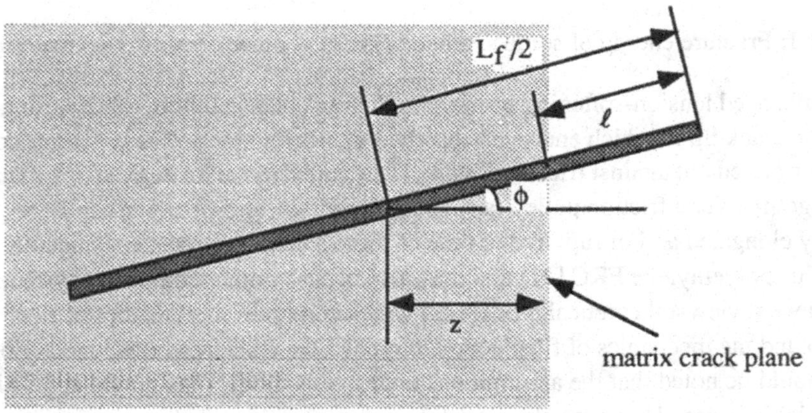

Fig. 3.2: A fiber crossing a matrix crack

For 3-D random orientation, the probability density function, $p(\phi)$, of the inclining angle ϕ is given by [42]

$$p(\phi) = \sin\phi \qquad for\; 0 \le \phi \le \pi/2 \tag{3.3}$$

The problem of two-sided fiber pull-out at ϕ=0 illustrated in Figure 3.3a will be described in the following section. For now, we turn our attention to the relationship of the force-displacement (P-δ) curve of a fiber arbitrarily oriented to that of a fiber with ϕ=0.

For a fiber with the same embedded length ℓ and with the same end slippage distance (s(0)), but with a non-zero inclining angle ϕ (Figure 3.3b), the bridging force will be increased. However, the distribution of the fiber axial force inside the matrix will still be the same for these fibers, irrespective of their inclining angles (ϕ). Using a snubbing friction coefficient, f, Li et al [43] related the bridging force for angle pull-out to that for ϕ = 0:

$$P|_{\phi} = e^{f\phi} P|_{\phi=0} \tag{3.4}$$

where f is an interface material parameter to be determined experimentally by single fiber pull-out test at various angles. The snubbing friction coefficients for nylon and polypropylene have been determined to be 0.99 and 0.70 respectively [43].

During the loading phase ($dP/d\delta \ge 0$), the crack separation for inclined fibers can be calculated from that of ϕ = 0 using superposition:

$$\delta|_{\phi} = \frac{P|_{\phi} - P|_{\phi=0}}{E_f A_f} L + \delta|_{\phi=0} \tag{3.5}$$

where E_f is the fiber elastic modulus, A_f is fiber cross sectional area, and L is the total slippage of the fiber ends. During unloading after reaching the peak load, part of the elastic elongation of the longer fiber embedded segment is recovered. The amount of the recovery depends on the snubbing friction force. Since the recovery occurs only after the shorter fiber segment has begun to slip out (i.e. when $s_1(0) > 0$), and the magnitude of this recovery is in general much smaller than the slippage distance, the recovery during unloading is neglected for simplicity. By so doing, $\delta|_{\phi}$ can always be determined from $\delta|_{\phi=0}$ for the same ℓ and L from (3.5), without the need of calculating the pull-out response directly for each ℓ and f. Equations (3.4) and (3.5) provide the P-δ relation (for any ϕ) sought for.

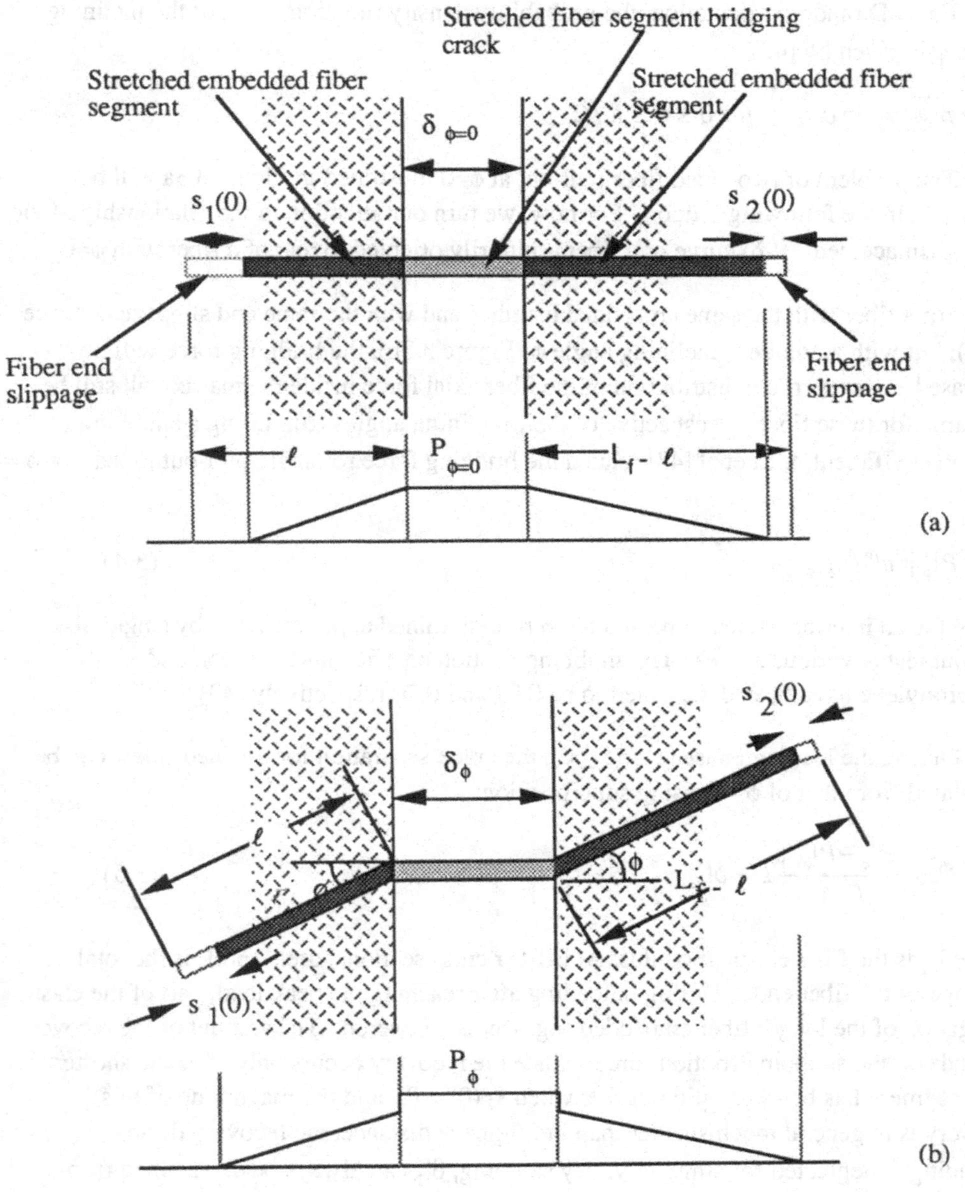

Fig. 3.3: (a) Two-sided fiber pull out for a fiber aligned in the direction of the tensile loading axis. The fiber has a length of L_f and a short embedded length of l, with end slippages $s_1(0)$ and $s_2(0)$. A bridging force P relates to a crack opening δ; (b) Two-sided fiber pull out for a fiber oriented at an arbitrary angle φ to the loading axis

To deduce the tension-softening curve for the composite, we may compute the traction transmitted across the matrix crack by integrating the force contributions from those fibers which are active in the bridging action. Thus for each crack opening δ, the composite bridging stress σ_b may be obtained from

$$\sigma_b = \frac{F}{A_c} = \frac{V_f}{A_f} \int_{z=0}^{L_f/2} \left[\int_{\phi=0}^{\arccos(2z/L_f)} P(\ell,\phi,\delta) p(\phi) d\phi \right] p(z) dz \qquad (3.6)$$

with p(z) and p(ϕ) given by (3.2) and (3.3). The integration limits in (3.6) exclude those fibers which do not cross the crack plane. The constituent parameters needed for (3.6) are (1) Fiber geometry -- diameter d_f and length L_f, (2) fiber properties -- elastic modulus E_f and fiber strength σ_f^u, (3) fiber volume fraction V_f, and (4) properties of fiber/matrix interface -- bond strength τ or slip-weakening/hardening law τ-s, and snubbing friction coefficient f. The fracture energy of the composite may be obtained by integration of (3.6) with respect to the crack opening. Equation (3.6) suggests that the $\sigma_b(\delta)$ scales with fiber volume fraction, and inversely with fiber diameter (since P varies linearly with d_f).

3.2 Single Fiber Pull-Out Model

The mechanics of pull-out of a single fiber embedded in a brittle matrix and aligned in the direction of tensile loading (Figure 3.3a) has been variously investigated, most notably in the form of shear lag models [e.g. 44, 45, 46, 47]. These models either assume that the debonded interface is traction free or is governed by a constant friction. In Wang et al [48,49], the model assumes a frictional bond at the fiber/matrix interface, neglecting any effect of an elastic bond due to expected low adhesion between polymer fibers and cement matrix. However, the frictional resistance is made to depend on the amount of slip so that either slip weakening or hardening phenomenon can be incorporated. In addition, the matrix is assumed to be much stiffer than the fiber and its deformation is neglected. The inclusion of slip-dependent frictional bond strength is motivated by observed fiber abrasion and the peculiar fiber pull out load-displacement curve measured for certain polymer fiber/cement matrix combinations.

The solution procedure of Wang et al [48,49] is illustrated in Figure 3.4 for the case of a slip-hardening frictional bond. In this case the short embedded end will be completely pulled out, whereas the long embedded end may have had some slippage. Geometric compatibility requires that $\delta = \delta_1 + \delta_2$ while equilibrium enforces the condition that $P_1 = P_2$. The latter condition implies that once the short fiber segment load P_1 decreases, the long fiber segment must also unload with partial retrieval of the slipped out segment back into the matrix, rather than following the complete direct pull-out curve indicated by the dashed line in Figure 3.4.

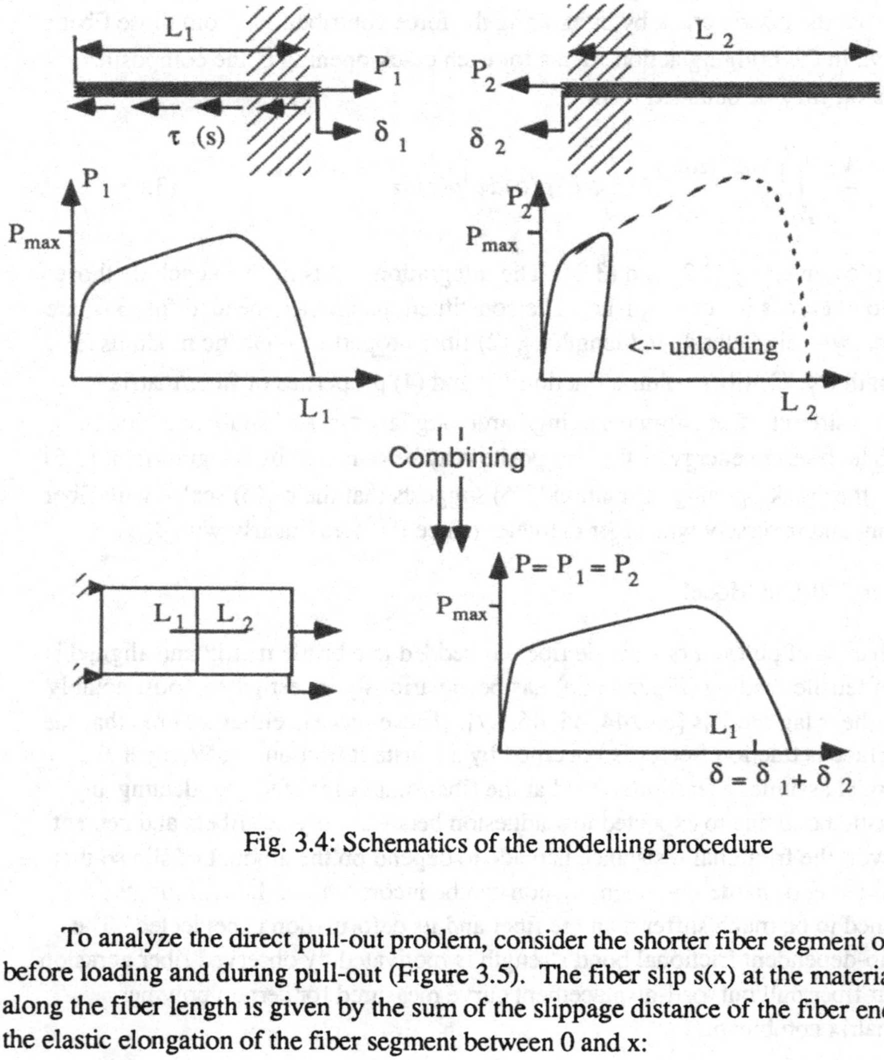

Fig. 3.4: Schematics of the modelling procedure

To analyze the direct pull-out problem, consider the shorter fiber segment of length L_1 before loading and during pull-out (Figure 3.5). The fiber slip s(x) at the material point x along the fiber length is given by the sum of the slippage distance of the fiber end s(0) plus the elastic elongation of the fiber segment between 0 and x:

$$s(x) = s(0) + \int_o^x \varepsilon(x)\,dx \tag{3.7}$$

where the axial strain ε(x) may be directly related to the axial force P(x):

$$\varepsilon(x) = \frac{4}{\pi d_f^2 E_f} P(x) \tag{3.8}$$

P(x) may be obtained from equilibrium considerations of a free body diagram of the fiber segment between 0 and x. Thus

$$P(x) = P_0 + \int_0^x \tau(x)\pi d_f [1+\varepsilon(x)]dx \tag{3.9}$$

and P_0 is a constant representing the fiber end anchorage effect when the fiber end slips ($P_0 = 0$ before complete debonding). The shear stress τ in general is a functional of local slip s(x) so that

$\tau(x) = \tau(s(x))$ for $x + s(x) < L_1$ (inside matrix) (3.10a)

$\tau(x) = 0$ for $x + s(x) \geq L_1$ (outside matrix) (3.10b)

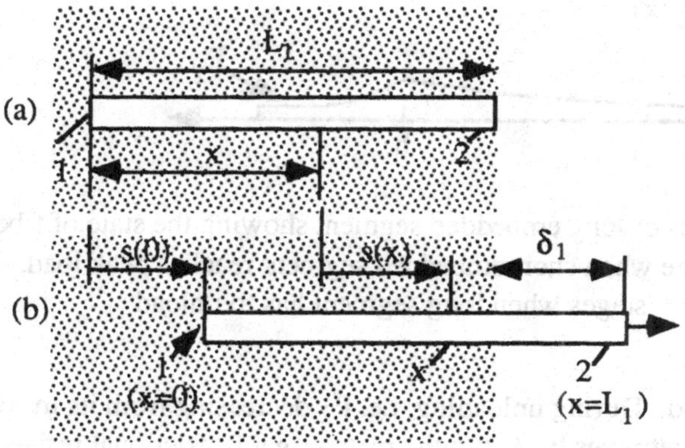

Fig. 3.5: Geometry (a) before and (b) during pull out of a fiber

Equations. (3.7) to (3.10) are in general coupled nonlinear equations and have to be solved numerically. At each loading stage, the load and the corresponding displacement at the exiting end of the fiber could be obtained from

$$P_1 = P(x = L_1) \tag{3.11}$$
$$\delta_1 = s(x = L_1) \tag{3.12}$$

To analyze the two sided pull-out problem, we now turn our attention to the behavior of the longer embedded fiber segment, illustrated in Figure 3.6. This fiber behaves exactly as described above for the shorter fiber segment, at least up to the maximum load P_{max}, after which unloading of this fiber occurs. Figure 3.6a illustrates the fiber condition at $P_2 = P_{max}$, and the slip distribution, the axial force and strain distribution at this state are labelled with a subscript m. In particular, the fiber axial force decreases from P_{max} at the exit end to

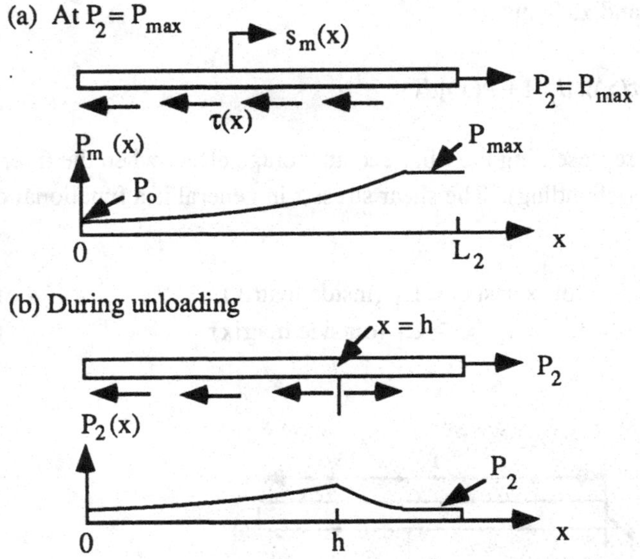

Fig. 3.6: Schematics of long embedded segment showing the state of fiber axial load distribution at (a) stage when short embedded segment reaches peak load; (b) subsequent stages when long segment has to unload

P_0 at the embedded end. During unloading, i.e. P_2 decreasing from its maximum value of P_{max}, part of the fiber retrieves back into the matrix against frictional resistance. Suppose the point x=h demarcates the boundary between the retrieving segment and the rest of the fiber which has not been affected by the unloading, then

$$P(x) = P_m(x) \qquad (x \leq h) \tag{3.13}$$
$$\varepsilon(x) = \varepsilon_m(x) \qquad (x \leq h) \tag{3.14}$$
$$s(x) = s_m(x) \qquad (x \leq h) \tag{3.15}$$

For x>h, the axial force is found from the equilibrium conditions as:

$$P(x) = P_m(h) - \int_h^x \tau(x)\pi d_f[1+\varepsilon(x)]dx \qquad (x > h) \tag{3.16}$$

and the slippage distance at x is given by the slippage distance when $P = P_{max}$ plus the slippage distance due to fiber elastic unloading:

$$s(x) = s_m(x) + \int_h^x [\varepsilon_m(h) - \varepsilon(x)]dx \qquad (x > h) \qquad (3.17)$$

Equations (3.8) and (3.10) are still applicable in describing the P-ε relation and the τ-s relation respectively for the retrieving fiber segment. Equations (3.13)-(3.17), together with (3.8) and (3.10) can be solved and the pull-out load and displacement at the fiber exit end can be determined:

$$P_2 = P(x = L_2) \qquad (3.18)$$
$$\delta_2 = s(x = L_2) \qquad (3.19)$$

By decreasing h from L_2 (no unloading when $P_2 = P_{max}$) until $P_2 = 0$, the P_2-δ_2 relationship for the unloading process can be obtained. The complete load versus crack separation curve P-δ is then calculated by adding the displacements of both short and long fiber segments together ($\delta = \delta_1 + \delta_2$) for the corresponding load ($P = P_1 = P_2$). This P-δ relationship is used (denoted as $P|_{\phi=0}$ and $\delta|_{\phi=0}$) in Equations (3.7) and (3.8) to compute the fiber load and crack separation for fibers with the same embedded length ($\ell = L_1$) and fiber length ($L_f = L_1 + L_2$) arbitrarily oriented to the loading axis.

Wang et al [48] represented the τ-s functional relationship in a quadratic form. They found that the model requires a slip-hardening representation in order to match data of certain synthetic fibers while a slip-weakening representation is required to match data of steel fibers. Figure 3.7 and 3.8 show the experimentally determined P-δ curves for a synthetic fiber and for a steel fiber pulled out from a cement matrix, and the corresponding model τ-s and predicted P-δ curves. The steel fiber pull-out data is from [50].

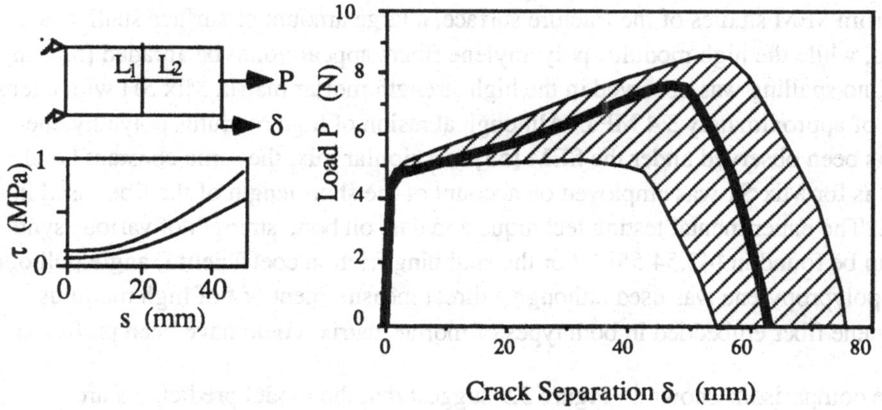

Fig. 3.7: Force-displacement relationship of single nylon fiber pulled out from a cement matrix. Experimental data (4 sets) fall in shaded area. Thick solid line is model prediction based on the slip-hardening curve shown in insert on left. Test configuration is also shown

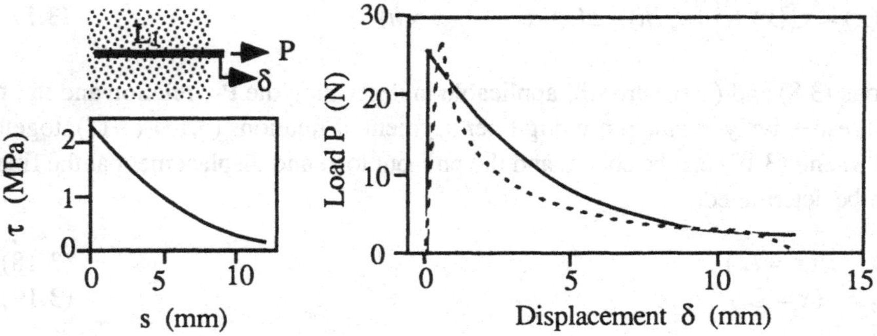

Fig. 3.8: Force-displacement relationship of single steel fiber pulled out from a cement matrix. Experimental data is shown as dotted line. Solid line is model prediction based on the slip-hardening curve shown in insert on left. Test configuration is also shown

3.3 Predicted Composite Tension-Softening Curve and Comparisons with Experimental Observations

As expected, the predicted composite tension-softening curves are sensitive to the nature of slip resistance (slip-hardening or slip-weakening), the fiber and matrix properties, and fiber geometry. A specific result was obtained for two mixes of high modulus polyethylene fiber reinforced mortar with material and geometric parameters given in Table 3.1. Both the model predicted curve and the experimentally determined curve based on a specially devised uniaxial testing technique [51,52] are presented in Figure 3.9. In Mix S1 the matrix material is a normal strength mortar with tensile strength of approximately 2.5 MPa. From SEM studies of the fracture surface, a large amount of surface spall was observed, while the high modulus polyethylene fibers appear not to be abraded [53]. In contrast, no spalling was observed in the high strength mortar matrix Mix SH with a tensile strength of approximately 3.4 MPa. Although abrasion of high modulus polyethylene fibers has been observed under the SEM [53] in a similar mix, the same constant bond strength as for Mix S1 was employed on account of the short length of the fiber used in the Mix SH. The experimental testing technique and data on bond strength of various synthetic fibers can be found in [43,54,55]. For the snubbing friction coefficient f, angle pull out data for polypropylene was used although a direct measurement of f of high modulus polyethylene fiber embedded in both types of mortar matrix would have been preferred.

The comparisons shown in Figure 3.9 suggest that the model predictions are reasonable, with the exception of the initial part of the curve for Mix S1 which may reflect the matrix spalling effect not accounted for in the model.

Table 3.1: Material and Geometry Parametric Values Used in the Prediction of the Tension-Softening Curves in Figure 3.9

Parameters	Mix S1	Mix SH
d_f (mm)	38	38
L_f (mm)	12.7	6.3
E_f (GPa)	120	120
σ_f^u (GPa)	2.6	2.6
V_f (%)	1	0.6
τ (MPa)	1.02	1.02
f	0.7	0.7

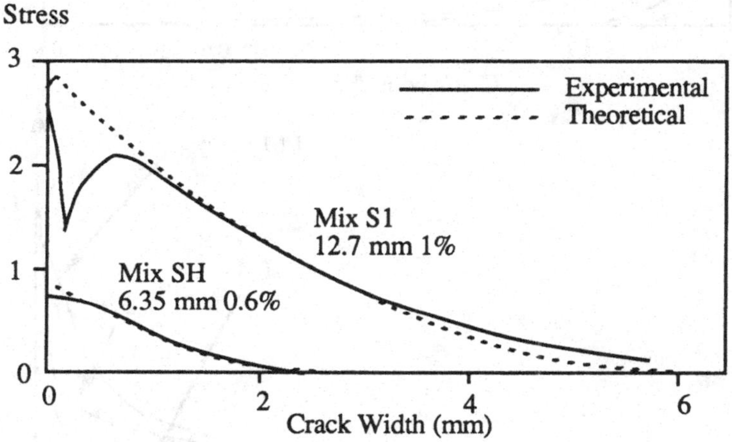

Fig. 3.9: Comparison of predicted and measured tension-softening curves for high modulus polyethylene fiber reinforced mortar. Mix S1 has a normal strength matrix. Mix SH has a high strength matrix

Li et al [42] found that the fiber length plays a significant role in controlling the composite behavior. When fiber length (for a given snubbing coefficient) is not long enough to cause fiber rupture, the fracture energy increases approximately quadratically with fiber length. However, long fiber length also implies more fiber ruptures and loss of reinforcement bridges. The result is a sharp drop shortly after peak load in the tension-softening curve and a corresponding reduction in fracture energy. Figure 3.10 shows the tension-softening curves schematically for three composites with everything identical except fiber length. The corresponding effect on the process zone bridging stresses is also illustrated.

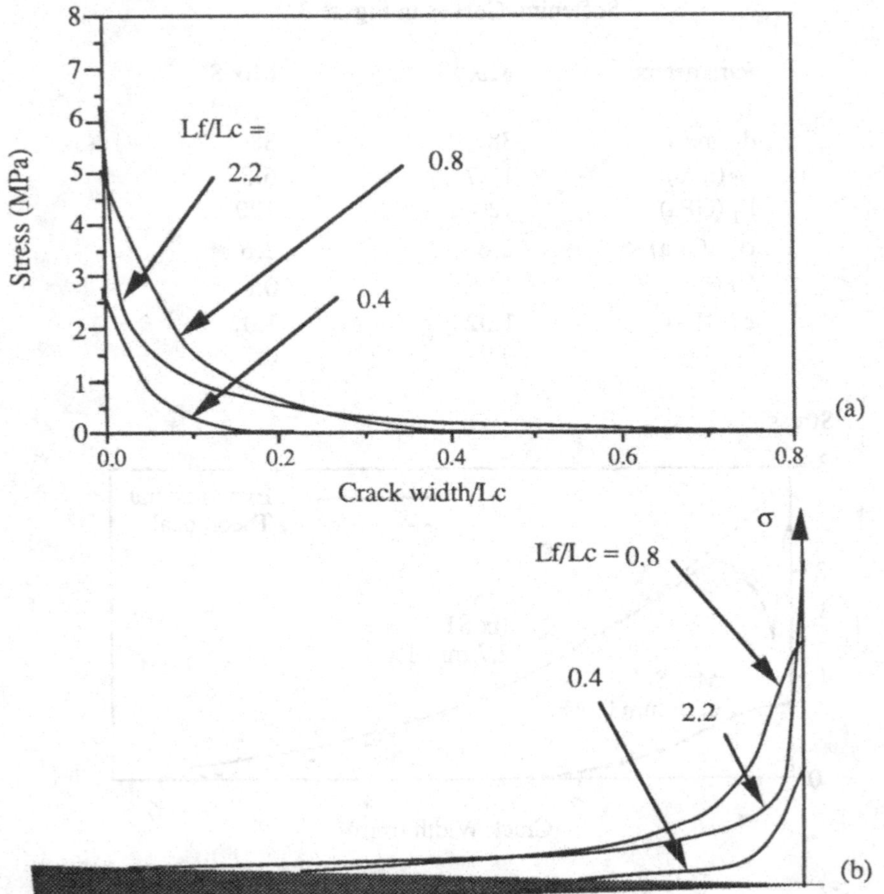

Fig. 3.10: Effect of fiber length on (a) the tension-softening curve and (b) the stress-distribution in the process zone. Fiber length is normalized by critical length L_c [43]

3.4 R-curve Studies

Because of the intimate relation between interfacial properties and the tension-softening curve, the relation between the tension-softening curve and process zone development, and the relation between process zone development and the R-curve (section 2.2), we now turn our attention to analyzing the R-curve of FRC in which fibers have been pretreated in order to change the interface properties. This discussion is based on a R-curve model for double cantilevered beam (DCB) specimen described in [56]. By treating the beams as undergoing pure bending, its compliance as a function of beam length (equal to the crack length) can be computed from elementary beam theory. This allows the energy release rate and the stress intensity factor K_a to be related to the applied loading, the crack length, the elastic modulus and Poisson's ratio, and the polar moment of inertia of the beam cross section. Similarly, the stress intensity factor, K_b, at the matrix crack tip due to fiber bridging stresses $\sigma_b(x)$ acting as a closing pressure was found to be

$$K_b = -\int_{a_0}^{a} \sigma_b(x) b(a-x)\left[bI\left(1-\nu^2\right)\right]^{-1/2} dx \tag{3.20}$$

where a_0 is the traction-free crack length. For equilibrium crack growth in the matrix, the total stress intensity factor, $K_a + K_b$, can be considered to be equal to the matrix toughness, K_m:

$$K_a + K_b = K_m \tag{3.21}$$

For a given load, solving (3.21) requires an iterative process to find the crack length, a. Once a is found, the corresponding composite fracture energy, G, can be calculated from

$$G = \int_{a_0}^{a} \sigma(x) \frac{\partial \delta(x)}{\partial x} dx + \frac{1-\upsilon^2}{E} K_m^2 \tag{3.22}$$

and the crack resistance, K_R, from $K_R = [EG/(1-\nu^2)]^{1/2}$. Note that when the process zone becomes fully developed, K_R will reach its steady state value, and the corresponding G is given by

$$G = \int_{0}^{\delta^*} \sigma(\delta) d\delta + \frac{1-\upsilon^2}{E} K_m^2 \tag{3.23}$$

where the integral part is simply the area under the σ_b-δ curve and $\sigma_b(\delta)=0$ when $\delta > \delta^*$.

This model was used to study R-curve measurement of Visalvanich and Naaman [57,58] in steel FRC tapered DCB. Typical values of E=21.5 GPa, ν=0.18, and K_{IC}=1 MPa $m^{1/2}$ for the matrix and experimentally determined σ_b-δ curve reported in [58] were used for the calculation. Depending on the choice of I_{min} or I_{max}, which correspond to the smallest and largest cross-sections of the DCB respectively, in the calculation, different R-curves were obtained. Their experimental data fall essentially in between these two limiting curves (Figure 3.11). As expected, for small crack extension (Δa), the experimental data lie closer to the R-curve for I_{min}, and for large Δa, the data bend toward the curve for I_{max}.

The tension-softening curves for three composites reinforced with one percent nylon monofilaments was approximated as shown in Figure 3.12a based on single fiber pull-out test for three pretreated fibers [54]. Case 1 was prewashed in hot water to remove the oil finish and may have enhanced the abrasion effect. Case 2 was coated with a fluorocarbon mold release agent to reduce the frictional resistance against sliding of the interface. Case 3 was mechanically crimped by running the fiber between a set of gears. In this case the pull-out resistance is significantly improved and could lead to fiber rupture during the pull-out process. The predicted R-curve based on the theoretical model described above together with these bridging σ_b-δ curves are shown in Figure 3.12b. It is seen that in all cases, the bridging zone contribution to fracture toughness far exceeds that of the matrix toughness.

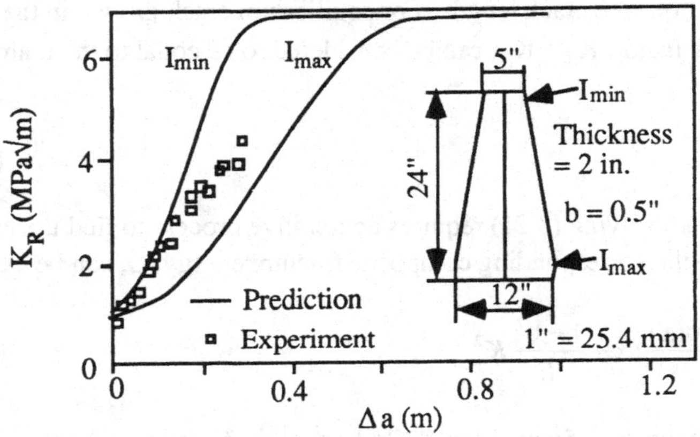

Fig. 3.11: Predicted and experimentally determined R-curve of steel FRC DCB

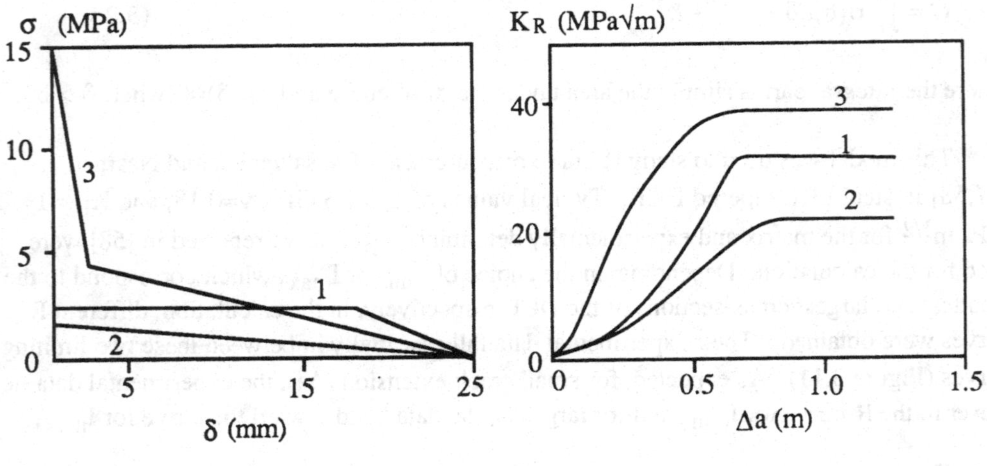

Fig. 3.12: (a) FRC σ–δ curves with fiber pre-treatment for 1) pre-washed, 2) fluorocarbon agent coated & 3) mechanically crimped, (b) corresponding predicted R-curves

Comparison between case 1 and 2 indicates that the surface finish can amount to a 50% change in fracture toughness. Case 3 suggest that mechanical interlock due to fiber crimps could lead to further increase in composite fracture toughness. In addition, the rise of apparent fracture toughness is much more rapid so that cracks could be arrested at an early stage before a long length is developed.

3.5 Concluding Remarks on Tension-Softening Models of FRC

This section discusses a theoretical model of post-peak behavior of short fiber reinforced brittle matrix composites with particular attention to the tension-softening behavior of synthetic fiber reinforced concrete, for which comparisons were made between experimental data and theoretical predictions. The model accounts for several physical processes related to the material structure and which appears to govern the tensile properties in such composites. These physical processes include the effect of fiber abrasion during fiber pull-out and result in a slip-hardening interfacial bond strength, as well as a snubbing effect exhibited during fiber withdrawal inclined to the matrix crack. The randomness of the fiber location and orientation is accounted for statistically. It is found that composite tension softening curve is strongly influenced by the slip-hardening behavior and fiber length.

A simple R-curve model is reviewed. This model was used to make predictions of R-curves based on various tension-softening relations. The effect of fiber treatment on R-curve behavior was discussed and found to play a significant role in determining composite toughness and the shape of the R-curve.

4. EXPERIMENTAL DETERMINATION OF TENSION-SOFTENING CURVES FOR QUASI-BRITTLE MATERIALS

In section 2, the post-cracking tension-softening behavior has been shown to control the growth of fracture in quasi-brittle materials. Section 3 presents a model of tension-softening curve for fiber reinforced concrete in relation to its failure mechanisms and phrased in terms of the material structure. In this section, we discuss methods for the experimental determination of tension-softening curves for quasi-brittle materials. Much of the examples are given for concrete and fiber reinforced concrete, but the test methods and particularly, the principals behind them are expected to be useful for other quasi-brittle materials such as rock and ceramics as well.

Two main classes of test are described. The first class involves the direct uniaxial test, and the second involves a special technique which employs non-linear elastic fracture theory to manipulate flexural type test data into tension-softening curves.

4.1 Uniaxial Tension Test

4.1.1 Stability and Other Considerations:

The uniaxial tension test is the most direct way of obtaining the uniaxial tensile behavior in any material. However, its application to obtaining the post-cracking tensile behavior have several demanding requirements which are difficult and not always met by measuring techniques. These requirements include [51]:

1. The testing fixture, including the testing machine, the loading grips, and the junctions between them, must be stiff so as to avoid unstable unloading after the specimen peak load is reached;

2. Misalignment of the specimen should not be introduced by the loading grips to avoid imposition of an unknown initial stress field on the specimen;

3. The testing fixture should have high rotational rigidity to prevent bending strains in the specimen and thus to ensure a uniform strain across the specimen.

The first requirement is related to the concept of stability. Figure 4.1 illustrates the chain of load transfer from machine to grips/fixtures to the specimen and finally to the fracture surface. One may conceptualize that the load is transferred through an effective spring with a certain compliance c (or inversely stiffness k = 1/c) contributed by a collective 'loading system' comprised of the testing machine, the loading grips, junctions between them as well as the volume of specimen material outside the fracture zone. During the

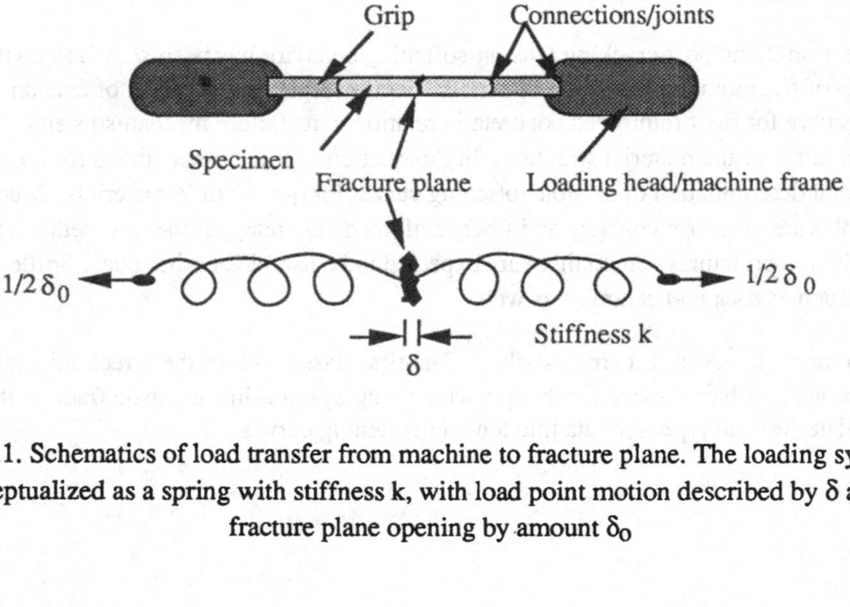

Fig. 4.1. Schematics of load transfer from machine to fracture plane. The loading system is conceptualized as a spring with stiffness k, with load point motion described by δ and the fracture plane opening by amount δ_0

tension-softening process as the crack plane opens, the load is continually reduced, and the loading system must unload to maintain equilibrium. This process may be represented graphically as following unloading lines with slopes determined by the loading system stiffness (Figure 4.2a). When the loading system is compliant, the unloading lines are closer to being flat on a plot of load (stress) versus displacement. When the loading system is stiff, the unloading lines are closer to being vertical. In the unloading stage, the constitutive relation of the fracture plane may be represented by its tension-softening curve. The load point displacements δ_o and fracture plane opening δ for each system are traced in Figure 4.2b, indicating a loss of stability when the unloading stiffness matches the slope of the tension-softening curve for a compliant system.

For the above reasons, attempts at stiffening machines have been made by several researchers. Petersson [59] used two parallel columns which were heated to act as a displacement controlled loading machine. Reinhardt [60] used electronic control to carry out cyclic loading by monitoring the load drop of the load cell. Gopalaratnam and Shah [61] used a electronic feedback system which monitor the crack opening. These techniques met with various degrees of success in making a stiff loading system. It should be mentioned that the load system also involves the specimen volume. Large specimen volume contributes to a reduced system stiffness and should therefore be avoided.

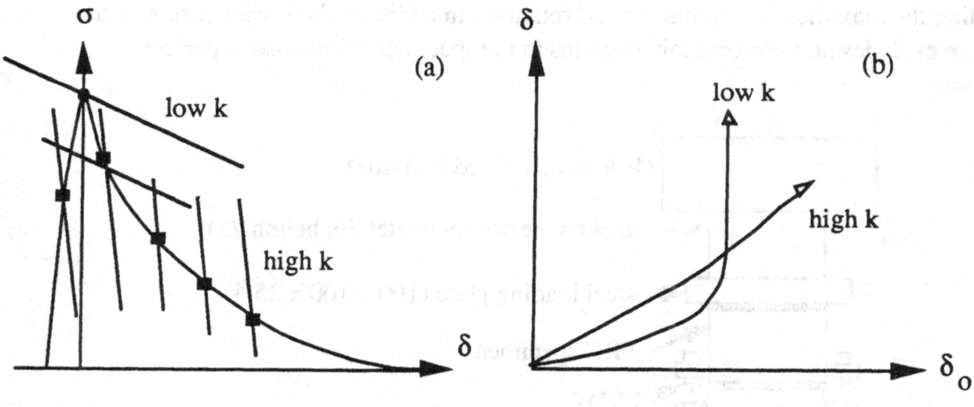

Fig. 4.2. Stability consideration during the unloading process. (a) Unloading lines with slope corresponding to spring stiffness k are superimposed on fracture plane tensile constitutive behavior. Points of equilibrium are indicated by square dots. A point of instability occurs for the spring with low k at the stage indicated by a circular dot. No instability occurs for the spring with high k. (b) Trace of load point displacement plotted against fracture plane opening. For the spring with low k, the fracture plane opening accelerates to infinity as instability is approached

The issues of system stiffness, system alignment and rotational rigidity are each discussed in [51]. Here we note that the problem of alignment often contributes to a reduction in measured tensile strength. This problem is particularly serious in brittle material sensitive to edge flaws which propagate across the specimen when bending stresses are introduced. Thus what is measured is really the residual strength of an unintentionally notched specimen under bending loads rather than the true tensile strength. In a tougher material such as composites with adequate fiber reinforcements, traction is maintained across an initiating crack, so that this problem may be expected to be less serious. In the following, we discuss an especially simple test technique used by Wang et al [51] for synthetic fiber reinforced concrete. Test methods and data for ceramics can be found in [62], for rocks in [62], for fiber reinforced thermoplastics in [4].

4.1.2 Test Set-up and Results:

The test set-up employs a 1331 servo-hydraulic Instron machine of 100 kN tension/compression capacity. The loading fixture consists of a pair of heavy steel plates tightly connected to the testing machine. One plate is bolted to the load cell and the other to the actuator piston, as illustrated in figure 4.3. The test specimen is glued to the loading plates with fast curing epoxy adhesive. By elimination of "soft" connections between specimen and machine, this set-up reduce the system compliance to a minimum as well as providing the maximum restrain from end rotation. In addition, the in-situ curing of the adhesive excludes non-uniform initial strains in the specimen with close to perfect alignment.

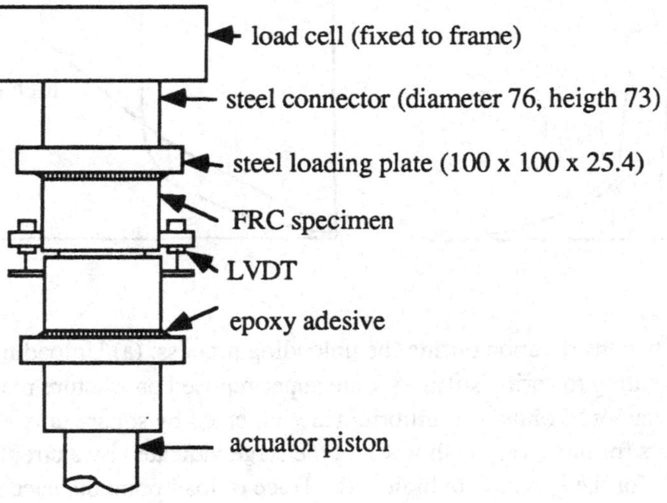

Fig. 4.3. Schematic illustration of loading fixture. Dimensions are in mm

Two LVDTs (linear variable differential transducers) were used to monitor the crack opening displacement, both with a displacement range of 5.08 mm. The LVDTs were mounted on two opposite sides of the specimen with aluminum holders glued to the specimen surface. The nominal measuring gage length was 12.7 mm. Signals of load, crack openings of the two sides, and the machine piston displacement were recorded by a micro computer. The test was performed at constant speeds of machine piston movement. Comparisons of the two LVDTs on either side of the specimen indicated no noticeable bending effect in the specimen. Details of specimen preparation and testing procedure can be found in [51].

After completing a uniaxial tension test, the data should be interpreted as shown in Figure 4.4. Correction for elastic behavior could be important for a material like concrete, but is likely to be unimportant for fiber composites. The tension-softening curves obtained using the method described above for several synthetic FRCs are given in Figure 4.5. Interpretations of the shape of these tension-softening curves in relation to material microstructures and failure mechanisms are given in [41]. Despite the successful use of this simple method in fiber reinforced concrete, its application to fiber reinforced high strength concrete and to mortar have met with difficulty, either due to inadequate machine stiffness or due to machine inaccuracy in displacement control, or both. An alternative method perhaps more suitable for quasi-brittle material with smaller material characteristic length l_{ch} than fiber reinforced concrete is described below.

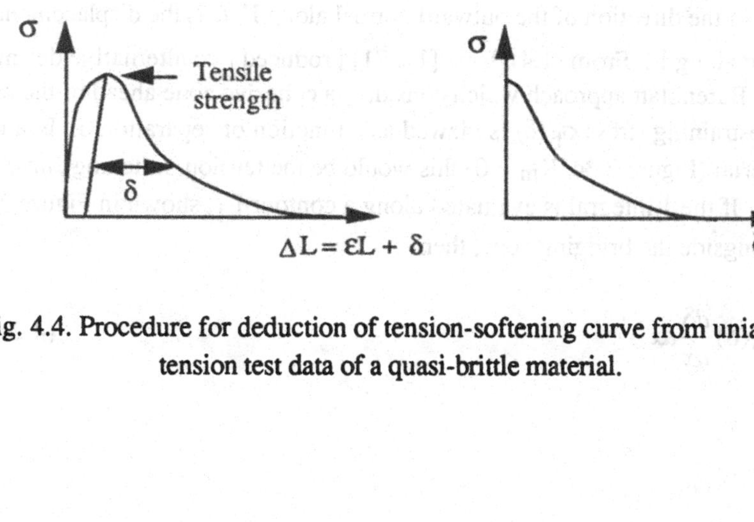

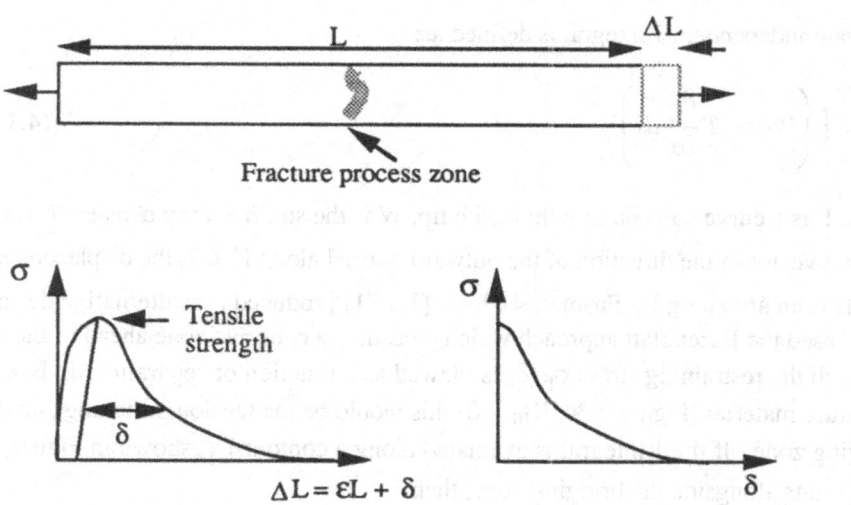

Fig. 4.4. Procedure for deduction of tension-softening curve from uniaxial tension test data of a quasi-brittle material.

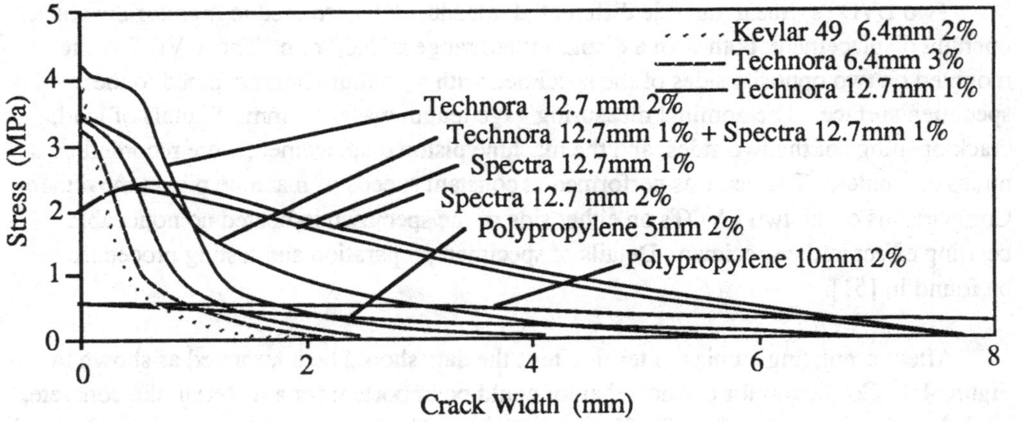

Fig. 4.5. Tension-softening curves for several mixes of synthetic FRC

4.2. Indirect Tension Test

The indirect tension test developed by Li and co-workers [64, 65, 66, 67, 68] for measurement of tension-softening curves of quasi-brittle materials, has so far been applied to mortar, fiber reinforced mortar and rock [69, 70]. Development of this method for ceramics is now ongoing.

4.2.1 Theoretical Considerations

The path independent J-integral is defined as:

$$J = \int_{\Gamma} \left(W dy - \underset{\sim}{T} \frac{\partial \underset{\sim}{u}}{\partial x} ds \right) \tag{4.1}$$

where Γ is a curve surrounding the notch tip, W is the strain energy density, $\underset{\sim}{T}$ is the traction vector in the direction of the outward normal along Γ, $\underset{\sim}{u}$ is the displacement vector and ds is an arc along Γ. From (1.4), Rice [17, 71] produced two alternative definitions of J. He used the Barenblatt approach which considers a cohesive zone ahead of the crack tip in which the restraining stress $\sigma_b(\delta)$ is viewed as a function of separation δ. In a tension-softening material (Figure 2.3c, $K_m = 0$) this would be the tension-softening curve in the bridging zone. If the J-integral is evaluated along a contour Γ_1, shown in Figure 2.4, which runs alongside the bridging zone, then

$$J = \oint_{\substack{bridging \\ zone}} \sigma_b(\delta) \frac{d\delta}{dx} dx \tag{4.2}$$

This definition may be interpreted as follows. If the crack opening at each point in the bridging zone increases by an amount dδ then the profile of the bridging zone boundary extends a distance dx. The quantity $\sigma(\delta)$dx is the force over each infinitesimal area and $\sigma(\delta)$ dx dδ is the energy absorbed during increased separation dδ. Thus (4.2) defines J as the rate of energy absorption with respect to bridging zone propagation. Equation (4.2) may also be expressed as

$$J = \int_0^{\delta_t} \sigma_b(\delta) d\delta \tag{4.3}$$

where δ_t is the separation at the physical crack tip. When δ_t reaches δ^* the real crack propagates and a critical J-integral value, J_c is reached

$$J_c = \int_0^{\delta^*} \sigma_b(\delta) d\delta \tag{4.4}$$

J_c is equal to the total area under the $\sigma_b - \delta$ curve and may be interpreted as the rate of energy absorption in the bridging zone with respect to crack tip propagation.

The second interpretation of J [17, 71] may be given as:

$$J = -\frac{\partial(PE)}{\partial a} \tag{4.5}$$

where PE is the potential energy of a body with crack length a. Thus J is equal to the rate at which the potential energy of a cracked specimen decreases as the crack propagates.

The basis of the indirect J-integral technique of finding the $\sigma_b - \delta$ relationship is first to determine J experimentally using (4.5) from which $\sigma_b(\delta)$ can be found by means of (4.3). Potential energy may be calculated simply from a load-displacement curve. However, since the crack tip position is difficult to locate accurately, it has so far not been possible to directly evaluate (4.5) by propagating a crack in a single specimen. One approximate procedure for getting around this problem is to use two cracked specimens identical in every respect except that for a slight difference in their initial crack lengths. If the load-load point displacement (P-Δ) curves are measured for each specimen, then the area A(Δ) between the two curves up to a load point displacement Δ represents the energy necessary to propagate the crack in this material for the distance Δa (the difference in crack lengths of the two specimens). Equation (4.5) may then be interpreted as:

$$J(\Delta) = \frac{A(\Delta)}{B(\Delta a)} \tag{4.6}$$

where B is the specimen thickness. If, during the experiment, the crack tip separation, δ, is also measured then it is possible to use the Δ–δ relationship to convert $J(\Delta)$ to $J(\delta)$. Differentiation of (4.3) then gives

$$\sigma_b(\delta) = \frac{\partial J(\delta)}{\partial \delta} \tag{4.7}$$

and the tension-softening curve may be determined from the slope of the $J(\delta)$ curve.

4.2.2 Numerical Verification of Test Technique

This method has been verified numerically for both beam bending and compact tension configurations. A. Hillerborg (private communications, 1985) provided verification by employing his fictitious crack model in a finite element scheme to simulate the load-load point displacement curves and load-crack tip separation curves of a pair of three-point bend specimens of slightly different crack lengths. He used an artificial bi-linear curve as input for the tension-softening behavior in the material ahead of the crack tips. The objective of the exercise was to extract the same curve using the indirect J-integral technique with his numerically derived 'test results'. The extracted curve essentially overlapped the assumed curve, thus verifying the theoretical basis. Reyes [72] used a boundary element method to carry out a similar procedure with a compact tension configuration. The load-load point displacement curves were first computed for the two specimens (not shown). Figure 4.6 shows the derived J-δ curve and a comparison of the input and calculated σ_b–δ curve. The results are again very encouraging.

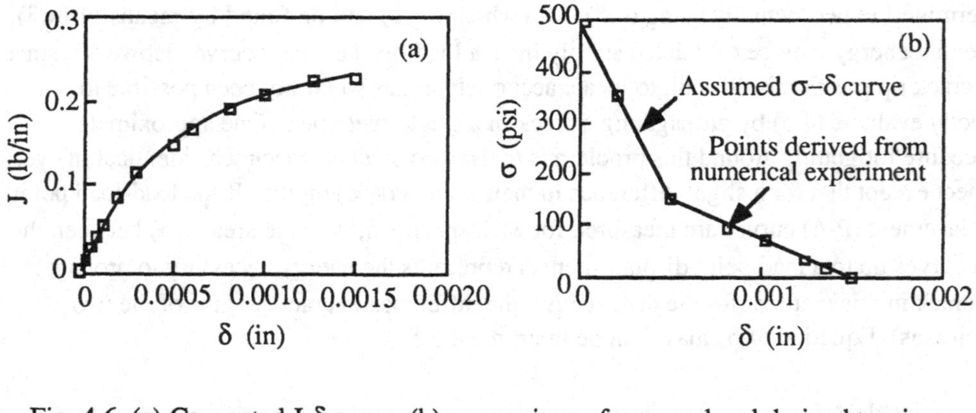

Fig. 4.6. (a) Computed J-δ curve, (b) comparison of assumed and derived tension-softening curve

4.2.3 Test Set Up and Results

This test method has been employed for the compact tension specimen and the four point bend specimen configurations (Figure 4.7). An Instron loading frame has been used. During the test, the load, load point displacement and crack tip separation are measured simultaneously. At the crack tip, a non-contacting proximity sensor capable of measuring up to .254 mm was used in parallel with an LVDT with a range of 5.08 mm (Figure 4.8). All signals were recorded by a Fluke data acquisition system and stored on a microcomputer for later analysis.

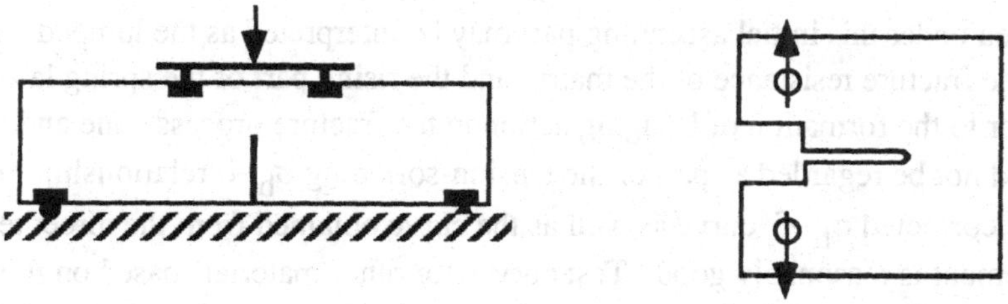

Fig. 4.7. Specimen configurations used with J-integral test

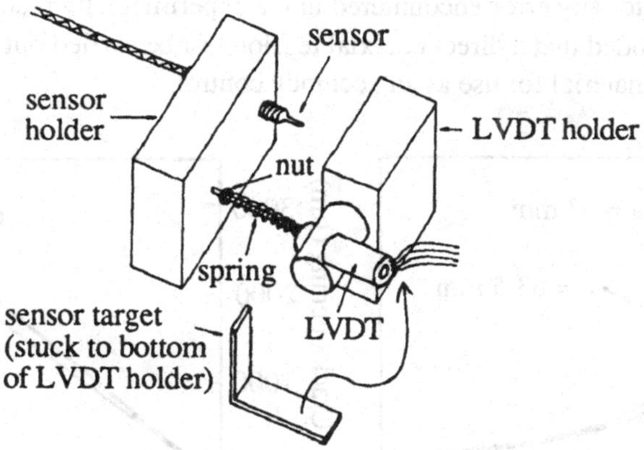

Fig. 4.8. LVDT and proximity sensors for measurement of crack tip opening displacements

To demonstrate the mechanics of the method, a set of experimental results for an aramid FRC is given and the deduced tension-softening curve is compared with a curve obtained from the direct tension test described in section 4.1.2 above. The specimen preparation details can be found in [67]. Figure 4.9 illustrates the load-load point displacement curves obtained using the four point bend test configuration. Figure 4.10

shows average crack opening values, δ_1, δ_2 expressed as a function of load point displacement Δ. δ_1 was measured at the initial crack tip, δ_2 at a point 6.35 mm (equal to Δa) above the crack tip, both in the short cracked beams. The crack separation used in the procedure described above is then obtain from the average $\delta = (\delta_1 + \delta_2)/2$. A J-Δ relationship was calculated using numerical integration. This result combined with the Δ-δ relation provides the J-δ curve, as shown in Figure 4.11. Numerical differentiation of the J-δ curve was achieved using Taylor expansions at five consecutive points J(δ-g), J(δ-h), J(δ), J(δ+j), J(δ+k) and solving for J'(δ). Figure 4.12 shows the deduced tension-softening curve. It shows an initial rise from $\sigma_b=0$ to $\sigma_b=f_t$ before descending back down to σ=0. The area under this initial ascending part may be interpreted as the lumped energy consumed by the fracture resistance of the matrix and the rising part of the spring law (see Figure 2.3) prior to the formation of bridging action in the fracture process zone and therefore should not be regarded as part of the tension-softening $\sigma_b-\delta$ relationship. Figure 4.13 shows the corrected $\sigma_b-\delta$ curve as well as the curve obtained from the direct tension test. The agreement is reasonably good. Test curves for other materials based on this method is shown in Figure 4.14. Further comparisons with direct tension-test results could be found in [66].

A major disadvantage of the method just described is the need for data differentiation which often exaggerates any error encountered in the experimentally recorded data. For this reason, it is recommended that a direct uniaxial tension test be carried out for the tensile strength of the same material for use as an accuracy control.

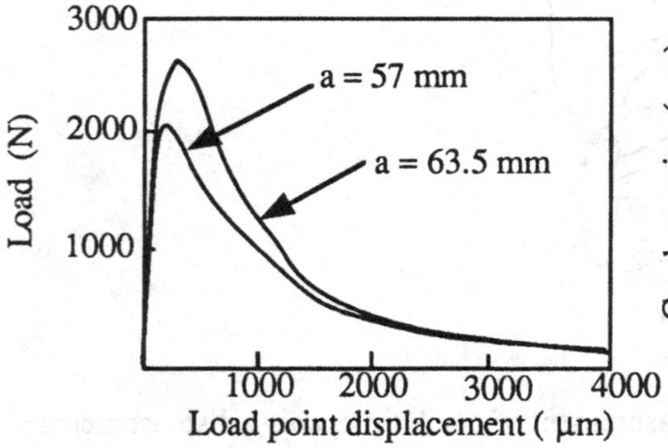

Fig. 4.9. Load vs load displacement

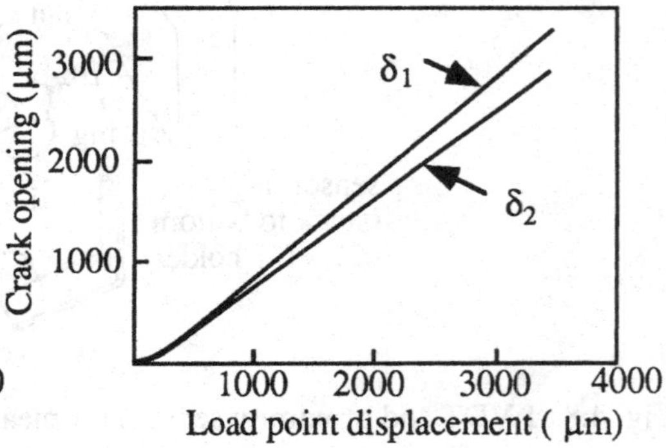

Fig. 4.10. Load displacement vs crack opening

4.2.4. Concluding Remarks on Tension-Softening Curve Measurements

Two types of experimental techniques for the determination of the tension-softening curves for quasi-brittle materials are described. Each has their own advantage and disadvantages. Results from the direct tensile test is easy to interpret, but the test is difficult to carry out especially for quasi-brittle material with high strength or low l_{ch}. An alternative test method using non-linear fracture mechanics theory could be used to indirectly determine the tension-softening curve. This method has been numerically and experimentally verified for several quasi-brittle material and for several testing configurations.

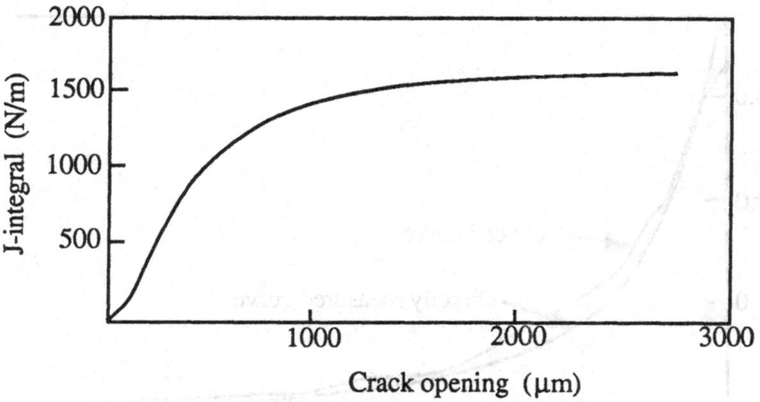

Fig. 4.11. J-integral versus crack opening curve

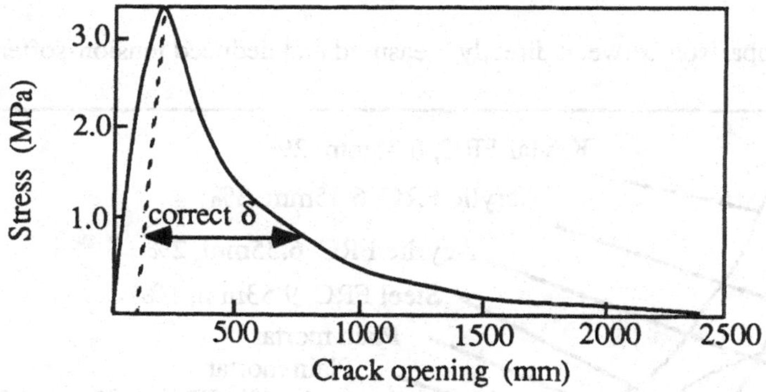

Fig. 4.12. Deduced tension softening curve

5. FINAL REMARKS

This chapter presents the theory of non-linear fracture mechanics of inhomogeneous quasi-brittle composites. Non-linear spring laws are used to describe the closing pressure acting on the crack flanks in the crack tip process zone. It is shown that such spring laws control the development of the fracture process, leading to the possibility of large scale 'yielding', and invalidating the use of linear elastic fracture mechanics.

From a materials performance point of view, it is suggested that the bridging spring action can substantially elevate the material toughness, thus motivating the attempt to relate material structure to the tension-softening spring law. While such models may provide a rational foundation and potential of systematically tailoring the material structure to improved material performance, much more research is needed and expected.

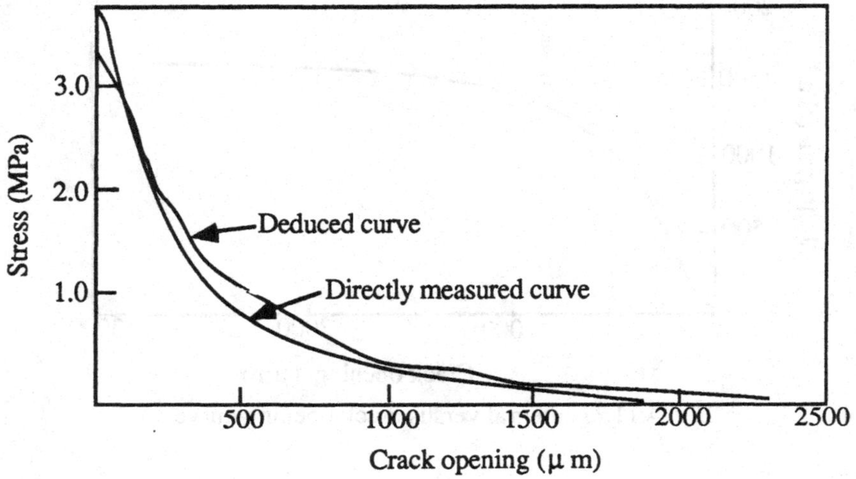

Fig. 4.13. Comparison between directly measured and deduced tension-softening curves

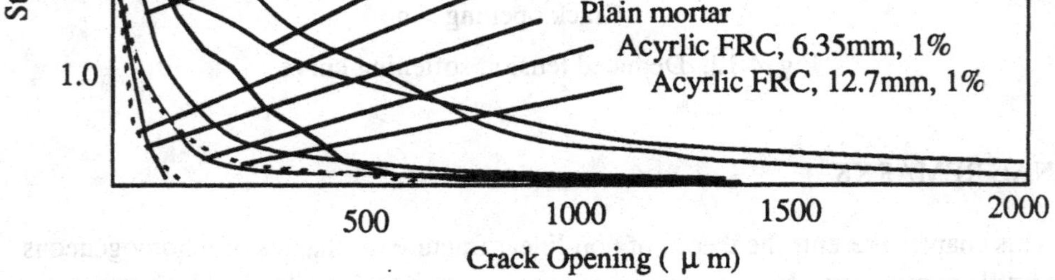

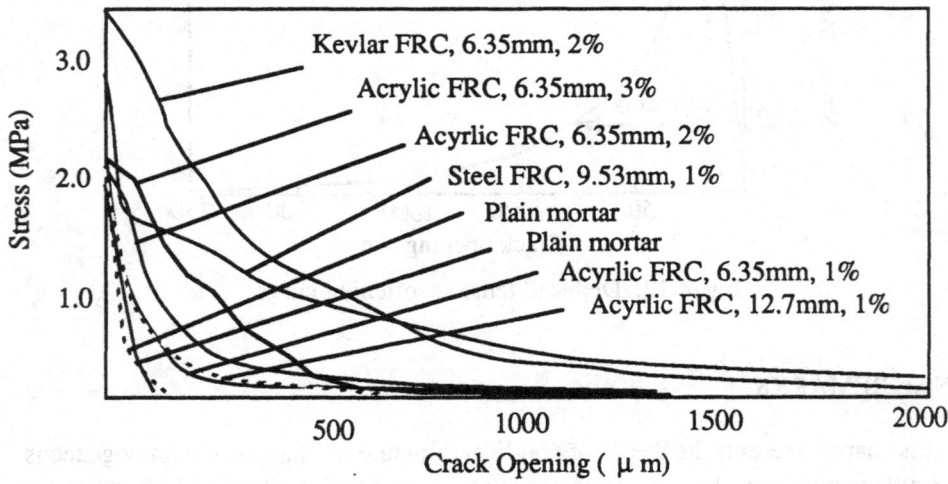

Fig. 4.14. Tension-softening curves of several quasi-brittle materials measured by the indirect J-integral based technique

Finally, the need to have a simple means of measuring the spring law for a given quasi-brittle material system cannot be overstated. Such development should help in providing constitutive relationship for modelling structural behavior on the one hand, and also assist in evaluating the post-cracking performance in inhomogeneous quasi-brittle materials.

ACKNOWLEDGEMENTS

This work is based on the research results of several projects which have been made possible by the generous support of the United States National Science Foundation, the United States National Aeronautics and Space Adminstration and the Shimizu Corporation. Helpful discussions with many individuals, particularly my colleague S. Backer, and former graduate students E. Green, J. Huang, Y. Kaneko, C. Leung, Y. Wang, and R. Ward are especially acknowledged.

REFERENCES

1. Diamond, S. and A. Bentur: On the Cracking in Concrete and Fiber Reinforced Cements, in: Application of Fracture Mechanics to Cementitious Composites (Ed. S.P. Shah), Martinus Nijhoff Publ., 1985, 87-140.

2. Li, V.C., S. Backer, Y. Wang, R. Ward, and E. Green: Toughened Behavior and Mechanisms of Synthetic Fiber Reinforced Normal Strength and High Strength Concrete, in: Fiber Reinforced Cements and Concretes, Recent Developments (Eds. R.N. Swamy and B. Barr), Elsevier Applied Science, London, 1989, 420-433.

3. Mandell, J., Darwish, and F. McGarry: Fracture testing of injection-molded glass and carbon fiber-reinforced thermoplastics, ASTM STP 734 (1981), 73-90.

4. Jang and Liu: Fracture behavior of short-fiber reinforced thermoplastics, I: crack propagation mode and fracture toughness, J. Appl. Poly. Sci., (1985), 3925-3942.

5. Becher, P.F., C.H. Hsueh, P. Angelini, and T.N. Tiegs: Toughening behavior in whisker-reinforced ceramic matrix composites, J. Am. Soc., 71, 12 (1988), 1050-1061.

6. Ruhle, M., B.J. Dalgleish, and A.G. Evans: On the toughening of ceramics by whisker, Scrip. Metal., 21 (1987), 681-686.

7. Evans, A.: The New High Toughness Ceramics, SB report, 87.

8. Mindess, S.: The Application of Fracture Mechanics to Cement and Concrete: A Historical Review, in: Fracture Mechanics of Concrete (Ed. F.H. Wittmann), Elsevier Publ., 1983, 1-30.

9. Swanson, P.L.: Crack-interface Traction: A Fracture-Resistance Mechanism in Brittle Polycrystals, in: Advances in Ceramics, Am. Ceram. Soc., Columbus, OH, 1986.

10. Swanson, P.L., C.J. Fairbanks, B.R. Lawn, Y.W. Mai, and B.J. Hockey: Crack-interface grain bridging as a fracture resistance mechanism in ceramics: I, experimental study on alumina, J. Am. Ceram. Soc., 70, 4 (1987), 279-289.

11. Sakai, M., J. I. Yoshimura, Y. Goto, and M. Inagaki: R-Curve behavior of a polycrystalline graphite: microcracking and grain bridging in the wake region, J. Am. Ceramic Soc., 71, 8 (1988), 609-616.

12. Reichl, A. and R.W. Steinbrech: Determination of crack-bridging forces in alumina, J. Am. Ceram. Soc., 71, 6 (1988), 299-301.

13. Kramer, E.J.: Microscopic and Molecular Fundamentals of Crazing, in: Crazing in Polymers, Advances in Polymer Science 52/53, (Ed. H.H. Kausch), Springer-Verlag, 1983, 1-56.

14. Budiansky, B.: Micromechanics II, in: Proceedings of Tenth U.S. Congress of Applied Mechanics, 1986.

15. Rose, L.R.F.: Crack reinforcement by distributed springs, J. Mech. Phys. Sol., 35 (1987), 383-405.

16. Marshall, D.B., and A.G. Evans: The influence of residual stress on the toughness of reinforced brittle materials, Material Forum, 11 (1988), 304-312.

17. Rice, J. R.: Mathematical Analysis in the Mechanics of Fracture, in: Fracture: An Advanced Treatise, 2, Academic Press, (1968), 191-311.

18. Rose, L.R.F.: Toughening due to crack-front interaction with a second-phase dispersion, Mech. of Mat., 6 (1987), 11-15.

19. Rice, J. R.: Crack Fronts Trapped By Arrays of Obstacles: Solutions Based on Linear Perturbation Theory, in: Analytical, Numerical, and Experimental Aspects of Three Dimensional Fracture Processes, AMD-91, (Eds. A.J. Rosakis, K. Ravi-Chandar, Y. Rajapakse), ASME, 1988, 175-184.

20. Irwin, G.R.: Analysis of stresses and strains near the end of a crack traversing a plate, J. Applied Mechanics, 24 (1957), 361-364.

21. Barenblatt, G.I.: Mathematical Theory of Equilibrium Cracks in Brittle Fracture, in: Advances in Applied Mechanics, VII, Academic Press, New York, 1962.

22. Dugdale, D.S.: Yielding of steel sheets containing slits, J. Mechanics and Physics of Solids, V.8, 1960.

23. Marshall, D.B. and B.N. Cox: A J-Integral method for calculating steady-state matrix cracking stresses in composites, Mechanics of Materials, 7 (1988), 127-133.

24. Li, V.C. and E. Liang: . Fracture process in concrete and fiber reinforced cementitious composites, ASCE J. of Engineering Mechanics, 112, 6 (1986), 566-586.

25. Palmer, A.C. and J.R. Rice: The Growth of Slip-Surfaces in the Progressive Failure of Overconsolidated Clay Slopes, Proceedings of the Royal Society of London, A332, (1973), 572-548.

26. Leung, C.K., and V.C. Li: Reliability of First-Cracking Strength for Short-fiber Reinforced Brittle Matrix Composites, In: Ceramics Eng. Sci. Proc., 9/10, 1989, 1164-1178.

27. Marshall, D.B. and A.G. Evans: Tensile Strength of Uniaxially Reinforced Ceramic Fiber Composites, in: Fracture Mechanics of Ceramics, (Eds. R.C. Bradt, A.G. Evans, D.P.H. Hasselman and F.F. Lange), 7 , Plenum Press, New York (1986), 1-15.

28. Hillerborg, A., M. Modeer, and P.E. Petersson: analysis of crack formation and crack growth in concrete by means of fracture mechanics and finite elements, Cement and Concrete Research, 6, (1976), 773-782.

29. Ingraffea, A. and W.H. Gerstle: Nonlinear Fracture Models for Discrete Crack Propagation, in: Applications of Fracture Mechanics to Cementitious Composites, NATO ASI Series, 94 (Ed. S.P. Shah) Martinus Nijhoff Publ., 1985, 247-286.

30. Ingraffea, A.: Theory of Crack Initiation and Propagation in Rock, in: Fracture Mechanics of Rock, (Ed. B. Atkinson), Academic Press, 1987, 71-110.

31. Mai, Y. and B. Lawn: Crack-interface grain bridging as a fracture resistance mechanism in ceramics: ii theoretical fracture mechanics model, J. Am. Ceram. Soc. 70, 4 (1987), 289-294.

32. Swanson, P.L.: Tensile fracture resistance mechanisms in brittle polycrystals: an ultrasonics and in-situ microscopy investigation, J. Geophysical Res., 92, (1987),

8015-8036.

33. Wecharatana, M. and S.P. Shah: Slow crack growth in cement composites, J. Stru., ASCE, 108, 6, (1982), 1400-1413.

34. Cook, R.: Transient Fracture Resistance in the Weak Toughening Limit, in: International Congress of Fracture, Proceeding Vol. 4, (Eds. Salama, Ravi-Chandar, Taplin, and Rao), 1989, 2747-2756.

35. Francois, D.: Fracture and Damage Mechanics of Concrete, in: Application of Fracture Mechanics to Cementitious Composites, (Ed. S.P. Shah) Martinus Nijhoff Publ., 1985, 141-156.

36. Takahashi, H. and H. Abé: Fracture Mechanics Applied to Hot, Dry Rock Geothermal Energy, in: Fracture Mechanics of Rock, (Ed. B. Atkinson), Academic Press, 1987, 241-276.

37. Ingraffea, A.R., K.L.Gaunsallus, J.F. Beech and P.P. Nelson: A Short Rod Based System for Fracture Toughness Testing of Rock, in: Chevron-Notched Specimens: Testing and Stress Analysis, ASTM STP 855, Louisville, (1983), 152-166.

38.Li, V. C., K. Chong, and H. H. Einstein, Tension Softening and Size Effects on Fracture Toughness Determination of Geomaterials, in: Fracture of Concrete and Rock, (Eds. Swartz and S. Shah), Springer-Verlag, New York, 255-264, 1989.

39. Li, V.C.: Mechanics of Shear Rupture Applied to Earthquake Zones, in Fracture Mechanics of Rock, (Ed. B.K. Atkinson), Academic Press, 1987, 351-432.

40. Bazant, Z.P., S. Sener and P.C. Prat: Fracture Mechanics Size Effect and Ultimate Load of Beams Under Torsion, in: Application of Fracture Mechanics to Concrete, Ed. V.C. Li and Z. Bazant, ACI STP, 1990, 171-178.

41. Wang, Y., V.C Li, and S. Backer: Tensile failure mechanisms in synthetic fiber reinforced mortar, submitted to the J. Materials Science (1990).

42. Li, V.C., Y. Wang, and S. Backer: A statistical-micromechanical model of tension-softening behavior of short fiber reinforced brittle matrix composites, submitted for publication, J. Mech. and Phys. of Solids (1990).

43. Li, V.C., Y. Wang, and S. Backer: Effect of inclining angle, bundling, and surface treatment on synthetic fiber pull-out from a cement matrix, to appear in Composites (1990).

44. Greszczuk, L. B.: Theoretical studies of the mechanics of the fiber-matrix interface in composites, in ASTM STP 452 (1969), 42-58.

45. Takaku, A. and R.G.C. Arridge: The effect of interfacial radial and shear stress on fiber pull-out in composite materials, Journal of Physics D: Applied Physics, 6 (1973), 2038-2047.

46. Lawrence, P.: Some theoretical considerations of fiber pull out from an elastic matrix, J. Mat.Sci.7, (1972) 1-6.

47. Gopalaratnam, V.S., and S.P. Shah: Tensile failure of steel fiber-reinforced mortar, ASCE J. of Engineering Mechanics, 113, 5, (1987), 635-652.

48. Wang, Y., V.C Li, and S. Backer: Modeling of fiber pull-out from a cement matrix, International J. of Cement Composites and lightweight Concrete, 10, 3, (1988), 143-149.

49. Wang, Y., V.C. Li, and S. Backer: Analysis of synthetic fiber pull-out from a cement matrix, Material Research Society Symposium Proceeding, 114, (1988), 159 -165.

50. Naaman, A., and S.P. Shah: Pull-out mechanism in steel fiber reinforced concrete, J. Struc., ASCE, 102, ST8, (1976), 1537-1548.

51. Wang, Y., V.C. Li, and S. Backer: Experimental determination of tensile behavior of fiber reinforced concrete, to appear in ACI Materials Journal (1990).

52. Wang, Y., V.C. Li, and S. Backer: Tensile properties of synthetic fiber reinforced mortar, to appear in J. of Cement Composites and Lightweight Concrete (1990).

53. Li, V.C., and E. Green: Tensile behavior of synthetic fiber reinforced high strength mortar and concrete, in preparation (1990).

54. Wang, Y., S. Backer, and V.C. Li: An experimental study of synthetic fiber reinforced cementitious composites, Journal of Materials Science, 22, (1987), 4281-4291.

55. Wang, Y., S. Backer, and V.C. Li: A special technique for determining the critical length of fiber pull-out from a cement matrix, J. of Materials Science Letters, 7 (1988), 842-844.

56. Li, V.C., Y. Wang and S. Backer: Effect of Fiber-Matrix Bond Strength on The Crack Resistance of Synthetic Fiber Reinforced Cementitious Composites, in: Bonding in Cementitious Composites (Eds. S. Mindess and S. Shah), MRS Symposia 114: 1988, 167-174.

57. Visalvanich K. and A.E. Naaman: In: Fracture Mechanics Methods for Ceramics, Rocks, and Concrete, ASTM STP 745, (Ed.S.W. Freiman and E.R. Fuller) 1981, 141-156.

58. Visalvanich, K. A.E. Naaman: Fracture Model for fiber reinforced concrete, ACI J. March-April, 1983, 128-138.

59. Petersson, P.E.: Crack Growth And Development Of Fracture Zone In Plain Concrete And Similar Materials. Div. of Building Materials, Lund Institute of Technology, Sweden. Report TVBM-1006, 1981.

60. Reinhardt, H.W.: Fracture Mechanics Of An Elastic Softening Material Like Concrete, Heron, Delft University of Technology, 29:2, 1984.

62. Marshall, D.B. and A.G. Evans: Failure mechanisms in ceramic-fiber/ceramic matrix composites, J. Am. Ceram. Soc., 68, 5 (1985) 225-31.

61. Gopalaratnam, V.S. and S.P. Shah,. Softening response of concrete in direct tension, J. Amer. Concrete Inst. 82:310 (1985).

63. Labuz, J.F., S.P. Shah and C. H. Dowding: Measurement and description of the tensile fracture process in rock, ASCE J. Eng. Mech., (1989).

64. Li, V.C.: Fracture Resistance Parameters For Cementitious Materials And Their Experimental Determination, in: Application of Fracture Mechanics to Cementitious Composites, (Ed. S.P. Shah) Martinus Nijhoff Publ., 1985, 431-452.

65. Li, V.C., C.M. Chan, and C.K.Y. Leung: Experimental determination of the tension softening relations for cementitious composites, Cement and Concrete Research, 17 (1987) 441-452.

66. Leung, C.K.Y. and V.C. Li: Determination of fracture toughness parameter of quasi-brittle materials with laboratory-size specimens, J. of Materials Science, 24, (1989) 854-862.

67. Li, V.C. and R. Ward: A Novel Testing Technique for Post-Peak Tensile Behavior of Cementitious Materials, in: Fracture Toughness and Fracture Energy -- Test Methods for Concrete and Rock, (Ed. Mihashi), in press, A.A.Balkema Publishers, Netherlands, 1989.

68. Chong, K.P., V.C. Li and H. Einstein: Size effects, process zone, and tension softening behavior in fracture of geomaterials", in press, Int'l J. of Engineering Fracture

Mechanics, 1989.

69. Chong, K.P., K.D. Basham, and D.Q. Wang: Fracture parameters derived from tension-softening measurements using semi-circular specimens, in: Fracture Toughness and Fracture Energy -- Test Methods for Concrete and Rock, (Ed. Mihashi), in press, A.A.Balkema Publishers, Netherlands, 1989.

70. Hashida, T.: Tension-softening curve measurements for fracture toughness determination in granite, in: Fracture Toughness and Fracture Energy -- Test Methods for Concrete and Rock, (Ed. Mihashi), in press, A.A.Balkema Publishers, Netherlands, 1989.

71. Rice, J.R.: A path independent integral and the approximate analysis of strain concentrations by notches and cracks, J. Applied Mechanics, 35 (1968) 379-386.

72. Reyes, O.M.L.: Numerical Modelling Of Fracture Propagation In Tension Softening Materials, M.S. Thesis, MIT, Cambridge, MA, (1987).

INTRODUCTION TO LINEAR FRACTURE MECHANICS

M. Matczynski
Polish Academy of Sciences, Warsaw, Poland

ABSTRACT.

On the basis of two-dimensional linear theory of elasticity in Muskhelishvili's formulation, some notions of the linear elastic fracture mechanics are introduced. The notions are used to present several fundamental fracture criteria proposed by A. A. Griffith, G. R. Irwin, G. I. Barenblatt and D. S. Dugdale.

1. TWO-DIMENSIONAL PROBLEM OF A CRACK.

The simplest mathematical model of a crack used in the analysis of the fracture process is based upon two-dimensional elastostatics.

In order to construct such a model assume the unbounded elastic body to contain a cut in plane x_1-x_3, (Fig.1). The cut is infinite in the direction of x_3-axis, finite and rectilinear in the direction of variable x_1 of the Cartesian coordinate system x_i (i=1,2,3). Moreover, assume the body to be subject to plane state of stress or strain, or anti-plane state of strain. Consequently, in view of the fact that the displacements, strains and stresses are functions of two variables only, say x_1 and x_2, the mathematical model of a crack is represented by a finite, rectilinear cut in plane x_1-x_2.

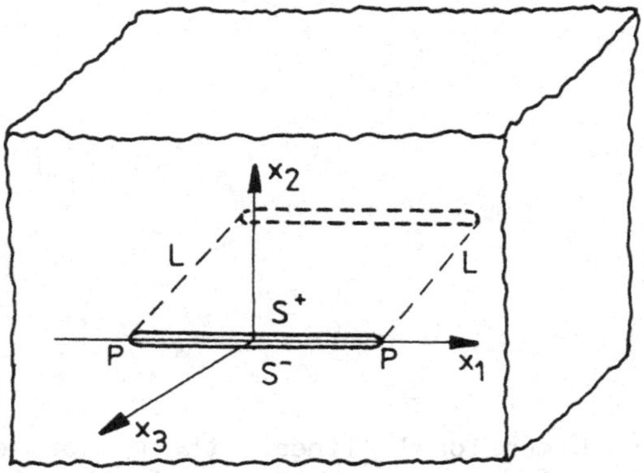

Fig. 1: L - edge (front) of a crack, P - tip of a crack, $S^{\pm}$ - upper and lower surfaces of a crack, respectively.

Introduce now the basic notions and definitions used in fracture mechanics, and in the linear elastic fracture mechanics (LEFM) in particular.

1.1 Plane state of stress or strain.

Let an unbounded elastic body be subject to a plane state of stress or strain and contain a crack (cut in the sense of the description introduced above) of length 2a. The crack is assumed to be stress free, the body being loaded by tractions $\underline{T}$ applied to points outside the crack . The state of displacements and stresses is denoted by (u_i, σ_{ij}), (Fig.2a)[1].

By means of the superposition principle we can represent this basic problem by a sum of two separate problems:

I. Problem of a continuous plane loaded by the given load $\underline{T}$ with $(\mathring{u}_i, \mathring{\sigma}_{ij})$ standing for displacements and stresses, respectively, (Fig.2b).

---

[1]Here and in what follows (unless stated otherwise) symbol $f^{\pm}$ denotes the value of function $f(x_1, x_2)$ on the upper S^+ and lower S^- surfaces of the crack, respectively.

II. Problem of a cracked plane with nonhomogeneous boundary conditions given on the crack surface where the displacements and stresses are denoted by (u_i, σ_{ij}), respectively, (Fig.2c). The boundary conditions are constructed on the basis of the solution of the first problem, (Fig.2b). We assume that stress vector components on the crack surface are opposite to the corresponding stresses evaluated from the first problem. In addition we also assume that the stresses σ_{ij} vanish at infinity.

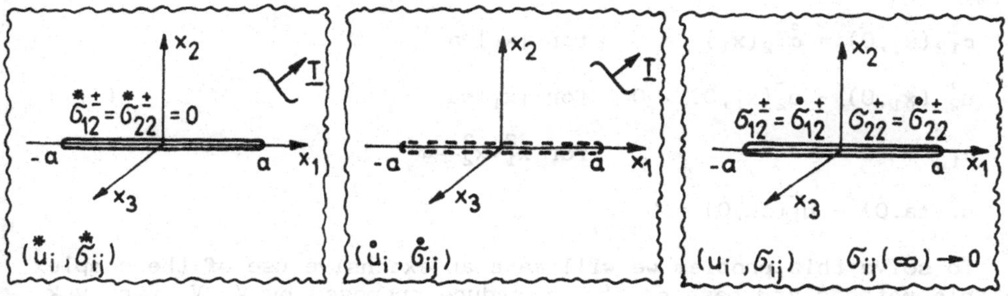

a) b) c)

Fig. 2.

Once the solutions of both problems are known, the solution of the basic problem (Fig.2a) will be represented by their sum and

$$(\overset{*}{u}_i, \overset{*}{\sigma}_{ij}) = (\overset{\circ}{u}_i, \overset{\circ}{\sigma}_{ij}) + (u_i, \sigma_{ij})$$

Assuming that the solution of the first problem is known, we will focus our attention on the second problem.

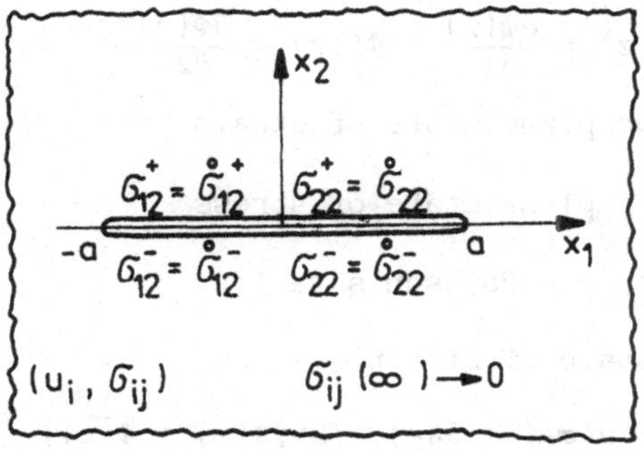

Fig. 3.

Let us find the distribution of displacements and stresses in the infinite elastic plane containing a single, straight crack of length 2a. We assume, contrary to the situation shown in Fig.2c, that the crack surfaces are loaded by an arbitrary non-equilibrated system of loads. Let the displaçements and stresses be denoted by (u_i, σ_{ij}), respectively, and let $\overset{\circ}{\sigma}{}^{\pm}_{ij}$ $(i,j=1,2)$ denote the given functions defined on the upper and lower surfaces of the crack. In addition, we assume that the stress components σ_{ij} vanish at infinity, (Fig.3).

From the point of view of mechanics, it is clear that the displacements have to be continuous at the crack tips and the crack surfaces cannot overlap. Assuming in addition that the crack surfaces do not meet each other at any point except the crack tips, the boundary conditions of the problem can be written as

$$\begin{aligned} &\sigma_{22}^{\pm}(x_1,0) = \overset{\circ}{\sigma}{}_{22}^{\pm}(x_1) && \text{for } |x_1|<a \\ &\sigma_{12}^{\pm}(x_1,0) = \overset{\circ}{\sigma}{}_{12}^{\pm}(x_1) && \text{for } |x_1|<a \\ &u_2^{+}(x_1,0) - u_2^{-}(x_1,0) > 0 && \text{for } |x_1|<a \\ &\sigma_{ij} \longrightarrow 0 && \text{for } x_1^2+x_2^2 \longrightarrow \infty \\ &u_i^{+}(\pm a,0) - u_i^{-}(\pm a,0) = 0 \end{aligned} \tag{1.1}$$

To solve this problem we will make an extensive use of the complex variable methods and base on the procedure proposed by V. V. Panasyuk, M. P. Savruk and A. P. Datsyshin in [1] and [2].

According to Muskhelishvili [3], displacements and stresses may be represented by two sectionally holomorphic functions $\varphi(z)$ and $\psi(z)$ of the complex variable $z=x_1+ix_2$, and

$$\begin{aligned} &\sigma_{11} + \sigma_{22} = 2[\Phi(z) + \overline{\Phi(z)}] \\ &\sigma_{22} - \sigma_{11} + 2i\sigma_{12} = 2[\bar{z}\Phi'(z) + \Psi(z)] \\ &2\mu(u_1 + iu_2) = \kappa\varphi(z) - z\overline{\Phi(z)} - \overline{\psi(z)} \end{aligned} \tag{1.2}$$

where

$$\Phi(z) = \frac{\partial\varphi(z)}{\partial z}, \quad \Psi(z) = \frac{\partial\psi(z)}{\partial z}, \quad \Phi'(z) = \frac{\partial\Phi(z)}{\partial z}$$

$$\kappa = \begin{cases} 3 - 4\nu & \text{for plane state of strain} \\ \dfrac{3-\nu}{1-\nu} & \text{for plane state of stress} \end{cases}$$

μ - Lame constant, ν - Poisson's ratio.

In the case of plane state of strain

$$\varepsilon_{33} = 0 \text{ and } \sigma_{33} = \nu(\sigma_{11} + \sigma_{22}) = 2\nu[\Phi(z) + \overline{\Phi(z)}]$$

while in the case of plane state of stress

$$\sigma_{33} = 0 \text{ and } \varepsilon_{33} = \frac{2\nu}{E}(\sigma_{11} + \sigma_{22}) = \frac{4\nu}{E}[\Phi(z) + \overline{\Phi(z)}]$$

where E stands for Young's modulus.

Since we are going to solve the problem by means of

Muskhelishvili's method, let us represent the boundary conditions (1.1) in a complex form.
Let

$$2p(x_1) = \overset{o+}{\sigma}_{22} + \overset{o-}{\sigma}_{22} - i(\overset{o+}{\sigma}_{12} + \overset{o-}{\sigma}_{12}) \quad \text{for } |x_1|<a$$
$$2q(x_1) = \overset{o+}{\sigma}_{22} - \overset{o-}{\sigma}_{22} - i(\overset{o+}{\sigma}_{12} - \overset{o-}{\sigma}_{12}) \quad \text{for } |x_1|<a \tag{1.3}$$

and let

$$[f(x_1,0)] = f^+(x_1,0) - f^-(x_1,0)$$

denote the discontinuity (jump) of function $f(x_1,x_2)$ across the crack surface.

Taking into account these relations, the boundary conditions (1.1) can be finally read as

$$\begin{aligned}
&\overset{+}{\sigma}_{22} + \overset{-}{\sigma}_{22} - i(\overset{+}{\sigma}_{12} + \overset{-}{\sigma}_{12}) = 2p(x_1) && \text{for } |x_1|<a\\
&[\sigma_{22}] - i[\sigma_{12}] = 2q(x_1) && \text{for } |x_1|<a\\
&[u_2] > 0 && \text{for } |x_1|<a\\
&\sigma_{ij} \longrightarrow 0 && \text{for } x_1^2+x_2^2 \longrightarrow \infty\\
&[u_i(\pm a,0)] = 0 &&
\end{aligned} \tag{1.4}$$

According to [1], the solution of the problem formulated above consists of two parts. The first part concerns a certain auxiliary problem, the solution of which will be used in the final, second part.

To introduce the auxiliary problem let us suppose, for the time being, that the discontinuities of displacements across the crack surface are known and let

$$[u_1(x_1,0)] + i[u_2(x_1,0)] = \frac{i(\kappa+1)}{2\mu}\, g(x_1) \quad \text{for } |x_1|\le a \tag{1.5}$$

where $g(x_1)$ is a complex function of real variable x_1 and, according to $(1.4)_5$, $g(\pm a)=0$.

According to this assumption, the boundary conditions of the auxiliary problem can now be formulated as

$$\begin{aligned}
&[\sigma_{22}] - i[\sigma_{12}] = 2q(x_1) && \text{for } |x_1|<a\\
&\frac{\partial}{\partial x_1}([u_1] + i[u_2]) = \frac{i(\kappa+1)}{2\mu}\, g'(x_1) && \text{for } |x_1|<a\\
&[u_2] > 0 && \text{for } |x_1|<a\\
&\sigma_{ij} \longrightarrow 0 && \text{for } x_1^2+x_2^2 \longrightarrow \infty\\
&g(\pm a) = 0 &&
\end{aligned} \tag{1.6}$$

where $q(x_1)$ is a given function as a part of the boundary condition (1.4) and $g(x_1)$ is assumed to be known.

From the relations (1.2) it follows that

$$\begin{aligned} \sigma_{22} - i\sigma_{12} &= \Phi(z) + \Omega(\bar{z}) + (z - \bar{z})\,\overline{\Phi'(z)} \\ 2\mu \frac{\partial}{\partial x_1}(u_1 + iu_2) &= \kappa\Phi(z) - \Omega(\bar{z}) - (z - \bar{z})\,\overline{\Phi'(z)} \end{aligned} \tag{1.7}$$

where

$$\Omega(z) = \bar{\Phi}(z) + z\bar{\Phi}'(z) + \bar{\Psi}(z) \tag{1.8}$$

with the notation

$$\bar{F}(z) = \overline{F(\bar{z})}.$$

Assuming that

$$(z - z)\Phi'(z) = 0 \qquad \text{for } x_2 \longrightarrow \pm 0$$

and observing that

$$\Omega(\bar{z}) = \Omega^{\mp}(x_1) \qquad \text{for } x_2 \longrightarrow \pm 0$$

the boundary conditions (1.6), due to the relations (1.7), can be written in terms of the limiting values of complex potentials $\Phi(z)$ and $\Omega(z)$ along the crack surface; after some manipulations we obtain

$$\begin{aligned} &\Phi^{+}(x_1) - \Phi^{-}(x_1) = iQ(x_1) && \text{for } |x_1|<a \\ &\Omega^{+}(x_1) - \Omega^{-}(x_1) = i[Q(x_1) + 2iq(x_1)] && \text{for } |x_1|<a \\ &\{\Phi(z), \Omega(z)\} \longrightarrow 0 && \text{for } |z| \longrightarrow \infty \end{aligned} \tag{1.9}$$

where

$$Q(x_1) = g'(x_1) - \frac{2i}{\kappa+1}\, q(x_1) \tag{1.10}$$

Equations (1.9) define the Hilbert problem concerning sectionally holomorphic functions $\Phi(z)$ and $\Omega(z)$ with the line of discontinuity $\{|x_1|<a, x_2=0\}$, [4].

To solve this problem let us recall that, following Muskhelishvili [4], if the function $f(x_1)$ satisfies certain continuity conditions along the segment $\{|x_1|<a, x_2=0\}$ and if the complex function $F(z)$ is defined by the Cauchy integral

$$F(z) = \frac{1}{2\pi i}\int_{-a}^{a} \frac{f(t)}{t - z}\, dt$$

then the limiting values of the function $F(z)$ along the segment

$\{|x_1|<a, x_2=0\}$ are given by the following Plemelj formulas:

$$F^+(x_1) - F^-(x_1) = f(x_1)$$
$$F^+(x_1) + F^-(x_1) = \frac{1}{i\pi}\int_{-a}^{a} \frac{f(t)}{t - x_1}\,dt \tag{1.11}$$

Keeping in mind the requirement that functions $\Phi(z)$ and $\Omega(z)$ should vanish at infinity, by virtue of Eqs.(1.9) and the Plemelj formulas (1.11), they can be represented by the following Cauchy integrals:

$$\Phi(z) = \frac{1}{2\pi}\int_{-a}^{a} \frac{Q(t)}{t - z}\,dt$$
$$\Omega(z) = \frac{1}{2\pi}\int_{-a}^{a} \frac{Q(t) + 2iq(t)}{t - z}\,dt \tag{1.12}$$

Knowing functions $\Phi(z)$ and $\Omega(z)$, the state of displacements and stresses, according to relations (1.2), will be established provided both functions $q(x_1)$ and $g'(x_1)$ are known. As a matter of fact, function $g'(x_1)$ is unknown, but satisfying the remaining part of the boundary conditions (1.4) and using the already derived solution (1.12), we can determine function $g'(x_1)$, and eventually solve the main problem defined by the boundary conditions (1.4).

Expressing the limiting values of stresses σ_{22} and σ_{12} in terms of complex potentials $\Phi(z)$ and $\Omega(z)$, the boundary condition $(1.4)_1$ can be written in the form

$$\Phi^+(x_1) + \Phi^-(x_1) + \Omega^+(x_1) + \Omega^-(x_1) = 2p(x_1) \quad \text{for } |x_1|<a \tag{1.13}$$

According to Eqs.(1.11) and (1.12), the limiting values of functions $\Phi(z)$ and $\Omega(z)$ along the crack surface satisfy the following equations:

$$\Phi^+(x_1) + \Phi^-(x_1) = \frac{1}{\pi}\int_{-a}^{a} \frac{Q(t)}{t - x_1}\,dt \quad \text{for } |x_1|<a$$

$$\Omega^+(x_1) + \Omega^-(x_1) = \frac{1}{\pi}\int_{-a}^{a} \frac{Q(t) + 2iq(t)}{t - x_1}\,dt \quad \text{for } |x_1|<a$$

Introduction of these expressions into (1.13) leads to the singular integral equation of Cauchy type [4]

$$\frac{1}{\pi}\int_{-a}^{a}\frac{Q(t)+iq(t)}{t-x_1}\,dt = p(x_1) \qquad \text{for } |x_1|<a$$

the solution of which is given by

$$g'(x_1) = -\frac{1}{\pi\sqrt{a^2-x_1^2}}\left[\int_{-a}^{a}\frac{\sqrt{a^2-t^2}}{t-x_1}\,p(t)dt + C\right] - \frac{i(\kappa-1)}{\kappa+1}\,q(x_1) \tag{1.14}$$

where C is an arbitrary complex constant.
To determine the constant C we recall that, according to $(1.6)_5$

$$\int_{-a}^{a} g'(x_1)dx_1 = 0$$

which with the help of formulas

$$\int_{-a}^{a}\frac{dt}{(t-x)\sqrt{a^2-x^2}} = \begin{cases} 0 & \text{for } |x|<a \\ \dfrac{\text{sign}(x)}{\sqrt{a^2-x^2}} & \text{for } |x|>a \end{cases}$$

$$\int_{-a}^{a}\frac{dt}{\sqrt{t^2-x^2}} = \pi \tag{1.15}$$

leads to

$$C = -\,iA \quad \text{where} \quad A = \frac{\kappa-1}{\kappa+1}\int_{-a}^{a} q(t)dt \tag{1.16}$$

Finally using the formula

$$\int_{-a}^{a}\frac{dt}{(t-z)\sqrt{a^2-t^2}} = -\frac{\pi}{\sqrt{z^2-a^2}} \qquad \text{for } z \notin \{|x_1|\le a, x_2=0\} \tag{1.17}$$

and bearing in mind Eqs.(1.10),(1.12),(1.14) and (1.16), the complex potentials $\Phi(z)$ and $\Omega(z)$ (under the condition that z does not coincide with any point of the crack surface) will be described by

$$\Phi(z) = \frac{1}{2\pi\sqrt{z^2-a^2}}\left[\int_{-a}^{a}\frac{\sqrt{a^2-t^2}}{t-z}\,p(t)dt - iA\right] + \frac{1}{2\pi i}\int_{-a}^{a}\frac{q(t)}{t-z}\,dt$$

$$\Omega(z) = \Phi(z) - \frac{1}{i\pi}\int_{-a}^{a}\frac{q(t)}{t-z}\,dt \qquad (1.18)$$

where A is given by (1.16).

Introduction of (1.18) into (1.8) and (1.2) solves our problem and enables us to determine the displacements and stresses at any point of the plane, (Fig.3).

1.2 Anti-plane state of strain.

According to Fig.1, we assume that the only non-vanishing displacements component is the displacement u_3 in the direction of the x_3-axis. Assuming that the displacement is a function of x_1 and x_2 only, the non-vanishing deformations and stresses are

$$\varepsilon_{i3} = \frac{\partial u_3}{\partial x_i}, \quad \sigma_{i3} = \mu\varepsilon_{i3} \quad (i=1,2) \qquad (1.19)$$

In absence of the body forces and by virtue of the equation of equilibrium it is readily seen that displacement u_3 is a harmonic function; thus it can be treated as a real or imaginary part of an analytic function f(z) of the complex variable $z=x_1+ix_2$.
Assuming that

$$\mu u_3(x_1,x_2) = \mathrm{Re}\{f(z)\} \qquad (1.20)$$

and bearing in mind (1.19), the representation (1.20) leads directly to the relation expressing stresses σ_{i3} in terms of the real and imaginary parts of the analytic function

$$F(z) = \frac{\partial f(z)}{\partial z} = f'(z) \qquad (1.21)$$

and

$$\sigma_{13} - i\sigma_{23} = F(z) \qquad (1.22)$$

Approaching a problem of crack we will follow the procedure used in the previous section and will solve a problem of a crack with the non-equilibrated load applied to its surfaces.

Let $\overset{o}{\sigma}{}^{\pm}_{23}$ be the prescribed functions defined on the upper and lower surfaces of the crack and let us assume that stresses σ_{i3} vanish at infinity, (Fig.4).

Introducing the following notation

$$2p(x_1) = \overset{o}{\sigma}{}^{+}_{23} + \overset{o}{\sigma}{}^{-}_{23} \quad \text{for } |x_1|<a \qquad (1.23)$$

$$2q(x_1) = \overset{\circ+}{\sigma}_{23} - \overset{\circ-}{\sigma}_{23} \qquad \text{for } |x_1|<a \tag{1.23}$$

the boundary conditions of the problem can be written as

$$\begin{aligned} &\overset{+}{\sigma}_{23} + \overset{-}{\sigma}_{23} = 2p(x_1) && \text{for } |x_1|<a \\ &\overset{+}{\sigma}_{23} - \overset{-}{\sigma}_{23} = 2q(x_1) && \text{for } |x_1|<a \\ &\sigma_{13} \longrightarrow 0 && \text{for } x_1^2+x_2^2 \longrightarrow \infty \\ &[u_3(\pm a,0)] = 0 \end{aligned} \tag{1.24}$$

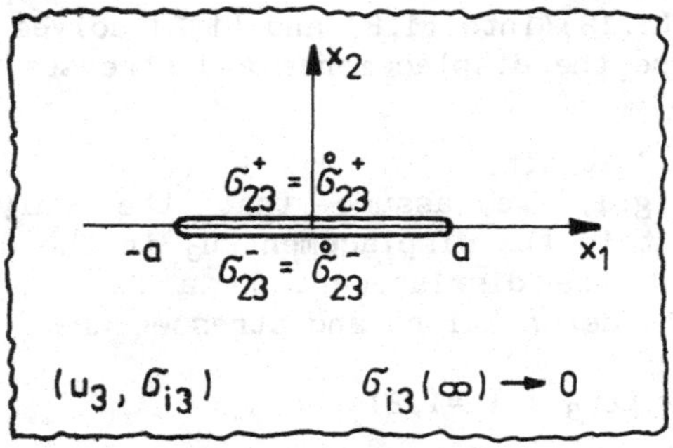

Fig. 4.

Introducing the auxiliary problem, let us suppose that, for the time being, the discontinuity of the displacement u_3 across the crack surface is known, and let

$$[u_3(x_1,0)] = \frac{2}{\mu}\, h(x_1) \qquad \text{for } |x_1| \le a \tag{1.25}$$

where according to $(1.24)_4$, $h(\pm a)=0$.
Bearing in mind Eq. (1.25), the boundary conditions of the auxiliary problem can be stated as follows

$$\begin{aligned} &[\sigma_{23}] = 2q(x_1) && \text{for } |x_1|<a \\ &\frac{\partial}{\partial x_1}[u_3] = \frac{2}{\mu}\, h'(x_1) && \text{for } |x_1|<a \\ &\sigma_{13} \longrightarrow 0 && \text{for } x_1^2+x_2^2 \longrightarrow \infty \\ &h(\pm a) = 0 \end{aligned} \tag{1.26}$$

Observing that

$$\frac{\partial u_3}{\partial x_1} = \frac{1}{2\mu}\,[F(z) + \overline{F(z)}] \quad , \quad \sigma_{23} = -\frac{1}{2i}\,[F(z) - \overline{F(z)}] \tag{1.27}$$

the boundary conditions (1.26) can be rewritten in terms of the limiting values of the potential F(z) along the crack surfaces what, after some transformation, leads to the following relations:

$$F^+(x_1) - F^-(x_1) = 2[h'(x_1) - iq(x_1)] \qquad \text{for } |x_1|<a$$

$$F(z) \to 0 \qquad \text{for } |z|\to\infty$$

The latter relations define the Hilbert problem concerning the function $F(z)$ with the line of discontinuity $\{|x_1|<a, x_2=0\}$ and, according to the Plemelj formulas (1.11), the function $F(z)$ can be represented by the following Cauchy integral:

$$F(z) = \frac{1}{i\pi}\int_{-a}^{a} \frac{h'(t) - iq(t)}{t - z}\,dt \tag{1.28}$$

To define the function $h(x_1)$ let us take into account the still unused boundary condition $(1.24)_1$, which may be written in terms of the limiting values of the potential $F(z)$

$$F^+(x_1) + F^-(x_1) - \overline{F^+(x_1)} - \overline{F^-(x_1)} = -4ip(x_1) \quad \text{for } |x_1|<a \tag{1.29}$$

On the other hand, by virtue of (1.11) and (1.28)

$$F^+(x_1) + F^-(x_1) = \frac{2}{i\pi}\int_{-a}^{a} \frac{h'(t) - iq(t)}{t - x_1}\,dt$$

This result together with the relation (1.29), leads to

$$\frac{1}{\pi}\int_{-a}^{a} \frac{h'(t)}{t - x_1}\,dt = p(x_1) \qquad \text{for } |x_1|<a$$

Making use of the boundary condition $(1.26)_4$ and of the formulas (1.15), solution of the last equation will be given by

$$h'(x_1) = -\frac{1}{\pi\sqrt{a^2 - x_1^2}}\int_{-a}^{a} \frac{\sqrt{a^2 - t^2}}{t - x_1}\,p(t)dt$$

Substitution of this equation into (1.28) leads, with the help of (1.17), to the final result

$$F(z) = \frac{1}{i\pi\sqrt{z^2 - a^2}}\int_{-a}^{a} \frac{\sqrt{a^2 - t^2}}{t - z}\,p(t)dt - \frac{1}{\pi}\int_{-a}^{a} \frac{q(t)}{t - z}\,dt \tag{1.30}$$

The formula (1.30) together with (1.20) and (1.22) solves the problem of the distribution of the displacement u_3 and stresses σ_{13} in the plane, (Fig.4).

1.3 Displacements and stresses near the tips of a crack.

From the analysis of Eqs.(1.2), (1.18) or (1.22), (1.30) it

follows that in the neighbourhood of the crack tips the stresses may become infinite, the order of infinity being 1/2. This type of singularity occurs in all solutions concerning elastic media with cracks, both finite and infinite, and is independent of the type of loading (not necessarily purely mechanical) applied to the body and producing stresses.
From the physical point of view such result is, obviously, nonrealistic and makes it necessary to introduce certain additional postulates into the classical elasticity theory, e.g. the suitable fracture criteria. We will return to this problem in the following sections.

Let us now use the solutions concerning unbounded medium with a crack to introduce one of the basic notions of the fracture theory, that is the stress intensity factors.

To this end introduce local Cartesian reference frames $x_i'(i=1,2,3)$ and polar coordinate systems r,θ centered at the crack tips, and assume that $z'=x_1'+ix_2'$, (Fig.5).

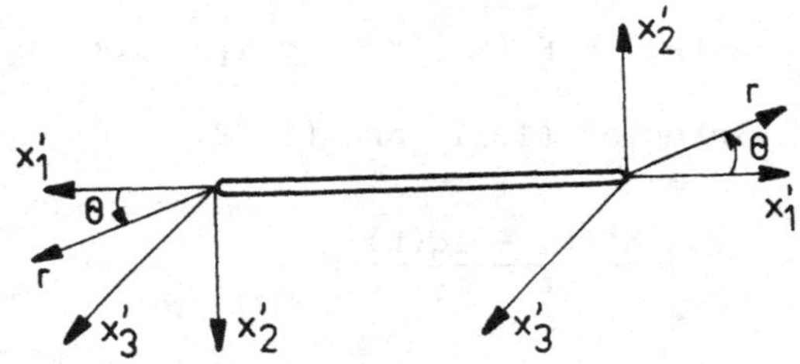

Fig. 5.

Investigation of the behaviour of the stresses and potentials $\Phi(z)$ and $F(z)$ in these coordinate systems enables us to prove, on the basis of Eqs.$(1.7)_1$, (1.18), that in the case of plane state of stress or strain

$$\sqrt{2\pi r}\,(\sigma_{22} - i\sigma_{12}) = K_I^{\pm} - iK_{II}^{\pm} \qquad \text{for } r\to 0 \text{ and } \theta=0$$
$$2\sqrt{2\pi z_1}\,\Phi(z_1) = K_I^{\pm} - iK_{II}^{\pm} \qquad \text{for } |z_1|\to 0 \tag{1.31}$$

where

$$K_I^{\pm} - iK_{II}^{\pm} = -\frac{1}{\sqrt{\pi a}}\left[\int_{-a}^{a}\sqrt{\frac{a\pm t}{a\mp t}}\,p(t)dt \pm \frac{i(\kappa-1)}{\kappa+1}\int_{-a}^{a}q(t)dt\right] \tag{1.32}$$

functions $p(t)$ and $q(t)$ being given by Eqs.(1.3).

In the case of anti-plane state of strain Eqs.(1.22), (1.30) yield the relations

$$\sqrt{2\pi r}\,\sigma_{23} = K_{III}^{\pm} \qquad \text{for } r\to 0 \text{ and } \theta=0$$
$$i\sqrt{2\pi z_1}\,F(z_1) = K_{III}^{\pm} \qquad \text{for } |z_1|\to 0 \tag{1.33}$$

where

$$K^{\pm}_{III} = -\frac{1}{\sqrt{\pi a}}\int_{-a}^{a}\sqrt{\frac{a \pm t}{a \mp t}}\,p(t)dt \tag{1.34}$$

function p(t) being defined by $(1.23)_1$.

Relations (1.31), (1.33) are of universal character, while Eqs. (1.32), (1.34) are true only in the case of a plane containing a straight crack of length 2a in the plane stress or strain cases (Fig. 3) or in anti-plane strain case (Fig. 4).

Coefficients $K^{\pm}_{I}$, $K^{\pm}_{II}$, $K^{\pm}_{III}$ play a very important role in the analysis of the fracture process and are called stress intensity factors at the crack tips, superscripts "-" and "+" referring to the left and right-hand crack tips, respectively. Subscripts I, II, III are connected with three possible crack deformation modes presented in Fig. 6.

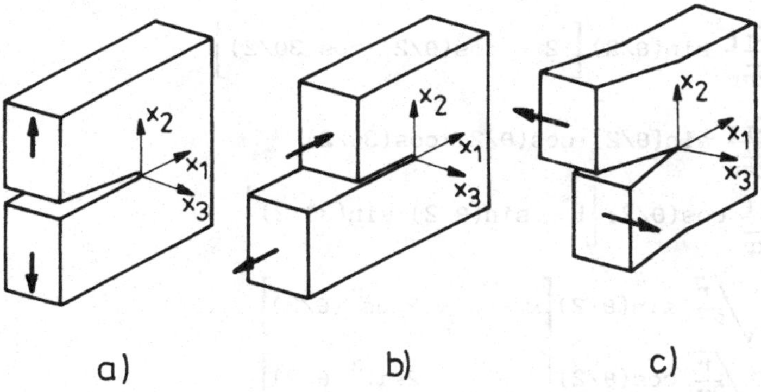

Fig. 6.

In Mode I (opening mode) $K_I \neq 0$, $K_{II} = K_{III} = 0$ and the displacement vector suffers a jump on the crack surface in the direction of x_2 axis, (Fig. 6a).

In Mode II (sliding mode) $K_{II} \neq 0$, $K_I = K_{III} = 0$ and the displacement vector component u_1 is discontinuous in the plane of the crack, (Fig. 6b).

In Mode III (tearing mode) $K_{III} \neq 0$, $K_I = K_{II} = 0$ and the displacement vector component u_3 is discontinuous in the plane of the crack, (Fig. 6c).

Analysis of the displacement and stress distributions in local polar coordinate system r, θ (Fig. 5), makes it possible to demonstrate [1] by means of the formulas (1.2), (1.31) and (1.33) that in the vicinity of crack tips, i.e. for $|z_1/a| \ll 1$ the displacements and stresses corresponding to various crack deformation modes may be represented by the following asymptotic formulas:

Mode I

$$\sigma_{11} = \frac{K_I}{\sqrt{2\pi r}} \cos(\theta/2)\left[1 - \sin(\theta/2)\cdot\sin(3\theta/2)\right]$$

$$\sigma_{22} = \frac{K_I}{\sqrt{2\pi r}} \cos(\theta/2)\left[1 + \sin(\theta/2)\cdot\sin(3\theta/2)\right]$$

$$\sigma_{12} = \frac{K_I}{\sqrt{2\pi r}} \sin(\theta/2)\cdot\cos(\theta/2)\cdot\cos(3\theta/2)$$

$$u_1 = \frac{K_I}{2\mu}\sqrt{\frac{r}{2\pi}} \cos(\theta/2)\left[\kappa - 1 + 2\sin^2(\theta/2)\right]$$

$$u_2 = \frac{K_I}{2\mu}\sqrt{\frac{r}{2\pi}} \sin(\theta/2)\left[\kappa + 1 - 2\cos^2(\theta/2)\right]$$

ModeII

$$\sigma_{11} = \frac{K_{II}}{\sqrt{2\pi r}} \sin(\theta/2)\left[-2 - \cos(\theta/2)\cdot\cos(3\theta/2)\right]$$

$$\sigma_{22} = \frac{K_{II}}{\sqrt{2\pi r}} \sin(\theta/2)\cdot\cos(\theta/2)\cdot\cos(3\theta/2) \qquad (1.35)$$

$$\sigma_{12} = \frac{K_{II}}{\sqrt{2\pi r}} \cos(\theta/2)\left[1 - \sin(\theta/2)\cdot\sin(3\theta/2)\right]$$

$$u_1 = \frac{K_{II}}{2\mu}\sqrt{\frac{r}{2\pi}} \sin(\theta/2)\left[\kappa + 1 + 2\cos^2(\theta/2)\right]$$

$$u_2 = \frac{K_{II}}{2\mu}\sqrt{\frac{r}{2\pi}} \cos(\theta/2)\left[1 - \kappa + 2\sin^2(\theta/2)\right]$$

Mode III

$$\sigma_{13} = -\frac{K_{III}}{\sqrt{2\pi r}} \sin(\theta/2)$$

$$\sigma_{23} = \frac{K_{III}}{\sqrt{2\pi r}} \cos(\theta/2)$$

$$u_3 = \frac{2K_{III}}{\mu}\sqrt{\frac{r}{2\pi}} \sin(\theta/2)$$

Symbols K_I, K_{II}, K_{III} represent here the stress intensity factors at the left or right-hand crack tip.

Equations (1.35) are universal in the sense that they remain valid for bounded or unbounded elastic solids and for cracks of arbitrary shape. However, stress intensity factors K_I, K_{II}, K_{III} are linear functions of the loads applied and are linear functions of elastic constants of the material, and depend on the geometry of the body and of the crack itself; they represent the stress measure in the neighbourhood of the crack tips for the particular problem considered and have the dimension of stress × length$^{-1/2}$.

The stress intensity factors are usually determined on the basis of analytical or numerical solutions of the corresponding boundary-value problems of elasticity. Depending on the geometry of the body and the crack and on the type of loading, such solutions are normally found by the complex variable methods or by means of the integral transforms.

Extensive discussion of the mathematical methods used in the analysis and numerous examples of solutions may be found in [5, 6, 7, 8, 9]; accurate or approximate values of the stress intensity factors are listed e. g. in [10, 11, 12].

1.4 J-integral.

In the mechanics of fracture (not necessarily limited to LEFM) an important role is played by the so-called J-integral. This notion was (sometimes under different names) introduced to the fracture analysis independently by several authors: J. D. Eshelby, J. L. Sanders, G. P. Cherepanov and J. R. Rice, in papers [13], [14], [9] and [15]. J-integral may also be viewed as a direct conclusion of one of the three conservation principles formulated by W. Günther [16] and extensively analyzed by J. K. Knowles and E. Sternberg [17].

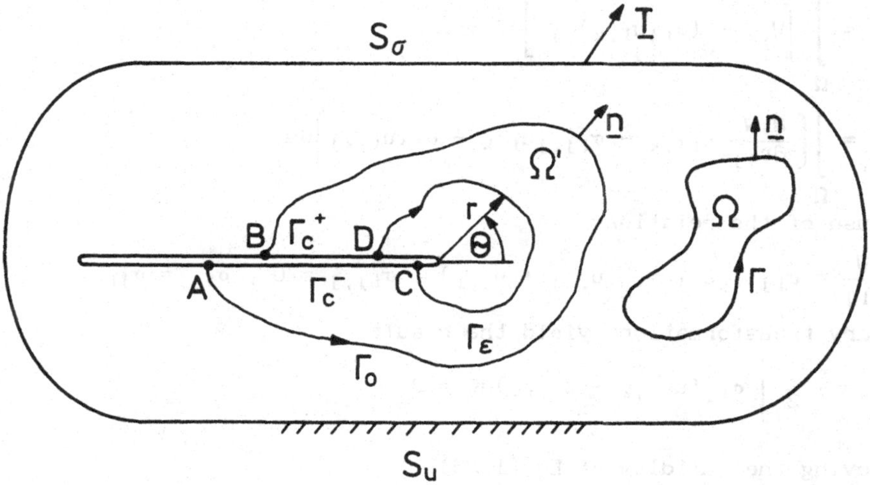

Fig. 7.

In the case of two-dimensional elasticity this principle may be formulated as follows: consider a two-dimensional, homogeneous elastic body V bounded by curve S. Disregard the body forces and assume that at one portion of the boundary S_σ, tractions $\underline{T}$ are prescribed, while at the remaining portion of the boundary, S_u, displacements are given, (Fig.7).

Let u_i, σ_{ij}, ε_{ij} represent the components of displacements, stresses and infinitesimal strains, respectively; let W denote the strain energy density define

$$W = W(\varepsilon_{ij}) = \int_0^{\varepsilon_{ij}} \sigma_{ij} d\varepsilon_{ij}$$

Then, for each closed curve Γ located within V and containing no singular points, defects or material inhomogeneities, the following condition is satisfied

$$\int_\Gamma (W n_k - T_i u_{i,k}) d\Gamma = 0 \tag{1.36}$$

Here k=1,2, $T_i = \sigma_{ij} n_j$ is the traction vector on Γ, $\underline{n}$ is the unit outward normal vector on Γ, and a subscript preceded by a comma indicates partial differentiation with respect to the corresponding Cartesian coordinate; summation over repeated indices is implied.

In order to prove Eq. (1.36), denote by Q_k the right-hand integral in (1.36) and assume curve Γ to represent the boundary of region $\Omega \subset V$, (Fig. 7). Then, on the basis of the Green-Gauss theorem

$$Q_k = \int_\Omega \left[W_{,k} - (\sigma_{ij} u_{i,k})_{,j} \right] d\Omega$$

$$= \int_\Omega \left(\frac{\partial W}{\partial \varepsilon_{ij}} \varepsilon_{ij,k} - \sigma_{ij,j} u_{i,k} - \sigma_{ij} u_{i,kj} \right) d\Omega$$

Making use of the relations

$$\frac{\partial W}{\partial \varepsilon_{ij}} = \sigma_{ij} \ , \quad \varepsilon_{ij} = \frac{1}{2}(u_{i,j} + u_{j,i}) \ , \quad \sigma_{ij,j} = 0 \ , \quad \sigma_{ij} = \sigma_{ji}$$

elementary transformations yield the result

$$Q_k = -\frac{1}{2} \int_\Omega \sigma_{ij} (u_{i,jk} - u_{j,ik}) d\Omega = 0$$

thus proving the validity of Eq. (1.36).

In the case when the region bounded by Γ contains a singular point, conservation law (1.36) is, in general, not satisfied and $Q_k \neq 0$ since then the Green-Gauss theorem has a different form.

Let us consider such a case and assume the curve Γ bounding region $\Omega' \subset V$ to contain one of the tips of a plane crack with stress free surfaces, (Fig. 7). Curve Γ is divided into four parts Γ_o, $\Gamma_c^{\pm}$, and Γ_ε; sense of the unit vector normal to Γ_o will be assumed to be positive if it is directed outwards the region Ω', and at the remaining curves if it is directed inwards this region. Under such assumptions region Ω' does not contain the crack tip and the conservation law (1.36) is satisfied, so that

$$\int_{\Gamma_\varepsilon} (Wn_k - T_i u_{i,k})d\Gamma = \int_{\Gamma_o} (Wn_k - T_i u_{i,k})d\Gamma - \int_{\Gamma_c} (Wn_k - T_i u_{i,k})d\Gamma \tag{1.37}$$

where $\Gamma_c = \Gamma_c^+ + \Gamma_c^-$.

Introducing the notation

$$J_k = \int_{\Gamma_\varepsilon} (Wn_k - T_i u_{i,k})d\Gamma \tag{1.38}$$

and making use of the fact that $T_i=0$ on $\Gamma_c^\pm$, relation (1.37) may be rewritten in the form

$$J_k = \int_{\Gamma_o} (Wn_k - T_i u_{i,k})d\Gamma - \int_{\Gamma_c} Wn_k d\Gamma \tag{1.39}$$

It follows that

$$J_1 \equiv J = \int_{\Gamma_o} (Wn_1 - T_i u_{i,1})d\Gamma$$

$$J_2 = \int_{\Gamma_o} (Wn_2 - T_i u_{i,2})d\Gamma - \int_{\Gamma_c} Wn_2 d\Gamma \tag{1.40}$$

where contour Γ_o is an arbitrary contour with the origin and end located on the lower and upper crack surfaces, respectively. J_1 is usually called (after Rice) the J-integral.

On the other hand, J_k is defined by the formula (1.38). In view of the fact that contour Γ_ε is an arbitrary contour surrounding the crack tip let us assume, for the sake of simplicity of further derivations, that it is a circular arc centered at the crack tip; let r, θ denote the polar coordinates of that contour, (Fig.7). Assume, moreover, that the material considered is linearly elastic ($W = \frac{1}{2}\sigma_{ij}\varepsilon_{ij}$), and the crack deforms simultaneously according to all three deformation modes I, II and III.

Making the additional assumption that the region bounded by Γ_ε is small enough ($r \rightarrow 0$, Fig.7), the integrands appearing in Eq.(1.38) may be replaced with the corresponding asymptotic expansions (1.35) what, after certain transformations and rearrangements, leads to the following formula [18]:

$$J_k = -\frac{\kappa+1}{8\pi\mu}\int_{\Gamma_\varepsilon}\left(\frac{\partial F_k}{\partial x_1}\frac{\partial x_1}{\partial \Gamma} + \frac{\partial F_k}{\partial x_2}\frac{\partial x_2}{\partial \Gamma}\right)d\Gamma$$

$$- \frac{1}{2\pi\mu}\int_{\Gamma_\varepsilon}\left(\frac{\partial G_k}{\partial x_1}\frac{\partial x_1}{\partial \Gamma} + \frac{\partial G_k}{\partial x_2}\frac{\partial x_2}{\partial \Gamma}\right)d\Gamma \qquad (1.43)$$

Here

$$2F_1 = K_I^2(\theta - \sin\theta\cdot\cos\theta) + K_{II}^2(\theta + \sin\theta\cdot\cos\theta) - 2K_IK_{II}\cos^2\theta$$

$$2G_1 = K_{III}^2\cdot\theta$$

$$2F_2 = K_I^2\cos^2\theta + K_{II}^2(2\ln\frac{r}{a} + \sin^2\theta) - 2K_IK_{II}(\theta + \sin\theta\cdot\cos\theta)$$

$$2G_2 = K_{III}^2\ln\frac{r}{a} \qquad (1.42)$$

$$x_1 = r\cos\theta\ ,\ x_2 = r\sin\theta\ ,$$

$$\frac{\partial x_1}{\partial\Gamma} = \cos(n,x_2)\ ,\ \frac{\partial x_2}{\partial\Gamma} = -\cos(n,x_1)$$

while 2a denotes crack's length.

The integrands appearing in (1.41) are total differentials what implies that the corresponding integrals are path-independent and equal the differences of the values of F_k and G_k taken at the end and origin of the contour Γ_ε, (Fig.7),

$$J_k = -\frac{\kappa+1}{8\pi\mu}\left[F_k(C) - F_k(D)\right] - \frac{1}{2\pi\mu}\left[G_k(C) - G_k(D)\right]$$

Finally, making use of this result and the relations (1.42) and (1.40), integrals J_1 and J_2 are expressed in the form

$$J_1 \equiv J = \int_{\Gamma_o}(Wn_1 - T_iu_{i,1})d\Gamma = \frac{\kappa+1}{8\mu}(K_I^2 + K_{II}^2) + \frac{1}{2\mu}K_{III}^2$$

$$J_2 = \int_{\Gamma_o}(Wn_2 - T_iu_{i,2})d\Gamma - \int_{\Gamma_c}Wn_2d\Gamma = -\frac{\kappa+1}{4\mu}K_IK_{II} \qquad (1.43)$$

From the relations derived it follows that the values J_k are completely determined by the corresponding stress intensity factors at the crack tip. On the other hand, the value of J is expressed by an integral taken along an arbitrary path Γ_o surrounding the crack tip. This property of J-integral is most important and is widely utilized in fracture analysis: the stress intensity factors may be determined by evaluating the contour integral what necessitates the knowledge of the displacement and strain field not in the entire region of the body considered but along the integration path only. Proper selection of the contour renders this evaluation (analytical, numerical or based on experimental measurements) much simpler. Physical interpretation of the J-integral will be discussed in Sec.3.3.

2. GRIFFITH'S BRITTLE FRACTURE CRITERION.

A classical approach to the fracture process was presented in 1921 by A. A. Griffith, [19]. This approach is based on energy considerations and on the fundamental assumption that during the entire fracture process the material remains linearly elastic outside the crack tips. This type of fracture is called brittle fracture (throughout the process the Hooke's law holds true).

2.1 Energy balance of a body with crack.

Consider a linearly elastic medium of volume V bounded by surface S, (Fig.8). Let the body contain a crack of length L, its surface S_L being stress free. Assume that the body is loaded by constant tractions T_i applied at portion S_σ of the bounding surface S, at the remaining portion S_u of that surface displacements $\overset{o}{u}_i$ being prescribed. For the sake of simplicity, action of body forces will be disregarded.

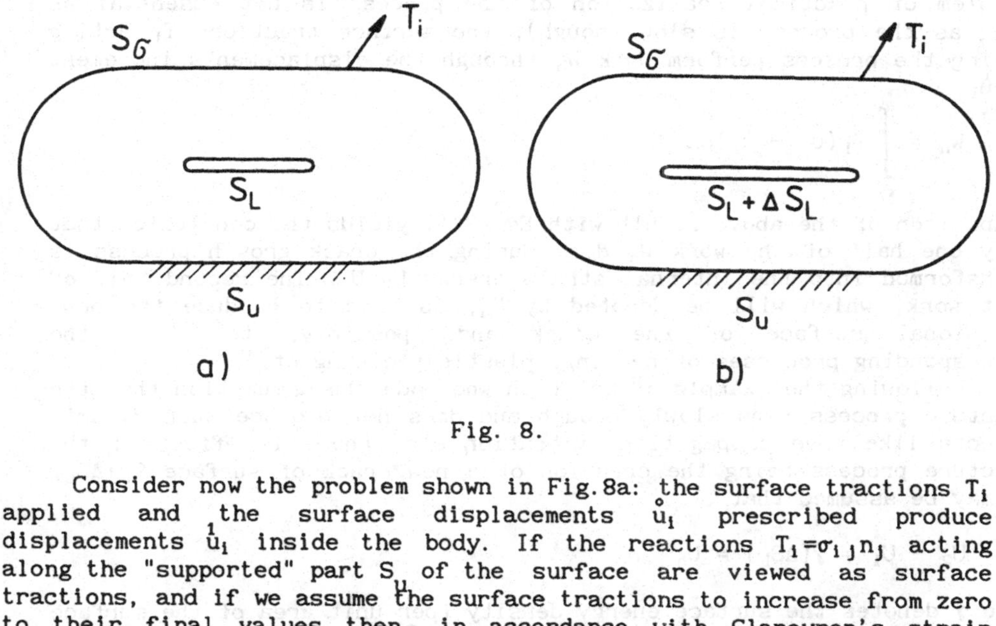

Fig. 8.

Consider now the problem shown in Fig.8a: the surface tractions T_i applied and the surface displacements $\overset{o}{u}_i$ prescribed produce displacements $\overset{1}{u}_i$ inside the body. If the reactions $T_i = \sigma_{ij} n_j$ acting along the "supported" part S_u of the surface are viewed as surface tractions, and if we assume the surface tractions to increase from zero to their final values then, in accordance with Clapeyron's strain energy theorem, the internal strain energy U_1 in System I will be defined by the relation

$$U_1 = \frac{1}{2} \int_S T_i \overset{1}{u}_i \, dS$$

If the body under consideration contained the crack $S_L + \Delta S_L$ (System II, Fig.8b) from the very beginning then, assuming its surface to be stress free and using the same assumptions as those made in connection with the problem shown in Fig. 8a, displacements $\overset{2}{u}_i$ would appear in the body; the internal strain energy U_2 corresponding to System II would then take the form

$$U_2 = \frac{1}{2} \int_S T_i \overset{2}{u}_i \, dS$$

what means that the strain energy increment corresponding to the passage from System I to II would be equal to

$$U_2 - U_1 = \frac{1}{2} \int_S T_i (\overset{2}{u}_i - \overset{1}{u}_i) dS \tag{2.1}$$

Consider now again the System I (Fig. 8a) and assume the surface S_σ to be loaded by surface tractions T_i of values equal to the final values of the previously considered loads, the body being already deformed under their action. If the size of the crack increases slowly from its original value S_L to $S_L+\Delta S_L$ under the assumption that the crack surfaces remain stress free during the crack growth process (the problem of practical realization of the process is not essential as long as the process is slow enough), the surface tractions T_i acting during the process perform work W_h through the displacements increment $\overset{2}{u}_i - \overset{1}{u}_i$,

$$W_h = \int_S T_i (\overset{2}{u}_i - \overset{1}{u}_i) dS$$

Comparison of the above result with Eq. (2.1) yields the conclusion that only one half of the work W_h done during the crack growth process is transformed into the internal strain energy U_2-U_1. The second half of that work, which will be denoted by U_h, is used to produce the new, additional surface of the crack and, possibly, to start the corresponding processes of heating, plastic yielding etc.

Following the example of Griffith who made the assumption that the fracture process runs slowly enough and does not produce such dynamic effects like wave propagation, vibration etc. the sole effect of the fracture process being the creation of a new crack of surface $S_L+\Delta S_L$, it may be assumed that

$$U_2 - U_1 = \gamma(\Delta S_L) = U_h$$

Here γ denotes the surface energy density (per unit area of the surface of the solid) of dimension force $\times$ length^{-1}, and U_h is called the energy of the newly created crack surface.

2.2 Griffith's fracture criterion.

Making use of the results presented in the preceding section let us define, following A. A. Griffith, the generalized potential energy for the body containing a stress free crack S_L,

$$\mathfrak{P}^* = U - \int_{S_\sigma} T_i u_i \, dS + U_h \qquad (2.2)$$

where

U - internal strain energy of the body with crack S_L,

$\int_{S_\sigma} T_i u_i \, dS$ - work done by surface tractions T_i through displacements u_i in the body with crack s_L,

$U_h = \gamma S_L$ - surface energy of crack S_L.

The minimum potential energy theorem for a body with a crack states that, among all geometrically admissible displacements u_i and additional surfaces S_L, the energetically privileged system is that which leads to the minimum value of the potential energy $\mathfrak{P}^*$. The state of the crack determined in that manner is stable if further growth of the crack by ΔS_L does not reduce the value of $\mathfrak{P}^*$

The type of equilibrium of the crack depends on the behaviour of the variation of the potential energy $\mathfrak{P}^*$ corresponding to small surface increments δS_L, and

1. If $\delta\mathfrak{P}^* > 0$ for $\delta S_L > 0$, the crack is stable and has no tendency to grow;
2. If $\delta\mathfrak{P}^* = 0$ for $\delta S_L > 0$, equilibrium of the system is unstable, and
   a) if $\delta^2\mathfrak{P}^* > 0$, further growth of the crack is possible under simultaneous slow growth of the load,
   b) if $\delta^2\mathfrak{P}^* \leq 0$, further growth of the crack is unstable, i.e. rapid propagation of the crack takes place;
3. If $\delta\mathfrak{P}^* < 0$ for $\delta S_L > 0$, the crack is unstable and propagation of the crack follows, leading to total failure of the system or to a new equilibrium position.

Vanishing of the variation of the potential energy $\mathfrak{P}^*$

$$\delta\mathfrak{P}^* = 0 \quad \text{for } \delta S_L > 0 \qquad (2.3)$$

called the Griffith brittle fracture criterion, determines the critical parameters, such as loads or crack dimensions, which make the crack unstable.

Let us assume that the surface tractions T_i increase slowly from zero to their final values; according to Clapeyron's theorem, we obtain

$$2U = \int_S T_i u_i \, dS$$

and, on the basis of Eq.(2.2), the potential energy $\mathfrak{P}^*$ of a body with a crack takes the form

$$\mathfrak{P}^* = U_h - U$$

In such a case the Griffith's brittle fracture criterion, i.e. answer to the question whether the crack is stable or not, depends on the behaviour of the variation

$$\delta \mathfrak{P}^{*} = \delta(U_h - U) \quad \text{for } \delta S_L > 0 \qquad (2.4)$$

Let us observe the difference between the notion of equilibrium of the system "load - crack" and the notion of crack stability. It is known that the condition of equilibrium of a mechanical system requires its potential energy to reach a minimum, what is equivalent to the condition of vanishing of the first variation of the energy; type of the equilibrium depends on the sign of the second variation of the potential energy of the system. In the case of a body with a crack, the crack remains stable even if $\delta \mathfrak{P}^{*} \neq 0$, provided the positive increment of crack's surface ($\delta S_L > 0$) is accompanied by increasing potential energy of the system ($\delta \mathfrak{P}^{*} > 0$).

For instance, in the case of absence of external loads, it follows from Eq.(2.4) that $\delta \mathfrak{P}^{*} = \gamma(\delta S_L) > 0$ and the whole system is not in equilibrium, though the crack should be considered as stable. Functional $\mathfrak{P}^{*}$ would reach a minimum value if the crack disappeared, i.e. if S_L vanished.

Non-equivalence of both the notions is an obvious result of the assumed irreversibility of the crack growth process. The notions of equilibrium of the system "load - crack" and of the crack stability coincide only in the case of unstable equilibrium of the system and of a critical state of the crack, since then the crack looses its stability and exhibits tendency to propagate into the body.

3. IRWIN'S FRACTURE CRITERION.

In spite of the fact that Griffith's hypothesis is based on such global notions as internal strain energy, work done by external forces, its character is actually local, in the sense that its application requires the knowledge of the state of the body in the immediate vicinity of the crack tip. This fact was observed by G. R. Irwin who applied a different approach, though equivalent to that proposed by Griffith, and based it on the notion of the stress intensity factor at the crack tip.

3.1 Local character of the fracture process.

Consider again a linear elastic medium of volume V bounded by surface S; let the medium contain a crack of length L, its surface being free of stresses. Assume the body to be loaded by constant surface tractions T_i acting at portion S_σ, at the remaining part of the surface S_u displacements $\overset{o}{u}_i$ are prescribed. Assume, in addition, that the hypothetical surface ΔS_L which will be created as a result of future growth of the original crack S_L, is acted on by forces $T_i = -\Delta T_i$, (Fig.9a). The surface loads applied and the displacements $\overset{o}{u}_i$ prescribed produce displacements $\overset{1}{u}_i$, strains $\overset{1}{\varepsilon}_{ij}$ and stresses $\overset{1}{\sigma}_{ij}$ within the body.

Potential energy of System I represented in Fig.9a assumes the form

$$\mathfrak{P}_1 = \frac{1}{2}\int_V \overset{1}{\sigma}_{ij}\overset{1}{\varepsilon}_{ij}\,dV - \int_{S_\sigma} T_i \overset{1}{u}_i\,dS$$

Consider now the other System II shown in Fig. 9b. It is assumed here that the body is of the same dimensions and is loaded in the same manner as that shown in Fig. 9a, the only difference consisting in the surface of the crack, which is now greater by ΔS_L. The surface of the newly created crack $S_L+\Delta S_L$ is free of stresses.

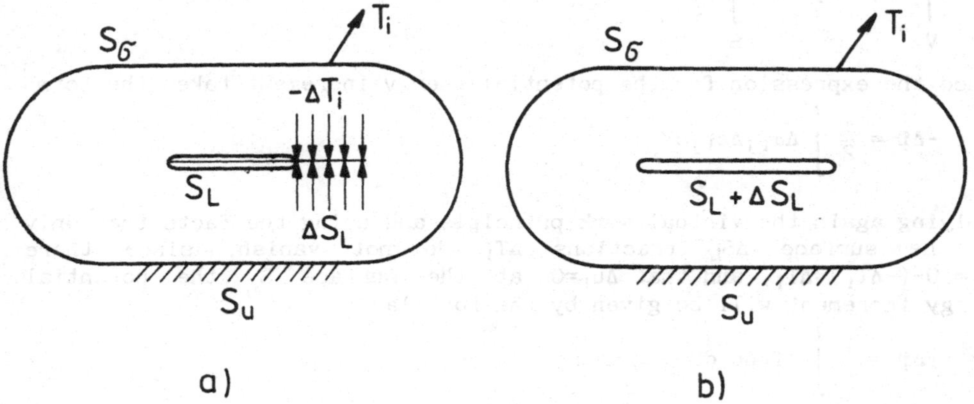

Fig. 9.

Crack surface increment ΔS_L and slow reduction of stresses on the additional surface from their original value $-\Delta T_i$ to zero introduce displacements u_i and stresses σ_{ij} into the body. Comparison with the corresponding magnitudes evaluated in System I makes it possible to calculate the increments

$$u_i = \overset{1}{u}_i + \Delta u_i\ ,\quad \varepsilon_{ij} = \overset{1}{\varepsilon}_{ij} + \Delta\varepsilon_{ij}\ ,\quad \sigma_{ij} = \overset{1}{\sigma}_{ij} + \Delta\sigma_{ij} \tag{3.1}$$

Potential energy of System II is written in the form

$$\mathfrak{P}_2 = \frac{1}{2}\int_V (\overset{1}{\sigma}_{ij} + \Delta\sigma_{ij})(\overset{1}{\varepsilon}_{ij} + \Delta\varepsilon_{ij})\,dV - \int_{S_\sigma} T_i(\overset{1}{u}_i + \Delta u_i)\,dS$$

The potential energy increment resulting from the transformation of the system from I to II is written in the form

$$-\Delta\mathfrak{P} = \mathfrak{P}_1 - \mathfrak{P}_2$$

$$= -\frac{1}{2}\int_V (\overset{1}{\sigma}_{ij}\Delta\varepsilon_{ij} + \overset{1}{\varepsilon}_{ij}\Delta\sigma_{ij} + \Delta\sigma_{ij}\Delta\varepsilon_{ij})\,dV + \int_{S_\sigma} T_i\Delta u_i\,dS \tag{3.2}$$

Taking into account relations (3.1) and the fact that

$$\sigma_{ij}\Delta\varepsilon_{ij} = \varepsilon_{ij}\Delta\sigma_{ij}$$

relation (3.2) assumes the form

$$-\Delta\mathfrak{P} = -\int_V \sigma_{ij}\Delta\varepsilon_{ij}dV + \frac{1}{2}\int_V \Delta\sigma_{ij}\Delta\varepsilon_{ij}dV + \int_{S_\sigma} T_i\Delta u_i dS$$

From the virtual work principle it follows that

$$\int_V \sigma_{ij}\Delta\varepsilon_{ij}dV = \int_{S_\sigma} T_i\Delta u_i dS$$

hence the expression for the potential energy increment takes the form

$$-\Delta\mathfrak{P} = \frac{1}{2}\int_V \Delta\sigma_{ij}\Delta\varepsilon_{ij}dV$$

Applying again the virtual work principle and using the facts that only on the surface ΔS_L tractions ΔT_i do not vanish, since there $\Delta T_i = 0-(-\Delta T_i)=\Delta T_i$, and that $\Delta u_i=0$ at the surface S_u, the potential energy increment will be given by the formula

$$-\Delta\mathfrak{P} = \frac{1}{2}\int_{\Delta S_L} \Delta T_i\Delta u_i dS$$

The stress vector increment ΔT_i appearing here represents a function which is continuous across the surface ΔS_L, contrary to the displacement vector increment which suffers a jump at that surface, hence

$$-\Delta\mathfrak{P} = \frac{1}{2}\int_{\Delta S_L} \Delta T_i[\Delta u_i]dS$$

Making use of the fact that $\Delta u_i = u_i - \overset{1}{u}_i$ and $[\overset{1}{u}_i]=0$ on ΔS_L let us note that $[\Delta u_i]=[u_i]$ and the expression describing the potential energy increment resulting from the system transformation I into II takes the final form

$$-\Delta\mathfrak{P} = \frac{1}{2}\int_{\Delta S_L} \Delta T_i[u_i]dS \tag{3.3}$$

Under assumption that the crack considered is plane and lies in the plane x_1-x_2, Eq. (3.3) may also be written as

$$-\Delta\mathfrak{P} = \frac{1}{2}\int_{\Delta S_L} \sigma_{i2}[u_i]dS \tag{3.4}$$

The last relation will now be used to determine the potential energy increment accompanying the passage from System I to System II due to all three possible crack deformation modes; further considerations will be confined to the Mode I crack deformation.

To this end let us assume that on the surface of the crack considered the displacement vector represents a discontinuous function in the direction of x_2-axis only. Consider a single, right-hand crack tip located at point (a,0), (Fig.2c) and assume that the crack surface increases by the length Δa in the direction of the x_1-axis. The corresponding potential energy increment is then found from Eq.(3.4),

$$-\Delta \mathfrak{P} = \frac{1}{2}\int_{a}^{a+\Delta a} \sigma_{22}(x_1,0;a)[u_2(x_1,0;a+\Delta a)]dx_1 \tag{3.5}$$

The notation used in the integrand indicates that stress $\sigma_{22}(x_1,0;a)$ corresponds to the crack with a tip at (a,0), while displacement $u_2(x_1,0;a+\Delta a)$ corresponds to the crack surface with a tip at point $(a+\Delta a,0)$, (Fig.10).

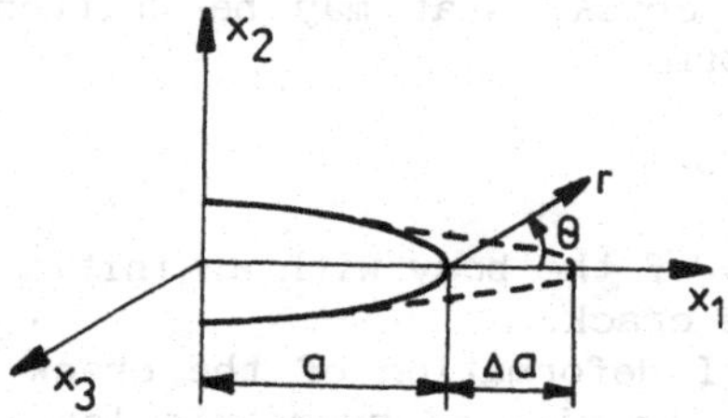

Fig. 10.

Moreover, assuming the crack extension Δa to be small enough ($\Delta a \to 0$), the accurate values of stress σ_{22} and displacement u_2 appearing in (3.5) may be replaced with their asymptotic expressions (1.35); in case of stress it should be assumed that $r=x_1$ and $\theta=0$, while in the case of displacement $r=\Delta a-x_1$ and $\theta=\pi$ (Fig.5). Then, on the basis of the aforementioned relations,

$$-\Delta \mathfrak{P} = \frac{(\kappa+1)K_I^2}{4\pi\mu}\int_{0}^{\Delta a}\sqrt{\frac{\Delta a-x_1}{x_1}}\,dx_1$$

Finally the conclusion follows that in the case of Mode I crack deformation, the potential energy increment due to infinitesimal right-hand tip displacement Δa may be written in the form

$$-\Delta \mathfrak{P} = \frac{(\kappa+1)K_I^2}{8\mu}\,\Delta a \tag{3.6}$$

Similar procedure applied to the remaining two crack deformation modes yields the following formulas:

$$-\Delta \mathfrak{P} = \frac{(\kappa+1)K_{II}^2}{8\mu}\,\Delta a \tag{3.7}$$

$$-\Delta \mathfrak{P} = \frac{(\kappa+1)K_{III}^2}{2\mu} \Delta a \tag{3.7}$$

In the case when the crack deforms according to all three deformation modes simultaneously, the potential energy increment produced by the horizontal crack extension Δa at one of its tips equals

$$-\Delta \mathfrak{P} = \left[\frac{\kappa+1}{8\mu} (K_I^2 + K_{II}^2) + \frac{1}{2\mu} K_{III}^2 \right] \Delta a \tag{3.8}$$

The fact that the energy increment $-\Delta \mathfrak{P}$ due to crack's elongation may be described by the stress intensity factors was used by G. R. Irwin to put the Griffith fracture criterion in a new form, more convenient in practical applications [20].

3.2 Griffith's fracture criterion in Irwin's formulation.

The condition of unstable equilibrium of a crack in Griffith's formulation is the condition of vanishing of the potential energy variation of the body with a crack, what may be written, in view of Eqs.(2.2) and (2.3), in the form

$$\delta(\mathfrak{P} + U_h) = 0 \quad \text{for } \delta S_L > 0 \tag{3.9}$$

Here $\mathfrak{P}$ is the potential energy of the body with an initial crack S_L and U_h - the surface energy of the crack.

In the case of the Mode I deformation of the crack $\{|x_1| \le a, x_2=0\}$ and under the assumption that, in view of symmetric loading, the crack will extend by equal distances at both tips, we obtain the result $U_h = 4\gamma(\Delta a)$ and condition (3.9) assumes the form

$$\lim_{\Delta a \to 0} \left(-\frac{\Delta \mathfrak{P}}{\Delta a} \right) = -\frac{\partial \mathfrak{P}}{\partial a} = 4\gamma$$

The above condition may be transformed by means of the relation (3.6) (to obtain the potential energy increment produced by symmetric extension of the crack at both ends, the value of $-\Delta \mathfrak{P}$ must be multiplied by 2) to yield

$$\frac{\kappa+1}{4\mu} K_I^2 = 4\gamma$$

The value of K_I obtained in this manner will be called the critical value of the stress intensity factor and denoted by K_{Ic}; thus, in case of brittle material

$$K_{Ic} = 4 \sqrt{\frac{\mu\gamma}{\kappa+1}} \tag{3.10}$$

Irwin's formulation of the unstable equilibrium condition of a crack states that the crack reaches this type of equilibrium at the instant when the stress intensity factor K_I reaches its critical value, i.e. when $K_I = K_{Ic}$.

For $K_I < K_{Ic}$ the crack is stable and does not exhibit the tendency to grow; if $K_I > K_{Ic}$, the growth of the crack becomes unstable and fracture occurs.

Parameter K_{Ic} characterizes the so-called fracture toughness of the material and may be determined experimentally.

3.3 Energy release rate.

In formulating the local fracture criterion Irwin introduced also the notion of the potential energy release rate. This parameter, which represents the decrease of potential strain energy in the process of crack growth, enables a deeper physical interpretation of the fracture hypothesis.

The potential energy release rate at a crack tip resulting from its propagation is represented by the following limit:

$$\lim_{\Delta a \to 0} \left(-\frac{\Delta \mathfrak{P}}{\Delta a} \right) = -\frac{\partial \mathfrak{P}}{\partial a} = \mathcal{G} \qquad (3.11)$$

It follows from Eqs.(3.8),(3.11) that when the crack deforms according to three deformations modes simultaneously, the energy release rate $\mathcal{G}$ produced by the crack growing in its plane at a single tip is completely determined by the corresponding stress intensity factors,

$$\mathcal{G} = \frac{\kappa+1}{8\mu} (K_I^2 + K_{II}^2) + \frac{1}{2\mu} K_{III}^2 \qquad (3.12)$$

The energy release rate $\mathcal{G}$ has the dimension of force × length^{-1} and is called sometimes the crack extension force; it may be understood as a force (per unit length of crack's increment) necessary for the crack's elongation.

On comparing the relations $(1.43)_1$ with (3.12) it is seen that the energy release rate $\mathcal{G}$ is identical with the J-integral, what adds to the physical sense of this magnitude; it means that the energy release rate may be determined by simple integration (according to the formula $(1.43)_1$) along an arbitrary contour Γ_o surrounding the crack tip, (Fig.7).

According to the energy balance discussed in Section 2.1, in the process of formation of a new crack surface, only half of the work W_h is being transformed into the strain energy. Let us now assume that the other half of that energy U_h is spent not only on the formation of a new surface of the crack but also on creation of plastic zones or other possible phase transformations in the region around the crack tip. Using relation (3.11), the condition of unstable equilibrium (3.9), formulated earlier for a perfectly brittle material, takes now the form

$$\mathcal{G} = \frac{\partial U_h}{\partial a} = \mathcal{G}_c \qquad (3.13)$$

Here $\mathcal{G}_c$ should be considered as a material constant representing the energy used for the formation of new crack surfaces and for creation of plastic zones around the crack tips. This type of crack stability condition applies to so-called quasi-brittle materials; it was originally proposed by G. R. Irwin [21] and E. Orowan [22].

If the crack is subject to the Mode I deformation and extends symmetrically at both ends, Eqs.(3.12),(3.13) yield the following crack stability condition:

$$\frac{1+\kappa}{4\mu} K_I^2 = \mathcal{G}_c$$

The critical value of the stress intensity factor obtained in this manner is given by the formula

$$K_{Ic} = 2\sqrt{\frac{\mu\mathcal{G}_c}{\kappa+1}} \qquad (3.14)$$

This means that K_{Ic} is a material constant, and both the crack stability conditions formulated on the basis of critical values of the energy release rate and of the critical stress intensity factor K_{Ic}, for brittle or quasi-brittle materials, are equivalent. In brittle materials $\mathcal{G}_c=4\gamma$ and Eq.(3.14) transforms into (3.10).

4. BARENBLATT'S FRACTURE CRITERION.

It was G. I. Barenblatt who in 1959 made the attempt to make the distribution of stresses and displacements around the crack tips more realistic (i.e. to avoid the stress singularities) [23]. He started from the assumption that stresses at crack tips in real bodies must be finite, and that from the physical point of view a sharp-edged crack cannot be deformed to an elliptical form. In formulating the theory, Barenblatt based his considerations, similarly to his predecessors, on the results of theoretical analysis of two-dimensional elasticity; however, he introduced three additional hypotheses the legitimacy of which cannot be verified, at least directly. To make the reasoning as clear as possible, the hypotheses will be introduced and discussed below in an order different from that used in the original paper [24].

The considerations will be simplified by discussing the problem of normal opening of a crack $\{|x_1|\le a, x_2=0\}$ in an infinite elastic medium. It will be moreover assumed that the crack surfaces are stress free, and the external load distribution is symmetric with respect to both axes x_1 and x_2 of the coordinate system.

Solution to this problem of theory of elasticity is known to lead to infinite stresses appearing at the crack tips. In order to eliminate the singularities Barenblatt assumed (Hypothesis I) the existence of cohesion forces between the opposite surfaces of the crack in the neighbourhood of both tips, the surfaces being attracted to each other in that region. It will be shown later that a direct result of the hypothetical forces of attraction consists in smooth closure of the crack surfaces at the tips forming a sharp edge.

The forces of cohesion may be treated initially as an additional, symmetric load applied to the crack surfaces; thus the corresponding solution may be obtained by means of the superposition principle and has the form of a sum of three following problems:

I. A load symmetric with respect to the axes x_1 and x_2 acting in the continuous elastic medium produces displacements $\mathring{u}_i$ and stresses $\mathring{\sigma}_{ij}$, (Fig.2a). Let us find the stress distribution along the x_1-axis. Owing to the symmetry of loading this distribution is given by

$$\mathring{\sigma}_{22}(x_1,0) = p(x_1)\ ,\ \mathring{\sigma}_{12}(x_1,0) = 0 \quad \text{for} \quad |x_1|<a$$

II. The medium under considerations contains a crack $\{|x_1|\le a, x_2=0\}$ and let $\overset{1}{u}_i$, $\overset{1}{\sigma}_{ij}$ denote the respective displacements and stresses in the medium, (Fig.2b). In absence of external loads, stresses should vanish at infinity, and on the crack surfaces the following conditions should be satisfied:

$$\overset{1}{\sigma}_{22}(x_1,0) = -p(x_1)\ ,\ \overset{1}{\sigma}_{12}(x_1,0) = 0 \quad \text{for } |x_1|<a$$

III. Let us assume again that the medium contains a crack $\{|x_1|\le a, x_2=0\}$ and let $\overset{2}{u}_i$, $\overset{2}{\sigma}_{ij}$ denote the respective displacements and stresses in the medium. In absence of external loads, stresses should vanish at infinity, and on the crack surfaces the following conditions should be satisfied:

$$\overset{2}{\sigma}_{22}(x_1,0) = g(x_1)\ ,\ \overset{2}{\sigma}_{12}(x_1,0) = 0 \quad \text{for } |x_1|<a$$

Here function $g(x_1)$ represents the forces of cohesion and thus it must be symmetric in x_1.

Assuming the displacement and stress distribution in the body without a crack as well as the cohesive forces distribution along the crack surfaces to be known, displacements and stresses on the surfaces of the crack or along its extension in Problems II and III are given by the corresponding relations (1.2). In the considerations to follow, asymptotic expressions describing the distribution of normal displacement of the surface and tensile stresses along crack's extensions will be sufficient.

Using relations (1.35) complemented by regular terms, displacement u_2 and stress σ_{22} in the vicinity of the crack tip are given by:
In Problem II

$$\begin{aligned} \overset{1}{u}_2(x_1,0) &= \frac{\kappa+1}{2\mu\sqrt{2\pi}} K_I\sqrt{a-x_1} + O[(a-x_1)^{3/2}] \\ \overset{1}{\sigma}_{22}(x_1,0) &= \frac{K_I}{\sqrt{2\pi}}\frac{1}{\sqrt{x_1-a}} - p(a) + O[(x_1-a)^{1/2}] \end{aligned} \tag{4.1}$$

In Problem III

$$\begin{aligned} \overset{2}{u}_2(x_1,0) &= \frac{\kappa+1}{2\mu\sqrt{2\pi}} K_c\sqrt{a-x_1} + O[(a-x_1)^{3/2}] \\ \overset{2}{\sigma}_{22}(x_1,0) &= \frac{K_c}{\sqrt{2\pi}}\frac{1}{\sqrt{x_1-a}} + g(a) + O[(x_1-a)^{1/2}] \end{aligned} \tag{4.2}$$

Here symbol $O(s^p)$ denotes the power series expansion starting from s^p, K_I and K_c are the stress intensity factors at the crack tips in the respective problems II and III; due to the symmetry of functions $p(x_1)$ and $g(x_1)$, the left and right-hand stress intensity factors are identical and according to (1.32) have the form

$$K_I = 2\sqrt{\frac{a}{\pi}}\int_0^a \frac{p(t)}{\sqrt{a^2-t^2}}\,dt, \quad K_c = -2\sqrt{\frac{a}{\pi}}\int_0^a \frac{g(t)}{\sqrt{a^2-t^2}}\,dt \tag{4.3}$$

In order to determine the displacement u and stress σ_{22} in the vicinity of the right-hand crack's tip corresponding to the initial problem (medium with a stress free crack), the expressions describing the displacement u_2 and stress σ_{22} around the tips in all three cases I, II, III must be added together.

Before summing up the solutions let us note that in Problem I (due to the symmetry of loading), displacement $\mathring{u}_2(x_1,0)=0$ and stress $\mathring{\sigma}_{22}(x_1,0)=p(x_1)$ is a continuous and differentiable function; its expansion into a power series in the neighbourhood of point $x_1=a$ is written in the form

$$\mathring{\sigma}_{22}(x_1,0) = p(a) + O(x_1-a)$$

Taking this into account, displacement u_2 and stress σ_{22} close to the crack tip due to the initial problem are evaluated by means of Eqs. (4.1), (4.2) and have the form

$$\begin{aligned} u_2(x_1,0) &= \frac{\kappa+1}{2\mu\sqrt{2\pi}}(K_I + K_c)\sqrt{a-x_1} + O[(a-x_1)^{3/2}] \\ \sigma_{22}(x_1,0) &= \frac{K_I + K_c}{\sqrt{2\pi}}\frac{1}{\sqrt{x_1-a}} + g(a) + O[(x_1-a)^{1/2}] \end{aligned} \tag{4.4}$$

From the results obtained it follows that in order to eliminate the σ_{22} - stress singularity at $x_1=a$, i.e. to secure finite values of the stress at the crack tip, the sum of both factors K_I and K_c must vanish, $K_I+K_c=0$; this means that the stress intensity factor produced by real loading of the body should be equal to the opposite value of the stress intensity factor produced by the forces of cohesion.

It directly follows from Eq. $(4.3)_2$ that the stress intensity factor K_c depends on the length of the crack (different lengths would lead to different K_c values). In order to eliminate this physical inconsistency, Barenblatt introduced two additional hypotheses on the basis of which factor K_c was found to represent a new physical constant, independent of the parameters characterizing the crack.

According to one of them, cohesive forces $g(x_1)$ act within a very small region of the crack, at a distance d from the tips. Consequently, identical forces of cohesion act at both the tips,

$$g(x_1) = \begin{cases} 0 & \text{for } 0<x_1<a-d \\ g(x_1) & \text{for } a-d<x_1<a \end{cases}$$

under the additional assumption that $d/a \ll 1$.

Relation $(4.3)_2$ yields now the following value of the stress intensity factor K_c:

$$K_c = -2\sqrt{\frac{a}{\pi}}\int_0^d \frac{G(t)}{\sqrt{t(2a-t)}}\,dt\,, \qquad G(t)=g(a-t)$$

Making use of the assumption that parameter $d/a \ll 1$, factor K_c may be written in the form

$$K_c = -\sqrt{\frac{2}{\pi}}\int_0^d \frac{G(t)}{\sqrt{t}}\,dt \tag{4.5}$$

It follows that K_c depends solely on d and on the cohesive forces distribution, provided d is small enough.

Assume now that the real load increases gradually to reach its final value. During such a process, the forces of cohesion which exist from the very beginning in small regions surrounding the crack tips (inter-molecular forces) also start to increase and counteract the tendency of the crack to open. The forces are assumed to reach a certain limiting value at which the crack starts to propagate. Denote this limiting value of cohesive forces $G(x_1)$ by $G_m(x_1)$, and the corresponding value of factor K_c by K_{cm}; from Eq. (4.5) it follows that

$$K_{cm} = -\sqrt{\frac{2}{\pi}}\,K$$

where K is defined by the formula

$$K = \int_0^d \frac{G_m(t)}{\sqrt{t}}\,dt \tag{4.6}$$

According to other hypothesis (to be formulated later), coefficient K represents a new material constant and is called by Barenblatt the modulus of cohesion.

From the relation $(4.4)_2$ it followed that the necessary condition for the stress σ_{22} to assume in the initial problem a finite value at the crack tip was the condition

$$K_I = -K_c \tag{4.7}$$

Let us note that gradual increase of load produces increasing values of the stress intensity factor K_I and of the cohesion forces which counteract the process of crack opening. Increasing forces of cohesion result in a simultaneous growth of the coefficient $-K_c$, so that condition (4.7) is satisfied at every stage of the process considered.

On the other hand, coefficient $-K_c$ cannot exceed the value $-K_{cm} = \sqrt{2/\pi}\ K$ since then the forces of cohesion would reach their maximum value, above which fracture occurs. It follows that Barenblatt's approach to the criterion of stability loss may be written in the form

$$\frac{1}{\sqrt{2\pi}} K_I = \frac{1}{\pi} K \tag{4.8}$$

Fulfillment of the above condition determines the critical state of the system with a crack (provided the stress intensity factor K_I is known), characterized by such critical parameters of the system as the load or crack's length; once the critical values are reached, the crack loses its stability.

Before approaching the formulation of the last Barenblatt's hypothesis (concerning the notion of cohesion modulus introduced above) let us observe that if the condition (4.7) is satisfied during the process of crack opening, Eq. $(4.4)_1$ yields the conclusion that displacement u_2 at the crack tip is of the order of $(a-x_1)^{3/2}$. This means that the crack surfaces close smoothly at the tips forming a sharp edge, and tangents to the crack surfaces at the tip become horizontal. For $a-d<x_1<a$, Eq. $(4.4)_1$ characterizes the form of a normal cross-section of the crack in the vicinity of its tips.

The last Barenblatt's hypothesis concerns the shape of cross-section of the deformed crack in the region surrounding the tips and may be formulated as following: the shape of the normal cross-section of a crack at its tip (as well as the local distribution of cohesion forces between the crack surfaces) in the critical state is independent of the loads applied to the body and is the same for a given type of material and under given physical conditions.

This hypothesis implies that during the process of crack propagation the tips of the crack are displaced but their shapes remain unchanged. Obviously such a situation is possible only if at least at one point of crack's surface the cohesion forces reach the maximum value (length d of the cohesion zone is independent of the loads applied to the body); in that sense the coefficient K defined by (4.6) may be treated as a new material constant.

On comparing the relations (4.8) with (3.10) (Irwin's formulation of the crack stability criterion) it is seen that constant K plays in the fracture process (Barenblatt's approach) the same role as the specific surface energy in the approach presented by Griffith. Detailed analysis of both approaches leads to a conclusion that the assumption

$$K = \sqrt{\frac{\pi}{2}} K_{Ic}$$

renders both the results equivalent.

5. DUGDALE'S MODEL OF A CRACK.

Another attempt to make the classical solution of the problem of stress distribution around the crack more realistic was made by D. S. Dugdale [25], who assumed the existence of narrow conical plastic zones at the front of the crack. The problem is reduced to purely elastic one by assuming uniform plastic stress distribution within the zones.

In order to formulate the basic ideas resulting from Dugdale's concept, consider a problem of an unbounded elastic medium containing a stress free crack $\{|x_1|\leq\ell, x_2=0\}$ loaded by external tractions $\underline{T}$, symmetric in x_1 and x_2; the crack is subject to Mode I deformation, (Fig. 2a). Moreover, assume at both crack tips the existence of two narrow plastic zones of lengths d, characterized by uniform distribution of tensile stresses, equal to the yield limit σ_T and entirely independent of the external load. Outside the zones the material is assumed to be elastic, but the stresses acting along the hypothetical crack of length $2a=2(\ell+d)$ cannot exceed the yield limit σ_T. Under such assumptions Dugdale's model is reduced to purely elastic problem of a crack of length 2a with an unknown parameter d representing the length of plastic zones, (Fig. 11); this parameter may be determined from the condition of finite stresses at the tips of the crack of length 2a

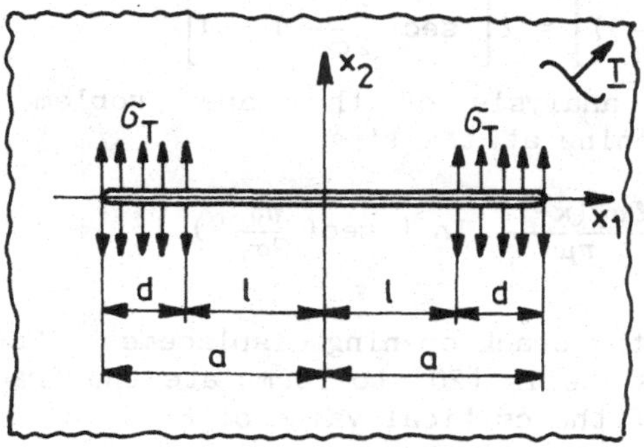

Fig. 11.

On repeating the reasoning used in formulating Barenblatt's criterion, solution of the problem stated may be represented in the form of the sum of solutions to the Problems I and II formulated in the preceding section; the boundary conditions of the problem II must now be replaced with the conditions

$$\overset{2}{\sigma}_{22}(x_1,0) = \begin{cases} -p(x_1) + \sigma_T & \text{for } -a<x_1<-\ell \\ -p(x_1) & \text{for } |x_1|<\ell \\ -p(x_1) + \sigma_T & \text{for } \ell<x_1<a \end{cases}$$

$$\overset{2}{\sigma}_{12}(x_1,0) = 0 \qquad \text{for } |x_1|<a$$

Displacement $\overset{2}{u}_2$ and stress $\overset{2}{\sigma}_{22}$ in the neighbourhood of the hypothetical crack of length 2a are now given by Eqs. (4.1), the stress intensity factors at the tips being found (cf. Eq. (4.3)$_1$) from the formula

$$K_I = 2\sqrt{\frac{a}{\pi}}\left[\int_0^a \frac{p(t)}{\sqrt{a^2-t^2}}\,dt - \sigma_T \operatorname{arc}(\cos\frac{\ell}{a})\right] \qquad (5.1)$$

where $a=\ell+d$.

Dugdale's postulate concerning finite values of stresses at the tips of the crack of length 2a is equivalent, as it follows from Eq. (4.1)$_2$, to the demand that the stress intensity factor given by

Eq. (5.1) should vanish, $K_I=0$, that is

$$\int_0^{\ell+d} \frac{p(t)}{\sqrt{(\ell+d)^2 - t^2}}\, dt - \sigma_T \text{arc}\left[\cos\left(\frac{\ell}{\ell+d}\right)\right] = 0 \qquad (5.2)$$

Solution of the Eq. (5.2) for the unknown parameter d yields the length of plastic zone at the front of the actual crack of length 2ℓ.

In the case when the unbounded elastic medium containing a crack $\{|x_1|\le\ell, x_2=0\}$ is subject to tensile stresses due to uniformly distributed loads applied at infinity and perpendicular to the crack surfaces, i.e. when $p(x_1)=\sigma$=const. then, on the basis of Eq. (5.2), length d of the plastic zones are found from the formula

$$d = a\left[1 - \cos\left(\frac{\pi\sigma}{2\sigma_T}\right)\right] = \ell\left[\sec\left(\frac{\pi\sigma}{2\sigma_T}\right) - 1\right]$$

A more detailed analysis of the same problem enables us to determine the crack opening at the tips,

$$\delta = [u_2(\pm\ell, 0)] = \frac{\ell\sigma_T(\kappa+1)}{\pi\mu} \ln\left[\sec\left(\frac{\pi\sigma}{2\sigma_T}\right)\right]$$

Parameter δ is called the crack opening displacement (COD) and the idea of it was used by A. A. Wells [26] to formulate the fracture criterion based on the concept of the critical value of δ.

REFERENCES:

1. Panasyuk, V. V., Savruk M. P. and Datsyshin A. P. : Stress distribution around cracks in plates and shells, (in Russian), Naukova Dumka, Kiev 1976.
2. Panasyuk, V.V., Savruk, M. P. and Datsyshin, A. P. : A general methods of solution of two-dimensional problems in theory of cracks, Engineering Fracture Mechanics, 9, (1977), 481-497.
3. Muskhelishvili, N. I.: Some basic problems of the mathematical theory of elasticity, Noordhoff, The Netherlands 1953.
4. Muskhelishvili, N. I.: Singular integral equations, Noordhoff, Groningen, The Netherlands 1953.
5. Sneddon, I. N. Lowengrub, M.: Crack problems in the classical theory of elasticity, John Wiley & Sons, Inc., New York, London, Sydney, Toronto 1969.
6. Methods of analysis and solutions of crack problems, in: Mechanics of fracture 1 (Ed. G. C. Sih), Noordhoff International Publishing, Leyden 1973.
7. Kassir, M. K. and Sih, G. C.: Three dimensional crack problems, in: Mechanics of fracture 2, (Ed. G. C. Sih), Noordhoff International Publishing, Leyden 1975.
8. Hahn, H. G.: Bruchmechanik, B. G. Teubner, Stuttgart 1976.
9. Cherepanov, G. P.: Mechanics of brittle fracture, McGraw-Hill International Book Company, New York 1979.

10. Tada, H., Paris, P. C. and Irwin, G. R.: The stress analysis of cracks handbook, Del Research Corporation, Hellertown, PA 1973.
11. Sih, G. C.: Handbook of stress intensity factors, Institute of Fracture and Solid Mechanics, Lehigh University, Bethlehem, PA 1973.
12. Rooke, D. P. and Cartwright, D. G.: Compendium of stress intensity factors, London, Hillingdon Press, Uxbridge 1976.
13. Eshelby, J. D., The continuum theory of lattice defects, in: Solid State Physics vol.3 (Ed. F. Sietz and D. Turnbull), Academic Press, New York, 1956.
14. Sanders, J. L.: On the Griffith-Irwin fracture theory, Journal of Applied Mechanics, 27, (1960), 352-353.
15. Rice, J. R.: A path independent integral and the approximate analysis of strain concentration by notches and cracks, Journal of Applied Mechanics, 35, (1968), 379-386.
16. Günther, W.: Über einige Randintegrale der Elastomechanik Abhandlungen, Braunschweiger Wissenschaftliche Gesellschaft, 14, (1962), 53-72.
17. Knowles, J. K. and Sternberg, E.: On a class of conservation laws in linearized and finite elasticity, Archive of Rational Mechanics and Analysis, 44, (1972), 187-211.
18. Matczyński, M., Sokołowski, M. and Zorski H.: Forces and moments on distributed defects and cracks, in: Defects and fracture (Ed. by Sih, G. C. and Zorski, H.), Martinus Nijhoff Publisher, The Hague, Boston, 1982, 109-119.
19. Griffith, A. A.: The phenomena of rupture and flow in solids, Philosophical Transactions, Royal Society of London, Series A 221, (1921), 163-198.
20. Irwin, G. R.: Analysis of stress and strains near the ends of a crack traversing a plate, Journal of Applied Mechanics, Trans. ASME, 24, (1957), 361-364.
21. Irwin, G. R.: Fracture dynamics, in: Fracturing of metals (Ed. F. Jonassen et. al.), American Society of Metals, Cleveland, 1947, 147-166.
22. Orowan, E.: Fracture and strength of solids, Report on Progress in Physics, Physics Society, London, 12, (1949), 185-232.
23. Barenblatt, G. I.: The formation of equilibrium crack during brittle fracture, Journal of Applied Mathematics and Mechanics, (Prikl. Mat. Mech.), 23, (1959), 434-444.
24. Barenblatt, G. I.: Mathematical theory of equilibrium cracks in brittle fracture, Advances in Applied Mathematics, VII, Academic Press, New York, (1962), 55-129.
25. Dugdale, D. S.: Yielding of steel sheets containing slits, Journal of the Mechanics and Physics of Solids, 8, (1960), 100-104.
26. Wells, A. A.: Unstable crack propagation in metals, cleavage and fast fracture, in: Proceedings of the crack propagation Symposium, Cranfield 1962 / England, 1962, 210-230.

ELASTO-PLASTIC FATIGUE CRACK GROWTH:
MATHEMATICAL MODELS AND EXPERIMENTAL EVIDENCE

J. C. Radon
Imperial College of Science and Technology, London, U.K.

Abstract

This paper describes some analytical models developed for the evaluation of the cyclic J-integral, ΔJ. The J-integral is usually obtained through involved graphical procedures with experimental load-load point displacement plots. Numerical and other procedures are even more time consuming. For typical specimens, such as C(T), TPB and DCB (double cantilever beam) the graphical procedures are much simpler. An easily applicable analytical formula for the cyclic J-integral, by means of which the integral could be calculated and which would be comprehensive enough to be applicable for the conditions of fatigue and creep would be particularly helpful. Such a formula may overcome the shortcomings of the graphical and other procedures which are laborious and time consuming, apart from involving considerable experimental work.

For developing the analytical model, the two halves of a specimen are considered as cantilevers built-in at the crack tip. Using non-linear beam theory the change in the distance between loading pins due to the elastic deformation, plastic deformation, cyclic creep and high temperature creep is evaluated. From this the load-displacement plot is analytically constructed. This gives the change in strain energy of the specimen after a number of cycles and, subsequently, ΔJ. Conditions for load cycling, strain cycling and cycling with incorporated dwell time are also considered.

Introduction

The successful application of linear elastic fracture mechanics (LEFM) analysis in the description of the brittle fracture behaviour of metals under monotonically increasing loads has led to a logical extension to the study of slow crack growth under both static and fatigue loads, commonly termed subcritical crack growth. Subcritical crack growth covers all processes in which a crack slowly propagates at a K level below the critical value K_c corresponding to unstable and catastrophic failure. These may be stress corrosion cracking (SCC), creep cracking, cyclic creep, fatigue crack propagation or any combination of these phenomena. Here, only the fracture mechanics approach to fatigue crack propagation will be considered.

It is known that the stress intensity factor, K, or the strain energy release rate, G, govern the intensity or magnitude of the stress field around the crack tip. Since crack extension occurs in this highly stressed region, it is reasonable to assume that the intensity of the crack tip stress field as represented by K should control the rate of crack extension. The cyclic crack growth rate, da/dN, may be a function of the range of the stress intensity factor, ΔK, in the loading cycle. ΔK is given by $\Delta K = K_{max} - K_{min}$, and $K_{max} = \sigma_{max} Y \sqrt{\pi a}$ and $K_{min} = \sigma_{min} Y \sqrt{\pi a}$, where K_{max} and σ_{max} are the maximum stress intensity factor and the maximum gross stress, respectively, and K_{min} and σ_{min} are the minimum values of the above variables in the loading cycle. The above expressions are the general equations for the stress intensity factor under mode I conditions, where a is the crack length and Y is a dimensionless factor, a function of the crack length and specimen geometry.

Cyclic crack growth (da/dN) will, of course, be dependent on many other variables, such as environment, frequency, mean stress, loading history, etc. The applicability of LEFM to the fatigue crack propagation problem is limited by the amount of plastic deformation in the specimen, component or in a structure. The specimen should be under linear elastic stresses and strains, apart from a very small plastic zone at the crack tip. However, there is some evidence, supported by the present work, that non-linear fracture mechanics can be successfully used to analyse fatigue crack growth data under elastic-plastic conditions.

Fatigue crack growth theories are numerous. A comprehensive description of the more relevant ones is available in the literature. Early theories did not use fracture mechanics analysis.

One of the first crack propagation laws to draw wide attention was proposed by Head. He employed a mechanical model which considered rigid-plastic work-hardening elements ahead of a crack tip and elastic elements over the remainder of the infinite sheet. The final equation obtained was:

$$\frac{da}{dN} = \frac{C\,\sigma^3\,a^{3/2}}{(\sigma_{ys} - \sigma)\,w_0^{1/2}} \tag{1}$$

where w_0 is the size of the plastic zone assumed constant during the crack propagation process, and σ_{ys} is the yield stress. C, m, n and q in the above equation and in other equations presented below are constants obtained experimentally and relate to any particular material investigated.

Many other equations of similar type to equation (1) have been constructed and are still in use.

It is usual to divide fatigue crack growth into three régimes or stages, as originally suggested by Forsyth (Figures 1 and 2).

Stage I includes nucleation and the early growth of fatigue cracks and is influenced by microstructural features and surface roughness. In this stage, the crack is short compared to both the plastic zone size and to microstructural units, and is therefore not readily modelled in terms of LEFM. Short fatigue cracks have been a major research topic during the last ten years, and some aspects of short crack behaviour will be discussed in a subsequent section. In stages II and III, the

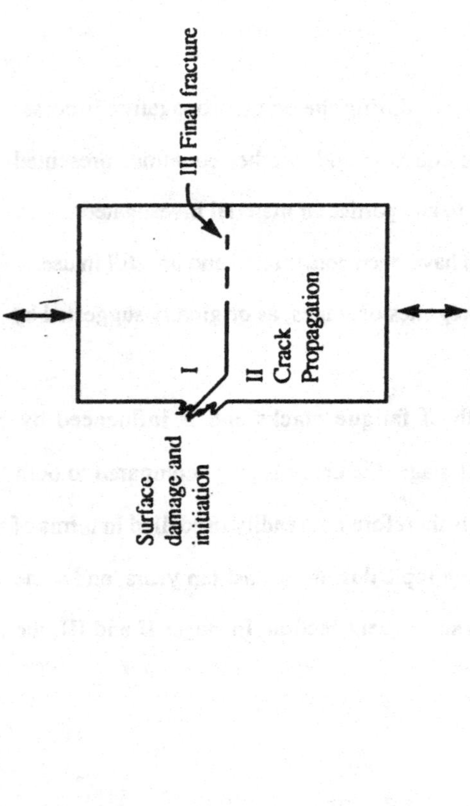

Fig. 1a. Schematic diagram of the different stagesin the fatigue failure process

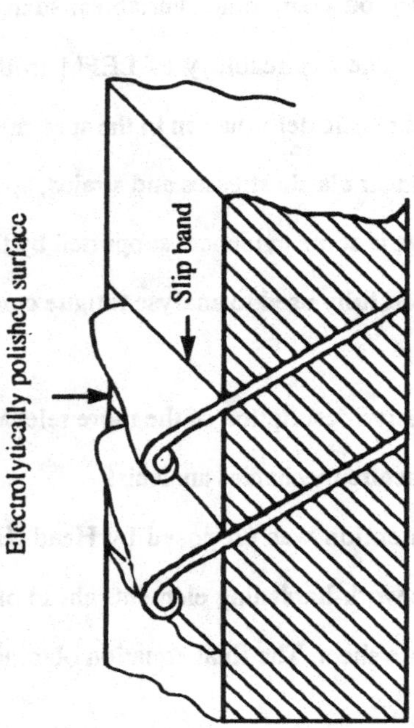

Fig. 1b. Slip-band extrusions

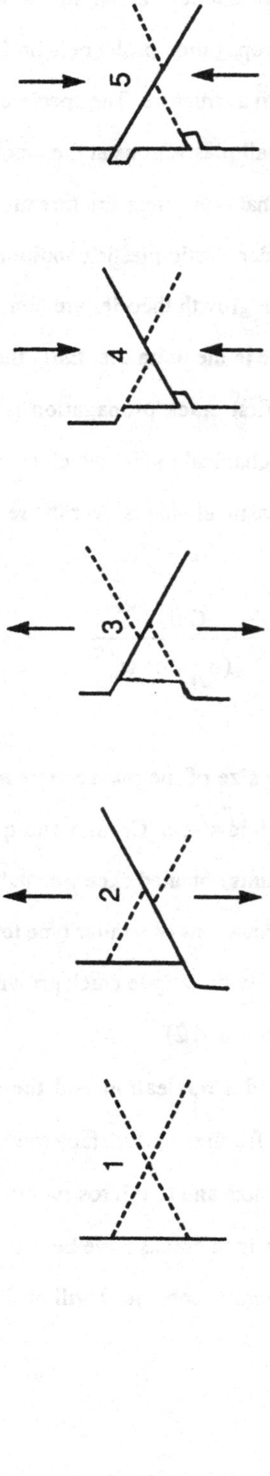

Fig. 1c. Cottrell and Hull's model for producing extrusions and intrusions

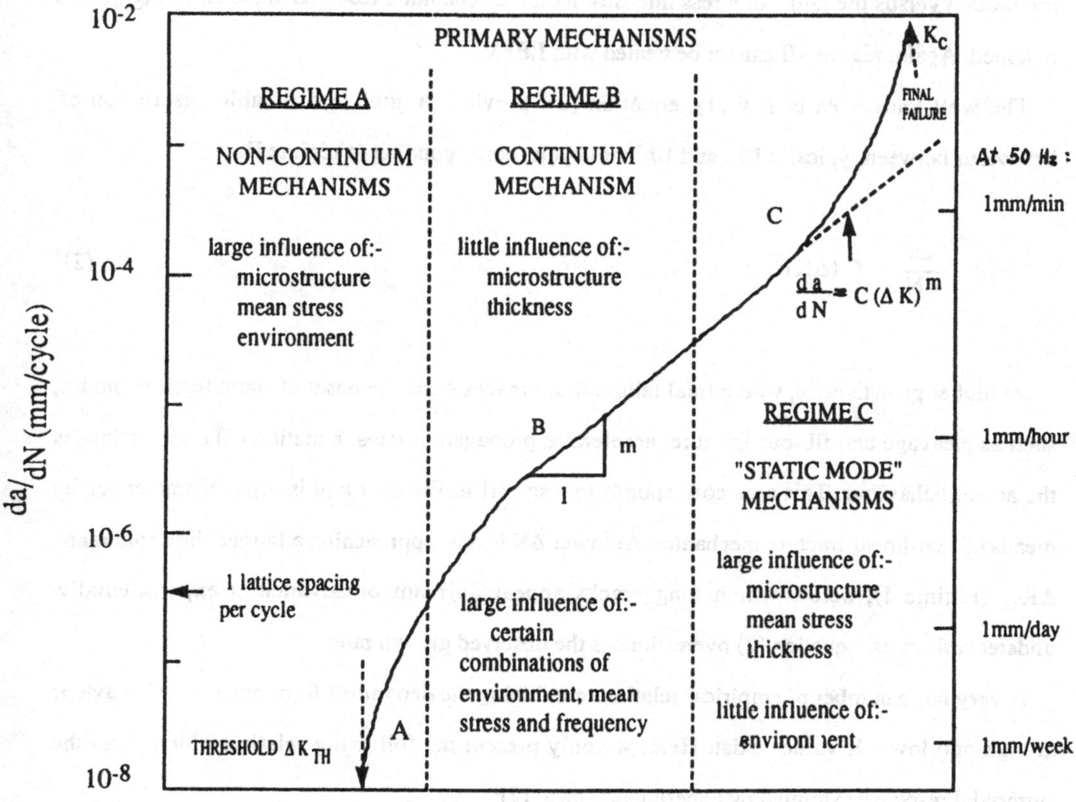

Fig. 2. The three regions of fatigue crack propagation behaviour and their corresponding mechanisms and characteristics.

crack is long enough to allow a continuum treatment. When measured fatigue crack growth rates are plotted versus the range of stress intensity factor, a schematic result as depicted in Figure 2 is obtained. Again, régime III cannot be treated with LEFM.

The well known Paris' law [1], equation (2), provides a good and reliable description of behaviour between typically 10^{-6} and 10^{-3} mm/cycle, corresponding to régime II:

$$\frac{da}{dN} = C\,(\Delta K)^{m} \tag{2}$$

At higher growth rates, where final failure is approached and the onset of static fracture modes, such as cleavage and fibrous fracture, accelerates propagation rates, equation (2) underestimates the actual behaviour. This area corresponds to stage III in Figure 1 and is often characterised by means of non-linear fracture mechanics. At lower ΔK levels, approaching a fatigue threshold range ΔK_{th} (régime I), below which long cracks appear dormant or advance at experimentally undetectable rates, equation (2) overestimates the observed growth rates.

A very large number of empirical relations combining the departures from power law behaviour at high and low ΔK values exist. Here, we only present the following relation which gives the sigmoidal response exhibited by experimental data, [2]:

$$\frac{1}{(da/dN)} = \frac{A_1(R)}{(\Delta K)^{n_1}} + A_2(R)\left\{\frac{1}{(\Delta K)^{n_2}} - \frac{1}{[K_c\,(1-R)]^{n_2}}\right\} \tag{3}$$

where $R = \sigma_{min}/\sigma_{max}$ is the stress ratio, $A_1(R)$ and $A_2(R)$ are functions which control the load ratio dependencies in régimes A and B in Figure 2, respectively, and n_1, n_2 and K_c are constants that can be obtained experimentally. The three terms in equation (3) correspond to the three régimes in Figure 2; transition régimes are modelled by a combination of two adjacent terms. The exponents n_1 and n_2 can be estimated from the slopes derived from a plot of the data in régimes A and B, respectively. The value of K_c, which is a fitting parameter to characterise the onset of instability, is also determined from a plot of the data. K_c should not be confused with the fracture toughness, K_{Ic}. As indicated by equation (3), the R-value dependence of da/dN differs for each of the three growth rate régimes. The effect is very strong in the threshold régime and at instability but rather weak in régime B. The general effect of R is to increase the growth rates at increased R-values.

A recent and more detailed description of empirical relations was given by Blom [3], who divided current fatigue research into studies of crack growth mechanisms and modelling methods of the actual propagation régime. In the following, attention will be concentrated on the growth under elasto-plastic conditions. In order to obtain a satisfactory prediction of the fatigue crack growth (FCG), other parameters, such as closure, residual stress, random loads, compressive stresses, size effects and geometries of short cracks, will have to be considered.

Fatigue Crack Growth at Large Plastic Strains

Fatigue crack growth can result from cycling with the maximum stress above or below the gross yield stress. The latter problem has been handled by applying LEFM and has been briefly reviewed in the previous section. With the approach of gross yielding, LEFM techniques are no longer applicable. There have been only a few studies of this aspect of fatigue crack propagation despite the engineering significance of fatigue crack growth resulting from plastic and strain cycling (low cycle fatigue crack propagation).

One of the first research studies in this field was performed by Laird and Smith [4]. Their observations on the nature of crack tip blunting and resharpening provided a better understanding of the low cycle fatigue crack propagation phenomenon. Later studies conducted by Crooker and Lange [5] have shown that for plate bend specimens containing an embedded surface flaw, the crack growth rate could be described by:

$$\frac{da}{dN} = C\,(\Delta\varepsilon_t)^q \tag{4}$$

The above equation described crack propagation under both elastic and plastic strain cycling.

McEvily [6] studied the low cycle fatigue of copper and used a strain concentration factor to describe his results in the form:

$$\frac{da}{dN} = C\,(\Delta\varepsilon_t \sqrt{a}\,)^q \qquad (5)$$

The parameter $\Delta\varepsilon_t\sqrt{a}$ was termed the strain-intensity factor.

Solomon [7] performed low cycle fatigue crack propagation tests in a low carbon steel, keeping the plastic strain range constant. The relationship obtained was:

$$\frac{da}{dN} = C\,a\,(\Delta\varepsilon_p)^q \qquad (6)$$

Tomkins [8] has proposed a model for crack propagation under the fully plastic situation of low cycle fatigue. The crack growth rate was expressed as a function of the cyclic strength coefficient K' and the cyclic strain hardening exponent n' in the form:

$$\frac{da}{dN} = \frac{\pi^2}{8\sqrt{2}}\left(\frac{K'}{2\sigma_{UTS}}\right)\frac{2\Delta\varepsilon_p^{(2n'+1)}}{(2n'+1)}\,a \qquad (7)$$

It can be seen that equation (7) has the same form as equation (5) and, after integration between the initial crack length and the final crack length for unstable fracture, gives the Coffin-Manson law.

Dover [9] reported results on crack growth rates in mild steel under constant COD cycling, and found that the crack growth rate was a power function of the hysteresis loop width and of the overall CTOD range.

The cyclic process of fatigue presents certain conceptual difficulties when analysed by the deformation theory of plasticity, on which the J-integral is based.

Irreversible deformation occurring during the unloading portion of a cycle makes application of the J-integral to fatigue theoretically invalid since the integral will no longer be path independent. As a consequence, several authors have developed arguments which attempt to suggest path independence during fatigue. In general, if one assumes that crack growth occurs only during the loading portion of the fatigue cycle, then it can be argued that damage occurring during unloading will automatically be reflected in the subsequent loading cycle. As a consequence, J might then be extended to fatigue where U is defined during the loading part of the fatigue cycle. Another argument proposed is that the material in the plastic zone has reached a stable condition represented by saturated hysteresis loops due to cyclic loading. Thus, there is no history-dependent behaviour. Yet another proposal is based on the assumption that each subsequent cycle is performed on a new, virgin specimen. This method will be discussed later.

Shortly after the first proposals by McEvily and Tomkins to implicate the strain range, Dowling and Begley [10] attempted analysis of fatigue data by using ΔJ instead of ΔK. Their tests were conducted under conditions of general yielding which gave a potential for unstable crack growth due to ratcheting under load control or a decrease in crack growth due to relaxation under deflection control. Consequently, tests were run under both deflection control and load control to a sloping line using compact tension specimens. At selected intervals during the tests, load versus deflection lines were plotted as hysteresis loops. Cyclic J values were then determined from areas under the rising portion of the load deflection curve, according to the earlier approximation proposed by Rice and given in equation (8), where B is the specimen thickness and b is the remaining uncracked ligament. The integral is represented as follows:

$$\Delta J = \frac{2}{B\,b}\int_0^{\delta_0} P\,d\delta \tag{8}$$

Equation (8) was obtained only for deeply cracked bend bars but is valid also for deep cracked compact tension specimens which are essentially bend type specimens. Dowling and Begley also chose to correct for crack closure by using only that portion of the area above a certain inflection point on the load-deflection curve. Note that ΔJ is related to ΔK in the linear elastic region, as shown in equation (9):

$$\Delta J = \frac{\Delta K^2}{E} \tag{9}$$

The use of ΔJ has significantly reduced scatter in the data and resulted in the extension of régime II or elimination of régime III. In short, the usual régime III behaviour has been mapped into a straight line, continuing régime II into the region where LEFM is not normally valid. Efforts by the authors to correlate the data by the usual linear elastic parameter, ΔK, proved unsuccessful. Dowling and Begley were unable to obtain any satisfactory correlation of data obtained for tests conducted under load control which was attributed to the influence of a mean value of J for which they could not account. They reasoned that crack growth occurs during the loading and evaluated the increase in J, that is ΔJ, for the loading portion of cycles on elastic-plastic specimens. Their tests were carried out under decreasing load conditions on compact tension and centre cracked specimens of A533-B steel. They then plotted the data in terms of da/dN versus ΔJ and found that the crack growth rates were related to ΔJ as:

$$\frac{da}{dN} = C\,(\Delta J)^m \tag{10}$$

At the same time, Branco [11] reported elastic-plastic crack growth data for BS15 mild steel and obtained a similar correlation, although his ΔJ values were determined under constant load cycling conditions. The procedure used in these tests will be discussed in detail at a later stage.

The J-integral has also been applied in analysing crack initiation and growth at notches. Lamba [12] showed that the J-integral, modified for cyclic loading, ΔJ, yields the same form for the strain concentration at a notch as the Neuber rule modified for the cyclic loading and concluded that the

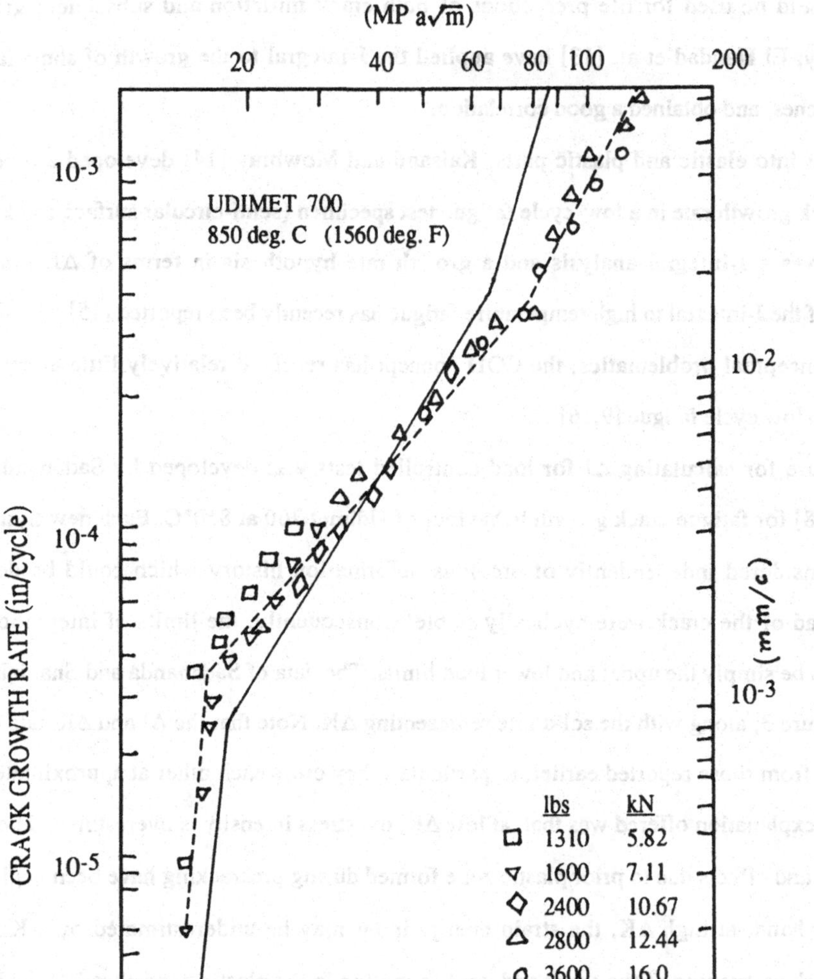

Fig.3. Fatigue crack growth data for Udimet 700 in terms of both ΔJ and ΔK (after Sadananda and Shahinian)

J-integral could be used for life predictions of both crack initiation and subsequent growth. Subsequently, El Haddad et al. [13] have applied the J-integral to the growth of short fatigue cracks at notches, and obtained a good correlation.

Dividing J into elastic and plastic parts, Kaisand and Mowbray [14] developed a model to describe crack growth rate in a low-cycle fatigue test specimen (semi-circular surface crack). The model involves a J-integral analysis and a growth rate hypothesis in terms of ΔJ. Another application of the J-integral to high temperature fatigue has recently been reported [15].

Due to conceptual problematics, the COD concept has received relatively little attention in application to low cycle fatigue [9,16].

A procedure for calculating ΔJ for load controlled tests was developed by Sadananda and Shahinian [18] for fatigue crack growth behaviour of Udimet 700 at 850°C. Each new hysteresis loop was considered independently of previous deformation history which could be done if material ahead of the crack were cyclically stable. Consequently, the limits of integration are considered to be simply the upper and lower load limits. The data of Sadananda and Shahinian are shown in Figure 3, along with the solid line representing ΔK. Note that the ΔJ and ΔK data differ substantially from those reported earlier, in particular, they cross each other at approximately 40 MPa√m. An explanation offered was that, at low ΔK, the stress intensity is overestimated because notch effects and effects due to prior plastic zone formed during precracking have been neglected. On the other hand, at high ΔK, the strain energy input may be underestimated by ΔK since processes, such as opening of the crack and void formation in the plastic zone ahead of the crack, are not included in the calculation of the stress intensity factor. By contrast, ΔJ is determined from the rising portion of the cyclic strain loop which is just the strain energy density input. As a consequence, ΔJ may be considered to represent, in a more realistic manner, the processes going on ahead of the crack.

Shao-Lun and Ji-Zhou [19] have studied a steel alloy using a wedge opening loading specimen and found results similar to Sadananda et al. in that there was a crossover between the lines for da/dN versus ΔK and ΔJ. However, Shao-Lun et al. suggest that the crossover is a result of the approximation used to calculate ΔJ and not a real phenomenon. It is suggested the error occurs in the LEFM region of $a/W < 0.53$, where the ΔJ approximation was used instead of the usual relation shown in equation (9).

Solomon [7,27] measured fatigue crack growth in 1018 steel under conditions of general yielding in cylindrical specimens which had been machined with a reduced section in one plane. Six different strain ranges were used ($\Delta\varepsilon_p = 0.001$ to 0.05). Test results were analysed in terms of:

$$\Delta(PK) = E\,(\Delta\varepsilon)\sqrt{a} \tag{11}$$

where $\Delta\varepsilon$ represents the total strain range and (PK) represents a pseudo stress-intensity. Minzhong and Liu [20] have recently analysed Solomon's data again in terms of ΔJ by analytically determining J for the specific specimen geometry using a finite element program. Because of Solomon's specimen geometry, Minzhong and Liu were able to assume plane stress loading. The finite element model gave results which were in excellent agreement with the experimental data and Minzhong and Liu conclude that fatigue crack growth correlates well with ΔJ, at least under plane stress conditions.

German, Delorenzi and Wilkening [21] also used a finite element technique to investigate cyclic versus monotonic J and conclude that cyclic behaviour shows a considerably smaller inelastic contribution than monotonic deformation. They find that ΔJ fully defines the damage done at the crack tip, but there is a distinct difference between load and deflection controlled tests which would not show up in monotonic J. At the present time, their model does not include crack closure.

El Haddad and Mukherjee [22], Tanaka, Hoshide and Nakata, and other workers [17,27] followed procedures similar to those of Dowling to evaluate J and correlate ΔJ with da/dN.

Low Cycle Fatigue (LCF) and Life Prediction

Apart from the work of Dowling and Begley [10] and Branco [11], Mowbray [24] and others developed simple models of fatigue crack growth using cylindrical specimens, and El Haddad [13] proposed a new concept of an effective crack length (now in general use), suggesting it could also be applied for very short cracks. This method was subsequently used by Bicego [25] for LCF life prediction. It is now accepted that very small cracks could be divided into two groups: microstructurally short cracks and mechanically small cracks. At low loads, i.e. in high cycle fatigue (HCF), microstructurally short cracks represent the major part of the life of a component, while mechanically small cracks are important at high loads, i.e. in the régime of LCF. Bicego analysed small fatigue crack growth in three different alloys. El Haddad's method was found particularly useful on the steel AISI304, for which Westinghouse and General Electric experimental data were available. Time effects, such as frequency changes or hold times, have not been included in the analysis and it is expected that some additional parameters, such as C^* or $\dot{J}$, would have to be used. The Bicego approach has a simple form of the crack growth, da/dN = function of ΔJ; the necessary constants are determined from the Paris equation in the elastic range.

Theoretical Considerations in the LCF Process

Consider three distinct regions of the stress and strain field surrounding the crack tip, as noted in Figure 4a. Region (1) can be viewed as an elastic field surrounding the crack tip where LEFM is applicable. Region (2) is an elastic-plastic field surrounding the crack tip where general yielding fracture mechanics (YFM) is applicable. Region (3) is a zone of intense deformation (currently under investigation). If zone (3) is comparatively small, then region (2) can be analysed using plasticity theories. To extend fatigue crack growth to the elastic-plastic régime, the J-integral will be used.

Based on an earlier analysis of Rice, and Rosengren and Hutchinson, McClintock derived the stress and strain distribution in terms of J-integral for power hardening material. Using this approach and applying a method suggested by Lamba [12], the strain distribution at the crack tip, under cyclic loading, can be expressed as:

$$\sigma\,(\theta,\,r)\,\alpha\,\left(\tfrac{1}{r}\right)^{n/n+1}$$

$$\varepsilon\,(\theta,\,r)\,\alpha\,\left(\tfrac{1}{r}\right)^{1/n+1}$$

w < r << planar dimensions

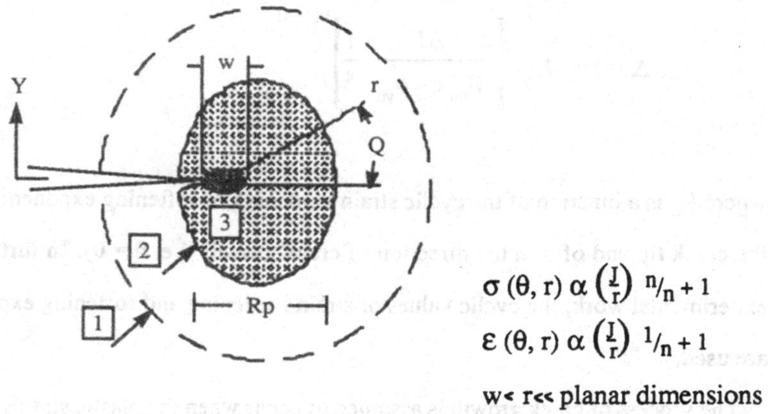

REGION 1 - AN ELASTIC FIELD SURROUNDING THE CRACK TIP
REGION 2 - AN ELASTIC-PLASTIC FIELD SURROUNDING THE CRACK TIP
REGION 3 - AN INTENSE ZONE OF DEFORMATION

Figure 4a. Crack tip stress and strain fields

increasing load — CRACK TIP CLOSED AT ZERO LOAD

CRACK TIP OPENS BUT STILL SHARP:
APPLIED LOAD BALANCING CLOSURE STRESS

$2r_c$ — CRACK TIP FULLY BLUNTED BUT NO CRACK
GROWTH

Δa — CRACK ADVANCES BY Δa WHEREVER THE CRACK
TIP STRAINS REACH THE FRACTURE ε_f

DIMPLES FORM IN FRONT OF THE CRACK TIP AND
GROW

DIMPLES COALESCE AND CRACK EXTENDS BY A
FURTHER Δa_{dt} DUE TO DIMPLE FORMATION AND
Δa_b — Δa_{dt} — DUCTILE TEARING

Total crack growth = $\Delta a_b + \Delta a_{dt}$

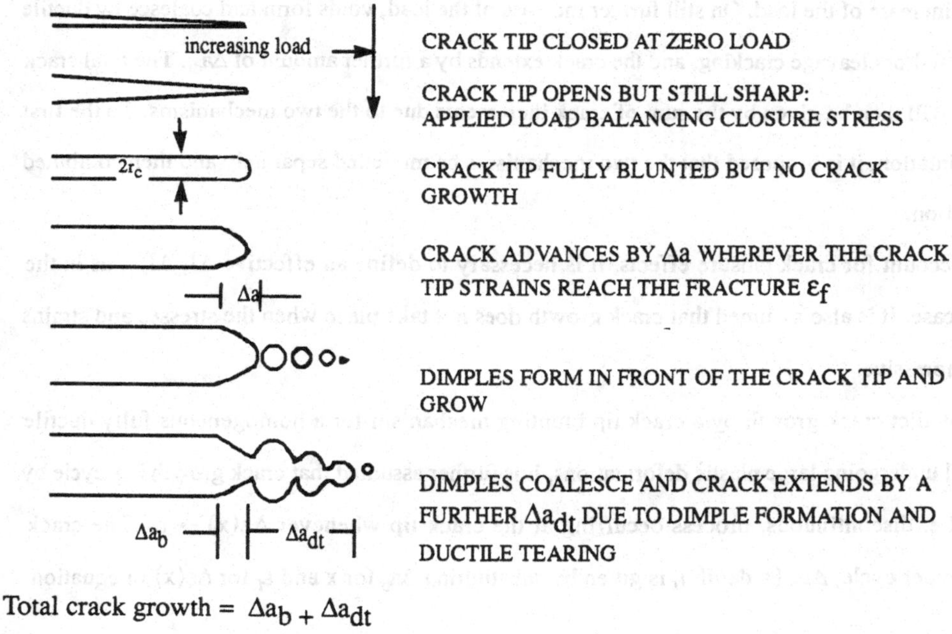

Fig. 4b. Crack tip deformations and crack growth under increasing load

$$\Delta\varepsilon(x) = 2\varepsilon_{yc}\left[\frac{\Delta J}{4I_{0n}\,\varepsilon_{yc}\,\sigma_{yc}}\,\frac{1}{x}\right]^{1/(1+n)} \tag{12}$$

where I_{0n} is a function of the cyclic strain hardening or softening exponent, n, the state of stress at the crack tip and of θ in the direction of crack growth (i.e. $\theta = 0$). In further analysis and in the experimental work, the cyclic values of strain hardening and softening exponents and coefficients are used.

The process of crack growth is assumed to occur when the plastic strains at the crack tip become large, i.e. of the order of the true fracture strain, ε_f. The crack tip of the originally sharp crack becomes blunted and its flanks, initially closed, will separate by a finite distance, yet remaining almost parallel. Further damage occurs only when this separation reaches a critical value, $2r_c$ (Figure 4b). The crack then advances by Δa, as a result of a crack tip blunting mechanism, on further increase of the load. On still further increase of the load, voids form and coalesce by ductile tearing and/or cleavage cracking, and the crack extends by a further amount of Δa_{dt}. The total crack growth will then be given by the sum of crack increments due to the two mechanisms. To the first approximation, it is proposed that the two mechanisms be modelled separately and then combined by addition.

To account for crack closure effects, it is necessary to define an effective ΔJ, ΔJ_{eff}, as in the elastic case. It is also assumed that crack growth does not take place when the stresses and strains are compressive.

To predict crack growth by a crack tip blunting mechanism for a homogeneous fully ductile material undergoing large plastic deformations, it is further assumed that crack growth is a cycle by cycle, i.e. discontinuous, process occurring at the crack tip whenever $\Delta\varepsilon(x) \rightarrow \varepsilon_f$. The crack advance per cycle, Δa_b (= da/dN), is given by substituting Δa_b for x and ε_f for $\Delta\varepsilon(x)$ in equation (12):

$$\frac{da}{dN} = \left(\frac{1}{\varepsilon_f}\right)^{n+1}\frac{(2\varepsilon_{yc})^n\,\Delta J_{eff}}{2\sigma_{yc}\,I_{0n}} - r_c \tag{13}$$

The value of r_c can be arrived at by allowing for da/dN → 0 at a critical value of ΔJ_{eff}, $\Delta J_{c,th}$, which is the critical threshold for a non-propagating crack (cf. $\Delta K_{c,th}$).

If the intense zone of deformation is large, additional crack growth will occur at fairly low strain ranges due to the void formation and coalescence ahead of the crack tip. Therefore, application of equation (13) is limited to those regions of LEFM and YFM where the intense zone of deformation, w, is small compared with the planar dimensions. However, when w is large, the expression (12) would underestimate the actual growth rate. At present, no simple analysis exists, except that by Rice and Johnson, who approximated deformation behaviour in this region using slip-line theory. The stress level at which voids start forming at the crack tip will depend on the material ductility and on the amount of second phase particles and inclusions present in the material. For an ideal homogeneous ductile material, voids may not form until after large plastic deformations. In this case, equation (13) would be applicable under both elastic and elastic-plastic conditions.

The lack of analytical solutions for zones of intense deformation may explain why fatigue crack growth models are frequently modified. An empirical factor in the form of $K_c/(K_c - K_{max})$ is often included in order to account for the increased growth rates that occur at high stress intensities due to void coalescence, creep and other processes. At high stress intensities, cyclic crack growth rates are found to correspond to the monotonic slow stable crack extensions. This observation suggests that the mechanism of crack growth is the same for both types of loading. To account for the static mode of crack growth superimposed in the 'pure' fatigue process, it is suggested that two approximations may be made to equation (13). The first of these approximations is achieved by multiplying the above equation by an empirically derived function of ΔJ_{eff} to fit the experimental data.

Alternatively, it is suggested that a double mechanism fatigue crack growth process may be assumed to take place in the following way:

At high stress intensities, the process is J_{max}-dependent and the crack growth may be estimated using crack growth resistance curves (R-curves). The R-curve may be approximated by:

$$\Delta a_R = C'(J_R)^{m'} \tag{14}$$

where C' and m' are material constants determined experimentally from the R-curve. It could be reasonably assumed that the crack tip blunting mechanism predominates in crack growth at low stress intensities, whereas void coalescence predominates at high stress intensities. Although there is, at present, no evidence available to show that these processes are additive, it is suggested that, at least at intermediate stress intensities, both mechanisms may occur simultaneously. Therefore, the total crack growth can be expressed by direct summation of equations (13) and (14) into equation (15). Including, then, experimental data for a low-alloy steel, we obtain:

$$\left(\frac{da}{dN}\right)_{total} = \frac{2^{1+n}\,(\Delta J_{eff} - \Delta J_{c,th})}{4 I_{0n}\, E^n\, \sigma_{yc}^{1-n}\, \varepsilon_f^{1+n}} + C'(J_{max})^{m'} \tag{15}$$

where J_R is now replaced by the maximum value of the applied ΔJ, which is $J_{max} = \Delta J/(1 - R)$. For $n = 0.19$, the value of $I_{0n} = 4.5\pi\,(1 + n) = 16.81$. For BS4360-50D steel, the constants are: $C = 7.5 \times 10^{-18}$ and $m = 2.6$, when da/dN is in mm/cycle and J is in N/m and $R < 0.1$.

Using the double mechanism model of equation (15), the predicted growth rates are compared with the experimental growth rates in Figure 5a. It will be noted that, although a very simplified crack growth model involving superposition of two growth mechanisms has been used, the prediction is good. Figure 5a indicates that, for BS4360-50D steel, the blunting mechanism is dominant at low growth rates $< 10^{-4}$ mm/cycle, whereas at high stress intensities void coalescence involving dimple formation and ductile tearing becomes the principal fracture mechanism. At intermediate growth rates, the two mechanisms seem to act simultaneously.

In Figure 5b, the data for the whole range of growth rates from threshold to fast fracture are presented. Equation (15) is seen to fit the data fairly well.

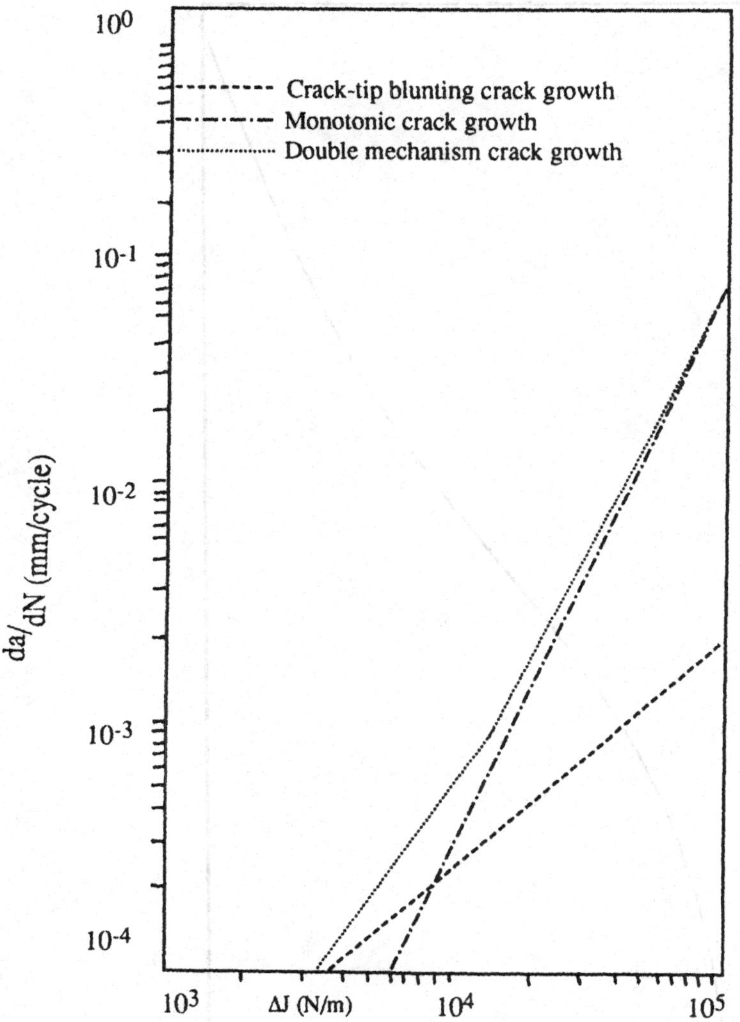

Fig. 5a. Modelling of elastic-plastic crack growth in BS4360-50D steel. For thickness 12, 25 and 50mm.

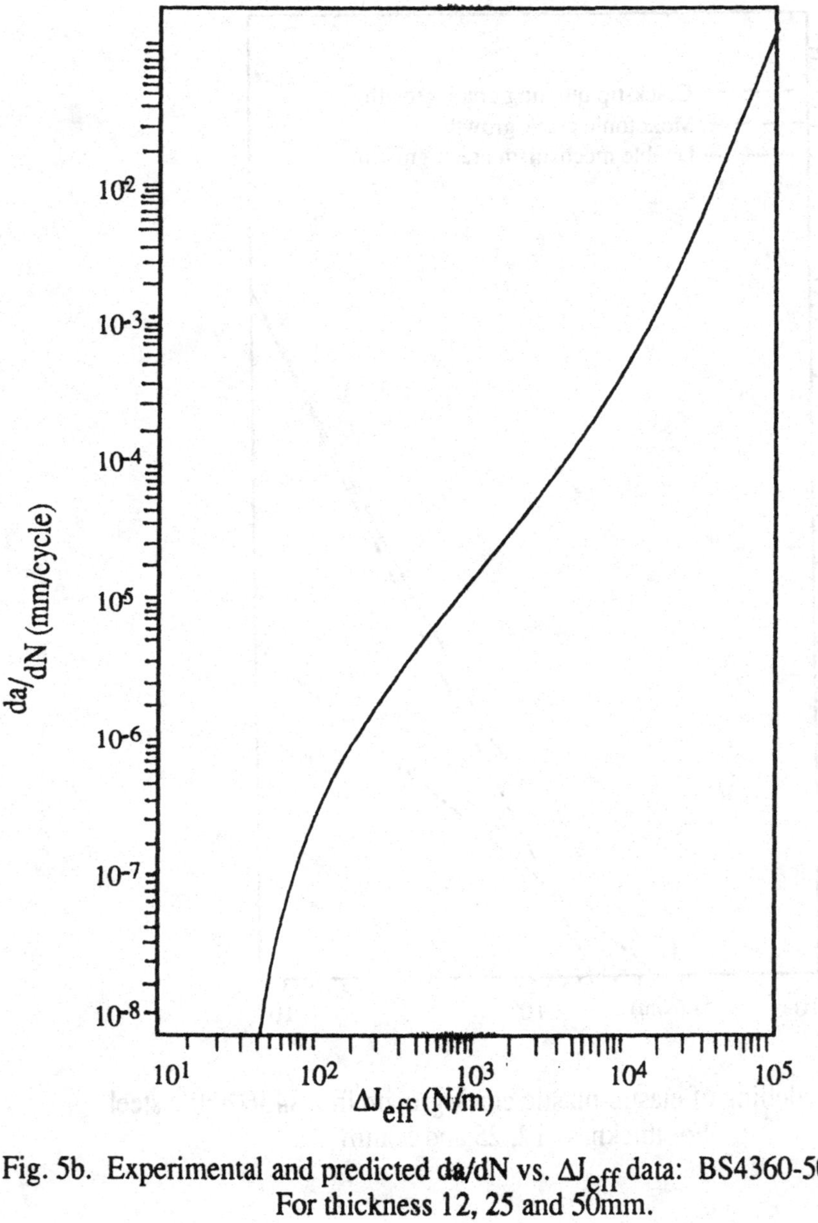

Fig. 5b. Experimental and predicted da/dN vs. ΔJ_{eff} data: BS4360-50D steel. For thickness 12, 25 and 50mm.

High Temperature Fatigue and the Rôle of Creep

Bowles [27] reported the recent research in the field of high temperature fatigue crack growth. The application of EPFM is particularly suitable because of the decreased yield strength and increased ductility at elevated temperatures. Furthermore, in general, the potential for time-dependent cyclic crack growth increases with increase in temperature. Sadananda and Shahinian [28] examined the effects of stress ratio and hold times on fatigue crack growth in Alloy 718 at 650°C and found that J_{max} provided a better correlation than ΔJ. They also raised the possibility of using the more complex rate-dependent integral, J*, for time-dependent crack growth. When both cycle-dependent and time-dependent processes are occurring, there may be synergistic effects requiring further research. Similar effects have been shown in our work on corrosion fatigue [34].

Okazaki et al. [29] examined the strain cycled fatigue of 304 stainless steel at 600°C using the strain range partitioning procedure, where the strain cycle was partitioned into creep and fatigue components. Strain range partitioning was first proposed by Manson in an attempt to separate the time-dependent and dynamic components of deformation associated with elevated temperature fatigue. Okazaki et al. have extended this procedure by applying the time-dependent J* to form a cyclic creep integral, ΔJ_c. The dynamic component of fatigue was evaluated by a cyclic component termed ΔJ_f. They found that the creep portion of crack growth correlated very well with ΔJ_c. The partitioning was further justified by the fact that the creep zone could be physically defined by Vickers hardness measurements. It is interesting to note that Okazaki found that the ratio $\Delta J_c/\Delta J_f$ decreased linearly with increasing crack length, suggesting that creep fatigue interactions may be less important in long cracks than in short cracks.

Recently, a number of models for fatigue-creep interaction has been proposed, and three basic approaches are currently favoured. These are:

- damage summation;
- frequency modified strain range; and
- strain range partitioning.

Some empirical models, tested on a range of materials and reported in the literature, are described below:

(a) Manson, Halford and Hirschberg [30]. This model is based on partitioning of the total strain range into certain components, each component contributing a fraction of the total life. Creep-fatigue is composed of fatigue, cyclic creep and creep-fatigue interaction components.

(b) Majumdar and Maiya [31]. This is an empirical model which gives the crack growth rate:

$$\frac{da}{dt} = A_t (\varepsilon_p)^m (\dot{\varepsilon}_p)^k . a \quad \text{(for tensile stress)} \quad \text{and}$$

$$\frac{da}{dt} = A_c (\varepsilon_p)^m (\dot{\varepsilon}_p)^k . a \quad \text{(for compressive stress)}$$

Time-dependent effects are included through the term $\dot{\varepsilon}_p$, the plastic strain rate. Though the authors have found good experimental corroboration, substantial differences have been recently reported.

(c) Plumtree and Lemaitre [32]. In this model, the damage fractions due to creep and fatigue are separately evaluated:

$$D_c = \text{damage due to creep} = 1 - [1 - \frac{N}{N_c}]^{1/(1+q')}$$

$$D_f = \text{damage due to fatigue} = 1 - [1 - \frac{N}{N_f}]^{1/(1+p)}$$

Total damage is obtained by summing up the two damages. The general applicability of the model has not yet been evaluated.

(d) Saxena et al. [33]. This is a semi-empirical model that focuses on the crack tip stress-strain conditions. Its basic assumption is that these conditions control creep-fatigue failures. The crack propagation rate is given as $da/dt = p (K^2/t)^m$. The parameter K^2/t characterises the time dependence of the stress-strain rate in the crack tip region. Janson, Weertman, Wareing, Min and Raj, and Challenger and others have also proposed models applicable in the creep-fatigue régime.

After achieving some success in extending the concepts of LEFM to fatigue and corrosion fatigue [34], creep crack growth rates in Al-alloy RR58, and later also in a mild steel, have been measured in this department using the stress intensity factor K. All three creep régimes have been investigated, but particularly encouraging results were obtained in the secondary region [35]. It may be appropriate to recapitulate some relevant results obtained from this work:

(1) The contoured DCB constant K specimen was used to give a constant creep crack growth rate over periods extending to 3000 hours.

(2) The effect of altering K at constant temperature suggested that creep crack growth rate was independent of previous history of K.

(3) The results showed that crack growth rate during the 'secondary' creep region could be expressed over the temperature range 100°C to 200°C for RR58 by:

$$\dot{a}_s \propto K^n e^{-Q/RT}$$

where n = 30 and Q = 18 kcal/mole. The value of n corresponds with the stress sensitivity of creep deformation for this material but Q is somewhat less than the value of 29 kcal/mole required to describe deformation in this temperature region.

(4) Because of the occurrence of a 'primary' region of crack growth at constant K, as yet not fully understood, it seemed unwise to measure the K sensitivity of crack growth rates on test piece configurations which did not impose K invariant with crack length, although the evidence available suggested the K dependence of primary and secondary crack growth could

be similar.

The above results indicated that the equation

$$\frac{da}{dt} = C \, . \, K^n$$

was only applicable when the amount of creep was confined to the crack tip region (small scale creep). In that situation, the remaining part of the test piece was behaving under linear elastic conditions. However, it was subsequently noted that data obtained in tests performed on the aluminium alloy RR58 at temperatures above 150°C did not correlate with the above equation. This was assumed to be caused by the large amount of creep obtained in the test piece (large scale creep).

To overcome this difficulty, other fracture mechanics parameters were proposed and, finally, a creep equivalent to the J-integral, $\dot{J}$, was derived. A reasonable correlation of this parameter with the crack growth rate data was then obtained.

The relationship of the J (or G) values and the crack growth rate is shown in Figure 25. In this graph, it was possible to correlate the small scale creep region (region I) with all the results available at 150°C, but only a few results at the higher temperatures corresponded to the low applied stresses. All the remaining data points obtained at 175°C and 200°C were included in the large scale creep region (region II). A further analysis of these tests may be found in [36].

The fatigue and creep interaction has also been investigated in this department by Dimopulos et al. [37], who used brittle and ductile low alloy steels and Ni-based superalloy AP1. The influence of the stress ratio R and of the changes in frequency were studied. At high frequencies, a transgranular fatigue failure was observed, while at low frequencies the time-dependent intergranular fracture dominated. The use of Paris' law during the fatigue process was acceptable and the effect of the R-ratio was correlated with the crack closure phenomenon. At low frequencies, the application of C* was found to be suitable and increasing creep ductility enhanced the resistance to creep-fatigue crack growth.

A standard form of Paris' law was used in the fatigue régime. However, in the creep régime, the crack growth was described as da/dt = function of C*. When these processes act together, both parts of the crack growth da/dN may be added, at least for the materials investigated. The prediction of the crack growth using this approach for ½ Cr Mo V steel at 565°C is shown in Figure 6 for a range of R-ratios at a constant value of K_{max}.

Subsequently, Nikbin [38] presented a model for growth rate expressed in terms of the damage accumulated in a process zone ahead of the crack tip. Also, an engineering creep crack growth assessment diagram was proposed which is independent of material properties (Figure 7). A relationship between creep crack growth rate and C* was developed.

Standard deeply-notched specimens used in the above investigations of the fatigue and creep processes were considered a well suited geometry for further research of cycling in the elasto-plastic régime. A large amount of data using three-point bend and compact type specimens is available in the literature. Those specimens were provided with a deep notch, a/W > 0.55. Other specimens used in the following analysis were of the double cantilever beam (DCB) type. This geometry was found very satisfactory in our earlier studies of fracture and fatigue in the LEFM régime [39].

Deeply-notched specimens were subjected to static and cyclic loading in various environments. Load-deflection plots could be experimentally and analytically determined using the non-linear beam theory. The J or ΔJ values were then calculated using standard formulae. All the partial deflections caused by the loading and environment, i.e. temperature and frequency, were included in the total deformation of the specimen arms. The tests using the DCB specimens are discussed next.

(1) Geometry of Double Cantilever Beam

Branco investigated the slow crack growth and, in [36], described a procedure for calculating the critical value of the stress intensity factor for unstable crack propagation, K_c. Using the contoured

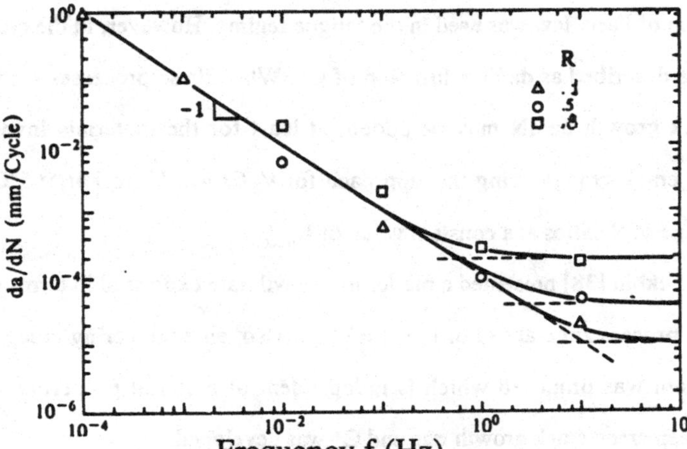

Fig. 6. Dependence of crack growth/cycle for 1/2CMV steel at 565° at a constant K_{max} of 30 MPa√m. [38]

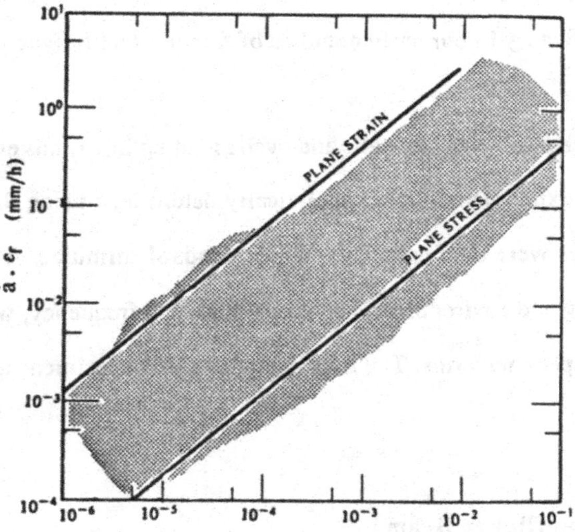

Fig. 7. Material independent engineering creep crack growth assessment diagram [38]

double cantilever beam specimen (a constant K specimen) [39], K_c was evaluated for various aluminium alloys within the range of applicability of LEFM. However, in the case of BS15 mild steel, this approach could not be used since the slow crack growth and an extensive plastic deformation occurred before sudden crack propagation. The method of calculating the J-integral and the application of EPFM will be described first.

The energy interpretation of J is illustrated in Figure 8 for the case of monotonic loading. Line OABC represents the load-deflection or load extension curve for a specimen in which the loading is linear elastic from O to A, work hardening occurs from A to B, and along BC the crack extends an amount da at a constant load P causing an increment of deflection dδ. This curve is typical for mild steel. For other metals, non-linear behaviour may be apparent from point O. The line OA_1C is the equivalent load-deflection curve for an initial crack length (a + da). The shaded area therefore represents the potential energy lost in propagating the crack from a to (a + da).

The potential energy, U, may be expressed as:

$$U = \frac{P_{ys}\,\delta_{el}}{2} + P_{pl}\,(\delta_{el} + \frac{\delta_{pl}}{(n+1)}) \tag{16}$$

Consequently:

$$J = -\frac{1}{B_n}[P\,(\frac{d\delta_{el}}{da} + \frac{d\delta_{pl}}{da}\frac{1}{(n+1)}) - P_{ys}\,(\frac{1}{2}\frac{d\delta_{el}}{da} + \frac{d\delta_{pl}}{da}\frac{1}{(n+1)}) - \frac{dP_{ys}}{da}\,(\frac{\delta_{el}}{2} + \frac{\delta_{pl}}{(n+1)})] \tag{17}$$

Provided that the boundary conditions are prescribed in terms of fixed loads, equation (17) allows the determination of the J values for any specimen geometry. The values δ_{el}, δ_{pl}, P_{ys} and their derivatives with respect to a may also be obtained at the loading points using the usual stress analysis methods.

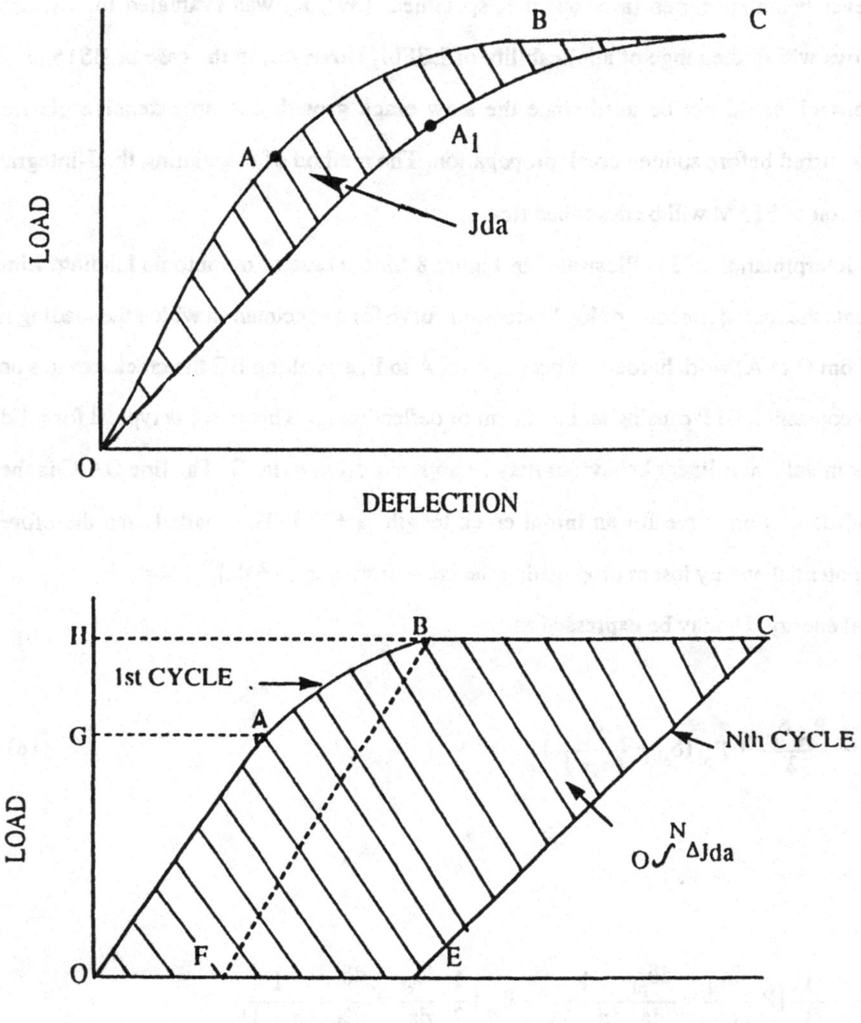

Fig.8 Energy Definition for Monotonic and Cyclic Loading

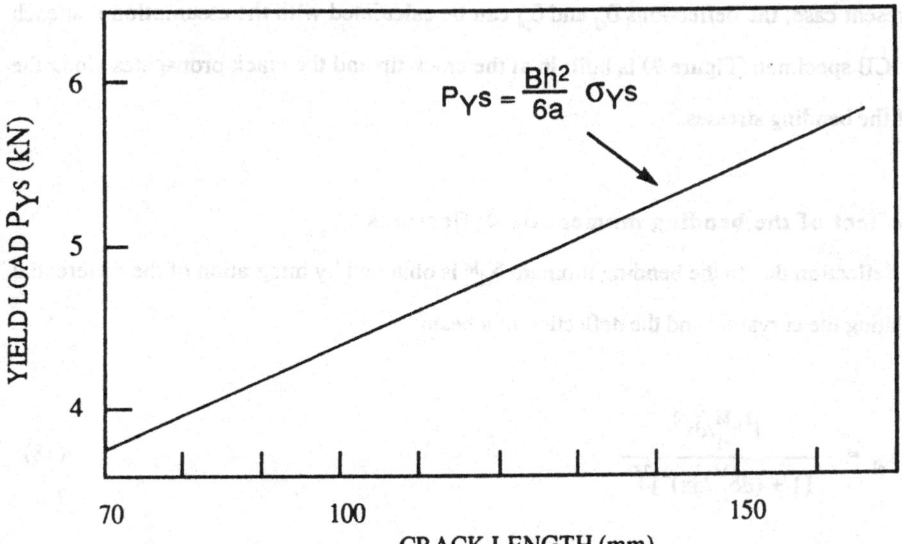

Fig. 9. Variation of the yield load with crack length. BS15 DCB specimen.

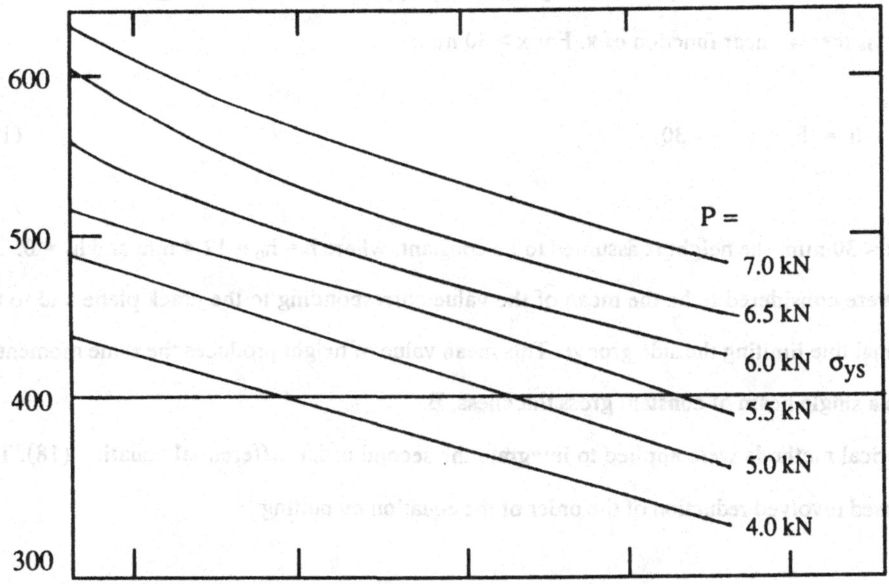

Fig. 10. Variation of the Maximum Bending Stress with Crack Length for Different Load Values. BS15. DCB Specimen.

In the present case, the deflections δ_{el} and δ_{pl} can be calculated with the assumption that each leg of the DCB specimen (Figure 9) is built in at the crack tip and the crack propagates along the direction of the bending stresses.

(A) The effect of the bending moment on deflections

The elastic deflection due to the bending moment δ_{el}^M is obtained by integration of the differential equation relating the curvature and the deflection of a beam:

$$k_{el} = -\frac{d^2\delta_{el}^M/dx^2}{[1 + (d\delta_{el}^M/dx)^2]^{3/2}} \tag{18}$$

where k_{el} is the elastic bending moment curvature.

The contour of the DCB specimen may be closely approximated by a straight line. The beam height, h, is then a linear function of x. For x > 30 mm:

$$h = h_0 + h_1 (x - 30) \tag{19}$$

and for x < 30 mm, the height is assumed to be constant, where $h = h_0 = 12.4$ mm and $h_1 = 0$. The heights were considered to be the mean of the value corresponding to the crack plane and to the longitudinal line limiting the side groove. This mean value of height produces the same moment of inertia as a single beam of constant gross thickness, B.

Numerical methods were applied to integrate the second order differential equation (18). The method used involved reduction of the order of the equation by putting

$$\frac{d\delta_{el}^M}{dx} = Z_1 \tag{20}$$

Equation (18) then becomes:

$$k_{el} = \frac{dZ_1/dx}{(1 + Z_1^2)^{3/2}} \tag{21}$$

Thus, equation (21) was integrated between 0 and the crack length x = a in a computer program using the modified Euler method and also the Runge-Kutta method, both methods yielding very similar results. Finally, integration of equation (20) was carried out in the same program using Simpson's rule with the Z_1 values previously obtained in the first integration. The accuracy of the numerical method depended on the step interval; after several trials, 0.5 mm was chosen. Crack length values ranged from 70 to 160 mm in 10 mm intervals.

The plastic deflection caused by the bending moment, δ_{pl}^{M}, was calculated by a similar process, the only difference being that the curvature equation used is that for a beam of a non-linear elastic material.

The yield load, P_{ys}, obtained from elementary beam theory, increased with crack length. The appropriate value of P_{ys} was substituted in the curvature equations before integration was carried out between 0 and a crack length, a.

Neglecting the crack tip singularity, the maximum bending stress in the beam is given for $\sigma_{max} > \sigma_{ys}$ as:

$$\sigma_{max} = \sigma_{ys} + \frac{2P_{pl}(2n + 1)}{B h^2 n} \tag{22}$$

In Figure 10, the curves for σ_{max} are plotted for six load values (4.5 to 7.0 kN). It can be seen that σ_{max} increases with applied load and decreases with crack length. With increasing load, linear elastic conditions are attained at increasing crack lengths. Plastic deformation will be obtained in the specimen for any monotonic loading to greater than 5.5 kN for crack lengths between 70 and

160 mm. For loads below 5.5 kN, the linear elastic stage is being obtained with increasing crack lengths as the load increases.

In Table 1 are the calculated values of δ_{el}^{M} and δ_{pl}^{M}; they show that δ_{el}^{M} increases with crack length while δ_{pl}^{M} decreases. This trend originates from an increase in P_{ys} with crack length and a corresponding decrease in P and σ_{max} (Figure 10).

(B) The effect of shear force and rotation at the crack tip on the deflections

The effect of the shear force on the deflections was also considered. The elastic shear force deflection, δ_{el}^{S}, is obtained by integrating the following equation:

$$\frac{d\delta_{el}^{S}}{dx} = \frac{(\tau_{yx}^{el})_{y=0}}{G'} \tag{23a}$$

where τ_{yx}^{el} is the linear elastic shear stress, and G' is the shear modulus. The plastic shear force deflection, δ_{pl}^{S}, is obtained by integrating the equation:

$$\frac{d\delta_{pl}^{S}}{dx} = D(\tau_{yx}^{pl})_{y=0}^{n} \tag{23b}$$

where D is a constant and is equal to A[$3^{\frac{n+1}{2}}$] ., and τ_{yx}^{pl} is the plastic shear stress.

Subsequently, the rotational effect at the crack tip was computed, assuming that the beam would rotate around the crack tip, with the end of the specimen being built in. It can be shown that this deflection is calculated as follows:

$$\delta^{R} = \frac{6\Delta P\, a^{2}(w - a)}{E\, B\, h_{u}^{3}} \tag{24}$$

TABLE 1

Theoretical and Experimental Deflection Values

(DCB Specimen - BS15)

Monotonic Loading P = 5.5kN

(All values in mm)

a	δ^{M}_{el}	δ^{M}_{pl}	δ_{M}	δ^{S}_{el}	δ^{S}_{pl}	δ^{R}	δ_{t}	δ_{exp}
70	1.94	5.37	7.31	0.06	0.09	0.41	7.87	7.924
80	2.65	5.15	7.80	0.08	0.09	0.47	8.44	-
90	2.98	4.3	7.28	0.09	0.10	0.51	7.98	-
100	3.87	3.43	7.30	0.09	0.11	0.53	8.03	8.145
110	4.23	2.30	6.53	0.10	0.11	0.54	7.30	-
120	4.96	1.35	6.31	0.11	0.12	0.56	7.10	-
130	5.34	0.99	6. 3	0.11	0.13	0.59	7.16	-
140	6.13	0.69	6.82	0.12	0.15	0.61	7.70	-
150	6.91	0.34	7.25	0.14	0.16	0.63	8.18	-
160	7.96	0.0	7.96	0.15	0.17	0.66	8.94	8.83

(C) The effect of high temperature on deflections

The deflection, δ_{cr}, caused by the influence of the environment (such as elevated temperatures) represents an amount of plastic strain under a steady load, described earlier. It becomes an important part of the total deformation at temperatures above 0.4 T_m and has to be included, particularly when cycling with incorporated hold times [37].

Deflections due to other environmental factors, such as corrosion or radiation effects, also have to be superposed.

The total theoretical deflection, δ_t, is then the superposition of all the calculated deflections:

$$\delta_t = \delta_{el}^M + \delta_{pl}^M + \delta_{el}^S + \delta_{pl}^S + \delta^R + \delta_{cr} + \delta_{cc} + \delta_{machine} \qquad (25)$$

A comparison between the computed bending moment deflections, δ^M (elastic + plastic), shear force deflections, δ^S (elastic + plastic), and rotational deflections is presented in Table 1. It may be seen that δ^S is between 2%-3% of δ^M and δ^R varies between 5%-8% of δ^M.

(D) Theoretical results for deflections, potential energy and J contour integral

Figure 11 shows the theoretical load-total deflection curves for three crack length values of 70, 100 and 130 mm and for an applied load of 6 kN.

Figure 12 also includes two theoretical load-deflection curves corresponding to an applied load of 7 kN and crack length values of 70 and 160 mm (maximum and minimum values of crack length).

Equation (16), with the appropriate values of δ_{el} and δ_{pl}, yields the potential energy, which is plotted against crack length in Figure 13. The amount of potential energy sufficient to cause linear elastic deformation for the same applied load has also been recorded and, as expected, the elastic-plastic potential energy is greater than the linear elastic counterpart. The difference between the two energies, which increases as load increases and decreases as crack length increases, is due to the increase in yield load with crack length which causes the plastic energy term to decrease with crack

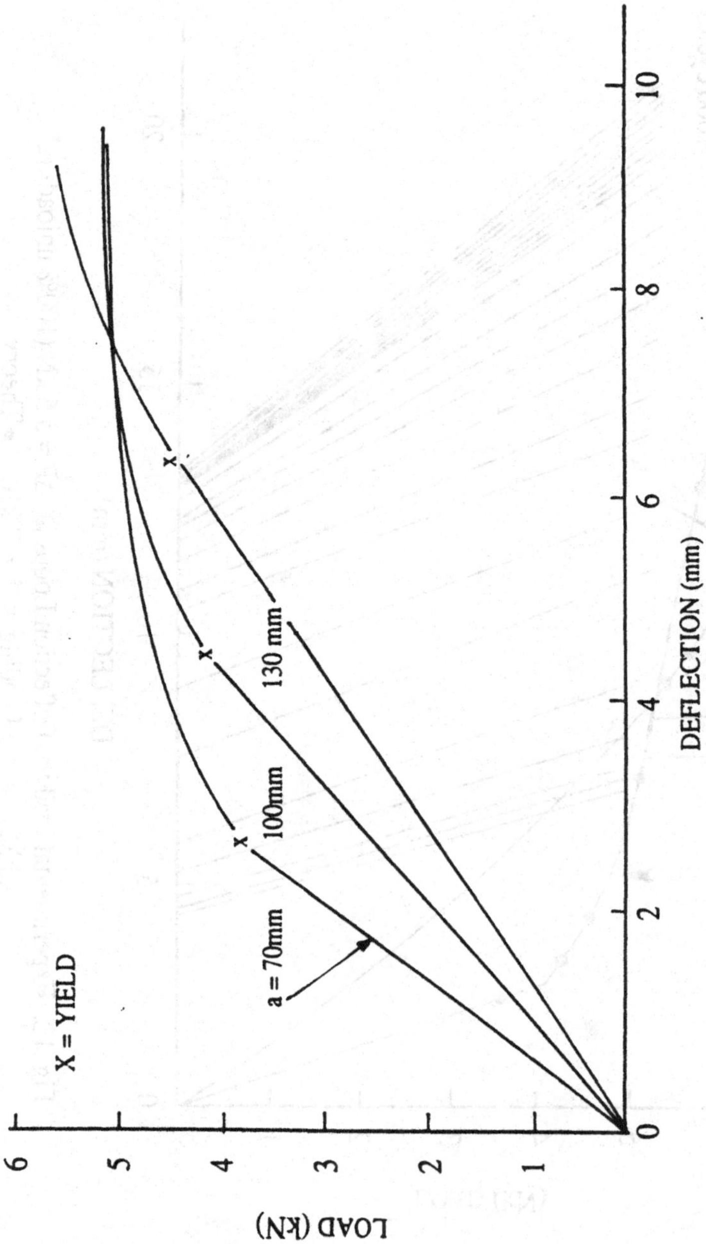

Fig. 11. Theoretical Load vs. Deflection Curves. P = 6 kN.

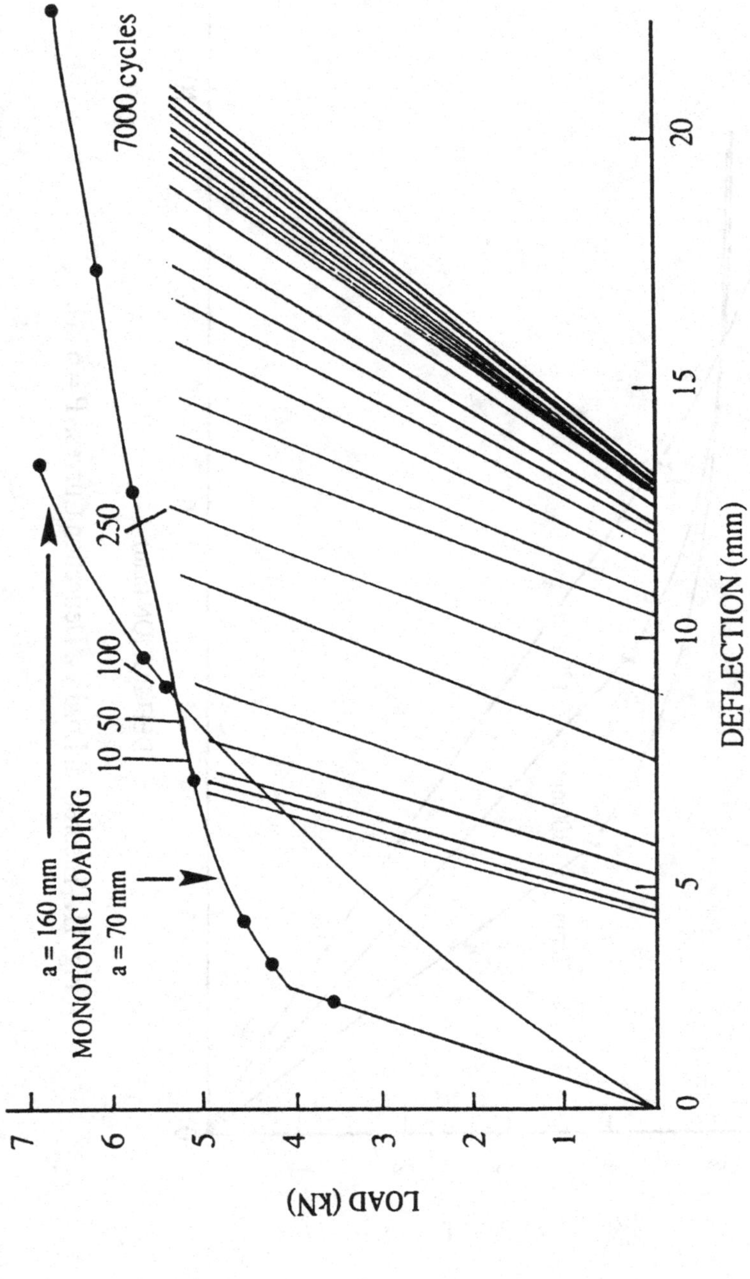

Fig. 12. Experimental load vs. Deflection Loops at $\Delta P = 5.5$ kN (100% unloading) and Monotonic Loading at $P = 7$ kN. ● Theory.

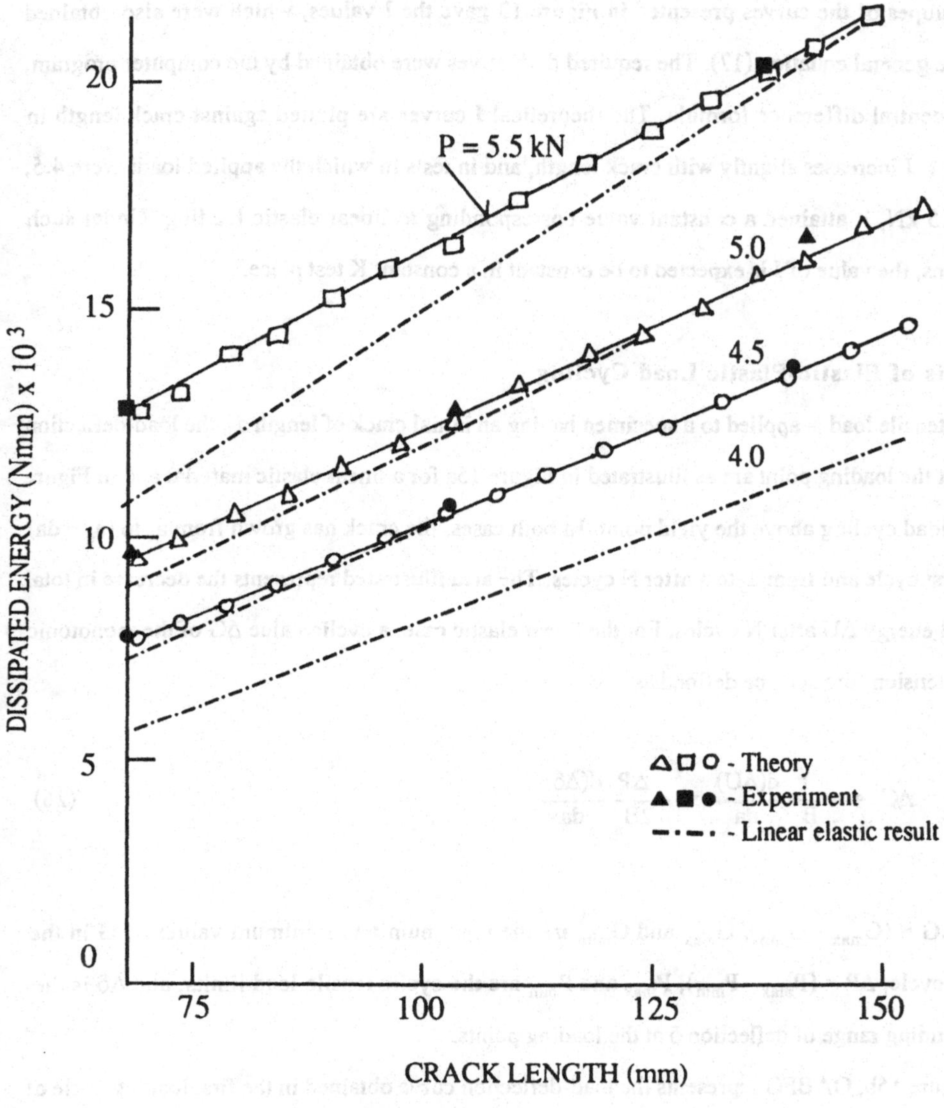

Fig 13. Dissipated Energy vs. Crack Length.

length.

The slopes of the curves presented in Figure 13 gave the J values, which were also obtained using the general equation (17). The required derivatives were obtained by the computer program, using a central difference formula. The theoretical J curves are plotted against crack length in Figure 14. J increases slightly with crack length, and in tests in which the applied loads were 4.5, 5 and 5.5 kN, it attained a constant value corresponding to linear elastic loading. Under such conditions, the value of J is expected to be constant in a constant K test piece.

Analysis of Elastic-Plastic Load Cycling

When a tensile load is applied to a specimen having an initial crack of length a_0, the load-deflection curves at the loading point are as illustrated in Figure 15a for a linear elastic material and in Figure 15b for load cycling above the yield point. In both cases, the crack has grown from a_0 to $(a_0 + da)$ in the first cycle and from a_0 to a after N cycles. The area illustrated represents the decrease in total potential energy ΔU after N cycles. For the linear elastic case, a cyclic value ΔG of the monotonic crack extension force can be defined as:

$$\Delta G = -\frac{1}{B_n}\frac{d(\Delta U)}{da} = -\frac{\Delta P}{2B_n}\frac{d(\Delta\delta)}{da} \qquad (26)$$

where $\Delta G = (G_{max} - G_{min})$, G_{max} and G_{min} are the maximum and minimum values of G in the loading cycle, $\Delta P = (P_{max} - P_{min})$, P_{max} and P_{min} are the cyclic tensile load limits, and $\Delta\delta$ is the corresponding range of deflection δ at the loading points.

In Figure 15b, OABFO represents the load-deflection curve obtained in the first loading cycle of an elastic-plastic material. The curve is linear elastic from O to A and plastic deformation occurs from A to B. This type of loading curve is typical for mild steel. For other metals, a non-linear curve starting at O may be more appropriate. BC represents the accumulated deflection caused by cyclic creep, or high temperature creep, and crack growth after N cycles at a constant load range,

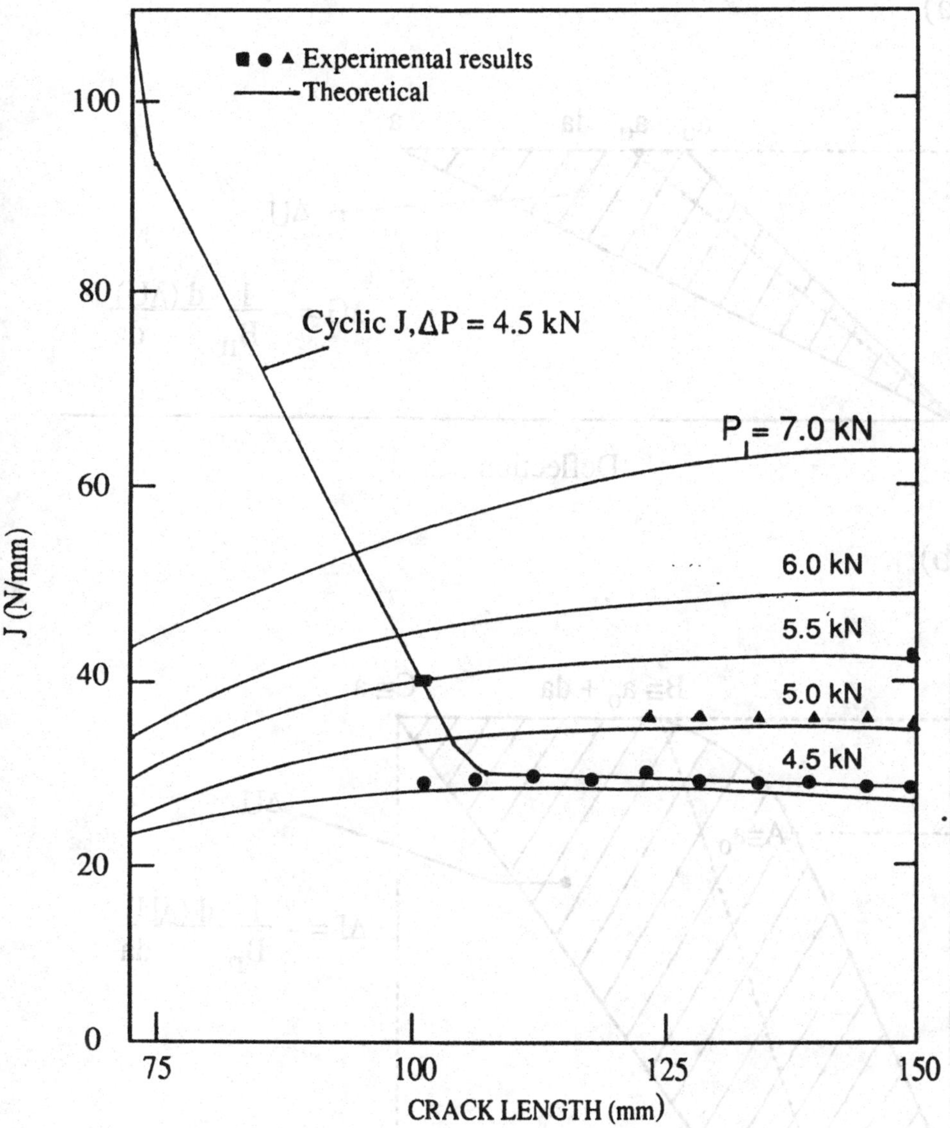

Fig. 14. J vs. Crack Length. BS15. DBC Specimen

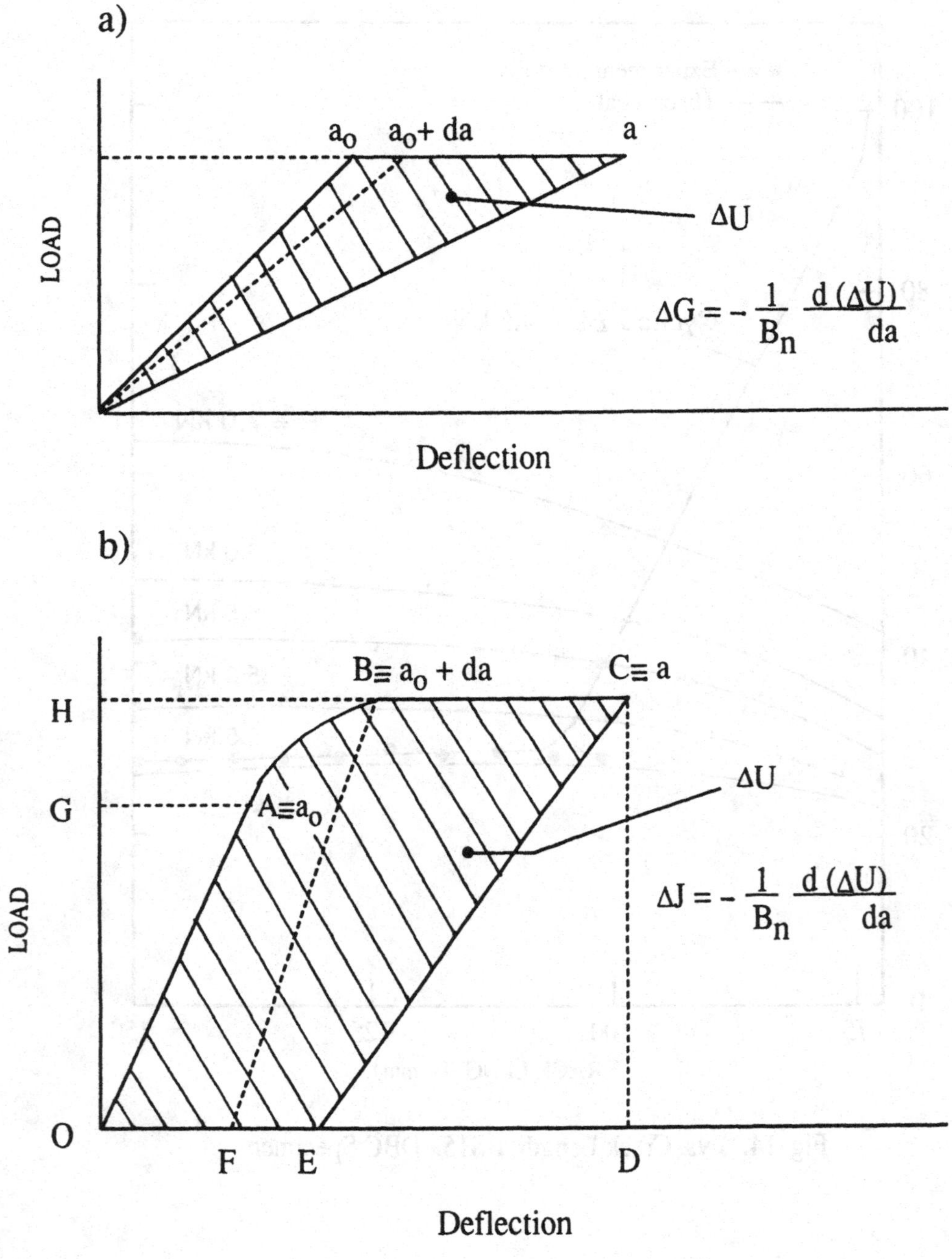

Fig 15. Operational Definitions of Cyclic ΔG and ΔJ after N Loading Cycles

ΔP.

The cyclic creep phenomenon has been extensively studied in the literature [40]. It consists basically of strain accumulation caused by load cycling at stress levels above the yield point. In general, the cyclic creep curve (accumulated strain against number of cycles) has a shape analogous to that of the ordinary creep curve. An initial primary stage is followed by a well-defined secondary stage with a constant cyclic creep strain rate, covering 80%-90% of the specimen life, and finally a very short tertiary stage near fracture. Consequently, the cyclic creep deflection represented by the quantity BC in Figure 15b is considered here to consist basically of the secondary stage, since experimental evidence indicates that the primary stage is virtually completed during the first few loading cycles and may be included in AB. The section AB will also include strain hardening (or softening).

The potential energy of the first cycle, ΔU_1, obtained by loading a specimen with crack length a_0 and zero initial deflection, is represented by the area OABHO in Figure 15b. The total accumulated potential energy after N cycles, ΔU_N, is given by the area OECHO. The energy ΔU_N is obtained by elastically loading from E to C a specimen having a crack length a and with a permanent initial deflection, indicated by OE; it is numerically equal to the real permanent deflection after N cycles. It should be pointed out that OE represents the accumulated (cyclic and high temperature) creep deflection after N cycles and, if crack growth had not occurred, the trace EC would have been parallel to BF. The elastic loading condition for EC is closely true since the yielding has been largely accomplished between points A and B.

Because only loading conditions were assumed, the energy interpretation of the J contour integral can be extended to cyclic loading of an elastic-plastic material to define a cyclic ΔJ value after N loading cycles as:

$$\Delta J = \frac{1}{B_n}\frac{d(\Delta U)}{da} \qquad (27)$$

where $\Delta U = \Delta U_N - \Delta U_1$.

It should be noted that, in the strict mathematical sense, the above definition and equation (27) are valid only if unloading phenomena are not considered (deformation theory of plasticity). However, the present approach may have more general applicability than the mathematical definition. The operational definition of ΔJ put forward is equivalent to saying that ΔJ defines the stress and strain fields near the crack tip during the loading half of the cycle.

This equivalent analysis does not take account of a slight overlapping effect (anelastic deformation) usually occurring between the unloading and loading lines on consecutive load-deflection loops. This overlapping is only noticeable in the first few loading cycles and is practically eliminated thereafter. Also, unloading might cause a yield stress increase due to cyclic work hardening which, again, is not considered here.

The second term in equation (27), $d(\Delta U)_1/da = 0$, because ΔU_1 depends on a_0 but not on a. From Figure 15, ΔU_N can be expressed as:

$$\Delta U_N = \Delta\delta_N \, \Delta P - \frac{\Delta P}{2} \Delta\delta_{el} \tag{28}$$

where $\Delta\delta_{el}$ is the elastic loading deflection ED, and $\Delta\delta_N$ is the total deflection HC at load P_{max}. Substitution of equation (28) into equation (27) leads finally to a generalised equation valid for any specimen geometry:

$$\Delta J = -\frac{1}{B_n}\left[\frac{d(\Delta\delta_N)}{da}\Delta P - \frac{\Delta P}{2}\frac{d(\Delta\delta_{el})}{da}\right] \tag{29}$$

The total deflection $\Delta\delta_N$ in equation (29) is obtained from a simplified type of equation (25) as follows:

$$\Delta\delta_N = \Delta\delta_1 + \Delta\delta_{cc} \tag{30}$$

where $\Delta\delta_1$ is the maximum deflection of the first 'specimen' (HB), or of the first 'cycle', and $\Delta\delta_{cc}$ is the accumulated deflection due to cyclic creep (or high temperature creep) and crack growth after N loading cycles (BC).

(A) Analysis for small deflections

Equation (29) may be used to compute ΔJ in any specimen geometry provided that the deflection or extension at the loading points is known under load-cycling conditions. Theoretically, $\Delta\delta_N$ and $\Delta\delta_{el}$ may be calculated for beam specimens using non-linear beam theory derived by Timoshenko. It is assumed here that the crack grows in the direction of the bending stresses, and the specimen is treated as a pair of variable-height beams built in at the crack tip, as mentioned earlier.

It is known from beam theory that $\Delta\delta_1$ may be expressed as:

$$\Delta\delta_1 = 2\int_0^{a_0} x\,k_0\,dx \tag{31}$$

where x is the coordinate along the beam length, and k_0 is the curvature caused by the maximum load at point B in the first cycle (Figure 15). Similarly, the expression for $\Delta\delta_{cc}$ is:

$$\Delta\delta_{cc} = 2\int_0^{a} x\,k_{cc}\,dx \tag{32}$$

where k_{cc} is the curvature caused by the secondary cyclic creep. Again, high temperature creep may be included here.

The beam height, h, was already given by equation (19) which was substituted in the curvature

equations for k_0 and k_{cc} and the integrations performed in a computer program using Simpson's rule. The step interval was 0.125 mm with $a_0 = 70$ mm and the values of the crack length, a, and number of cycles, N, were those obtained in the fatigue crack growth tests at 4.5 and 5 kN. The appropriate values of A and n obtained experimentally for BS15 mild steel were used in the curvature equation for $\Delta\delta_1$. The secondary cyclic creep law used for the evaluation of C_1 and m in the equation for k_{cc} has the form given by:

$$\varepsilon = C_1 N \sigma_{pl}^m \tag{33}$$

The theoretical curves of $\Delta\delta_N$ for load ranges $\Delta P = 4.5$ kN and 5 kN are presented in Figure 16. The total deflection $\Delta\delta_N$ increases with the applied load range and crack length. The cyclic creep deflection rate decreases with crack length as expected, since bending stress σ_{max} also decreases with crack length. Elastic conditions are, respectively, attained at points A and B in Figure 16, at which the crack length values are 105 and 125 mm. This happens when $\Delta P = P_{ys}$, at which point cyclic creep stops. Other appropriate ΔP values would have to be inserted for high temperature or environmental deformations.

(B) Analysis for large deflections

The analysis presented in the previous section is valid for small deflections only, i.e. when the appropriate curvature is assumed to be equal to the second derivative of the deflection. The exact analytical solution is provided by the solution of the differential equations [11]. In [26], a similar differential equation was used to calculate the elastic and plastic deflections for monotonic loading. Additionally, some rotation at the crack tip may occur. This latter effect has also been computed following the procedure explained in [23].

For larger load range values of 5.5, 6 and 6.5 kN, respectively, it is necessary to calculate $\Delta\delta_N$ by integration. The value of $\Delta\delta_1$ has already been calculated and the method employed to calculate $\Delta\delta_{cc}$ is basically the same as that for the monotonic loading case described earlier. Thus, the order

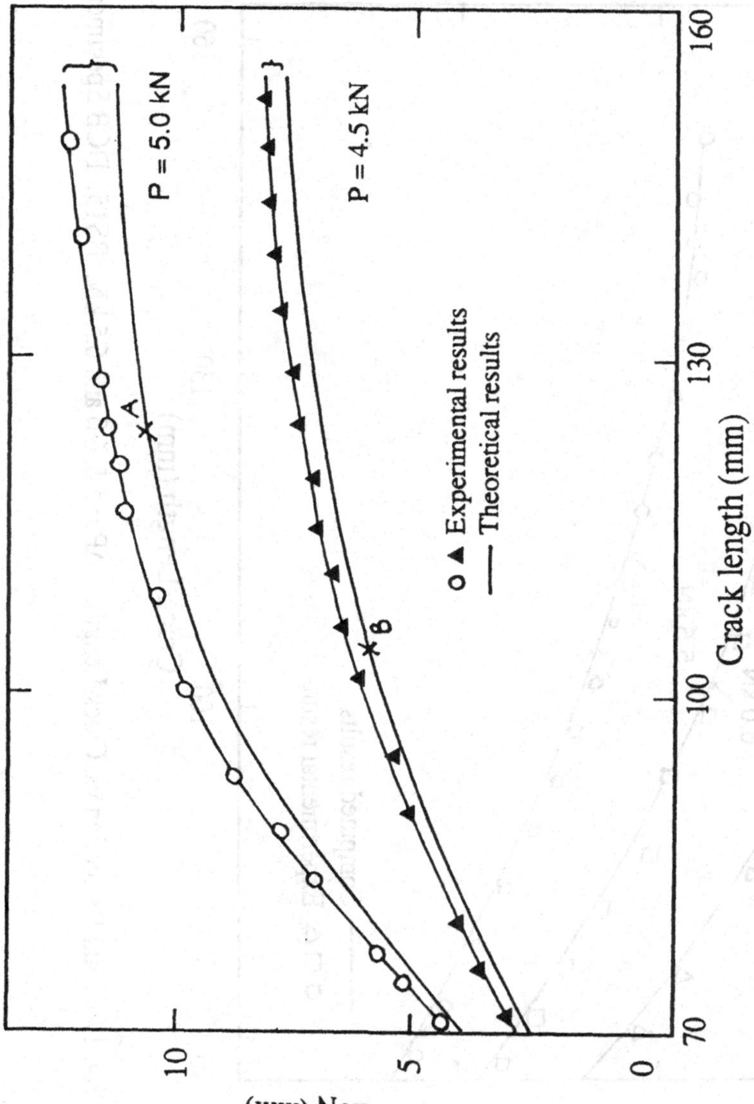

Fig. 16. Total Deflection vs. Crack Length. ΔP = 4.5 and 5.0 kN. BS15. DCB specimen.

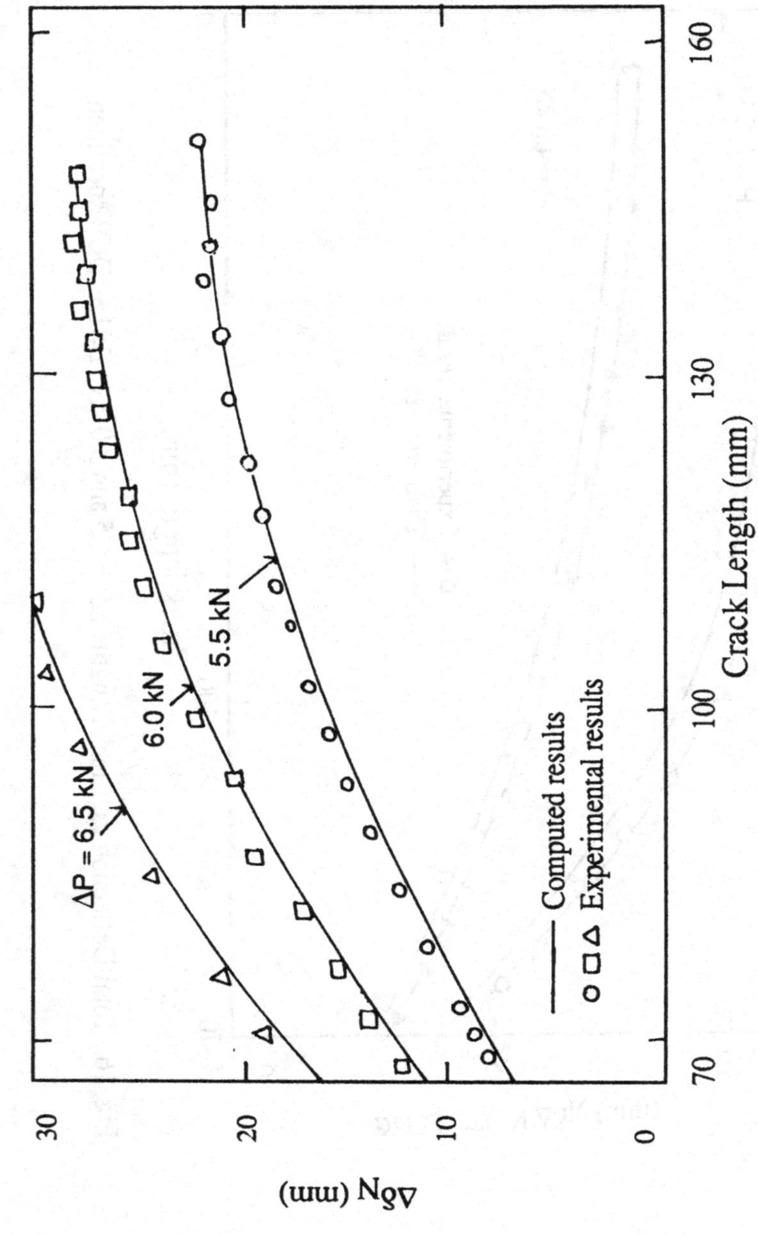

Fig. 17. Total Deflection vs. Crack Length. ΔP = 5.5, 6.0 and 6.5 kN. BS15. DCB Specimen.

of the differential equation has been reduced by putting

$$\frac{d\Delta\delta_{cc}}{dx} = Z_2 \tag{33a}$$

and the curvature equation is:

$$k_{cc} = \frac{dZ_2/dx}{(1 + Z_2^2)^{3/2}} \tag{33b}$$

Equation (33b) was integrated between 0 and the crack length x = a by a computer program applying the modified Euler method. Finally, integration of equation (33a) was carried out in the same program using Simpson's rule with the Z_2 values previously obtained in the first integration. The step interval chosen was 0.125 mm, and the values of the crack length and number of cycles were those obtained in fatigue crack propagation tests at the appropriate load range values.

The computed $\Delta\delta_N$ curves are plotted against crack length in Figure 17. $\Delta\delta_N$ values, which include the deflection δ^R caused by crack tip rotation, increase with the applied load range and crack length, showing the same trend as observed in Figure 16. Linear elastic conditions were only obtained for 5.5 kN at crack length values greater than 152 mm, since the two other load range values (6 and 6.5 kN) exceed the yield load of the specimen.

(C) Evaluation of $\Delta\delta_{cr}$

A number of useful equations is available for creep strain. These relate creep strain to the variables that control creep, like temperature, stress level and hold time. The following equation may be used here:

$$\varepsilon_{cr} = C\,\sigma^{q}\,t^{\beta}\,e^{-p/T} \tag{34a}$$

where ε_{cr} = creep strain after N cycles, t = creep time = $(N\, t_D)$ = (number of cycles elapsed) × (dwell time per cycle), and σ = maximum stress in the cycle that prevails during the dwell time, and the bending stress under the maximum load P_{max} of the fatigue cycle = $P_{max}(a)/(Bh^2/6)$. Making the necessary substitutions:

$$\varepsilon_{cr} = C\,(N\, t_D)^{\beta}\, e^{-p/T}\, \left[\frac{P_{max}(6a)}{B\,h^2}\right]^{q}$$

The effective load causing the creep is the maximum load in the cycle, P_{max}, which operates during the hold period of the trapezoidal load wave.

Analysing, as in cyclic creep, we obtain:

$$\Delta\delta_{cr} = 2[C\,(N\, t_D)^{\beta}\, e^{-p/T}]\left[\frac{P_{max}(2q+1)}{2q\,B}\right]\frac{a^{q+2}}{(q+2)\,(h/2)^{2q+1}} \quad (34b)$$

(D) The ΔJ results

Using linear elastic beam theory, the value of the derivative $d(\Delta\delta_{el})/da$ in equation (29) caused solely by the bending moment is given by:

$$\frac{d(\Delta\delta_{el})}{da} = \frac{24\Delta P}{E\,B}\,\frac{a^2}{(h_0 + h_1\, a)^3} \quad (35)$$

Furthermore, numerical differentiation of the $\Delta\delta_N$ values, equation (30), with respect to a, using a central difference formula inserted in the computer program, gave $d(\Delta\delta_N)/da$. Theoretical values of ΔJ were then calculated by substituting the numerical values of both terms, $d(\Delta\delta_N)/da$ and $d(\Delta\delta_{el})/da$, into the general equation (equation (29)). The theoretical ΔJ against crack length curve is shown in Figure 18 for the load range 5 kN, and in Figure 19 for load ranges 5.5, 6 and 6.5 kN. The value of ΔJ decreases with the crack length corresponding to the decrease in $d(\Delta\delta_N)/da$. For the load range ΔP = 5 kN, the value of J is constant for cracks longer than 115 mm, as expected (Figure 16).

(E) Fatigue crack propagation results

Figures 20 and 12 exemplify load-deflection loops obtained in fatigue crack growth tests at ΔP = 5 and 5.5 kN, respectively. The variation of crack length with number of cycles, presented in Figure 21, shows that for ΔP = 5 kN a constant crack growth rate region starts at a crack length of about 115 mm, whilst for ΔP = 4.5 kN it starts at approximately 100 mm. For shorter cracks, crack growth is faster and non-linear. In Figure 22, the linear region was not attained for load ranges of 6 and 6.5 kN. The load-deflection loops for linear el. unloading converge to point R.

The experimentally determined total potential energy, ΔU, which is the area within the load-deflection loops, is plotted against crack length in Figure 23 for the load ranges of 5.5 and 6 kN. The slopes of these curves are the experimental ΔJ values, which are superimposed on the theoretical curves in Figures 18 and 19.

The relationship between da/dN and ΔJ (or K^2/E, where K^2/E refers to the linear elastic results) is shown in Figure 24. The present experimental crack growth rate values under elastic-plastic cyclic conditions increased by about 2 orders of magnitude from 4.5×10^{-3} mm/cycle to 2.5×10^{-1} mm/cycle, and the maximum ΔJ reached 1100 N/mm.

Discussion

Both theoretical and experimental ΔJ values decrease with crack length and reach the linear elastic value ΔG (Figure 18). The crack growth rate also decreases (Figures 21 and 22), suggesting that the crack growth process may be controlled by ΔJ. The experimental values of ΔJ (Figure 18) are greater than the theoretical ones, differences ranging from 5% for short crack lengths to 10% for long crack lengths. These discrepancies may be explained in terms of deflection differences. The

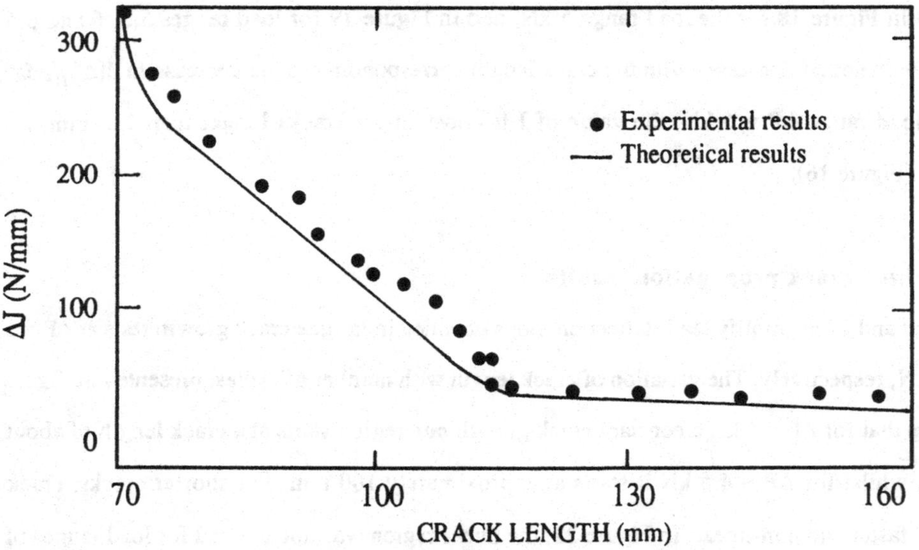

Fig. 18. J vs. Crack Length. ΔP = 5 kN. BS15. DCB Specimen.

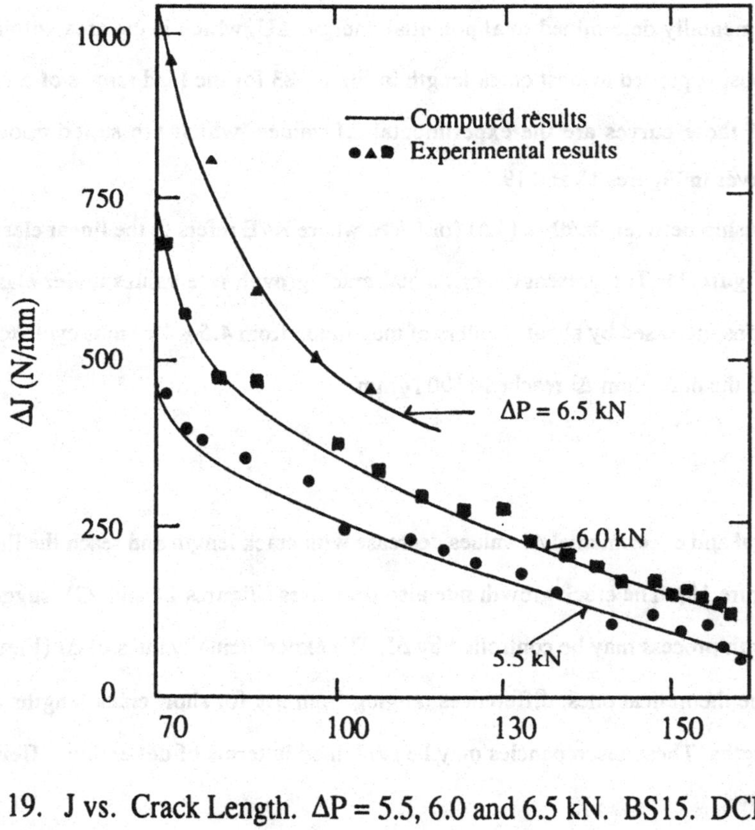

Fig. 19. J vs. Crack Length. ΔP = 5.5, 6.0 and 6.5 kN. BS15. DCB Specimen.

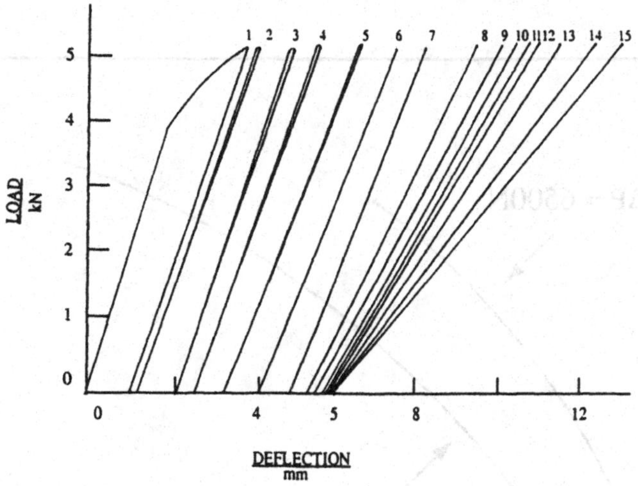

Loop	Cycle	Loop	Cycle	Loop	Cycle
1	1	6	600	11	2400
2	10	7	800	12	2800
3	100	8	1200	13	4000
4	200	9	1600	14	6000
5	400	10	2000	15	8000

Fig. 20. Load vs. deflection loops: P = 5 kN

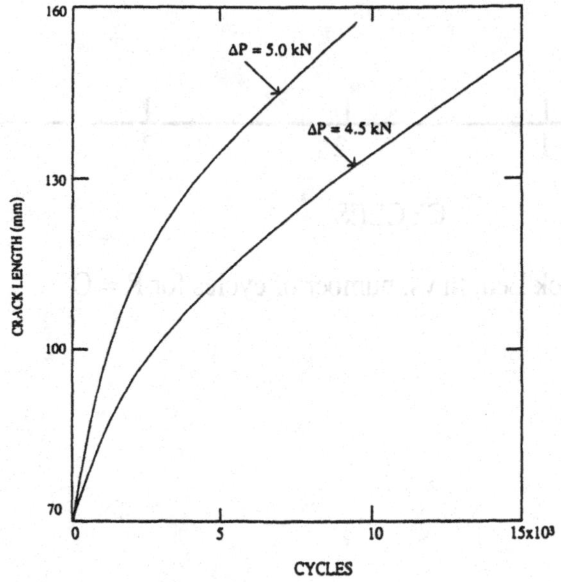

Fig. 21. Crack Growth for Load Cycling. BS15

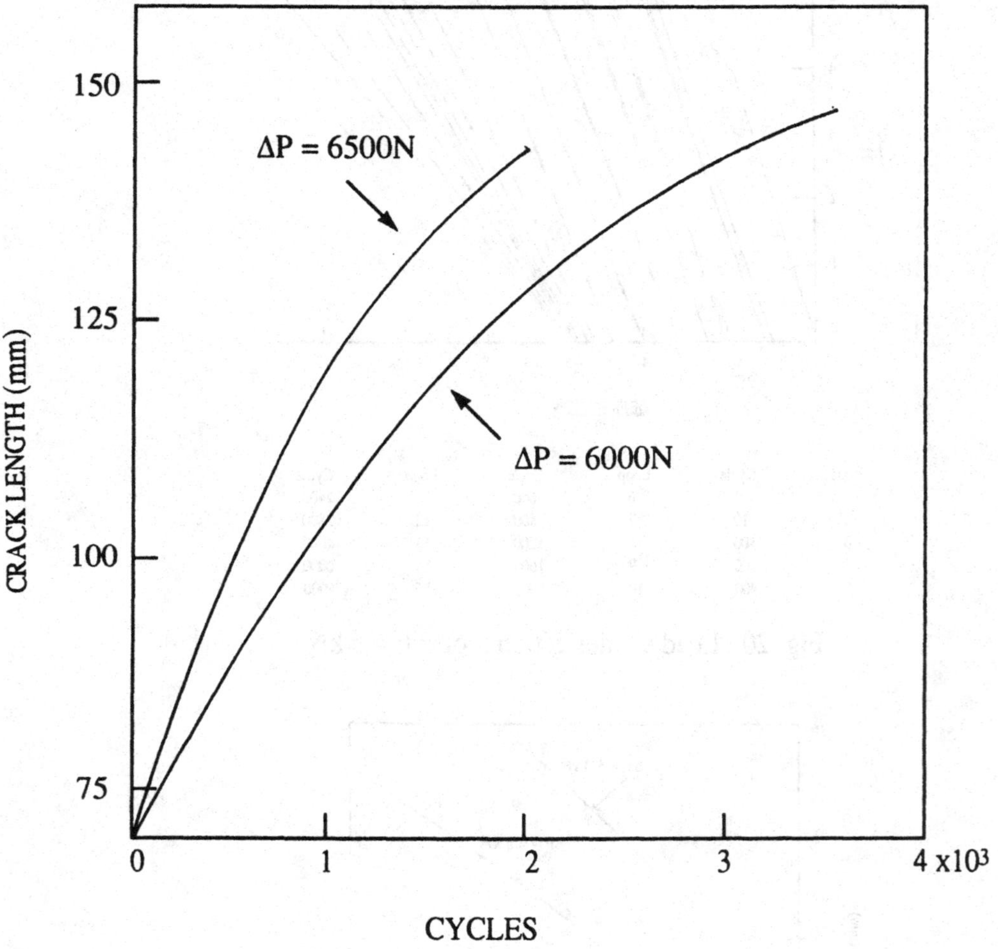

Fig. 22. Crack Length vs. number of cycles for R = O

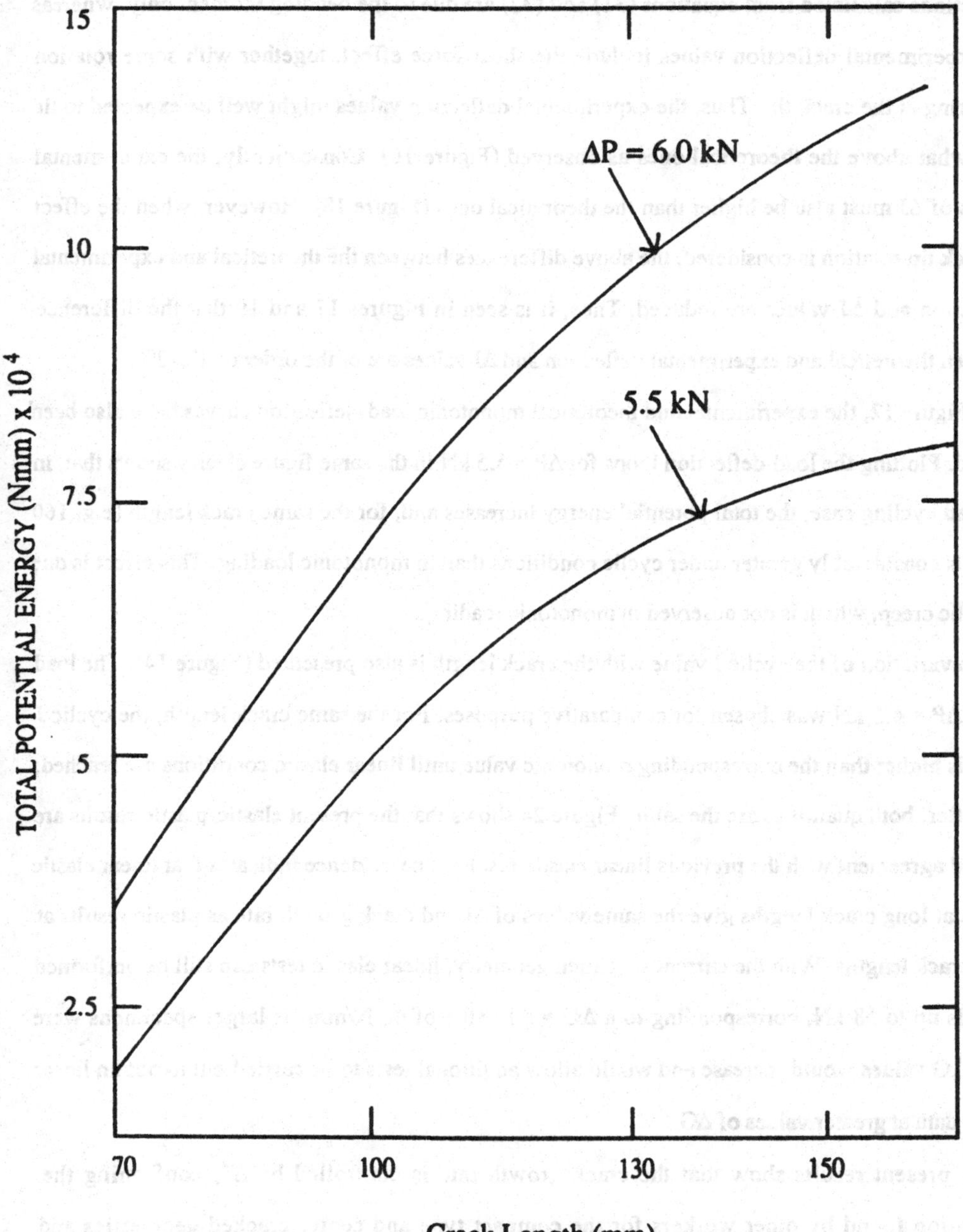

Fig. 23. Total Potential Energy vs. Crack Length. BS15

deflections calculated from equations (31) and (32) are due to the bending moment only, whereas the experimental deflection values include the shear force effect, together with some rotation occurring at the crack tip. Thus, the experimental deflection values might well be expected to lie somewhat above the theoretical ones as observed (Figure 16). Consequently, the experimental values of ΔJ must also be higher than the theoretical ones (Figure 18). However, when the effect of crack tip rotation is considered, the above differences between the theoretical and experimental deflection and ΔJ values are reduced. Thus, it is seen in Figures 17 and 19 that the difference between theoretical and experimental deflection and ΔJ values are of the order of 1%-2%.

In Figure 12, the experimental and theoretical monotonic load-deflection curves have also been plotted. Plotting the load-deflection loops for ΔP = 5.5 kN in the same figure clearly shows that, in the load cycling case, the total potential energy increases and, for the same crack length (e.g. 160 mm), is considerably greater under cyclic conditions than in monotonic loading. This effect is due to cyclic creep, which is not observed in monotonic loading.

The variation of the cyclic J value with the crack length is also presented (Figure 14). The load range ΔP = 4.5 kN was chosen for comparative purposes. For the same crack length, the cyclic J value is higher than the corresponding monotonic value until linear elastic conditions are reached; thereafter, both quantities are the same. Figure 24 shows that the present elastic-plastic results are in good agreement with the previous linear elastic results. The evidence indicates that linear elastic results at long crack lengths give the same values of ΔJ and crack growth rate as plastic results at short crack lengths. With the current specimen geometry, linear elastic tests can still be performed at loads up to 58 kN, corresponding to a $\Delta G = \Delta J$ value of 48 N/mm. If larger specimens were used, ΔG values would increase and would allow additional tests to be carried out to obtain linear elastic data at greater values of ΔG.

The present results show that the crack growth rate is controlled by ΔJ, confirming the correlation found by other workers for the compact type and centre cracked geometries and discussed later. A simplified relationship between the crack growth rate and ΔJ is similar to the equation da/dN = const. $(K_{max}^2 - K_{min}^2)$ developed in this department [34], despite the fact that ΔJ

was calculated by a different process and under different loading conditions. It can be shown that there is a direct relationship between ΔJ and the parameter $\lambda = K_{max}^2 - K_{min}^2$ correlating linear elastic fatigue crack growth data for a range of Al-alloys and steels.This relationship has the form:

$$\Delta J = \frac{\lambda}{E} \tag{36}$$

and by substitution:

$$\frac{da}{dN} = C_{pl}(\lambda)^n \tag{37}$$

Therefore, to the first approximation, a crack growth rate correlation with λ in the linear elastic range implies a similar correlation with ΔJ in the elastic-plastic range.

(2) Geometry of Plates in Tension

Here, again, the generalised equation (29) can be used and the total deflection $\Delta\delta_N$ (equation (30)) expressed as:

$$\Delta\delta_N = l_0(\varepsilon_{ys} + \varepsilon_{pl} + \varepsilon_{cc}) + \Delta\delta_{el} \tag{38}$$

where l_0 is the gauge length of the specimen or the length of the component. As previously, only a simplified type of equation (25) is used here.

Substituting the appropriate values into equation (29), we obtain:

$$\Delta J = l_0 \frac{\sigma A}{B} \frac{d}{da}[(\frac{\sigma}{k})^{1/n} + C_1 \Delta\sigma^m N] + \frac{Y^2 \sigma_{max}^2 \pi a}{E} \tag{39}$$

where B and W are the dimensions of the plate, cross-section A, n is a cyclic value of the strain hardening exponent, and κ is the strain hardening coefficient. The plastic component of the stress $\Delta\sigma = \sigma - \sigma_{ys}$. The first term of equation (39) does not depend on a; its derivative is zero. The crack growth rate in the plastic region is:

$$\frac{da}{dN} = C\,(\Delta J)^{\beta} \tag{40}$$

where C, C_1 and β are experimental constants.

(3) Geometry of Bend Specimens

The application of the generalised equation (29) and equation (30) is appropriate.

Large bending deflections $\Delta\delta_{cc}$ can be obtained from the cyclic creep curvature k_{cc} as follows:

$$k_{cc} = \frac{d^2(\Delta\delta_{cc})/dx^2}{[1 + (\frac{d\Delta\delta_{cc}}{dx})^2]^{3/2}} \tag{41}$$

The extension of the beam fibre located at a distance y_1 from the neutral axis is:

$$\varepsilon_{cc} = \frac{y_1}{\rho_{cc}} \tag{42}$$

where ρ_{cc} is the radius of curvature, equal to $1/k_{cc}$. Using now:

$$\sigma = k\,\varepsilon^n = k\,(\frac{y_1}{\rho})^n \quad \text{and} \quad \Delta\sigma = k\,(\frac{y_1}{\rho})^n - \sigma_{ys} \tag{43}$$

We obtain:

$$k_{cc} = \frac{N C_1}{y_1}[k(\frac{y_1}{\rho})^n - \sigma_{ys}]^m \tag{44}$$

In this equation, $1/\rho$ is expressed as a function of the bending moment M; a suitable function must be selected in the form:

$$k_{cc} = f(y_1) \tag{45}$$

For example, for $y_1 = W/2$, equation (44) becomes:

$$k_{cc} = \frac{2 N C_1}{W}[k(\frac{W}{2\rho})^n - \sigma_{ys}]^m \tag{44a}$$

and $\Delta\delta_{cc}$ is determined from equation (41).

The bending moment, M, for a beam of rectangular cross-section is:

$$M = 2B[\int_0^d \sigma_{y_1} dy_1 + \int_d^{W/2} \sigma_{y_1} dy_1] \tag{46}$$

and on appropriate substitution, we obtain:

$$M = 2B\{\frac{\sigma_{ys}^3 \rho^2}{3E^2} + \frac{k}{(n+2)\rho^n}[(\frac{W}{2})^{n+2} - (\frac{\sigma_{ys}\rho}{E})^{n+2}]\} \tag{46a}$$

giving the curvature ρ as a function of M; the value of k_{cc} can be obtained from equation (44a).

The values of Y in equation (39) are available for three-point bend, cantilever bending and for pure bending in the literature (cf.p.ex. ASTM E399).

Expressions for ΔJ

Using the generalised equation (29) and noting that $\Delta\delta_1$ is not dependent on a, equation (40) can be applied to eliminate the number of cycles N and, after some simplification, equation (29) becomes:

$$\Delta J = \frac{\sigma_{max} W^2 C_2}{6 L C \Delta J^{\beta}} + \frac{Y^2 \sigma_{max}^2 \pi a}{E} \tag{47}$$

which can be solved, by iteration, as $\Delta J = f(a)$.

In the case of three-point bend, the only modification needed would be to change ΔP to $2\Delta P$; thus, the factor 6 in the first term will become 3 and the geometrical factor Y in the second term will now be 2Y. Here, C_2 is in mm/cycle.

When applying bending moment M instead of the load P in a three-point bend specimen, the final form of equation (29) will be:

$$\Delta J = \frac{\sigma_{max} W^2 C_3}{6 C \Delta J^{\beta}} + \frac{Y^2 \sigma_{max}^2 \pi a}{E} \tag{48}$$

where the constant C_3 is in rad/cycle. The angular deflection term $d(\Delta\theta_1 + \Delta\theta_{cc})/da$ has replaced the linear extension term used originally in equation (30). Again, the expression θ_1 does not depend on a, and consequently $d(\Delta\theta_1)/da = 0$. Further, by the integration of equation (46a), the value of $\Delta\theta_{cc}$ was arrived at, and the solution of:

$$\Delta\theta_{cc} = \frac{d(\Delta\delta_{cc})}{da} \tag{49}$$

by integration resulted in the form:

$$\Delta\theta_{cc} = C_3 N \tag{50}$$

A similar method was then used for $\Delta\delta_{cc}$ and, by substitution, equation (48) was finally obtained.

Conclusions

Non-linear fracture mechanics parameters can be used to characterise the fatigue crack propagation phenomenon for metals in which extensive nominal plastic deformation occurs. In the present work, only elastic-plastic fatigue crack growth under load cycling has been studied. However, it is suggested that other processes, such as high temperature creep, may be included in the analysis.

On the basis of the energy interpretation of the J contour integral, the variation of the total potential energy with crack length was adopted as a definition of the cyclic value of J, ΔJ. An analytical method based on non-linear beam theory was derived to compute ΔJ values for situations in which cyclic creep occurs under tensile load cycling above the yield load. Application of the method to contoured DCB specimens showed that the derived relation for ΔJ is a function of load, crack length, number of cycles, the strain hardening or softening and creep properties (high temperature as well as cyclic) of the material. ΔJ was found to decrease with the crack length, reaching a constant value if linear elastic conditions were met.

Experimental investigation of fatigue crack growth phenomena in BS15, under conditions of cyclic tensile load and under elastic-plastic conditions, showed that the cyclic values of J determined from the variation of the total cyclic potential energy with the crack length are in good agreement with the theoretical values of ΔJ.

Elastic-plastic fatigue crack growth rates correlated with ΔJ and with previous linear elastic results, yielding an equation of the form:

$$\frac{da}{dN} = C_{pl} (\Delta J)^{n_1}$$

where C_{pl} and n_1 are constants.

The J contour integral was also successfully evaluated, both theoretically and experimentally, for the constant K contoured DCB specimen of BS15 under monotonic loading conditions. J increased with crack length until a constant value was attained for linear elastic loading. ΔJ values under load cycling are greater than the monotonic values, the difference being due to the increase in potential energy caused by cyclic creep in load cycling. A critical value for J for crack propagation, J_c, was obtained for this material using compact type and contoured DCB specimens.

Acknowledgement

The author wishes to thank C. M. Branco, J. M. Musuva and other members of the Fracture Group for carrying out the tests and numerical calculations.

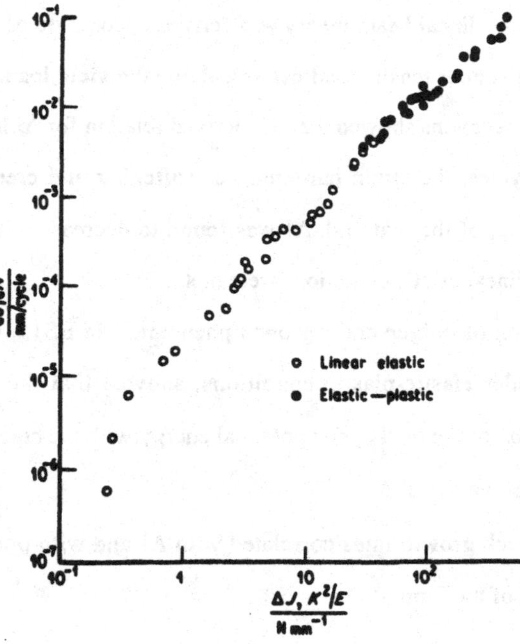

Fig. 24. Crack growth rate vs ΔJ (or K^2/E); R=0

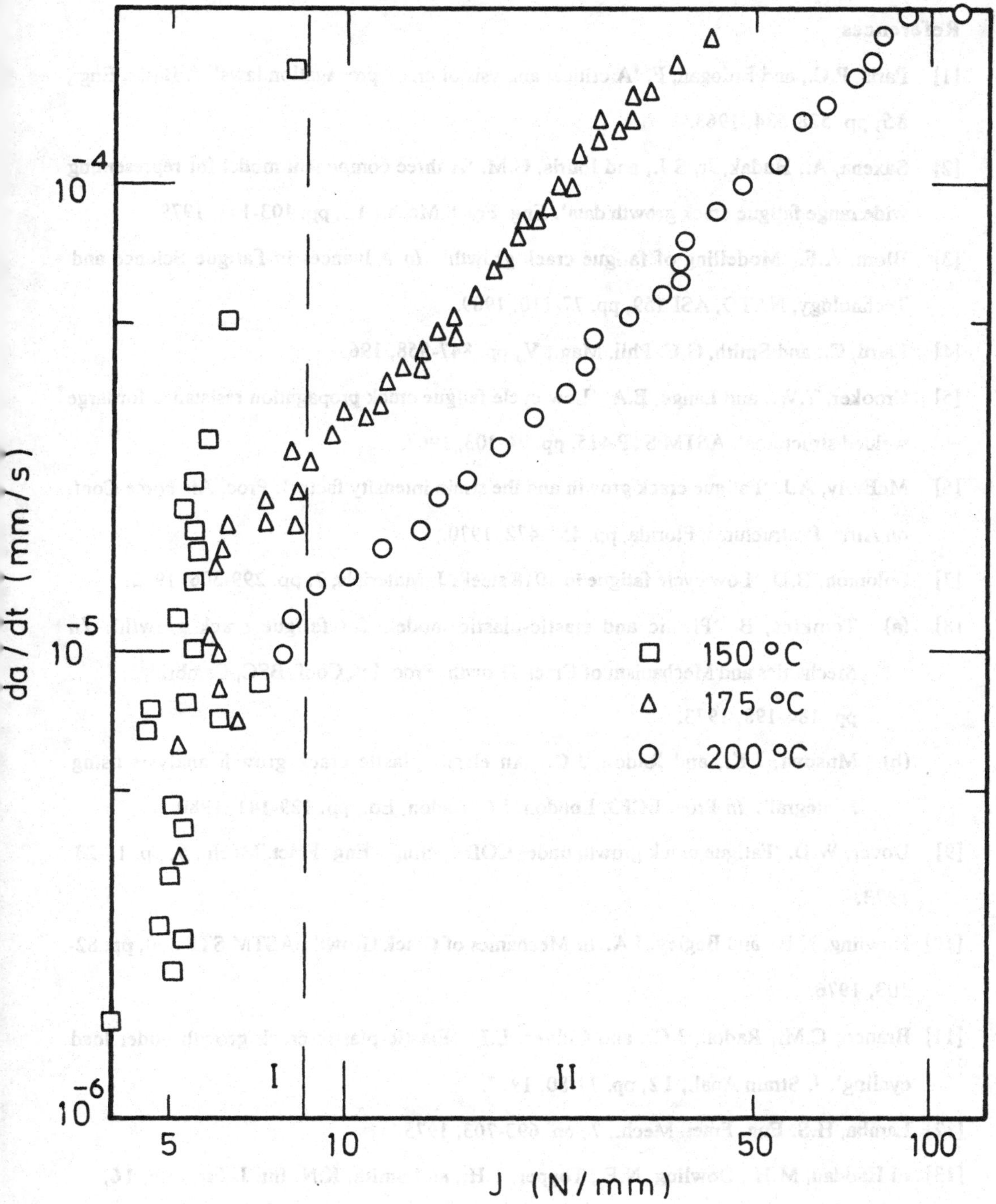

Fig. 25. Crack growth rate vs J. Thickness 25 mm.

References

[1] Paris, P.C., and Erdogan, F. 'A critical analysis of crack propagation laws'. J. Basic. Eng., 85, pp. 528-534, 1963.

[2] Saxena, A., Hudak, Jr, S.J., and Jouris, G.M. 'A three component model for representing wide range fatigue crack growth data'. Eng. Fract. Mech., 12, pp. 103-115, 1979.

[3] Blom, A.F. 'Modelling of fatigue crack growth'. *In* Advances in Fatigue Science and Technology, NATO, ASI 159, pp. 77-110, 1989.

[4] Laird, C., and Smith, G.C. Phil. Mag., V, pp. 847-858, 1961.

[5] Crooker, T.W., and Lange, E.A. 'Low cycle fatigue crack propagation resistance for large welded structures'. ASTM STP 415, pp. 94-103, 1967.

[6] McEvily, A.J. 'Fatigue crack growth and the strain intensity factor'. Proc. Air Force Conf. on Aircraft Structures, Florida, pp. 451-472, 1970.

[7] Solomon, H.D. 'Low cycle fatigue in 1018 steel'. J. Materials, 7, pp. 299-306, 1972.

[8] (a) Tomkins, B. 'Plastic and elastic-plastic models for fatigue crack growth'. *In* Mechanics and Mechanism of Crack Growth, Proc. Int. Conf. BSC, Cambridge, pp. 184-198, 1973.

(b) Musuva, J.K., and Radon, J.C. 'An elastic-plastic crack growth analysis using J-integral'. *In* Proc. ECF3, London, J.C. Radon, Ed., pp. 129-141, 1980.

[9] Dover, W.D. 'Fatigue crack growth under COD cycling'. Eng. Fract. Mech., 5, pp. 11-23, 1973.

[10] Dowling, N.E., and Begley, J.A. *In* Mechanics of Crack Growth, ASTM STP 590, pp. 82-103, 1976.

[11] Branco, C.M., Radon, J.C., and Culver, L.E. 'Elastic-plastic crack growth under load cycling'. J. Strain Anal., 12, pp. 71-80, 1977.

[12] Lamba, H.S. Eng. Fract. Mech., 7, pp. 693-703, 1975.

[13] El Haddad, M.H., Dowling, N.E., Topper, T.H., and Smith, K.N. Int. J. Fracture, 16, pp. 15-30, 1980.

[14] Kaisand, L.R., and Mowbray, D.F. J. Testing & Eval., 7, pp. 270-280, 1979.

[15] Douglas, M.J., and Plumtree, A. *In* ASTM STP 677, pp. 68-84, 1979.

[16] Radon, J.C., De Castro, P.M.S.T., and Culver, L.E. 'Crack growth in constant load and constant displacement'. Proc. ECF2, Darmstadt, pp. 311-336, 1979.

[17] Musuva, J.M., and Radon, J.C. 'Size effects and J-integral approach to low cycle fatigue'. Proc. Int. Symp. on Low Cycle Fatigue and Elastic-Plastic Behaviour, DVM, Stuttgart, Rie & Haibach, Eds., pp. 479-494, 1979.

[18] Sadananda, K., and Shahinian, P. Eng. Fract. Mech., 11, pp. 73-86, 1979.

[19] Shao-Lun, Liu and Ji-Zhou. Trans. ASME, J. Eng. Gas Turbines, 108, pp. 521-524, 1986.

[20] Minzhong, Zheng and Liu, H.W. Trans. ASME, J. Eng. Mats. & Technology, 108, pp. 198-205, 1986.

[21] German, M.D., Delorenzi, H.G., and Wilkening, W.W. Eng. Fract. Mech., 22, pp. 871-883, 1985.

[22] El Haddad, M.H., and Mukherjee, B. *In* Elastic-Plastic Fracture, ASTM STP 803, pp. 689-707, 1983.

[23] (a) Branco, C.M., Radon, J.C., and Culver, L.E. 'Fatigue crack growth in gross plasticity cycling'. ASM, Proc. 2nd Int. Conf. MBM, Boston, pp. 632-635, 1976.

(b) Srinivas, M. 'High Temperature Fatigue Crack Growth Studies in Elastic-Plastic Régimes'. PhD Thesis, IIT, Bombay, 1988.

[24] Mowbray, D.F. *In* ASTM STP 601, pp. 33-46, 1976.

[25] Bicego, V. 'Low cycle fatigue life predictions in terms of an EPFM small crack model'. Eng. Fract. Mech., 32, pp. 339-349, 1989.

[26] Radon, J.C., Branco, C.M., & Culver, L.E. 'Creep in low endurance fatigue of mild steel'. Int. J. Fracture, 13, pp. 595-610, 1977.

[27] Bowles, C.Q. 'The J-integral approach in elastic-plastic fatigue crack propagation'. *In* Advances in Fatigue, NATO, ASI 159, pp. 467-488, 1989.

[28] Sadananda, K., and Shahinian, P. *In* Fracture Mechanics, ASTM STP 700, pp. 152-163, 1980.

[29] Okazaki, M., Hattori, I., Ikeda, T., and Koizumi, T. Trans. ASME, **107**, pp. 346-355, 1985.

[30] Manson, S., Halford, G.R., & Hirschberg, H.H. NASA Report TMX-67838, 1971.

[31] Majumdar, and Maiya, N. ASME - MPC, pp. 323-341, 1976.

[32] Plumtree and Lemaitre. *In* Proc. ICF5, Advances in Fracture, pp. 2379-2384, 1982.

[33] Saxena, et al. ASTM STP 743, 1981.

[34] (a) Arad, S., Radon, J.C., and Culver, L.E. 'Growth of fatigue cracks in metals and polymers'. Eng. Fract. Mech., 6, pp. 195-211, 1974.

(b) Radon, J.C. 'Influence of environment on threshold in fatigue crack growth'. Metal Sci., 13, pp. 411-419, 1979.

[35] Kenyon, J.L., et al. 'An investigation of the application of fracture mechanics to creep cracking'. *In* Proc. Int. Conf. on Creep and Fatigue, Philadelphia, 1973, CP13, Paper C157/73, pp. 1-8, 1973.

[36] Branco, C.M., and Radon, J.C. 'Analysis of creep cracking by the J-integral concept'. *In* Proc. Int. Conf. on Engineering Aspects of Creep, Sheffield, Paper C210/80, pp. 43-48, 1980.

[37] Dimopulos, V., Nikbin, K.M., and Webster, G.A. 'Influence of cyclic to mean load ratio on creep/fatigue crack growth'. ASM 8620-001, 1986.

[38] Nikbin, K.M., Smith, D.J., and Webster, G.A. 'An engineering approach to the prediction of creep crack growth'. Trans. ASME, **108**, pp. 186-191, 1986.

[39] Radon, J.C., et al. 'Use of DCB tests in fracture studies'. Conf. on Practical Applications of Fracture Mechanics, I. Mech. E., Paper C7/71, pp. 48-55, 1971.

[40] Radon, J.C., and Oldroyd, P.W.J. 'Plasticity of steels in reversed cycling'. *In* Non-Linear Problems in Stress Analysis, P. Stanley, Ed., pp. 285-198, Applied Science Publishers, 1978.

ON J INTEGRAL AND J_{Ic} MATERIAL TESTING (WITH SPECIAL REFERENCE TO WELDED STRUCTURES)

S. Sedmak
University of Belgrade, Belgrade, Yugoslavia

ABSTRACT

The introduction of J integral for crack significance evaluation enabled the extension of fracture mechanics approach to wider range of mechanical behaviour of materials and structures. In addition to brittle fracture, controlled by plane strain fracture toughness, K_{Ic} and J_{Ic}, J integral can also desribe the crack behaviour in plastic range. Standardized methods ASTM E399, E813 and E1152, and JSME 001 are shortly described with comments on their application. Special attention is paid to J integral direct measurement method, that enabled the evaluation of crack behaviour not only on tensile panel specimens with surface crack, but also on the real structure such a pressure vessel. The crack behaviour in welded structure is strongly depended on structural heterogeneity of welded joints and the differences in mechanical properties of base metal, weld metal and heat-affected-zone.

The effect of weldment heterogeneity on its strength and ductility is studied on SM80P high-strength low-alloyed steel of 800 MPa ultimate tensile strength and three different welded joints, performed by sumerged are welding. The strain distribution in welded joint is depended on matching effect of weld metal, that was performed as normalmatched (slightly lower yield strength of weld metal compared to 800 MPa base metal), undermatched and overmatched. For the same weldments combinations, the effect of crack size and position on strength and ductility was analyzed and the benefits of overmatching effect for cracked weldment safety were clearly recoanized. Direct measurement of J integral for cracks positioned in weld meta and heat-affected-zone enabled the comparison with the same value on small three-point-bend specimens. Pop-in effect was observed in small specimens, with the crack tip in HAZ. Practical application of fracture mechanics parameters is demonstrated for residual strength evaluation of cracked pressure vessel by use of Ratvani, Erdogan and Irwin elasto-plastic model for crack driving force and comparison with materials J resistance cuives

INTRODUCTION

Crack driving force, G, and stress intensity factor, K, fracture mechanics parameters established by mathematical analysis of elastic cracked body [1,2], enabled the development and definition of basic property of cracked material, K_{Ic}, the plane strain fracture toughness [3,4]. In the same time, the introduction of crack opening displacement (COD) measurement [5] had open an additional practical approach for the analysis of crack behaviour in structural materials [6]. It had been recognized that requirements for valid testing of K_{Ic} [3] reduces its application only to the high-strength materials by the conditions

$$B, a, W - a \geq 2.5 \left(\frac{K_{Ic}}{R_{p0,2}}\right)^2 \quad (1)$$

B is specimen thickness, W its width, a crack length, $R_{p0,2}$ yield strength. Practical application of COD has been successfull in different situations [7,8,9], but in many cases this application is questionable because the mathematical analysis for an extension to plastically deformed regions is not available [10].

Path independent J integral is defined by Rice [11] as an energy criterion in linear-elastic fracture mechanics along an arbitrary path Γ (Fig. 1)

$$J = \int_\Gamma \left(W dy - \vec{T} \frac{\partial \vec{u}}{\partial x} ds \right) \quad (2)$$

Here, W is strain energy density, $\vec{T}$ is outward traction vector, $\vec{u}$ displacement vector and ds is an increment of the contour path. The introduction of J integral offerred further possibility in development of fracture mechanics testing. In linear elastic region J integral represents the crack driving force, G, and therefore, stress intensity factor, K_I, can be expressed by

$$K_I^2 = E' G_I = E' J_I \quad (3)$$

where E'= E for plane stress condition and E'= E/(1- ν^2) for plane strain condition; E is elasticity modulus, ν Poisson's ratio.

The relation

$$K_{Ic}^2 = \frac{E}{1-\nu^2} J_{Ic} \quad (4)$$

enables the indirect determination of K_{Ic} by J_{Ic} measurement [12], simplifying the application of linear elastic fracture mechanics in evaluation and testing of cracked material.

Tne extension of J integral into inelastic region is recently accepted in ASTM E 1152-87 Standard Test Method for Determining J-R Curves [13] for characterisation of the resistance of metallic materials to slow stable crack growth after initiation from a preexisting very sharp flaw.

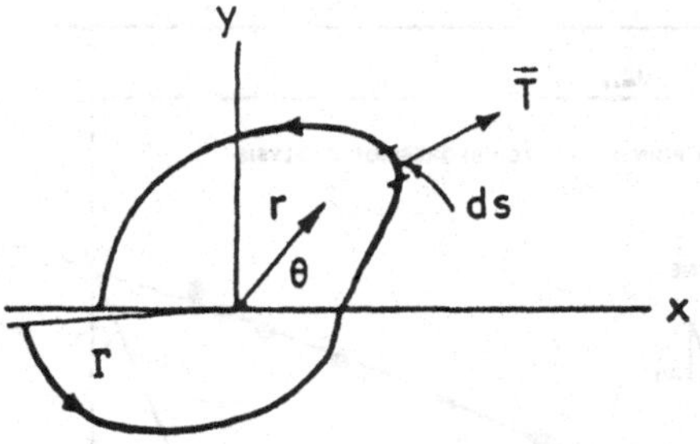

Fig. 1. The path of J integral: definition

2. J_{Ic} - A MEASURE OF FRACTURE TOUGHNESS

According to ASTM E 616-82 [14], fracture toughness is generic term for measures of resistance to extension of a crack.

It is generally accepted that fracture toughness reaches lower limiting value when sharp crack is exposed to severe tensile constraint, such that the state of stress near crack front approaches tritensile plane strain and the crack tip plastic region is small compared to crack size and specimen dimensions in constraint direction. The value K_{Ic} of plane-strain fracture toughness is critical (minimal) value of stress intensity factor K_I [2].

The resistance of material to crack growth could be expressed, probably in most convenient way, in the form of J-R curve [13], that is J dependance of crack extension Δ a. Two typical plots, given in Fig.2, are taken from ASTM E 813 (a) and ASTM E 1152 (b).

The explanation of basic points of J_{Ic} value determination could help for better understanding of processes on the crack tip and in its vicinity during loading process. This is of special interest when heterogeneous structures, such as weldments, are considered.

Different specimen shapes have been developed by fracture mechanics analysis [14]. Two of them, namely **S**ingle **E**dge **B**ending - SE(B) and **C**ompact version of SE **T**ension - C(T) are recommended in many standards [3,4,6,12,13] as most convenient. However, there are some differences in C(T) specimen shapes defined in (a) E 399 and (b) in both E 813 and E 1152, as presented in Fig. 3.

There are also some differences in definitions of SE(B) specimens for three-point bending (Fig. 4).

Very important requirement in specimen production is fatigue precracking from machined notch (Fig. 5), in order to assure the plane-strain conditions at the crack tip in reproducible way.

Typical load-displacement records for K_{Ic} determination are given in Fig. 6. In addition to the requirement (1), the recorded ratio P_{max}/P_Q must not exceed 1.1 for valid K_c test. P_Q is load for conditional K_Q, calculated from

$$K_Q = \frac{P_Q}{BW^{1/2}} f\left(\frac{a}{W}\right) \quad \text{for } C(T) \qquad K_Q = \frac{P_Q S}{BW^{3/2}} f\left(\frac{a}{W}\right) \quad \text{for } SE(B) \qquad (5)$$

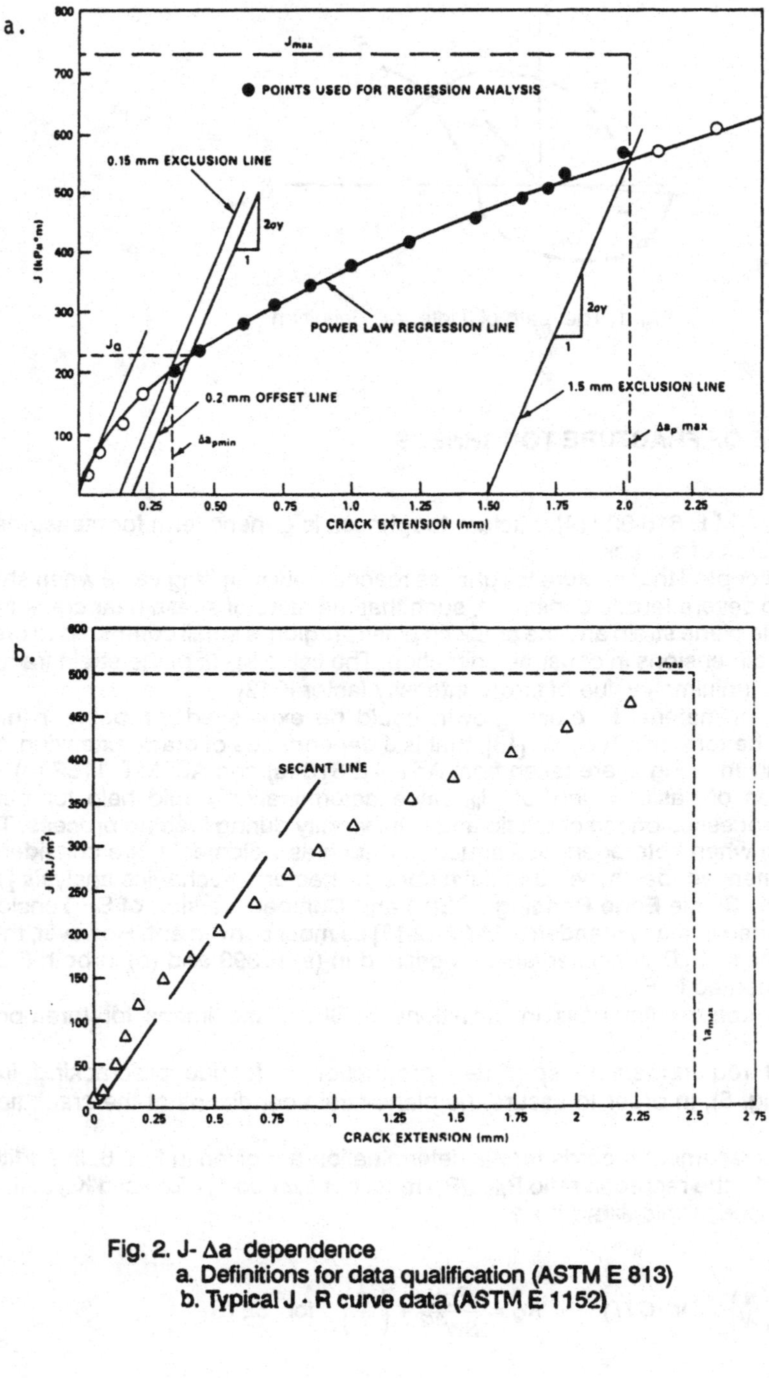

Fig. 2. J- Δa dependence
a. Definitions for data qualification (ASTM E 813)
b. Typical J - R curve data (ASTM E 1152)

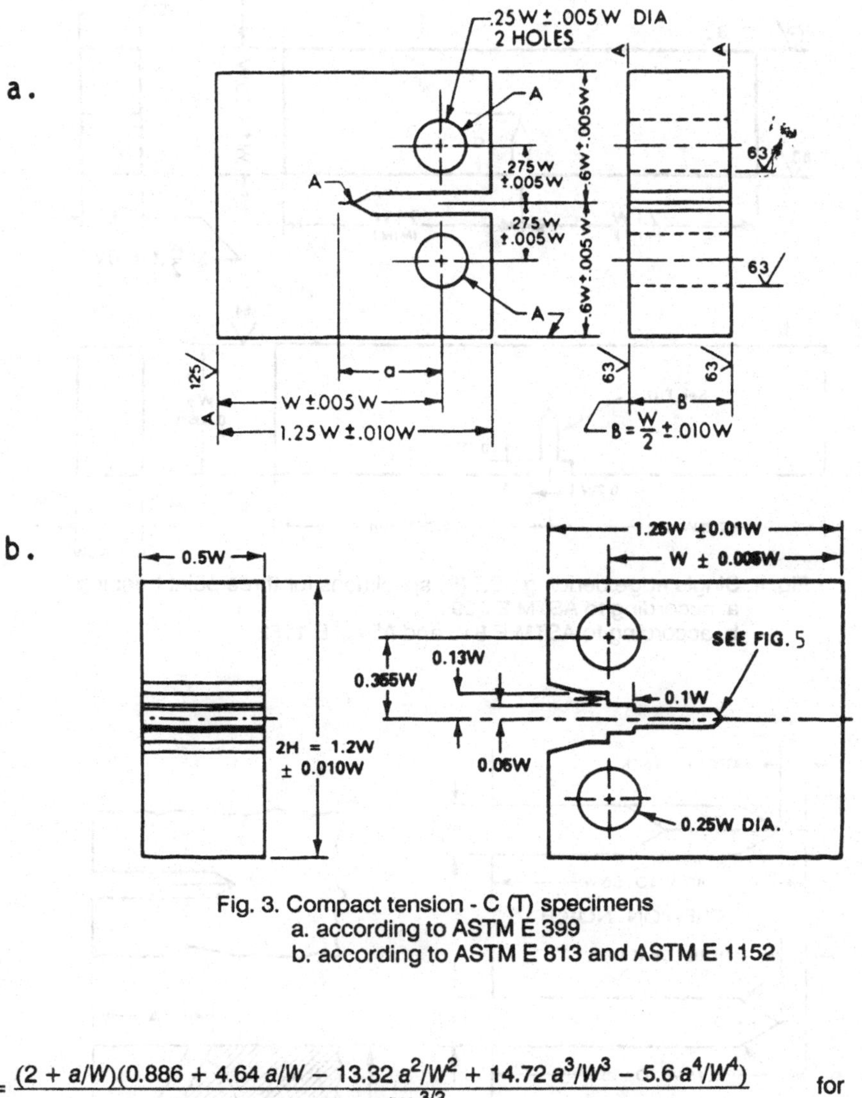

Fig. 3. Compact tension - C (T) specimens
a. according to ASTM E 399
b. according to ASTM E 813 and ASTM E 1152

where

$$f\left(\frac{a}{W}\right) = \frac{(2 + a/W)(0.886 + 4.64\, a/W - 13.32\, a^2/W^2 + 14.72\, a^3/W^3 - 5.6\, a^4/W^4)}{(1 - a/W)^{3/2}} \quad \text{for} \quad SE\,(B)$$

(6))

$$f\left(\frac{a}{W}\right) = \frac{3\,(a/W)^{1/2}\,\left[1.99 - (a/W)(1 - a/W)(1 - a/W)\,(2.15 - 3.93\, a/W + 2.7\, a^2/W^2)\right]}{2\,(1 + 2a/W)\,(1 - a/W)^{3/2}} \quad \text{for } C(T)$$

S is span for SE(B) specimen, usually S = 4W.

Two basic requirements for valid K_{Ic} correspond to constraint achieved by sufficient thickness of high strength material according to req. (1) or by brittle behaviour of tested

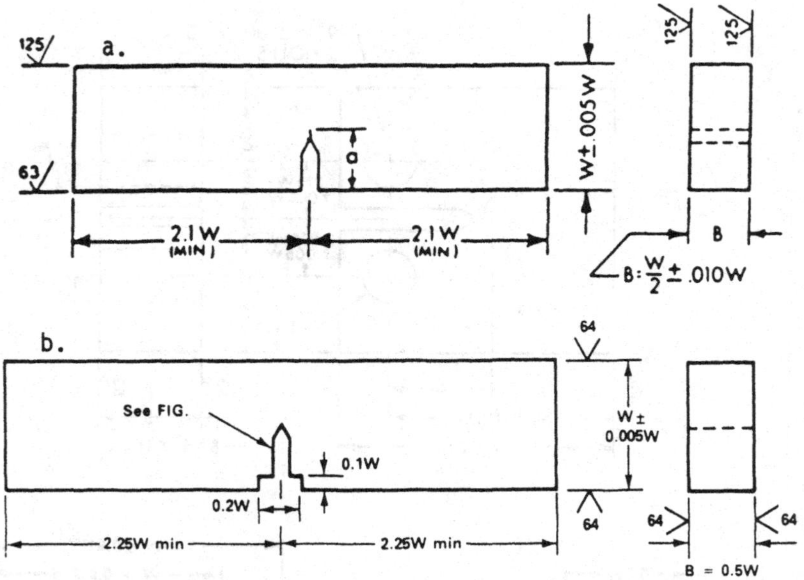

Fig. 4. Single edge bending - SE (B) specimens for three point bending
a. according to ASTM E 399
b. according to ASTM E 813 and ASTM E 1152

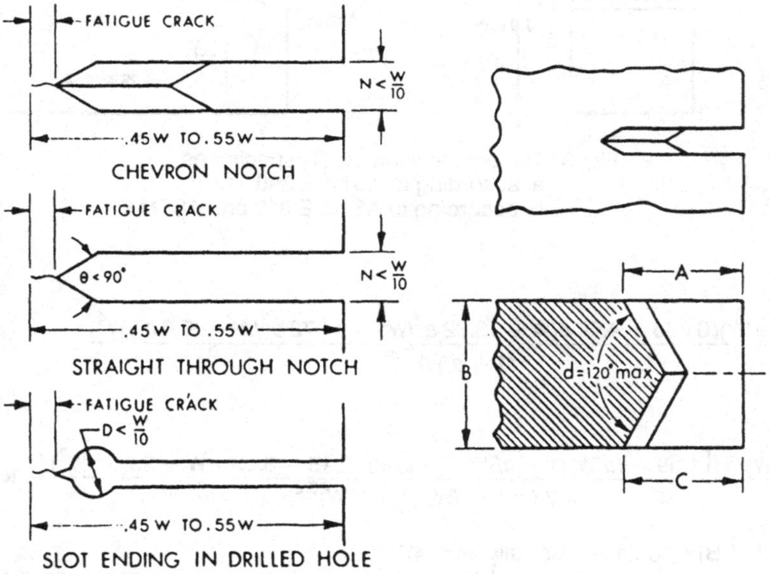

Fig. 5. Crack starter notch and fatigue crack configurations [3]

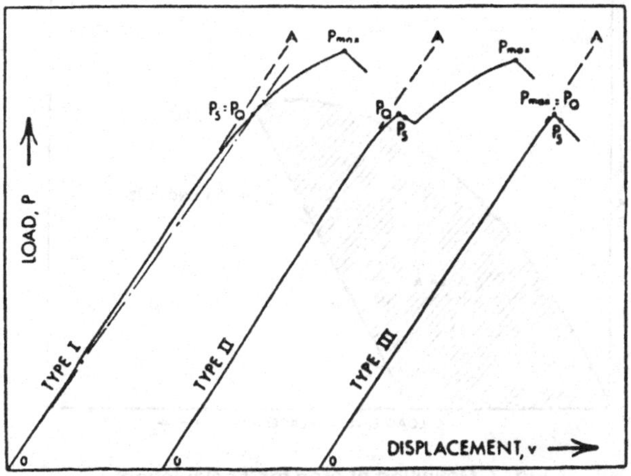

Fig. 6. Principal types of load-displacement records for K_{Ic} determination

material, expressed by requirement $P_{max}/P_Q < 1.1$. If these conditions are met, $K_{Ic} = K_Q$. It is to be mentioned that, generally, high-strength steels are brittle.

It is not posible to determine K_{Ic} for materials with low yield strength level, because specimens dimensions, required by (1) are to large, and the loading capacity of testing machine would not be sufficient for specimen fracturing. The application of J integral has extended the limits for valid determination of fracture toughness. The basis for this aplication is determination of J_Q value (Fig. 2a) from J-R curve under conditions given in E 813. That means that $J_Q = J_{Ic}$ if:

1. $B, b_o\ (= W - a) > 25 \dfrac{J_Q}{\sigma y}$ (7)

2. The slope of power law regression line, dJ/da, evaluated at Δa_Q is less than σ_y (σ_y is average of 0.2% offset tensile yield strength, $R_{p0,2}$ and ultimate tensile strength, R_m).

3. The additional four requirements, defined in E813 are met. It is necessary to say that the requirements in E 813 are less strict than those in E 399.

Original version of J integral testing is based on the definition that J integral is equal to the value obtained from two identical bodies with infinitesimally differing crack areas, each subjucted to load. Consequently, J is the difference in work per unit difference in crack area at a fixed value of displacement or, where appropriate, at a fixed value of load. The consequence is that for each point in Fig. 2a one test specimen is reqiured, and the crack length is allowed to extend to different final value at the end of test (Fig. 7). That means, for similar initial crack length in the specimens of the same size, the areas A must be different. It has been accepted that J value consists of elastic component J_{el} and plastic component J_{pl}.

$$J = J_{el} + J_{pl} \tag{8}$$

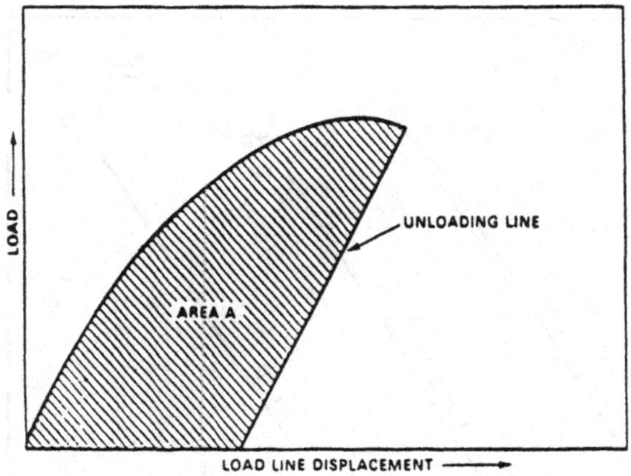

Fig. 7. Definition of area for J calculation

For J_{el} eq. (3) is valid, and K could be calculated from eq. (5) and (6). Plastic part corresponds to

$$J_{pl} = \frac{\eta \cdot A_{pl}}{B \cdot b_o} \tag{9}$$

where η = 2 for SE(B) specimen and $\eta = 2 + 0.522\, b_o/W$ for C(T) specimen; A_{pl} is area A defined in Fig. 7. With different crack extensions the residual compliance will be different and this is expressed by different slopes of unloading lines. This affects the value of A_{pl} (Fig. 7) and can be taken into account by the procedures, described in E 813 and E 1152.

Very expensive multispecimen tests for J_{Ic} evaluation is a severe limitation for practical use.

For this reason, ASTM E 1152 standard has accepted single specimen unloading technique for J-R curve determination, as it is presented by typical load-displacement records for ASTM A 516 Gr 70 steel and its submeged-arc-welded (SAW) weld metal in Fig. 8. It is to be mentioned here that J_{Ic} = 283 kPa $\sqrt{m}$ is obtained for A 516 Gr 70 steel and J_{Ic} = 100 kPa $\sqrt{m}$ for its SAW weld metal by straight line regression law according to ASTM E 813-81 [15].

Fig. 9 represents simplified behaviour of stressed crack. Initial part corresponds to blunting of sharp crack and material obeys linear relationship on J-R curve, defined by [16]

$$J = 2\,\sigma_y \Delta a \tag{10}$$

Apparent crack growth during blunting process will be over by formation of final stretch zone at the moment when the crack starts to grow by stable tearing. Corresponding point in J-R curve is an intersection of blunting line and stable tearing line, indicating onset of crack growth. J value in this specific point is J_{Ic}, a measure of fracture toughness.

Stretch zone width (SZW) is a basis for japanese standard JSME S001/81 [17]. Blunting line in this standard is defined by J-SZW relationship before stable crack growth (Fig. 10).

Two or more specimens are loaded to the specified displacement below crack onset, crack extension in blunting is marked by additional fatigue (Fig. 10a) and stretch zone width

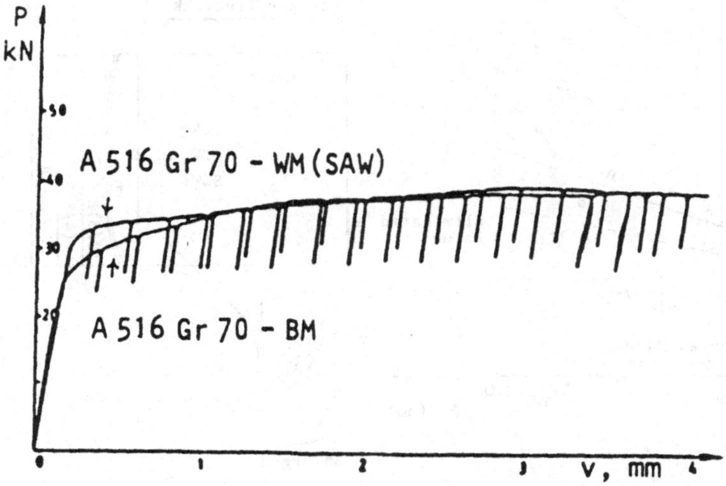

Fig. 8. Load-displacement records for single specimen J-R curves testing of ASTM A 516 Gr 70 steel and its submerged-arc-welded (SAW) weld metal

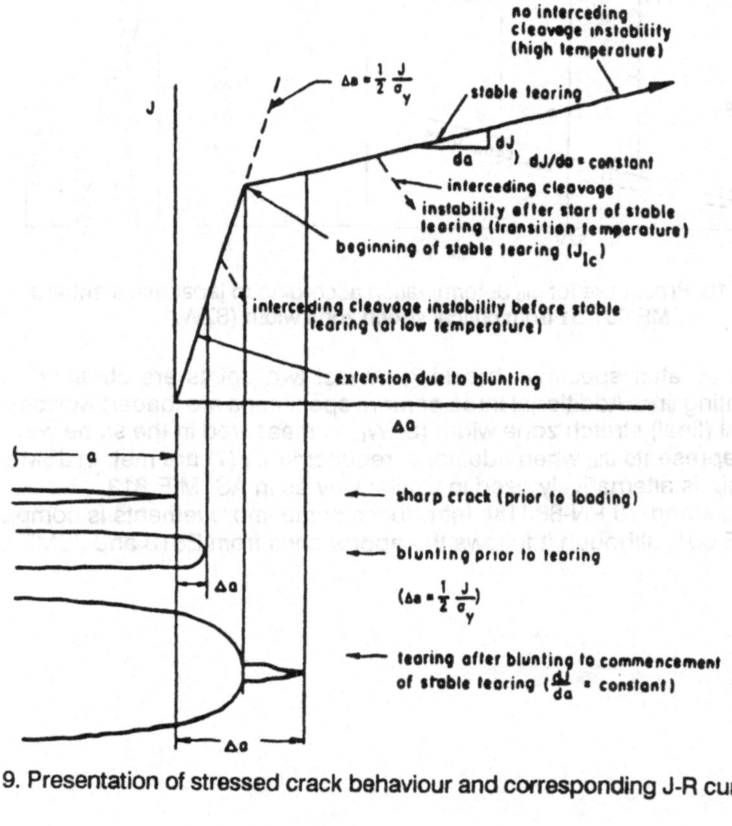

Fig. 9. Presentation of stressed crack behaviour and corresponding J-R curve

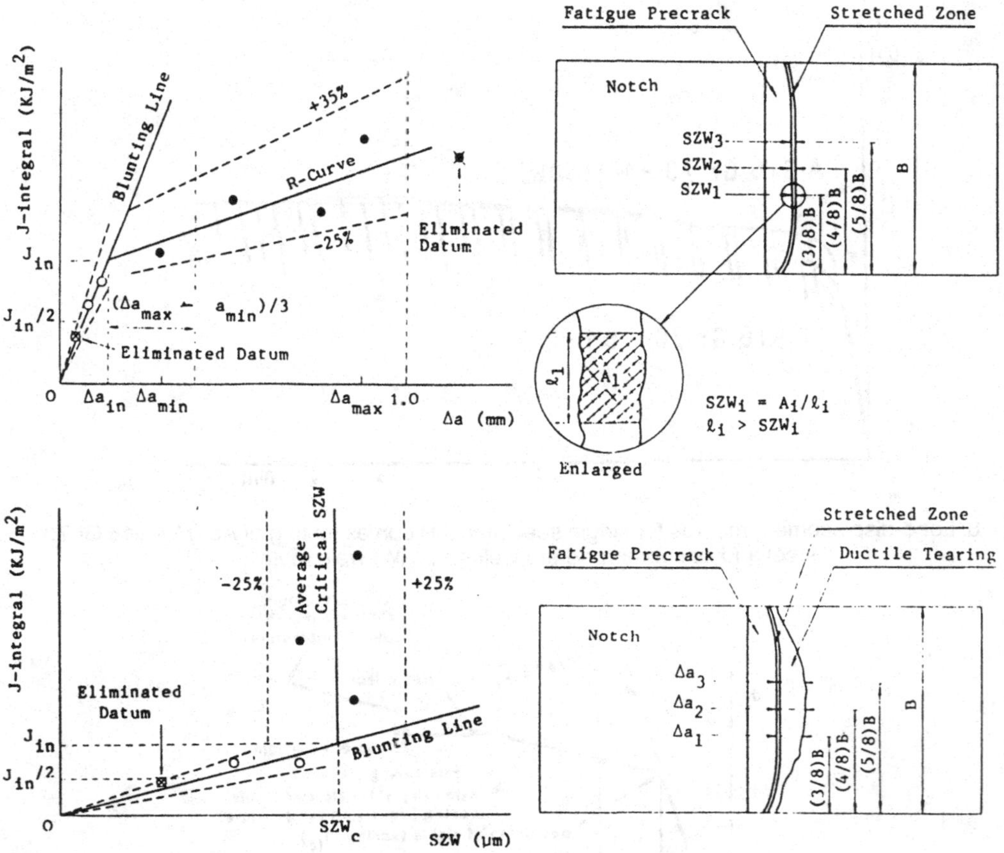

Fig. 10. Procedure for J_{Ic} determination according to japanese standard JSME 001/81 from critical strech zone width (SZW_c)

(SZW) is measured after specimen breaking. At last two points are obtained in this way, defining the blunting line. Additional three or more specimens are loaded well beyond crack onset and critical (final) stretch zone width (SZW_c) is measured in the same way. The point of intersection represents J_{Ic} when additional requirements (7) are met. In JSME 001/81 J- Δ a relationship is alternatively used in similar way as in ASTM E 813.

A new polish standard PN-88 [18] introduces some improvements is comparison to E 813 and JSME 001, although it follows the approaches from E813 and JSME S001.

3. EXPERIMENTAL METHOD FOR DIRECT EVALUATION OF THE J CONTOUR INTEGRAL

The selection of an appropriate contour around the crack and direct use of eq. (2) enabled the development of an experimental method for direct measurement of J integral, Read [19]. This contour is given in Fig. 11.

Required strain values during tension are followed from strain gauge (SG) readings. On this basis, strain energy density can be defined in the form

$$W = \int \sigma_{yy}\, d\varepsilon_{yy} \tag{11}$$

because on the free specimen surface along the contour DC and AB only the component σ_{yy} of stress tensor σ_{ij} exists. The contour segment BC, that intersects the specimen, has to be far enough from the crack tip, in order to avoid the singularity effect.

In the linear elastic range the relation $\sigma_{yy} = E \cdot \sigma_{yy}$ holds. Yield criterion is $E\,\varepsilon_{yy} \geq R_{p0,2}$. These two assumptions allowed the definition of bilinear relationship (Fig. 11-II)

$$W = \frac{1}{2}\sigma_{yy}\,\varepsilon_{yy} = \frac{1}{2}E\,\varepsilon_{yy}^2 \quad \text{for} \quad \varepsilon_{yy} \leq \frac{R_{p0.2}}{E}$$

$$(12)$$

$$W = \frac{R_{p0.2}^2}{2E} + R_{p0.2}\left(\varepsilon_{yy} - \frac{R_{p0.2}}{E}\right) \quad \text{for} \quad \varepsilon_{yy} > \frac{R_{p0.2}}{E}$$

The traction vector is zero on the free specimen surfaces along the segments AB and CD. In BC section (Fig. 11) partial derivates $\partial v/\partial x$ can be expressed in the form

$$\frac{\partial v}{\partial x} = \frac{v\,(C) - v\,(B)}{x\,(C) - x\,(B)} = \frac{v\,(C) - v\,(B)}{B} \tag{13}$$

where v(B) and v(C) are displacement in points B and C on smooth and cracked side, respectively, and B is specimen thickness. Traction vector on BC segment is

$$\vec{T_i} = \sigma_{ij} \cdot \vec{n_j} \tag{14}$$

with $\vec{n_j}$ denoting the unit vector of contour outer normal. Along BC segment $n_y = 1$ and $n_x = 0$, and, consequently, $T_y = \sigma_{yy}$ and $T_x = \sigma_{xy}$. Shear stress component σ_{xy} can be neglected in the remote section.

Due to the simmetry, one can write according to Fig. 11-I

$$J = 2\,(J_{DC} + J_{AB} - J_{BC}) \tag{15}$$

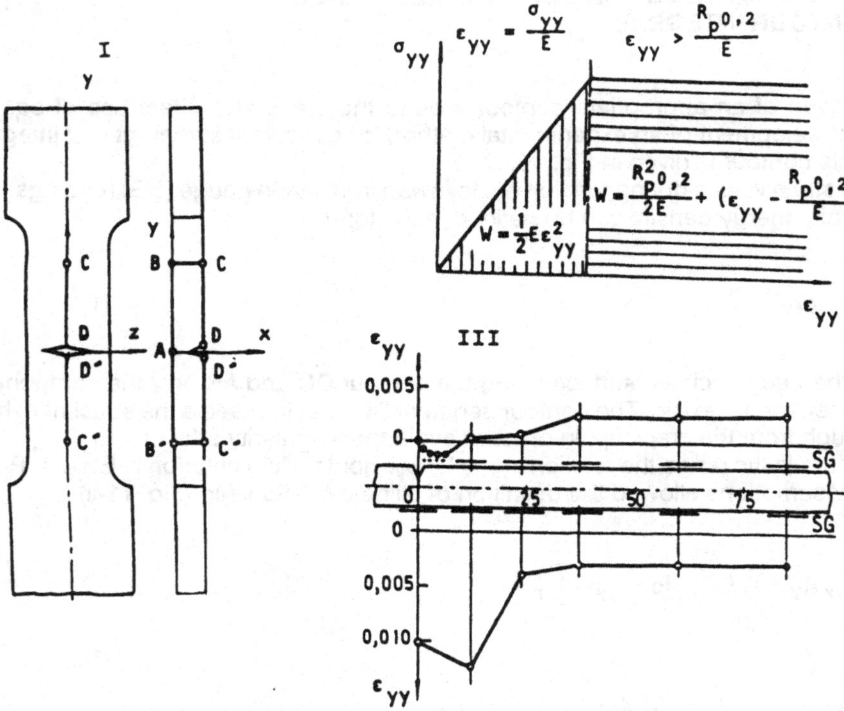

Fig. 11. Procedure for J integral direct evaluation
I Contour on tensile panel (ABCD)
II Simplified relationship σ_{yy} - ε_{yy}
III The distribution of ε_{yy} strain along segments on AB smooth side and CD on cracked specimen side

Strain distribution is determined from SG readings (Fig. 11-III), and from this the strain energy density W could be evaluated in both elastic and plastic regions. In this way the terms J_{CD} and J_{AB} are defined. The third term, J_{BC}, corresponding to the bend effect of traction vector $\vec{T}$, could be obtained by multiplying T_y ($\cong\sigma_{yy}$) with the displacement difference v(C) - v(B), that represents the value ($\partial\vec{u}/\partial x$) ds. The displacement v(B) on the smooth specimen side is available from strain distribution (Fig. 11-III) using numerical integration

$$v(B) = 2\int_A^B \varepsilon_{yy} dy \tag{16}$$

On the cracked specimen side the crack opening displacement (COD) has to be taken into account:

$$v(C) = 2\int_C^D \varepsilon_{yy}\, dy + COD \tag{17}$$

Crack opening displacement can be measured by clip gauge.

The basic relationship is a plot of load F vs. COD, obtained by successive loadings and partial unloadings. Typical plots could be of regular shape (as presented in Fig. 8 and Fig. 12a) or exhibited "pop-in" behaviour as it is shown in Fig. 12.b.

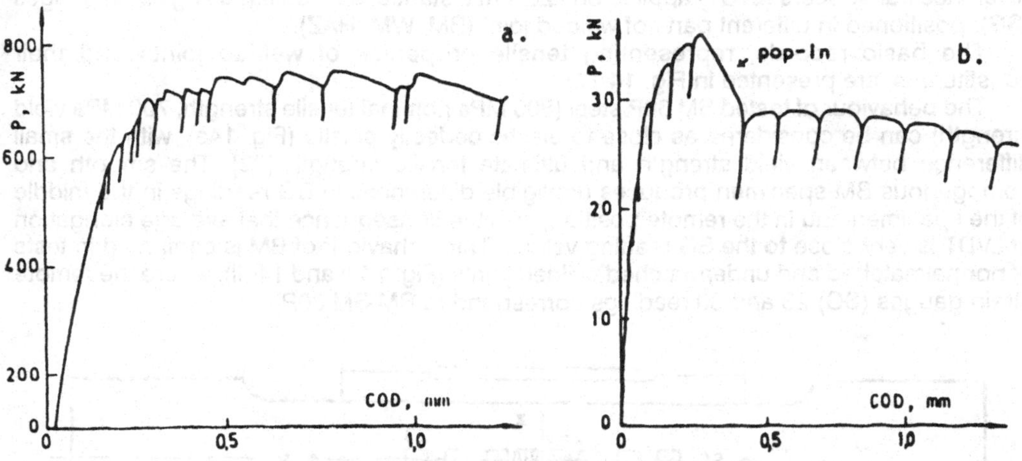

Fig 12. Typical plots load Fvs. COD, obtained by successive loadin loadings and partial unloadings of fracture pre cracked specimens
a. Regular shape, obtained with steel of 700 MPa yield strength class
b. "Pop-in" effect obtained in testing of specimen precracked in HAZ of weldment of same steel

4. WELDMENT HETEROGENEITY EFFECTS

The basic approach in welded structures design is that weld metal strength overmatches the strength of base (parent) metal (overmatching effect). That means,when the load reaches required level the yielding will start in base metal (BM) and the weld metal (WM) will start to yield only when the stress in BM reaches (by strain hardening) the level of WM yield strength. The behaviour of heat - affected-zone (HAZ) in stressed weldment is directed by its small volume entail and ctructural heterogeneity, as well as by the different mechanical properties of HAZ regions. A well produced welded joint, designed on overmatching basis, will be fractured by plastic collapse in BM when overloaded.

The overmatching can affect the occurance of cold cracks after welding in high-strength low-alloyed steels (700 MPa yield strength class) [20]. To avoid this cracking and to enable the defect free welding of structural steels of this class, the welding consumable has been designed in a way to produce WM with slightly lower strength properties compared to BM (undermatching effect). In such weldments WM starts to yield before BM. Only when the stress level in WM reaches (by strain hardening) the BM yield strength, the yielding in BM will occur.

In order to get a close insight in different behaviour of welded joint as whole and its constituents, tensile tests had been performed with the specimens (Fig. 13), taken from differently matched joints [21,22]. These tests indicate the basic response of welded joints to loading. The elongation had been measured globally, using three linear variable difference transducers (LVDT) applied on 228 mm distance, and locally, using strain gauges (SG), positioned in different parts of welded joint (BM, WM, HAZ).

The basic records, representing tensile properties of welded joints and their constituents, are presented in Fig. 14-15.

The behaviour of tested SM 80P steel (800 MPa nominal tensile strength, 700 MPa yield strength) can be considered as close to elastic-perfectly plastic (Fig. 14a), with the small difference between yield strength and ultimate tensile strength [22]. The smooth and homogenous BM specimen produces negligible differences in SG readings in the middle of the specimen and in the remote position, with the consequence that average elongation of LVDT is very close to the SG reading values. This behaviour of BM is confirmed in tests of normalmatched and undermatched welded joints (Fig. 14-II and 14-III), where the remote strain gauges (SG) 23 and 30 readings correspond to BM-SM 80P.

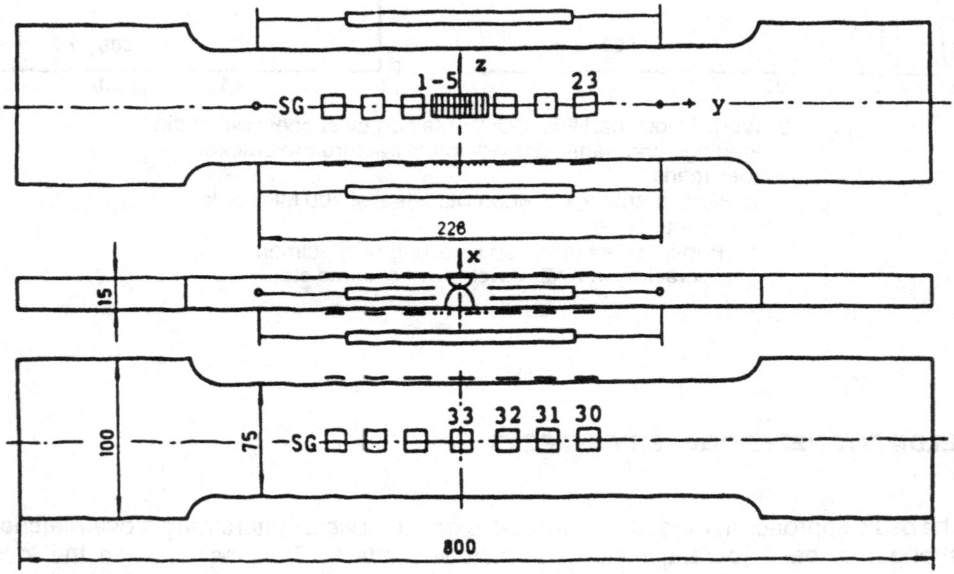

Fig 13. Tensile panel for tensile behaviour evaluation of welded joint and its constituents
Linear variable difference transducer - LVDT
Strain gauge - SG

Normalmatched welded joint (WM-N) exhibits slightly lower WM yield strength compared to BM (Fig. 14-II), designed in a way to produce yielding in BM after partial strain hardening of WM, so that further load increase should produce plastic strains in both WM and BM. The plot of stress-average strain relatioship that corresponds to the welded joints assembly, is positioned between the plots for BM and WM. The final fracture of specimen will happen in WM, because of its lower ultimate tensile strength. For an exaggerated undermatching effect (Fig. 14-III), the ultimate tensile strength of weld metal is clearly lower than yield strength of base metal SM 80P, and the later behaves elastically (remote points

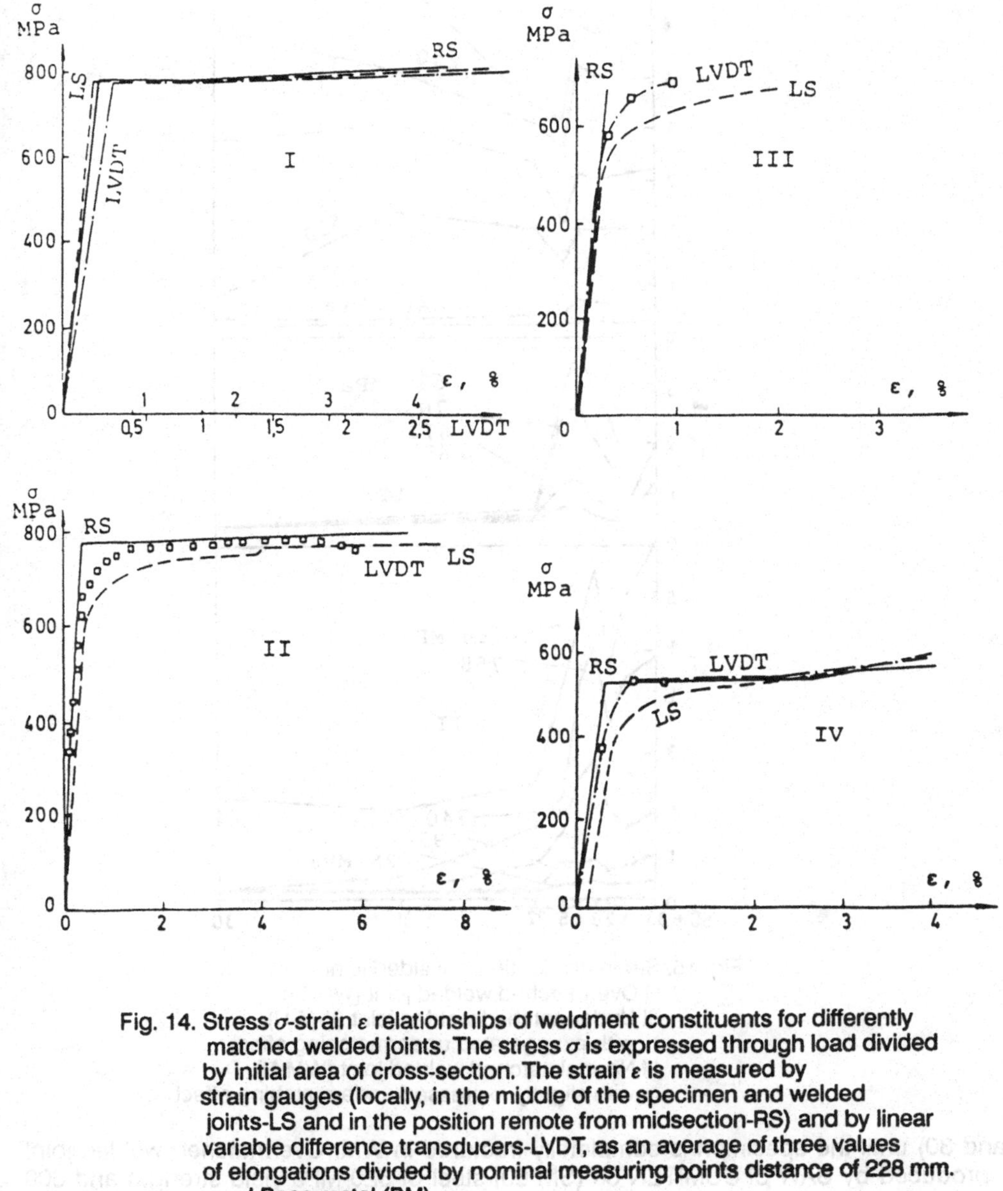

Fig. 14. Stress σ-strain ε relationships of weldment constituents for differently matched welded joints. The stress σ is expressed through load divided by initial area of cross-section. The strain ε is measured by strain gauges (locally, in the middle of the specimen and welded joints-LS and in the position remote from midsection-RS) and by linear variable difference transducers-LVDT, as an average of three values of elongations divided by nominal measuring points distance of 228 mm.
I Base metal (BM)
II Normalmatched welded joint (WM-N), with slightly expressed undermatching effect
III Undermatched welded joint (WM-U), with exaggerated undermatching effect
IV Overmatched welded joint (WM-O)

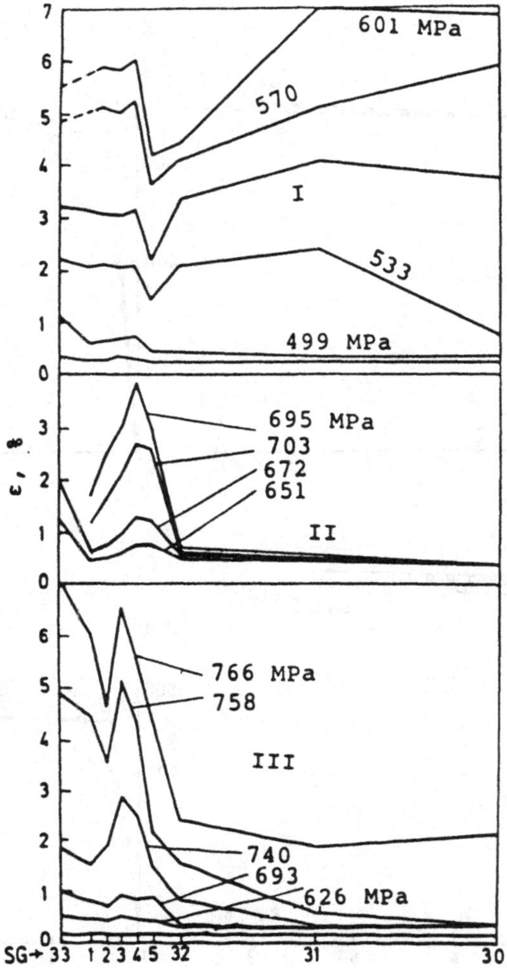

Fig. 15. Strain distribution in welded joint
I Overmatched welded joint (WM-O)
II Undermatched welded joint (WM-U), with exaggerated undermatching effect
III Normalmatched welded joint (WM-N), with slightly expressed undermatching effect

23 and 30) until the specimen is completelly fractured in WM. Overmatched welded joint was produced by SAW of SUMITEN 60 (SM 60) steel of 520 MPa yield strength and 600 MPa ultimate tensile strength class with US 80B wire, designed for slightly lower strength properties of WM when applied for SM 80P welding. The SM 60 steel response had been followed by SG 23 and SG 30 on the welded joint specimen, and again the stress-strain relationship is near to elastic-perfectly plastic behaviour (Fig. 14-IV) as in the case of SM 80P steel. Even in the overmatched welded joint WM starts to yield before BM and only when the yield point of BM is achieved by strain hardening of WM, the yielding would develop

in specimen as a whole. The specimen will be fractured in BM, since the hardening exponent is higher in WM, as well as its ultimate tensile strength.

Strain behaviour of weldments constituents is presented in the Fig. 15. For normalmatched welded joint (WM-N) of SM 80P steel, the larger strains are developed in WM (Fig. 15-III). In the heat-affected-zone, covered by SG1-SG5 the regions of large and constrained strains could be found, BM starts to yield (SG32-SG31-SG30) at the certain level of plastic strain in WM (SG33), corresponding to approximately 740 MPa. In the welded joint with exaggerated undermatsching effect (Fig. 15-II) large strains in HAZ could be again recognized, as well as the region of constrained strains. At the strain level of 2% SG33 failed, but before that the largest strain was read in this position. The strain in BM are in elastic range before the final fracture of the specimen, that occured in WM. The strain distribution in the case of overmatched weld metal is also irregular (Fig. 15-I), with the smallest strain in HAZ, and with the largest strain replaced from WM into BM at the stresses above 570 MPa.

Tensile testing of weldments have shown irregular distribution of elongation along different constituents. The strain distribution is strongly affected by the matching effect. The region of small strains has been found in HAZ in all considered specimens. Therefore, the microstrustural analysis of HAZ is important, because different structures, that could be expected in a phase transformation [23], are positioned in small region, and this inevitably affects the mechanical behaviour of welded joint.

5. THE EFFECT OF CRACK ON THE STRENGTH OF WELDED JOINT

The effect of crack on the strength of welded joint has been analyzed experimentally with the specimens, taken from BM SM80P and from all the three combinations of welded joints matching. The crack had been positioned in the middle of the specimen and weld metal [24]. The effect of crack (Fig. 16) has been analyzed in relationship between remote stress σ (the ratio of applied load and the initial area of cross-section) and average strain ε (measured by LVDT). The obtained results are presented in Fig. 17.

The plots in Fig. 17 show that cracks influence the strength and ductility of tested tensile panels in different way. The small crack (a = 2,5 mm) in base metal (Fig. 17-I), that reduces the cross-section area in about 2.5% does not change the basic relationship (remote stress - average strain) and the large crack (a = 5 mm), that reduces the cross-section area in about 13%, produces the minimal reduction in strength and significant reduction in ductility, The most expressed crack effect is found in undermatched welded joint (WM-U) (Fig. 17-III). The loading capacity of this specimen is lower than the load corresponding to the residual ligament area. Therefore the elongation at fracture represents only a small portion of the value, that corresponds to the smooth specimen (no crack); this is clearly expressed in the case of large crack. In the normalmatched specimen the loading capacity approximately corresponds to the area of residual ligament, although the ductility of the specimen is significantly reduced (Fig. 17-II).

In the overmatched welded joint (WM-O) the crack existance does not effect the stress σ–strain ε relationship in the testing range of 0.02 average strain (Fig. 17-IV), that is about seven times higher than yield strain.

It is possible to conclude that the crack in an undermatched welded joint reduces both the strength and ductility, the later one more significantly. This reduction of ductility, under specific condition, can cause premature failure of the structure.

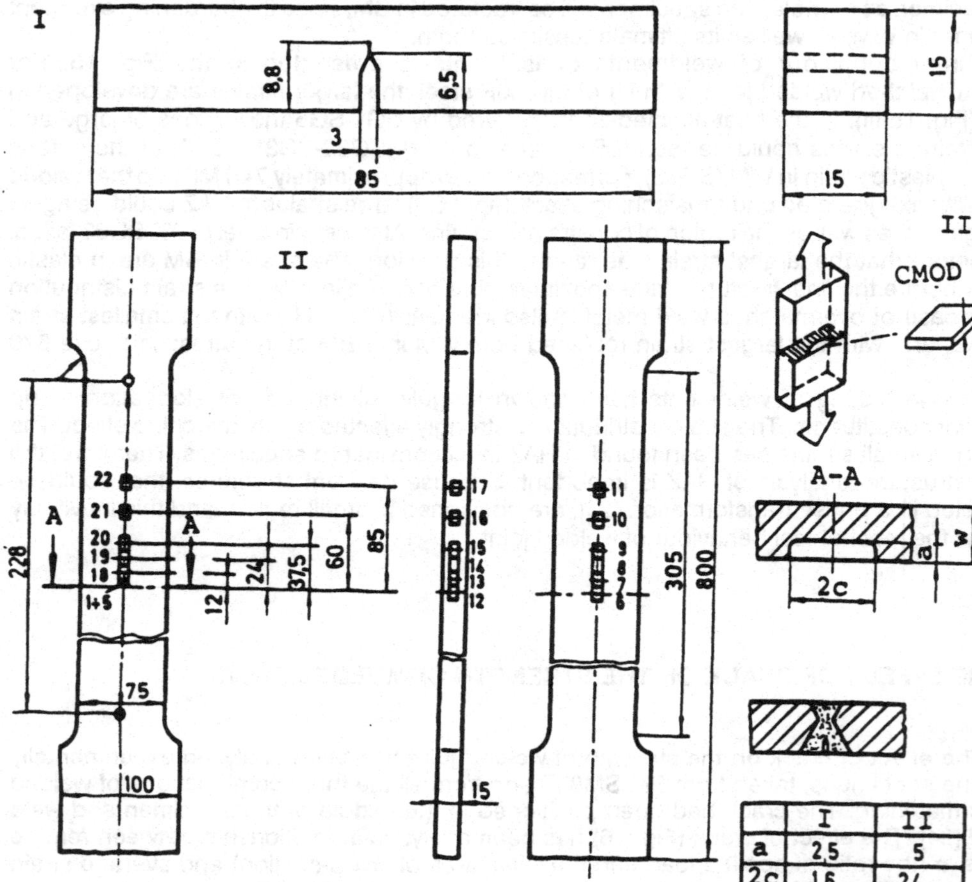

	I	II
a	2,5	5
2c	16	24

Fig. 16. Shape and dimensions (in mm) of cracked specimens
I Three-points bend specimen
II Tensile panel
III The data of cracks in tensile panels
1-5 Small strain gauges
6-22 Strain gauges of normal size

The special attention should be paid in this case to the stress concentration at the crack tip, that contributes to the stress triaxiality and brittle fracture. On the other hand, stress concentration increases the load capacity of material in the vicinity of crack tip, since the cross-section area, due to constrained condition, is only slightly reduced before the fracture occurs.

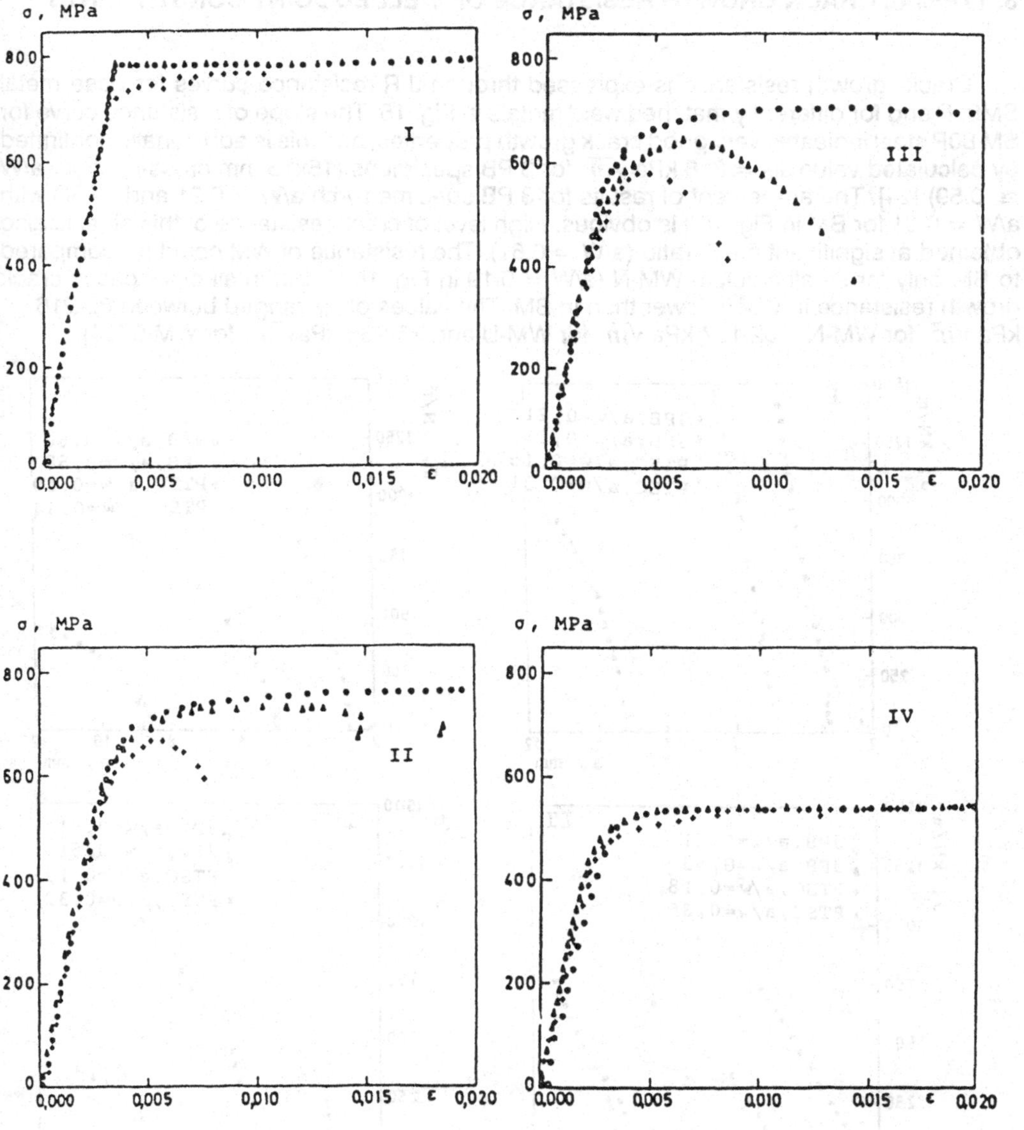

Fig. 17. The relation ship between average stress (applied load divided by initial cross-section area) and average strain (measured by 3 LVDT on measuring distance 228 mm) for specimens with no crack (o), small crack (Δ) and large crack (+).
I Base metal (BM)
II Normalmatched weld metal (WM-N)
III Undermatched weld metal (WM-U)
IV Overmatched weld metal (WM-O)

6. TYPICAL CRACK GROWTH RESISTANCE OF WELDED JOINT CONSTITUENTS

Crack growth resistance is expressed through J-R resistance curves for base metal SM80P and for differently matched weld metals in Fig. 18. The slope of resistance curve for SM 80P steel indicates very good crack growth properties, and this is additionally confirmed by calculated value J_{Ic} = 316 kPa $\sqrt{m}$ for 3 PB specimens (15x15 mm cross-section, a/W ≅ 0.59) [24]. The agreement of results for 3 PB specimen with a/W = 0.21 and PTSC with a/W = 0.31 for BM in Fig. 18-I is obvious. High level of crack resistance of this steel is also obtained at significant crack ratio (a/W = 0.61). The resistance of WM could be compared to BM only for small crack in WM-N (a/W = 0.19 in Fig. 18-II), but in all other cases crack growth resistance in WM is lower than in BM. The values of J_{Ic} ranged between 80-216 kPa $\sqrt{m}$ for WM-N, 102-137 kPa $\sqrt{m}$ for WM-U and 75-238 kPa$\sqrt{m}$ for WM-0 [24].

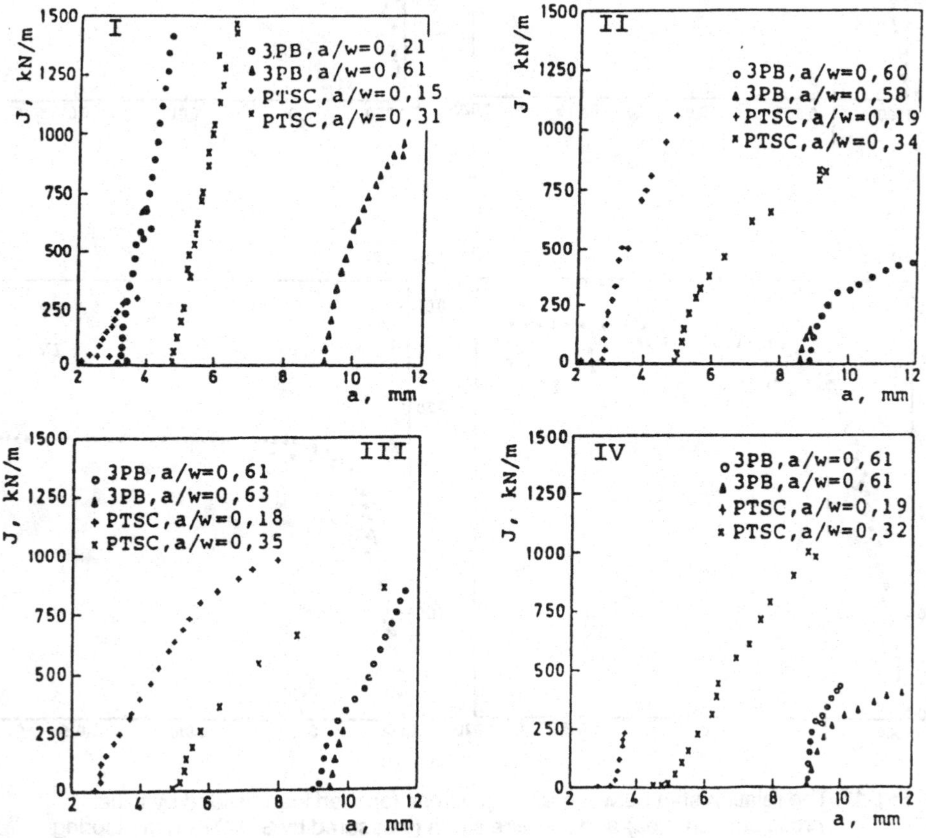

Fig. 18. Crack growth resistance curve J-R

3PB - three-point bend specimen, TPSC - tensile panel with surface crack

I Base metal SM 80 (BM) II Normalmatched weld metal (WM-N)

III Undermatched weld metal (WM-U) IV Overmatched weld metal (WM-O)

Significant differences in crack growth resistance of heat-affected-zone are caused by heterogeneity of microstructures and by the actual position of crack tip (Fig. 19). Mechanical behaviour of materials on different sides of HAZ is different (Fig. 15) and asymetrical distribution of stresses and strains affects the crack behaviour. For this reason the J integral path has to be instrumented completely as it is presented in Fig. 11 for DCBAB'C'D'. J integral behaviour in tensile panels with the crack in HAZ is presented in Fig. 20. The J-R curves for 3 PB specimens are presented in Fig. 21. The behaviour of cracked HAZ in PTSC (Fig. 20) can be considered as similar to cracked BM (Fig. 18), but this is not the case for 3 PB specimens. Pop-in effect, presented in Fig. 12b, produced in 3 PB specimens with low a/W ratio (0.21 and 0.20) irregular shape of J-R curve (Fig. 21), reducing crack growth resistance.

The method of J integral direct evaluation enables the analysis of effects of asymetrical stress and strain distribution on crack growth in HAZ by separation of terms along part ABCD and AB'C'D' (Fig. 11-I). The symetry is disturbed when crack is positioned in HAZ (and this is most common case for real welded structures), because the crack growth is controlled by BM on one crack side and WM on the other side. Recently performed analysis revealed that the effect of strain distribution in both normalmatched and overmatched weldments is partially covered by traction vector effect in a way that J integral can be evaluated as for the symetrical problem.

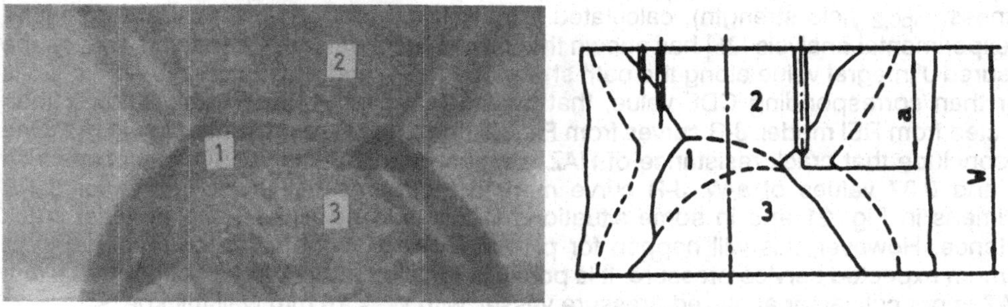

Fig. 19. Macrostructure of welded joint (a) with the crack tip positions in HAZ (b)

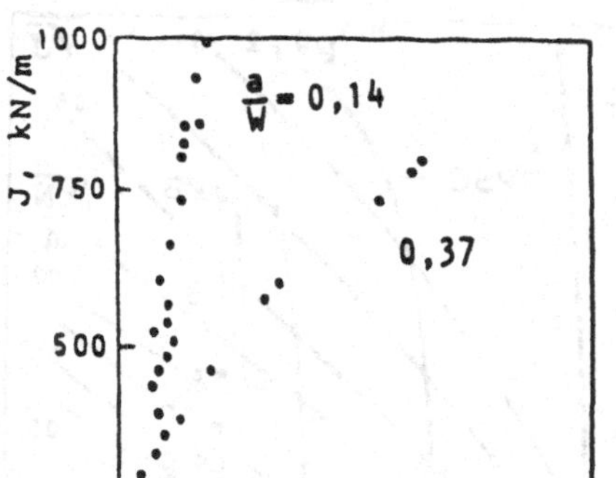

Fig. 20. Tensile panel J-R curves for HAZ of normalmatched weldment of SM 80P steel

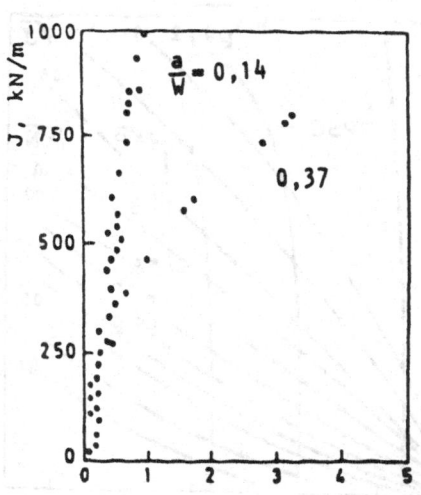

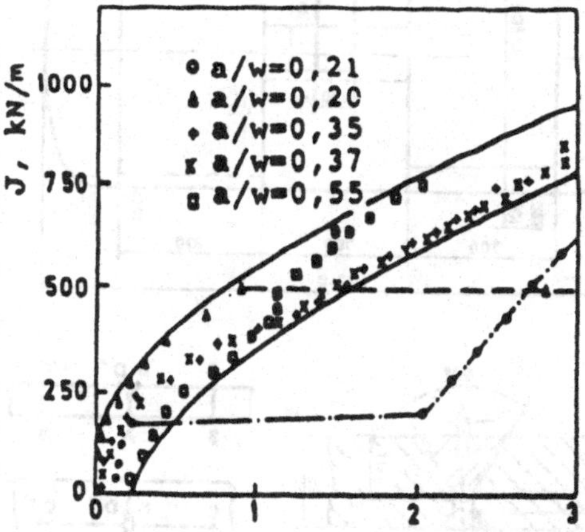

Fig. 21. 3PB specimen J-R curves for normalmached weldment of SM 80P steel

7. THE APPLICATION OF J-R CURVE AND J_{Ic} FOR CRACK SIGNIFICANCE ANALYSIS IN WELDED STRUCTURE

The value of fracture toughness K_{Ic}, determined from eq. (4) can be used directly for crack significance evaluation according to relation

$$K_{Ic} = \sigma\sqrt{a_c}\, f\left(\frac{a}{W}\right) \tag{18}$$

from which critical crack size a_c can be found for known values of acting stress σ and crack shape function f(a/W).

It is also possible to use J-R curve for crack significance analysis. For this purpose, a model for crack driving force (CDF) evaluation in structure will be applied and obtained set of CDF curves compared to material crack resistance (J-R curves) for different a/W values. For a pressure vessel, containing the crack in WM (Fig. 22) a set of CDF curves is presented in Fig. 23 for different $pR/WR_{p0,2}$ values (p is pressure, R inner vessel radius, W wall thickness, $R_{p0,2}$ yield strength), calculated from Ratwani-Erdogan-Irwin (REI) model [25]. The experimental analysis [26] has shown that model is conservative enough, because the measured J integral value along the path shown in Fig. 22 and presented by A in Fig. 23 is lower than corresponding CDF value, that means real crack driving force is lower than predicted from REI model. J-R curves from Fig. 20 and 21 are are presented in Fig. 23. One can conclude that crack resistance of HAZ is higher than CDF for TPSC specimens with 0.14 and 0.37 values of a/W. J-R curve marked LB represents lower bound for 3 PB speçimens in Fig. 21 and in some situations CDF could be higher than material crack resistance. However, this will happen for $pR/WR_{p0,2} \cong .0.7$, the value corresponding to maximum expected service pressure. It is possible to conclude that crack of depth 8,8 mm in HAZ is not critical for analyzed pressure vessel, with W = 16 mm wall thickness.

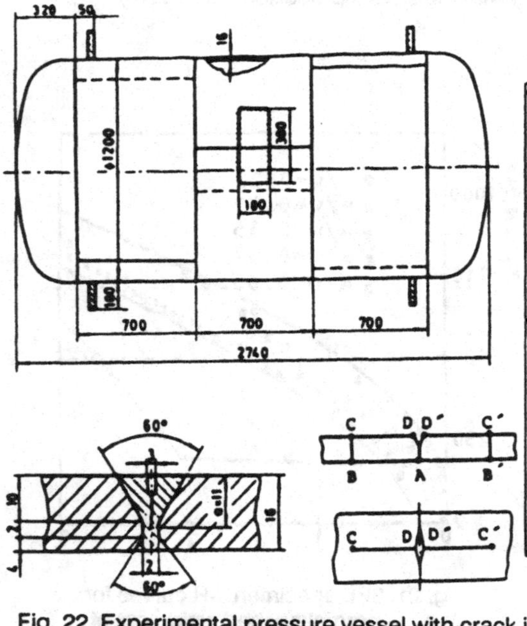

Fig. 22. Experimental pressure vessel with crack in WM and path for J integral direct measurement

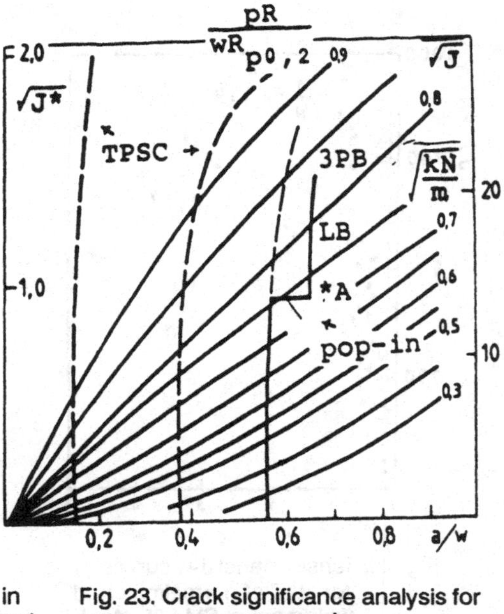

Fig. 23. Crack significance analysis for pressure vessel

REFERENCES

1. Irwin,G. R.: "Fracture", in Encyclopedia of Physics, ed. Flugge, S. Volume VI: Elasticity and Plasticity, Springer Verlag, Berlin, 1958.
2. "Fracture toughness testing and its application", Special Technical Publication, STP381, ASTM, Philadelphia, 1965.
3. ASTM E 399-83: Standard method of test for plane-strain fracture toughness of metallic materials, ASTM Book of Standards 03.01.ASTM, Philadelphia, 1988.
4. BS 5447: Methods for plane strain fracture toughness testing, British Standard Instution, London, 1979.
5 Burdekin, F.M.,Stone, D.E.W.: "The crack opening displacement approach to fracture mechanics in yielding" J Strain Analysis, 1, 1966, 145 - 153.
6. BS 5762: Methods for crack opening displacement (COD) testing, British Standard Instution, London, 1980.
7. Harrison, J.D.: "The "state-of-the-art" in crack tip opening displacement (CTOD) testing and analysis", Part 3: Application of CTOD approach, Metal Construction, Nov. 1980.
8. Reed, R.P.,Kasen, M.B., McHenry, H.I., Fortunko, C.M., Read, D.T.: "Fittness-for-service criteria for pipeline girth welds quality", Bulletin 296, Welding Research Council. New York, July 1984.
9. McHenry, H.I., Shives, T.R., Read, D.T., McColskey, J.D. Brady, C.H., Purtcher, P.T.: "Examination of a pressure vessel that ruptured at the Chicago Refinery of the Union Oil Company on July 23, 1984.", NBSIR 86-3049, National Bureau of Standards, Boulder, Colorado, 1986.
10. "Safety of pressure vessels weldments", discussions of R. Labbens and F.M. Burdekin, in Modern aspects of design and production of pressure vessels and penstocks, ed. S. Sedmak, Gosa-Faculty of Technology and Metallurgy Belgrade, 1982. (in Serbo-Croatian), 287 - 290.
11. Rice, J.R.: "A path independent integral and the approximate analysis of strain concentration by notches and cracks", J. Applied Mechanics, June 1986, 379 - 386.
12. ASTM E 813-81: Standard test method for J_{Ic}, a measure of fracture toughness, ASTM Book of Standards 03.01. ASTM, Philadelphia, 1988.
13. ASTM E 1152-87: Standard test method for determing J-R curves, ASTM Book of Standards 03.01. ASTM Philadelphia, 1988.
14. ASTM E 616-82: Standard teminology relating to fracture testing, ASTM Book of Standards 03.01. ASTM Philadelphia, 1988.
15. Sedmak, S.: "The evaluation of welded joints properties by testing of cracked specimens", in Fracture mechanics of weldments, ed. S. Sedmak, Gosa-Faculty of Technology and Metallurgy, Belgrade, 1985. (in Serbo-Croatian), 287 - 306.
16. Paris, P.C., Tada, H., Ernst, H.A.,Zahoor, A.: "An initial experimental investigation of tearing instability theory", in Elastic-plastic fracture, STP 668, ASTM, Philadelphia, 1979, 251 - 266.
17. Miyamoto, H., Kobayashi, H., Ohtsuka, N.: "Standard method of test for elastic-plastic fracture toughness J_{Ic} recommended in Japan", in ICM 4 Proceedings, Volume 2, Pergamon Press, Stockholm, 1983, 747 - 753.

18. PN-88 metals: Test method for J_{Ic}, a measure of fracture toughness, Polski Komitet Normalizacij Miar Jakoszi, 1988.

19. Read, D.T.: "Experimental method for direct evaluation of J-contour integral", in Fracture Mechanics Fourteenth Symposium, Vol.II, Testing and Applications, STP 791, ASTM, Philadelphia, 1983, 199 - 213.

20. Satoh, K., Toyoda, M.: "Joint strength of heavy plates with lower strength weld metal", Welding Journal Research Supplement, September 1975.

21. Sedmak, S., Petrovski, B.: "The experimental analysis for differently matched weld metals", in JOM3, Proceedings of the International Conference Joining of Metals, Helsingor, 1986, 309 - 314.

22. Sedmak, S., Petrovski, B., Graovac, B.: "The effect of weldment heterogeneity on specimen behaviour in tensile testing", in Proceedings of International Symposium Weldability and Test Methods, Vranje, 1985. (in Serbo -Croatian), 131 - 143.

23. Bozic, B., Novovic-Simovic, N., Vidojevic, N., Kerecki, R.: "The influence of heating temperature and cooling rate on the microstructure and hardness of the low carbon high strength CrNiMnMo steel", in Bulletin Academie Serbe des sciences et des art, Tome XCIII, Classe des science technique, No 23, Beograd, 1987.

24. Read, D.T., Petrovski, B.: "Elastic-plastic fracture of high strength low alloyed steel welded tensile panels with surface flaw", in Prospect of development and application of fracture mechanics, ed. S. Sedmak, Gosa -Faculty of Technology and Metallurgy, Belgrade, 1987. (in Serbo-Croatian), 119 - 139.

25. Ratwani, M., Erdogan, F., Irwin, G.: "Fracture propagation in cylindrical shell containing an initial flaw", Lehigh University, Betlehem, 1974.

26. Sedmak, S., Petrovski, B.: "Application of direct measurement of J integral on a pressure vessel with axial notch", in Fracture Mechanics: Eighteenth Symposium, STP 945, ASTM, Philadelphia, 1987, 730 - 740.

EXPERIMENTAL INVESTIGATION OF FAILURE MECHANISMS IN DUCTILE MATERIALS

W. Szczepinski
Polish Academy of Sciences, Warsaw, Poland

ABSTRACT

Theoretical and experimental study of interaction between variously oriented voids and cracks in a plastic medium is presented in an attempt to obtain a deeper insight into the processes of ductile failure in metals. Theoretical analysis is based on the slip-line technique. Experiments were performed with the use of specimens made of ductile metals. In these specimens systems of holes or slits simulating voids and cracks were prepared. Experimental results substantiate theoretical analysis and demonstrate the interaction between softening and hardening effects during the process of ductile failure in metals.

1. INTRODUCTION

The mechanics of ductile fracture is very complex and still fully not examined. Depending on the temperature, the rate of deformation and the structure of the metal various mechanisms may be responsible for the softening and fracture. Coalescence of voids, interaction between variously oriented cracks leading to local internal microdecohesion may contribute to the progressing process of ductile fracture. In polycrystalline metals intercrystalline void formation and intergranular sliding may lead to the softening and finally to the ductile fracture of the aggregate. These phenomena cannot be analysed in terms of the classical fracture mechanics based on the assumption of the elastic model of the material (Fig.1a). The model of an elastic-plastic material (Fig.1b) is more adequate to analyse mechanisms of ductile fracture. Numerous problems of this kind have been solved with the use of various numerical techniques. On the other hand the possibilities of application of the theory of plasticity based on the assumption of a rigid-plastic model of the material (Fig.1c) to the analysis of the phenomena of ductile fracture have been relatively unexplored.

In each theory of fracture the existence of certain systems of defects is assumed. These defects may be formed during the solidification of metals, during the manufacturing process (mainly cold working operations) or may be formed during the deformation process preceding the failure. However, it should be noted that for the rigid-perfectly plastic model of the material (Fig.1c) theoretical solutions may be obtained leading to the separation of the material without any assumption of the existence of internal defects. This will be shown here on an elementary example of a bar pulled in tension under plane strain conditions.

Initial configuration is shown in Fig.2. In order to ensure the existence of plane strain conditions the width of the bar in the direction perpendicular to the plane of the figure is assumed to be very large. In other words a sheet of the initial thickness c_0 is pulled in tension. Initial configuration of slip-lines is presented in Fig.2a. Advanced stage of deformation is shown in Fig.2b. This well known solution represents a particular case of a more general solution to the problem of plastic yielding of bars weakened by a pair of V-notches given by LEE [1]. If the total angle of the notch is 2β, then the angle α of the inclination of the deformed straight boundary is

$$\alpha = \arctan\frac{(1+cos\beta)^2}{(2+cos\beta)sin\beta}.$$

The problem presented in Fig.2 may be treated as a limit case of a V-notch with the total angle $2\beta = 180^0$. Then from the above formula we obtain $\alpha = arctan 0.5 = 26^0 34$ as shown in Fig.2b. This well known solution is described here because we shall need it for the analysis of failure of bars with certain systems of cracks.

According to this slip-line solution if the elongation of the bar is ΔL then its thickness in the neck reduces to the value $c = c_0 - \Delta L$ (cf.Fig.2b). Thus the

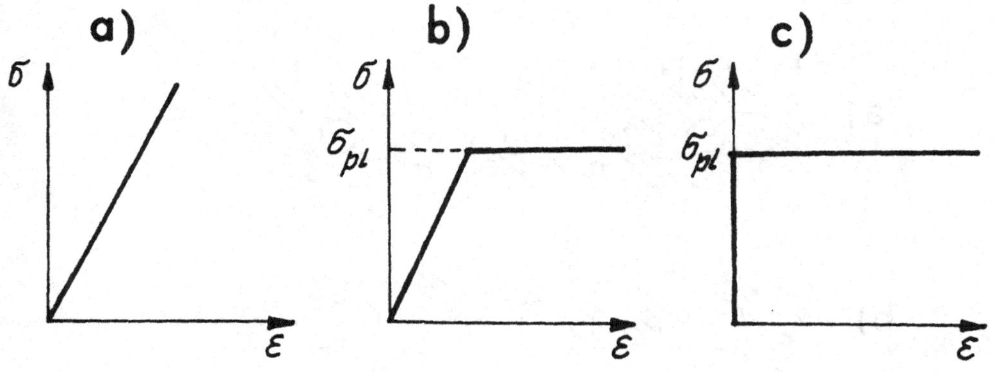

Fig.1, Basic stress-strain relations in theoretical models

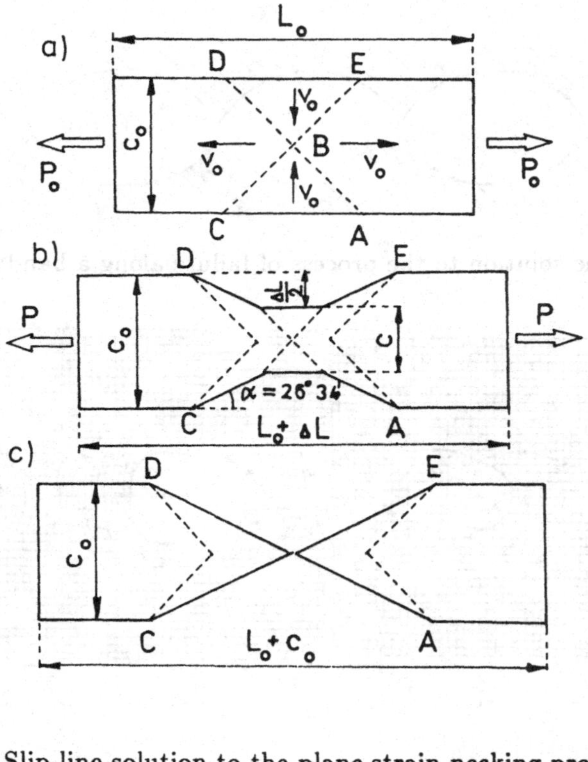

Fig.2, Slip-line solution to the plane strain necking process.

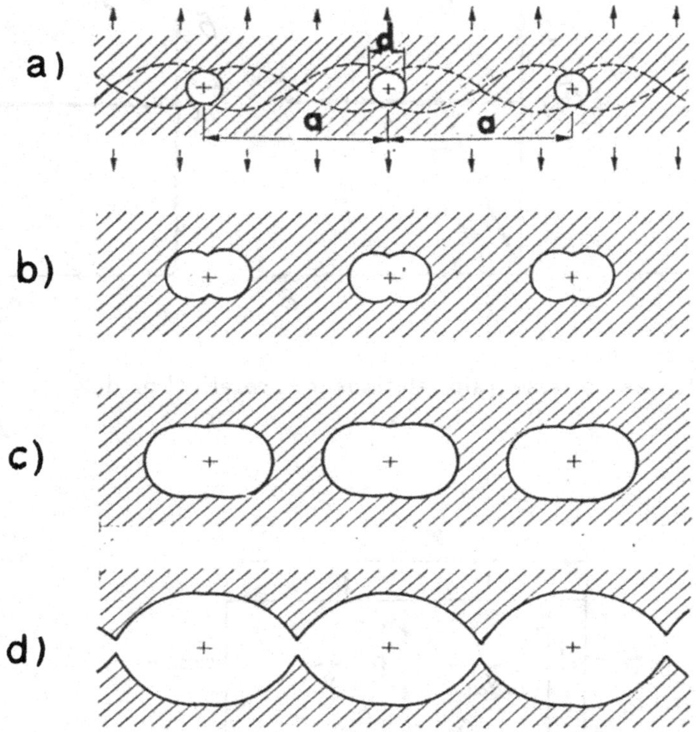

Fig.3, Slip-line solution to the process of failure along a band of holes.

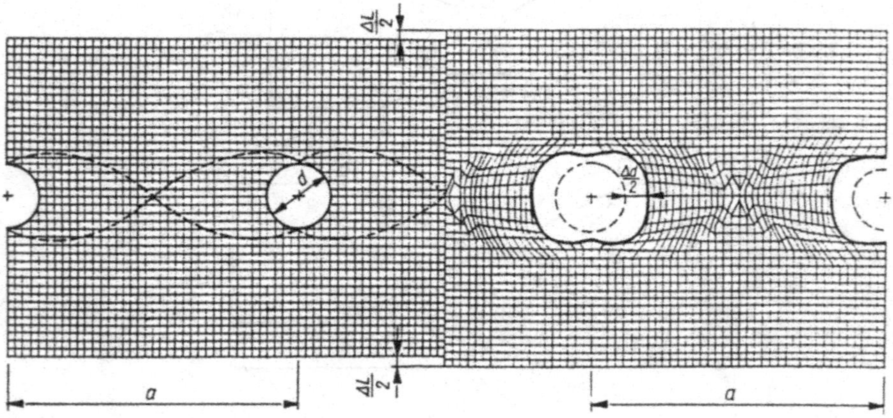

Fig.4, Deformation pattern at the stage of deformation shown in Fig.3b.

The left-hand side of Fig.4 shows the initial configuration with the square grid assumed to exist in the non-deformed body. On the right the initial stage of deformation (cf. GARR et al.[13]) with the strong strain concentration along the band of holes has been presented. For the relatively small total elongation of the body $\triangle L$ shown in the figure the conventional lateral strain within the band is as large as $\triangle d/(a-d) \sim 0.22$. This stage of deformation has also been shown in Fig.3b. Fig.3c (cf. WANG [14]) shows how quickly the process of the hole growth advances, leading finally to the separation of the body along the band of holes as shown in Fig.3d. The separation of the body occurs at rather small total elongation $\triangle L$. Thus assuming the rigid-ideally plastic model of the material, we obtain the final result very close in effect to the brittle fracture of the body.

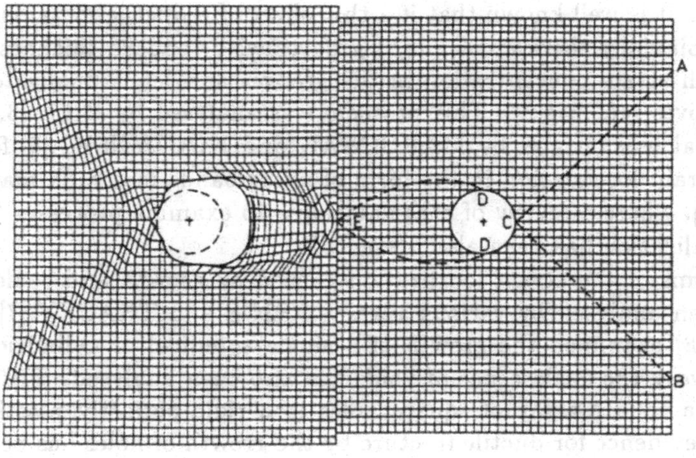

Fig.5, Slip-line solution for a bar with two holes.

Fig.5 shows the theoretical plane strain slip-line solution to the problem of the growth of two holes prepared in a rigid-ideally plastic bar of the finite width b. The initial configuration of slip-lines is shown on the right-hand side of the figure. The left-hand side shows an advanced stage of the process of the hole growth and demonstrates the strain concentration in the central part between the holes. The deformation pattern in the ligament between the holes is the same as that shown in Fig.4 (cf.[13]). The conventional lateral strain in the ligament between the holes is in the particular stage of deformation shown in the figure approximately twice as large as the conventional lateral strain in the outer strips. Further stages of the deformation process are presented in Fig.6 (cf. [14]).

Fig.6a represents the initial configuration and Fig.6b corresponds to the stage previously shown in Fig.5. It can be seen that for the relatively small total elon-

conventional lateral strain in the neck may be written as

$$\epsilon_c = -\frac{\Delta L}{c_0}.$$

Note that this theoretical solution reducing the thickness to zero is of practical significance. Such a failure mode is observed in some metals tested at high temperatures and also in some metals (gold, pure lead, aluminium - see e.g.[2]) tested at the room temperature.

2. ON THE ROLE OF STRAIN CONCENTRATION

Strain concentrations play an important role in the mechanics of ductile fracture in metals. It is well known that if a thin sheet of a ductile metal is stretched uniaxially or biaxially there appear in it grooves with reduced thickness. This is connected with strain concentration in the grooves and leads finally to fracture along the grooves (cf. Fig.2). This problem was analysed by HILL [3,4] for the case of uniaxial tension and by MARCINIAK and KUCZYŃSKI [5] for biaxial stretching. Strain localization in the form of shear bands in a bulk material was analysed in [6], where a review of other attempts to examine processes leading to the strain localization has been also presented.

Mechanisms of the strain localization caused by growth and coalescence of voids have been analysed by various authors. RICE and TRACEY [7] and McCLINTOCK [8] examined the growth of a single spherical or cylindrical cavity. In the latter work the coalescence of voids has also been analysed. THOMASON [9] has taken a square array of square holes in a rigid-perfectly plastic matrix. Experimental evidence for ductile fracture by the growth of holes has been known for a number of years (cf. [8] and [10]).

The problem of interaction between regularly distributed voids was examined by means of the slip-line technique by NAGPAL et al. [11], in order to obtain approximate equations for fracture by the development of localized flow at the front of a crack. Following previous paper (SZCZEPIŃSKI [12]) we shall use the slip-line technique for the study of evolution of strain localization.

As an introductory example let us analyse the process of ideally ductile failure of a rigid-ideally plastic body weakened by a single row of circular holes as shown in Fig.3a. The problem will be examined under plane strain conditions. The presence of even very small holes causes strain localization along the band of holes. The initial slip-line system consisting of logarithmic spirals is shown in Fig.3a. For the assumed ratio of dimensions $\frac{a}{d} = 4.75$, where a is the spacing between the holes and d stands for the diameter of the holes, the slip-lines do not overlap. The theoretical analysis presented below is kinematically admissible only, since the extension of the slip-line field into the rigid region is not known.

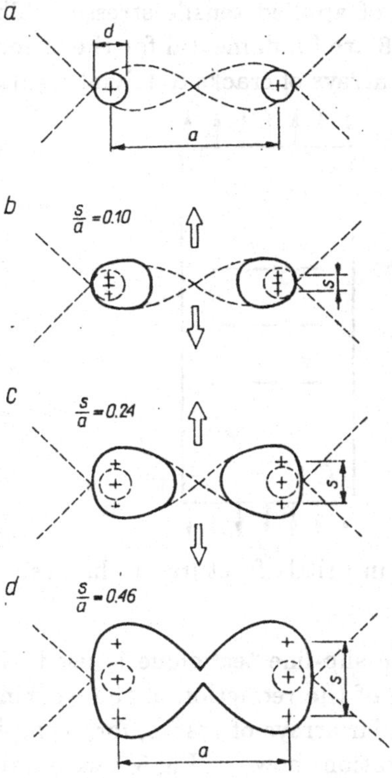

Fig.6, Slip-line solution to a problem of coalescence of holes.

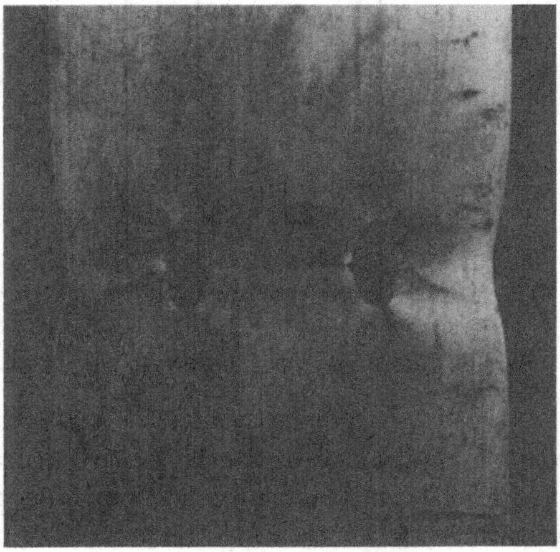

Fig.7, Fracture in the specimen with holes pulled in tension.

gation of the bar equal to $s = 0.46a$ the width of the ligament between the holes has been reduced to the point. Thus the theoretical analysis of the process of ideally plastic deformation leads in this case to the local internal separation of the material. The successive appearance of such local fractures in the body with random distribution of voids may be interpreted as the evolution of the damage process.

Fig.7 shows the specimen with two holes made of a ductile Al-2% Mg aluminium alloy after plastic deformation. The dimensions of the specimen correspond to the theoretical solution discussed above. Owing to the limited ability of the material used in the test to deform plastically fracture occured at certain stage of deformation in the region of strain concentration.

3. DUCTILE FAILURE IN THE PRESENCE OF LINEAR DEFECTS

Assume that in the rigid-perfectly plastic material there exists a system of cracks oriented perpendicularly to the direction of applied tensile stresses. The three configurations of such cracks shown in Fig.8 are fundamental for the theory of brittle fracture. However, they lead for regular arrays of cracks to trivial results,

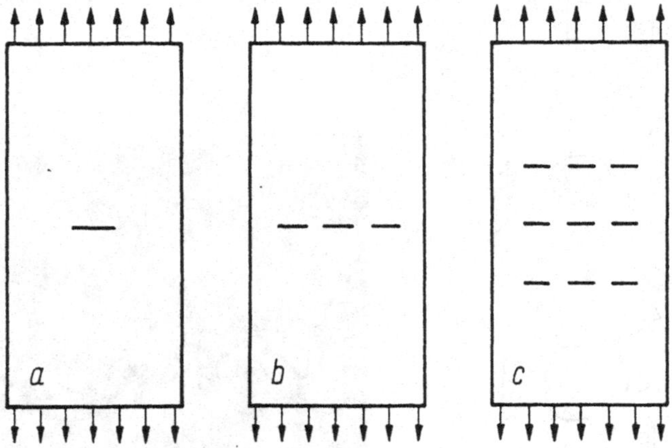

Fig.8, Basic configurations of cracks analysed in brittle fracture mechanics.

when a rigid-plastic material is assumed and the slip-line technique is used. In theoretical solutions we obtain a smooth process of the reduction of the working area between the cracks to the point. For nonregular arrays of cracks, for example for various spacings between cracks in the configuration shown in Fig.8b, we obtain the process of consecutive separation of ligaments between adjacent cracks.

Although we shall not discuss in detail the configurations of cracks shown in Fig.8, in view of their limited significance in the theory of ductile failure, some schemes of deformation will be shown and briefly discussed.

Usually the damage parameter in the material undergoing deformations is defined as the relative reduction of the working area. If for example inside the material there exists a band of uniformly distributed cracks of the length a and the spacing s between them so defined damage parameter may be written as

$$\omega_0 = \frac{a}{s}. \tag{3.1}$$

However, kinematically admissible deformation mechanism resulting from the slip-line solution shown in Fig.9 indicates, that plastic yielding of the ligaments B-A between the cracks begins when stresses in them reach the value $p^+ = \sigma_{pl}(1 + \frac{\pi}{2})$, much larger than the yield locus of the material σ_{pl}. Since we do not know the extension of the slip-line field into the rigid regions the value p^+ constitutes only the upper bound on the unknown exact value of tensions p.

A lower bound on stresses p can be calculated for example by using statically admissible stress field shown in Fig.10.

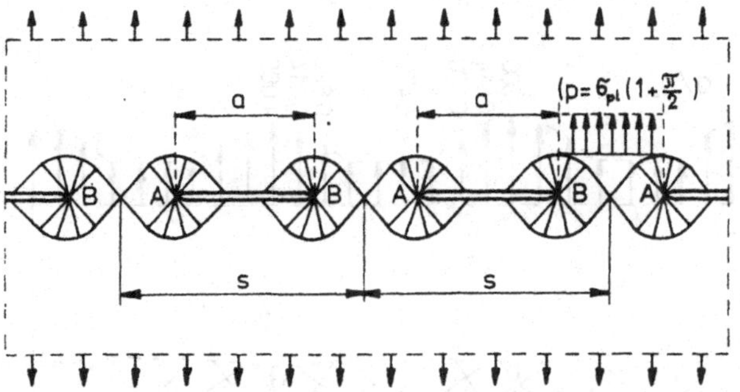

Fig.9, Kinematically admissible mechanism of plastic failure for a body with a band of cracks.

This field is composed of fields given by BISHOP [45] for notched bars. The width of these fields should be equal to the spacing s. This condition determines the value of the angle γ. Thus the lower bound on the stresses p in ligaments B-A is $p^- = \sigma_{pl}(1+\gamma)$.

The resultant limit force per unit thickness acting on the sector of the length s is $P^* = (s-a)p$, while the analogous limit force for the material without cracks is $P_0 = s\sigma_{pl}$. Now the damage parameter may be defined as

$$\omega^* = 1 - \frac{P^*}{P_0} = 1 - (1 - \frac{a}{s})\frac{p}{\sigma_{pl}}. \tag{3.2}$$

In the particular case presented in Figs.9 and 10 we have $a/s = 0,565$ and $\gamma = 30^0$. By substituting the upper and lower bounds on p we obtain that

$$0,118 \leq \omega^* \leq 0,227,$$

while from the formula (3.1) we get much larger value $\omega_0 = 0,565$.

This simple example shows that when there exist linear defects in the material the damage parameter not always can be defined simply as the relative reduction of the working area, as it is for example in the case shown in Fig.11.

The left-hand figure shows the initial configuration along with the lines of velocity discontinuity A-C and B-C. The deformation consists in the rigid block motion. The incipient deformation pattern is shown in the right-hand figure. Since in this slip-line solution we have in the two triangles ABC the state of uniaxial tension with stresses equal to σ_{pl} the damage parameter is equal to the relative reduction of the working area

$$\omega^* = \frac{b}{d}.$$

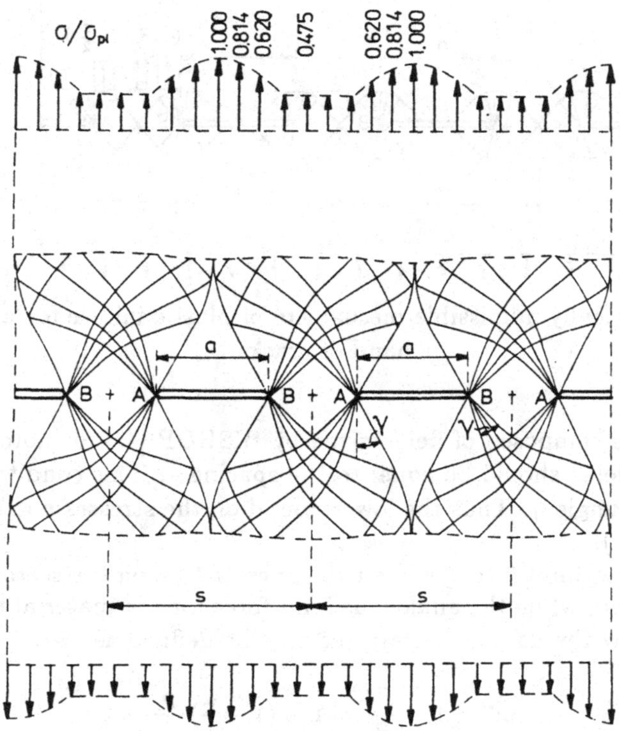

Fig.10, Statically admissible plastic stress field for a body with a band of cracks.

A regular array of linear defects (cracks), for which the damage parameter should be calculated also from formula (3.1), was discussed in the previous work [35].

Thus it can be seen that according to the slip-line solutions the load carrying capacity of ligaments between the cracks may assume various values depending of the mode of interaction between the cracks. If the width of the ligament is c this carrying capacity per unit thickness may in extreme cases be equal to $P = c\sigma_{pl}$, as in the case shown in Fig.11, or even $P = (1 + \frac{\pi}{2})c\sigma_{pl}$ as in certain cases similar to that shown in Fig.9. Important cases of a pair of cracks parallel to the direction of the tensile stresses are shown in Fig.12. Hypothetical processes of forming such system of cracks and the deformation modes taking place in a plastic body in

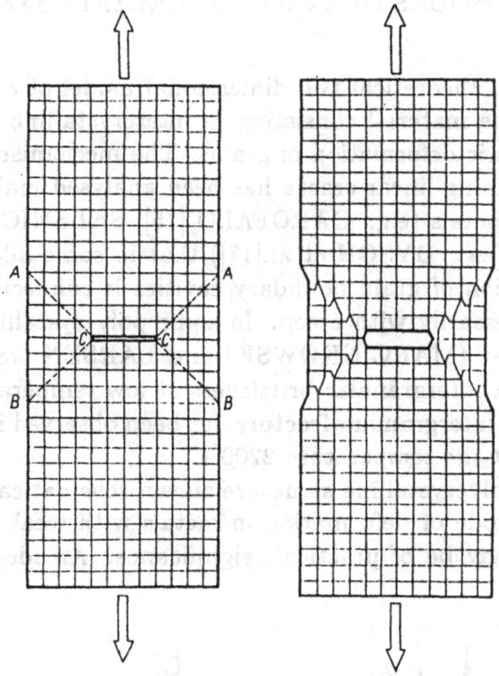

Fig.11, Slip-line solution to the problem of plastic failure of the strip with a single crack.

which such system exist will be discussed in the followig sections.

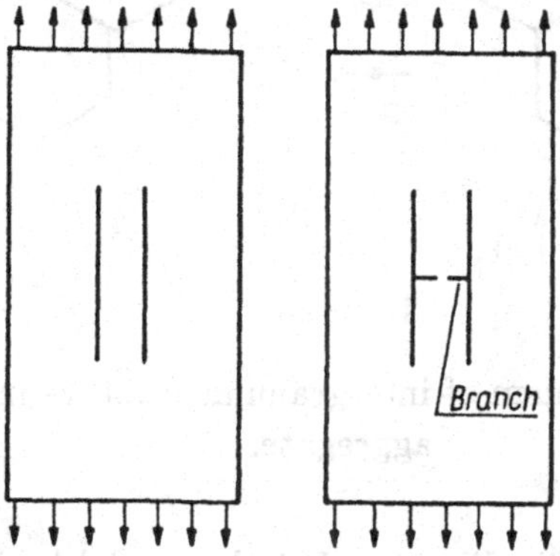

Fig.12, Configurations of cracks parallel to the direction of the tensile loading.

4. HYPOTHETICAL MODES OF FORMING CERTAIN SYSTEMS OF CRACKS

Consider at first a theoretical two-dimensional model of a deformation process in a polycrystalline material consisting in intercrystalline sliding connected with simultaneous plastic deformation of grains. The mechanism of intergranular fracture leading to forming linear cracks has been analysed mainly in papers devoted to the creep of metals (e.g. GAROFALO [15], SKLENICKA et al.[16]). It is observed, however, (e.g. DYSON et al.[17]) that in some alloys under certain conditions the generation of grain boundary cavities is connected with the plastic strain and not necessarily with creep. In some polycrystalline ordered alloys fracture is intergranular (MARCINKOWSKI and LARSEN [18]). Also vacuum melted iron can exhibit intergranular brittleness at low temperatures (RELLICK and McMAHON [19]). Intergranular fracture has been observed in tungsten tested under simple tension at the temperature 2200^0C.

The model of a polycrystalline structure shown schematically in Fig.13 presents a hypothetical mode of deformation in metals with weak grain boundaries which in some cases may be of practical significance. An idealized initial con-

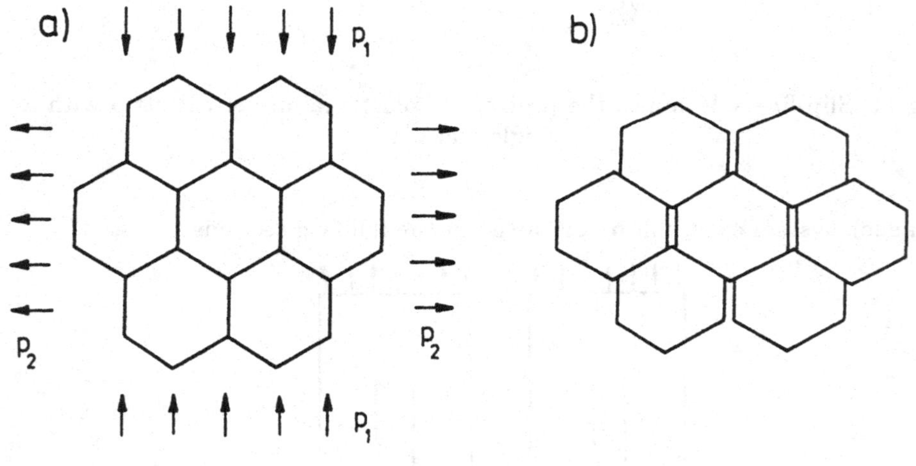

Fig.13, Idealized mechanism of intergranular fracture in a polycrystalline aggregate.

figuration of grains is shown in Fig.13a. Let the model be loaded by the stresses p_1 and p_2 such that deformation of the type shown in Fig.13b takes place. The

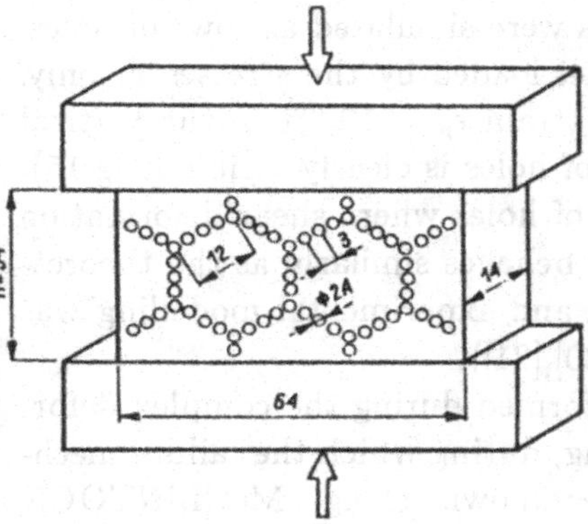

Fig.14, Experimental setup for modelling intergranular fracture.

Fig.15, Specimen used for modelling intergranular fracture after compression test.

grains are deformed plastically. However, part of the deformation of the aggregate is caused by sliding of hexagons along the inclined contact surfaces. Along the vertical contact surfaces the gaps are formed. This process of separation along these surfaces is irreversible, while along other boundaries the adhesion was not damaged by the sliding process. For more information see Appendix.

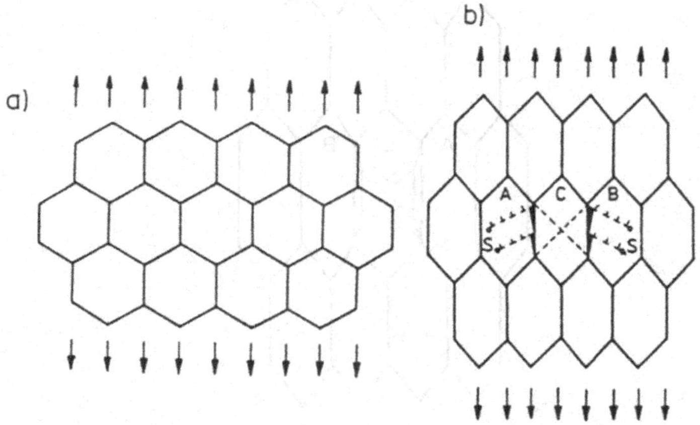

Fig.16, Formation of cracks along grain boundaries.

In order to demonstrate how the gaps may be formed a simple experimental test simulating real process has been performed. Specimen of the shape shown

in Fig.14 made of a ductile Al-2%Mg aluminium alloy has been compressed as shown in the figure. The weak grain boundaries were simulated as rows of holes. Such a test corresponds to the theoretical model loaded by the stresses p_1 only. The specimen was deformed to the permanent strain $\epsilon_p = 10.3\%$ in the vertical direction. Total separation along vertical rows of holes is clearly visible (Fig.15). There is no decohesion along the inclined rows of holes where shear deformation only is observed. Thus our experimental model behaves similarly as the theoretical model. Extensive study of such theoretical and experimental modelling was presented in previous papers (SZCZEPIŃSKI [20],[21]).

In the model discussed above cracks were formed during the complex deformation history preceding the final tensile loading, during which the failure mechanism begins to develop. Consider now another known (cf.e.g. McCLINTOCK and ARGON [22] or HIRTH and LOTHE [23]) mechanism of forming cracks along the grain boundaries during the process of uniaxial tension without the preceding loading history. An idealized model of a polycrystal (Fig.16a) is pulled in tension and its grains are elongated plastically to the state shown in Fig.16b. Assume that in some grains (say in grains A and B) there exist dislocation sources S producing dislocations during the deformation process. Accumulation of dislocations at the grain boundaries leads to the formation of cracks along

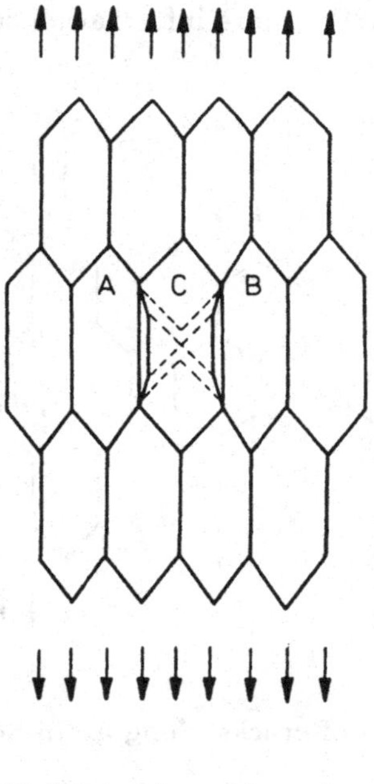

Fig.17, Internal necking of a grain.

these boundaries. Thus at a certain level of plastic deformation cracks separating central grain C from grains A and B appear at their boundaries.

Suppose now that the materials with cracks formed previously in this or other way during the preceding loading history is loaded at high temperature by tension in such a manner that cracks are parallel to the direction of loading. After certain amount of uniform plastic deformation the grain C, separated from grains A and B by a pair of cracks, begins to neck down independently according to the mechanism shown in Fig.2. The advanced stage of the local necking is presented schematically in Fig.17.

It is not excluded that for certain metals the internal necks may be formed even at the room temperature. This supposition seems to be confirmed by experiments in macroscale discussed in next sections.

Another mechanism of forming elongated defects parallel to the direction of tensile loading is shown in Fig.18, which presents a deformed specimen made of a ductile aluminium alloy. Before the tension test two small circular holes were drilled in the specimen. Then the specimen was pulled in tension. It can be

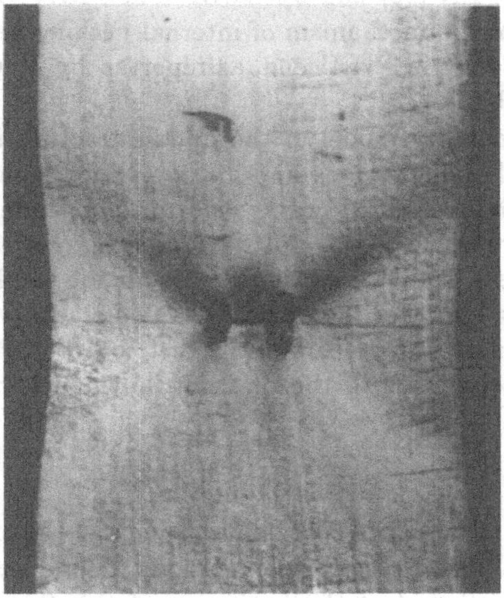

Fig.18, Elongation of circular holes and fracture between them.

seen that during the initial stage of plastic deformation the holes were strongly elongated forming a configuration similar to a pair of cracks parallel to the direction of tensile loading. During the advanced stage of deformation a neck is formed in the ligament between the deformed holes leading finally to the internal fracture seen in the figure.

5. INTERNAL MICRONECKING

Cracks parallel to the direction of tension may radically affect the mechanism of ductile fracture. Note that in the theory of brittle fracture cracks of such orientation do not play any significant role, contrary to cracks perpendicular to the tensile loading. In the mechanics of ductile fracture their role may be significant. For example ROGERS [24] compared the tensile ductility of two theoretical models. One is a cylindrical rod of a perfectly plastic material and the other is a cable of equal external diameter consisting of a bundle of cylindrical strands. Each of the strands behaves in the same manner as a single rod, necking down to a point. Thus total elongation of the rod just before separation into two pieces is many times larger than that for the cable. A similar mechanism of ductile fracture was discussed by BACKOFEN [25] and by HULT and TRAVNICEK [26]. The real physical significance of the mechanism of internal necking has been confirmed by RHINES who simulated it in Plasticene, as reported by McCLINTOCK [8]. See also for example MARTIN [27].

Consider an elementary model of the ductile fracture mechanism shown in Fig.19a for the case of a single crack and in Fig.19b for a pair of cracks. Each of the strips formed by the separating cracks deforms independently according to the necking mechanism shown in Fig.2. In the present case a rigid-plastic model of the material is assumed. Total separation occurs for the single crack model at the limit elongation $\triangle L^* = \frac{1}{2}c_0$. For the model with two cracks we have $\triangle L^* = \frac{1}{3}c_0$ as shown in Fig.19b. Generally for n uniformly distributed parallel cracks the limit elongation at which separation of the plane strain model into two parts occurs may be written as

$$\triangle L^* = \frac{1}{n+1}c_0.$$

Thus using the theory of ideally plastic deformation one obtains the failure accompanied by a very small total plastic deformation. This effect is similar to those analysed in papers mentioned at the beginning of this Section. The possibility of application of the slip-line technique to the analysis of this problem was mentioned by BROWN and EMBURY [28] and also in the recent book [29] edited by TREFILOV.

Consider now a more realistic case when the cracks are nonuniformly distributed. For the sake of simplicity we shall discuss at first the particular case of a pair of cracks. The initial configuration with the slip-line system is shown as the stage A_1 in Fig.20 (cf. SZCZEPIŃSKI [30], [31], [32]). For not too large total

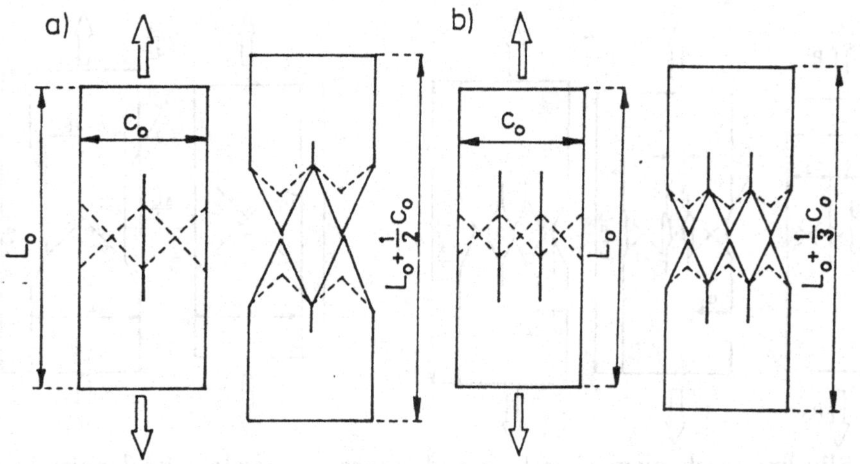

Fig.19, Slip-line mechanism of ductile failure in models with longitudinal cracks.

elongation $\triangle L$ of the model each of the strips separated by the cracks deforms independently according to the solution shown in Fig.2. A neck is formed in each of them as shown in Fig.20A_2. A strong plastic strain concentration is connected with such a mode of deformation. It is evident that in the elastic state no stress or strain concentration in the element with cracks forming the system such as in Fig.20A_1 appear under tensile loading.

In order to estimate the magnitude of the strain concentration consider the stage of deformation presented in Fig.20A_2. If the total elongation of the model is $\triangle L$, then the lateral shortening of all three strips separated by the slits will be equal to $-\triangle L$. This follows from the slip-line solution (cf. Fig.2). Thus the conventional lateral strain in central strip may be written as

$$\epsilon_{a_1} = -\triangle L/a_1,$$

and in the two outer strips it is equal to

$$\epsilon_{a_2} = -\triangle L/a_2.$$

Comparing both values the strain concentration factor f can be calculated

$$f = \epsilon_{a_1}/\epsilon_{a_2} = a_2/a_1.$$

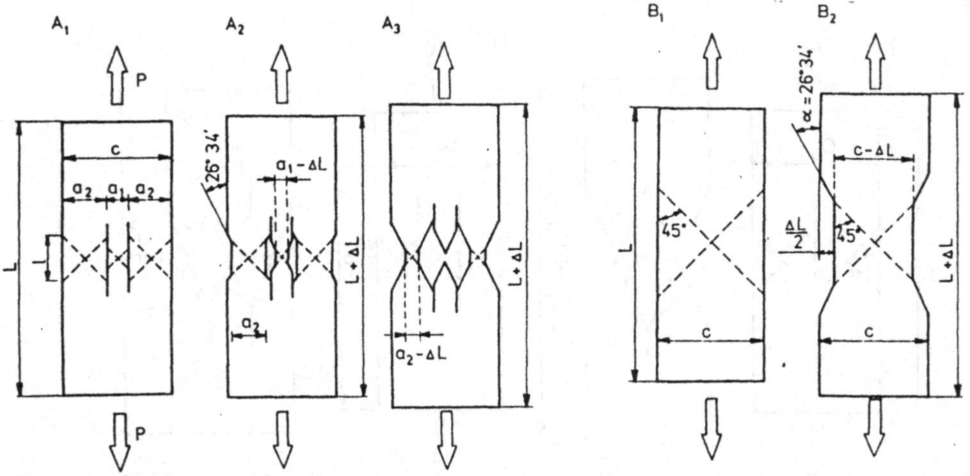

Fig.20, Slip-line mechanism of necking and failure in a model with longitudinal cracks.

This formula is valid only when the central strip is not separated into two parts as shown in Fig.20A_2.

Another definition of the strain concentration factor is obtained by comparing the conventional lateral strain in the narrowest strip of the model with that of the solid model without slits (Fig.20B_2). The strain concentration factor f^* defined in this manner can be calculated from the formula

$$f^* = \epsilon_{a_1}/\epsilon_c = c/a_1.$$

Such a strain concentration factor may reach in real situations very large values.

Since the analysis presented above is based upon the assumption of an idealized rigid-plastic material with no strain hardening, simple experiments were performed in order to ascertain whether the internal damage due to the local micronecking (cf. Fig.20A_3) occurs in real metals. Results of such experiments were published in [30] and [31]. Typical example is shown in Fig.21. The specimen of a ductile Al-2%Mg aluminium alloy was deformed plastically by uniaxial tension. Two slits were machined in the specimen according to the scheme presented in Fig.20A_1. Necking process is only slightly advanced in the two wide strips while the narrow central strip has ruptured. This simple experiment shows that the theoretical analysis based on the slip-line technique has a real practical significance. Thus the mechanism of local internal necking should be considered as one of possible factors responsible for the process of ductile fracture in metals.

Let us analyse the history of deformation in a plane strain model with two parallel cracks analogous to that shown in Fig.20A_1 and subjected to repeated

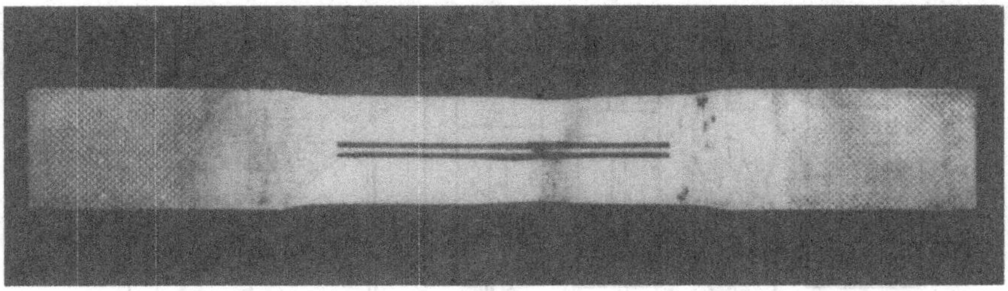

Fig.21, Deformation and internal fracture in a specimen with a pair of slits.

tension-compression loadings with the symmetrical strain amplitude. In Fig.22 are presented seven stages of deformation during the first two cycles of loading with the conventional strain amplitude $\pm\epsilon = \pm \triangle l^*/c_0$. Since in our model the plastic deformation is strongly localized, a conventional measure of strain $\epsilon = \triangle l/c_0$ is assumed, where $\triangle l$ is the total elongation of the model and c_0 stands for the initial width of the model.

As in the previous analysis, we assume that the material is rigid-plastic with no strain-hardening. Consecutive stages from 1 to 7 show how the internal structure of the model changes during the first cycle of loading. Stage 1 shows the initial configuration. Plastic yielding begins when the pulling force per unit thickness reaches the value $P^* = c_0\sigma_{pl}$. Stage 2 is analogous to the final stage shown in Fig.20A_3 and the total length of the model is $l_0 + \triangle l^*$. Between the stages 2 and 3, the model is compressed and the two separated parts of the central strip are flattened. This leads to the configuration shown as the stage 3 with an additional horizontal cracks between the two initial vertical cracks. The total length of the model has been restored to its initial value l_0. Between stages 3 and 4 the model is compressed up to the conventional strain $\epsilon^* = -\triangle l^*/c_0$ by the constant stress equal to yield locus of the material σ_{pl}. Initial length l_0 is reduced to $l_1 = l_0 - \triangle l^*$ and spacings a_1 and a_2 increased to $a_1' = l_0 a_1/(l_0 - \triangle l^*)$ and $a_2' = l_0 a_2/(l_0 - \triangle l^*)$, respectively. Also the total width increases from c_0 to $c_1 = l_0 c_0/(l_0 - \triangle l^*)$. Thus the total force at the end of this stage is equal to $P = -\sigma_{pl} c_1$. When, subsequently the tensile stress is applied, the two outer strips of the width a_2' begin to neck. This is illustrated in Fig.22a by the stage 5 showing the situation for the conventional strain equal to $+\epsilon^*$. For the following compression the length of the model is reduced again to l_1 (stage 7). The intermediate stage of deformation corresponding to the initial length l_0 is shown as the stage 6. Note that the configuration shown as the stage 7 is identical with that shown as the stage 4. Fig.22b shows how the stress-strain loop changes after the first cycle of loading.

The change of the internal structure after the first cycle of loading may be

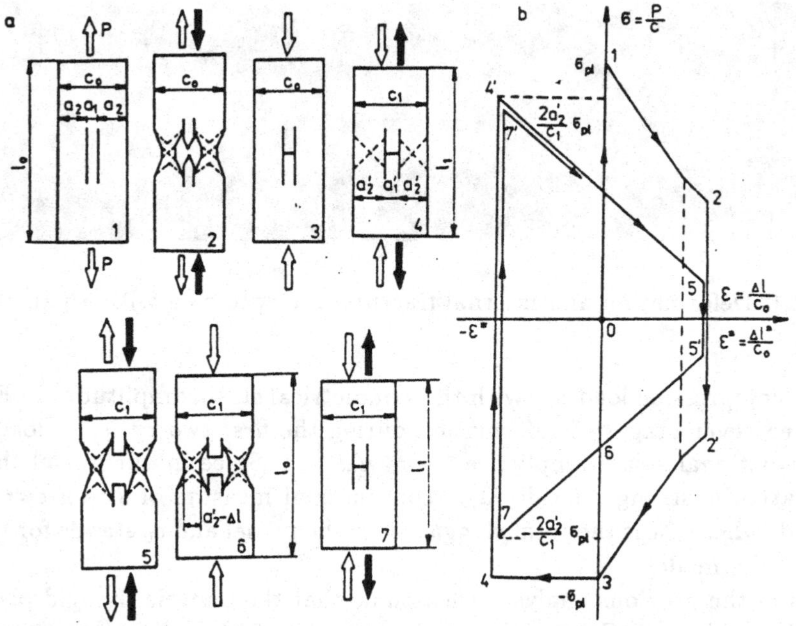

Fig.22, Cyclic loading of a theoretical model of a body with two longitudinal cracks.

interpreted as a damage parameter. In our model the separated portion of the cross-sectional area of the width a_1' may be used as a measure of damage. Referring to Fig.22a the damage parameter may be defined as

$$\omega = a_1/c_0 = a_1'/c_1. \tag{5.1}$$

Note that this value of the damage parameter is valid also in the case when the deformation process is finished at the stage 3.

Assume that the deformation process is finished at the moment which in Fig.22 is shown as the stage 6. Thus the final length of the model is equal to the initial length l_0, while its width increased to the new value c_1. A cavity of complex shape is formed inside the model and the damage parameter defined as the relative reduction of the cross-sectional area may be written as

$$\omega = (a_1 + 2\Delta l^*)/c_1. \tag{5.2}$$

This simple analysis shows how the material may be weakened by the deformation process if initially it has imperfections in the form of microcracks.

Fig.23 illustrates evolution of damage in a model with a number of cracks with different spacings between them. Such a progressing process of damage was examined experimentally.

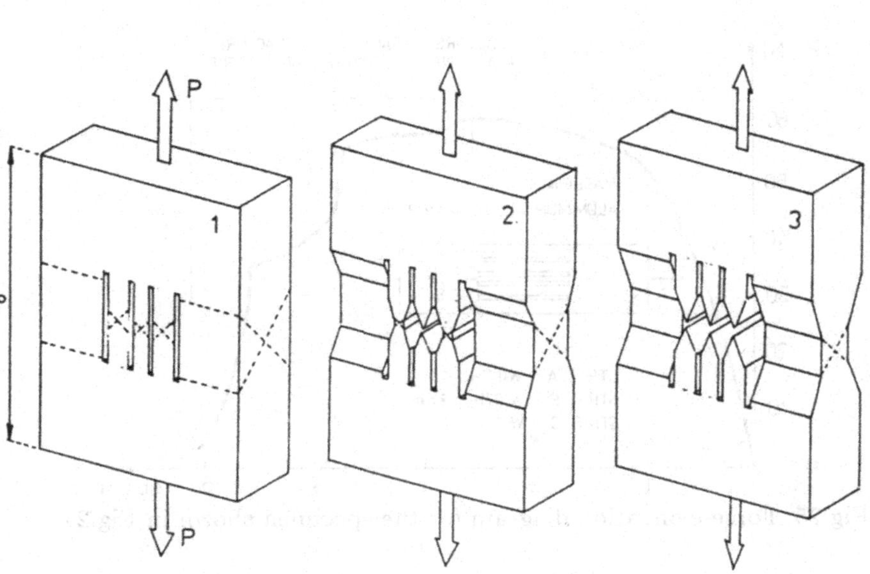

Fig.23, Slip-line mechanism of necking and ductile failure in a model with four longitudinal cracks.

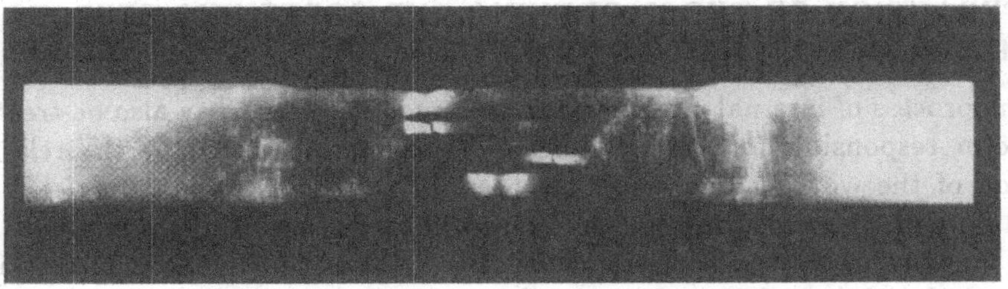

Fig.24, Deformation and internal fracture in a specimen with four longitudinal slits.

Fig.24 shows a plastically deformed specimen with four machined narrow slits. The dimensions of the strips between these slits are shown in the inset in Fig.25. The thickness of the specimen was 15 mm. Thus the conditions of plastic deformation were close to those of plane strain. The specimen was made of a ductile Al-2% Mg aluminium alloy [32].

The force-elongation diagram is shown in Fig.25. In the final sector of the diagram, there appear sharp steps corresponding to the formation of necks in the

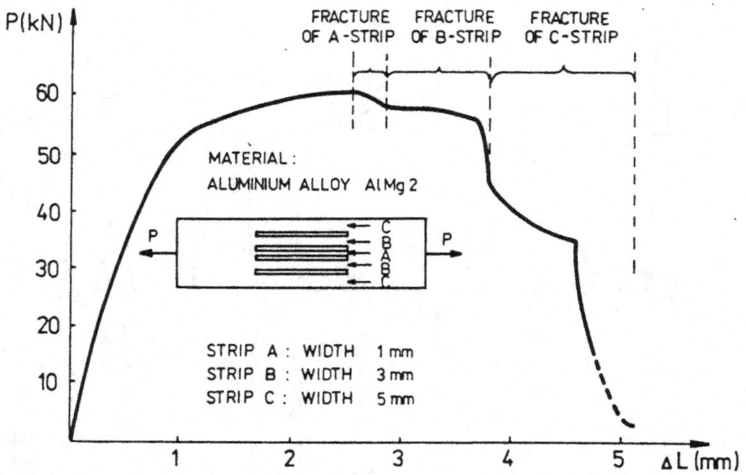

Fig.25, Force-elongation diagram for the specimen shown in Fig.24.

consecutive strips between the slits. After the fracture of the central narrow strip A, the specimen still displays quite a great carrying capacity and ability to deform plastically. The second step corresponds to the fracture of strips B.

6. INTERACTION OF THE HARDENING AND SOFTENING EFFECTS IN THE MECHANICS OF DUCTILE FRACTURE

The process of internal micronecking presented in Fig.20 may also be treated as a factor responsible for the softening of the material caused by the relative reduction of the working cross-sectional area. The process of softening and its interaction with the simultaneous hardening will be discussed in this Section.

In the advanced stage of plastic deformation, the softening effects of the kind mentioned above may play an important role in the global behaviour of metals. The yield criterion should include not only the hardening parameters α, but also the softening parameters β connected with the reduction of the effective cross-sectional area. Generally such a yield condition may be written in the form

$$f(\sigma_{ij}, \alpha, \beta) = 0. \tag{6.1}$$

Now following previous works [34], [35] we will examine the model with two parallel cracks with short branches corresponding to the configuration shown on the right-hand side in Fig.12. These branches form the internal notch within the model.

Assuming the rigid-perfectly plastic material we may analyse the entire deformation process with the use of the slip-lines technique. The initial configuration of slip-lines is presented in Fig.26a. The two outer strips begin to neck according to the scheme shown in Fig.20. Theoretical solution to the process of deformation in the narrowest section of the central strip is identical with the solutions given by LEE [36] and LEE and WANG [37] for notched bars pulled in tension.

Figure 26b shows the intermediate stage of the plastic deformation process. Cavities of a complex shape have been formed on both sides of the central strip. This process leads finally to the total separation of the central strip (Fig.26c). In this slip-line solution, in which the strain hardening is neglected, the necking in the two outer strips begins simultaneously with the beginning of plastic yielding of the notched part.

In order to examine whether such a mechanism of internal fracture may occur in real ductile metals which display the strain hardening effect, a set of specimens of an Al-2% Mg aluminium alloy was prepared [35]. The specimens were machined from a rolled bar 14 mm thick. Their dimensions in milimeters are shown in Fig.27. Dimensions of all specimens were identical, except the notch width, which was different in consecutive specimens. Theoretical and experimental analysis (cf. DRUCKER [38], SZCZEPIŃSKI and MIASTKOWSKI [39]) indicates that for such a thickness of specimens the plane strain solutions constitute fairly good approximation. Thus the specimens may serve for the experimental testing of the theoretical solution.

The behaviour of the specimens under tension is different from that of the theoretical rigid-plastic model analysed in Fig.26. Both outer strips deform uniformly until a considerable amount of plastic deformation owing to the strain hardening of the material. Then the necks begin to form. This is demonstrated in Fig.28. The photographs show two stages of plastic deformation of the specimen with the initial notch width b=6 mm. In the left-hand figure the central notched strip has been ruptured while the two outer strips are deformed uniformly. The necks begin to form in them in a very advanced stage of deformation. This is shown in the right-hand photograph.

The two following figures show how the force-elongation diagram depends on the geometrical parameter defined as the notch width b. We may observe the sharp drop of the pulling force connected with the fracture of the notched central strip. After this sudden decrease of the force the carrying capacity of specimens increases owing to the strain hardening effect in the outer strips. Thus our idealized experimental model simulates the possible interaction between the hardening and softening parameters α and β appearing in the yield condition (6.1).

It can be seen that the recovery of the carrying capacity is full for deeply notched central strip and the maximum force is larger than that causing fracture in the notched part. The diagram in Fig.29 was obtained for the specimen with the notch width $b = 2mm$. In the case shown in Fig.30 the notch width was $b = 2.5mm$. For less deep notches ($b = 4mm$ and $b = 6mm$) this recovery was not

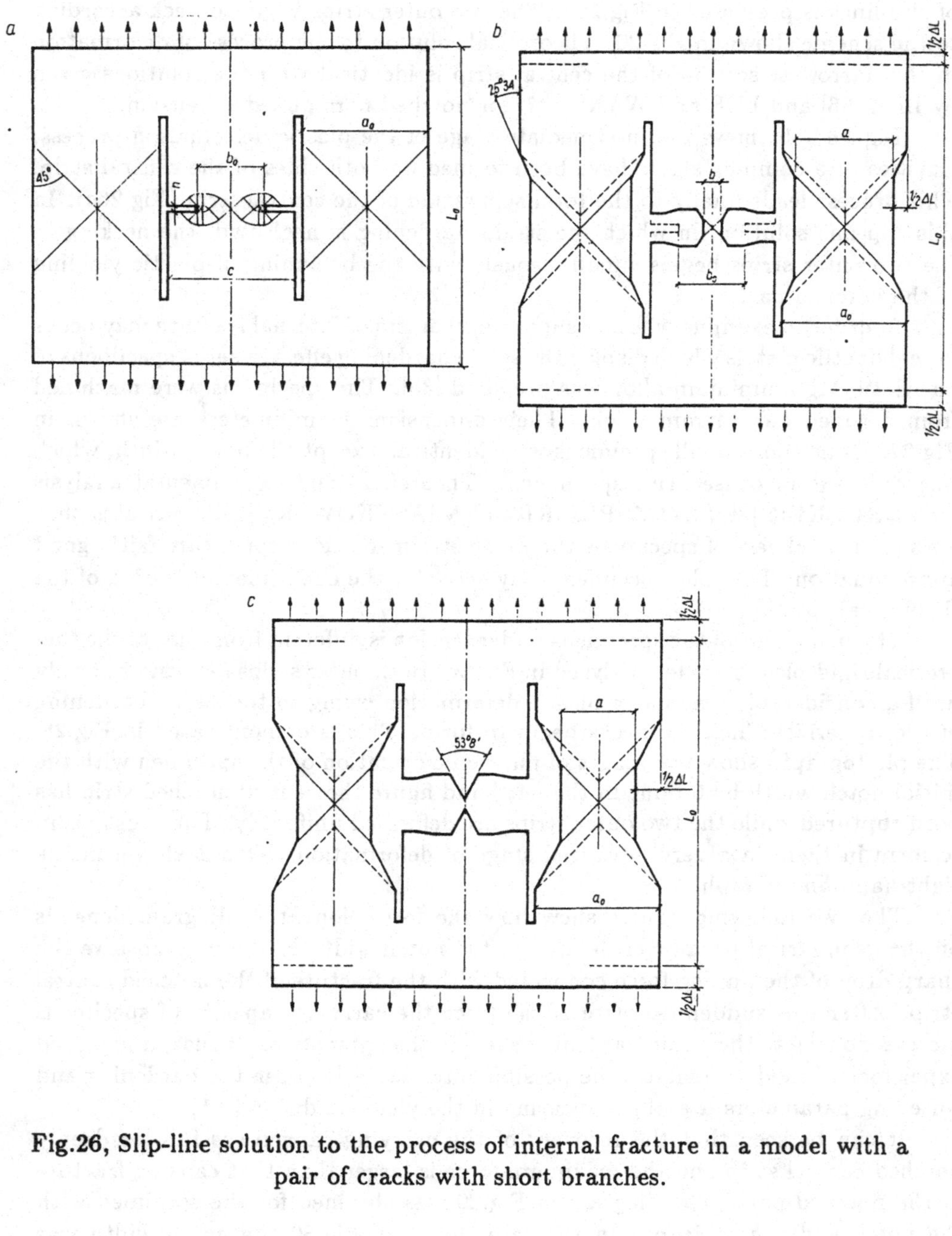

Fig.26, Slip-line solution to the process of internal fracture in a model with a pair of cracks with short branches.

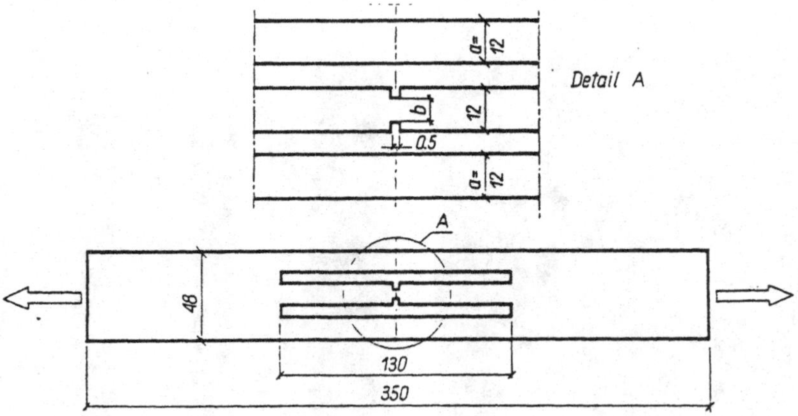

Fig.27, Specimen with a pair of slits with short branches.

full (details may be found in [35]).

The interaction between hardening and softening is even more visible in experimental models with several notched strips. In Fig.31 is presented a force-elongation diagram for a specimen with three smooth strips and two strips with notches of different width as shown in the inset. The specimen was made from the same material as the previous specimens. Two sharp steps can be seen in the initial portion of the force-elongation diagram. These steps are connected with the consecutive rupture of the notched strips. The hardening effect may be observed also on the short sector between the two steps.

Similarly as in the case shown in Figs.20 and 22, we may interpret the process of deformation of our experimental models as the evolution of the damage parameter. In the above mentioned case the damage parameter after necking the narrowest strip down to the point was simply defined as the relative reduction of the working cross-sectional area. In the case of the configuration of defects shown in Fig.12b, which is discussed here, the damage parameter after fracture of the notched central strip cannot be defined in such a simple manner. The carrying capacity of the notched bar pulled in tension is larger than that of a smooth bar with the same area of cross-section. This means, referring to Fig.26a, that the carrying capacity of the central strip is larger than that of the un-notched strip of the initial width b_0. In the theory of notched bars this increase of the carrying capacity is defined by the load factor

$$\kappa = P/P_0,$$

where in the case of plane strain notched bars $P_0 = b_0\sigma_{pl}$ is the yield load of a smooth bar of the width b_0 and P is the yield force per unit thickness resulting

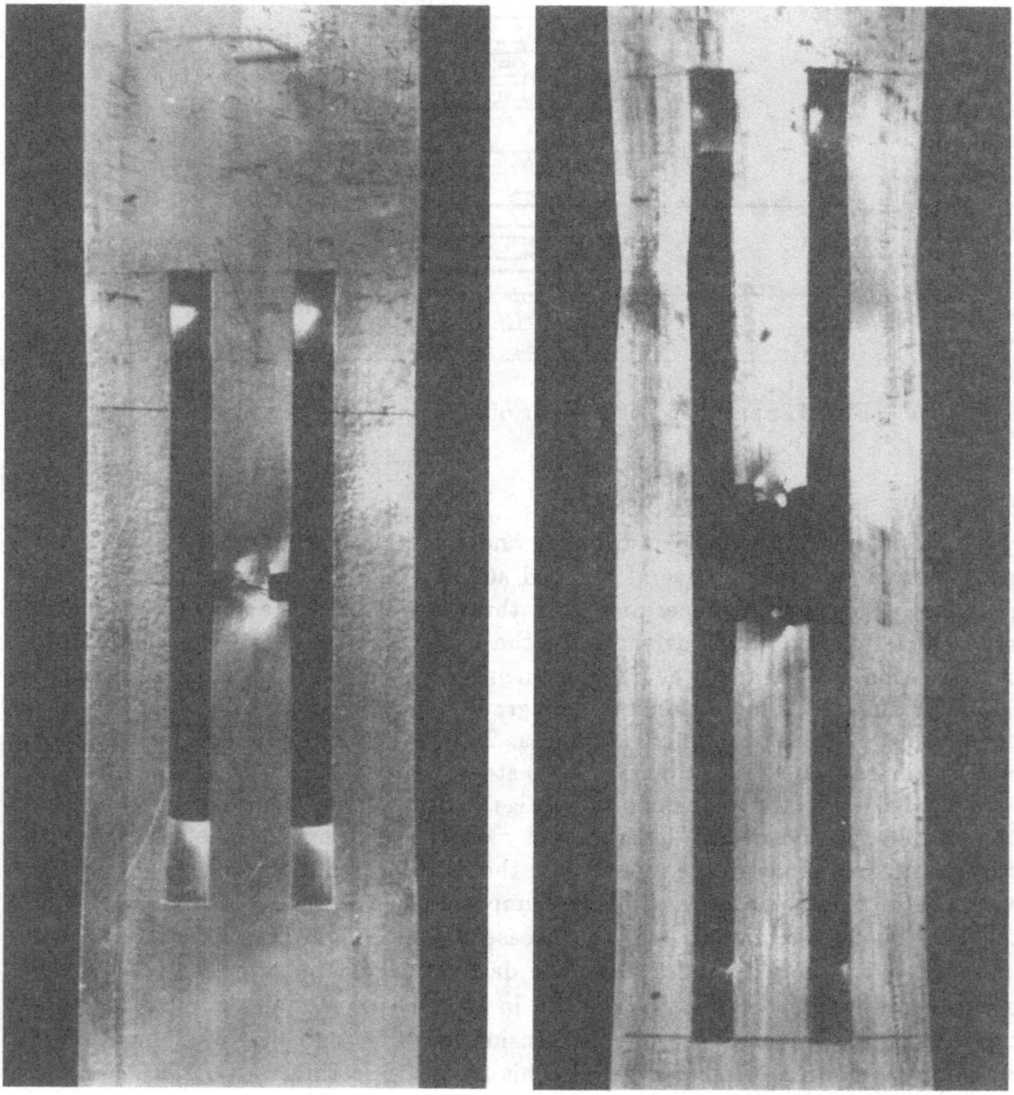

Fig.28, Two stages of deformation of the specimen shown in Fig.27.

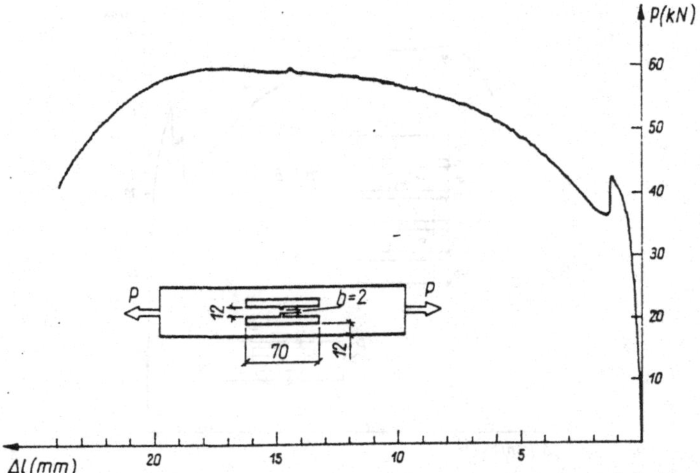

Fig.29, Force-elongation diagram for a specimen with two slits with branches ($b = 2mm$).

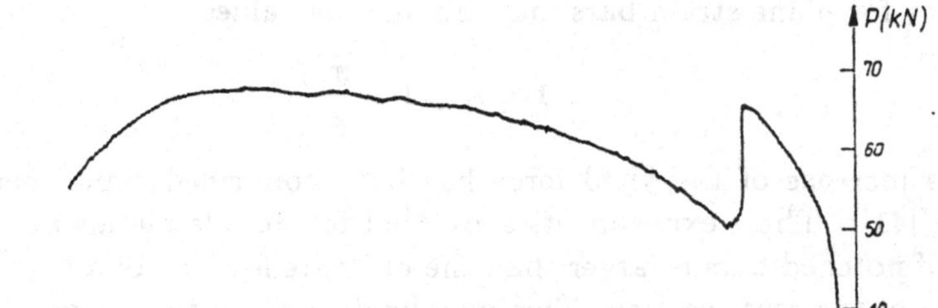

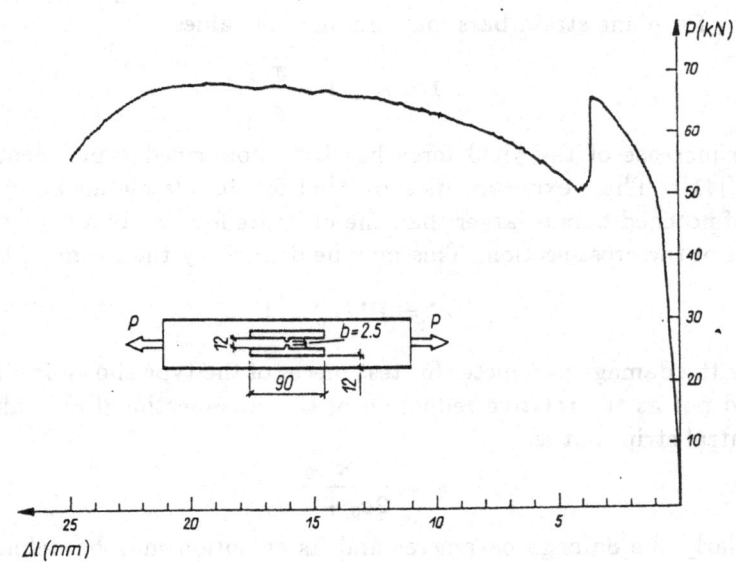

Fig.30, Force-elongation diagram for a specimen with two slits with branches ($b = 2.5mm$).

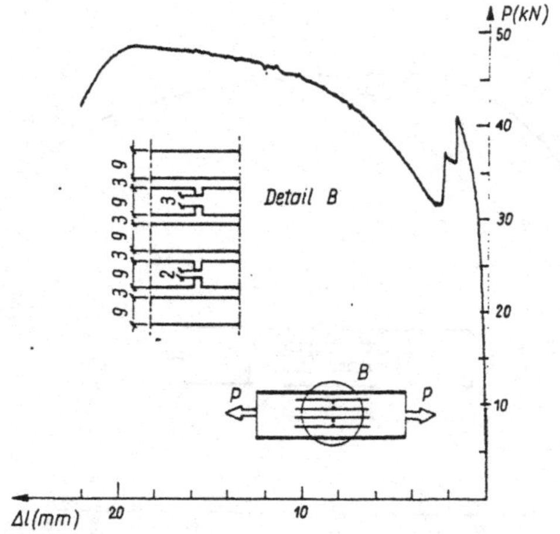

Fig.31, Force-elongation diagram for a specimen with four slits with branches.

from the slip-line solution. Depending on the shape of the notch and on the width of the bar outside the notch (on the width c in the case in Fig.26a), this theoretical load factor for plane strain bars may assume the values :

$$1 < \kappa \leq 1 + \frac{\pi}{2}.$$

This increase of the yield force has been confirmed experimentally (cf.e.g. [39], [40], [41]). These experiments show that for ductile metals also the ultimate load P^* of notched bars is larger than the ultimate load P_0^* of a smooth bar of the same area of the cross-section. This may be defined by the ultimate load factor

$$\kappa^* = P^*/P_0^* > 1.$$

Thus the damage parameter for test pieces of the type shown in Fig.27 should be defined not as the relative reduction of the cross-sectional area after fracture of the central strip, but as

$$\omega = \frac{\kappa^* b_0}{2a_0 + \kappa^* b_0}$$

Similarly the damage parameter and its evolution may be defined for specimens with several strips with notches of different width on the type shown in Fig.31. Note that analogous problem with defining the damage parameter was discussed in the case shown in Fig.9.

It is worth mentioning that the effect of a sudden drop of the pulling force occuring in diagrams presented in Figs.29, 30, 31 is of a similar type as that observed in rock-like materials. For these materials rheological models have been proposed which contain elastic springs and a fracturing constituent (see LIPPMANN [42]). By introducing sliding elements these models may be generalized also for problems considered here.

7. INTERACTION BETWEEN LINEAR DEFECTS

The role of strain concentration, discussed in previous Sections, is also significant in the presence of certain systems of linear defects existing in the material. Such linear defects may appear in the material as the result of the previous cold working operations or may be produced in the course of the process of plastic deformation leading finally to the ductile fracture. Consider a case of two cracks D-C and G-P-A oriented as shown in Fig.32, (cf.[44]).

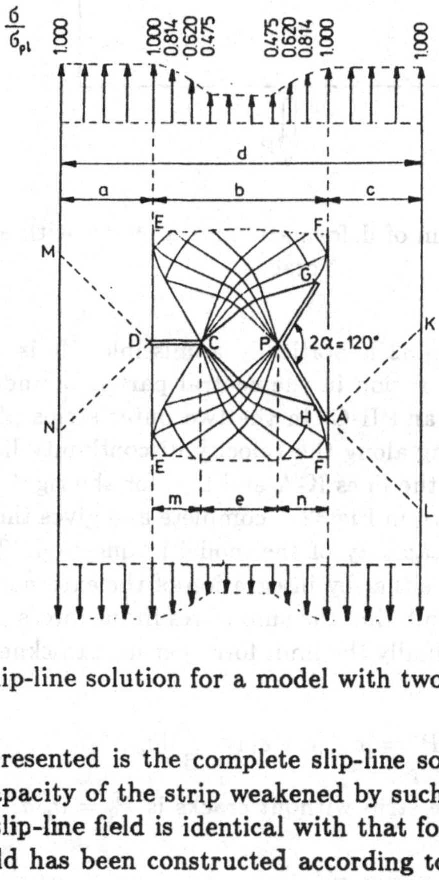

Fig.32, Complete slip-line solution for a model with two cracks D-C and H-P-G.

In the figure presented is the complete slip-line solution to the problem of the load carrying capacity of the strip weakened by such a system of cracks. The central part of the slip-line field is identical with that for a notched bar problem. This part of the field has been constructed according to the standard Bishoph's procedure [45]. In the two outer strips of the width a and c respectively, a state of uniaxial tensile stress equal to the yield point σ_{pl} has been assumed. The entire

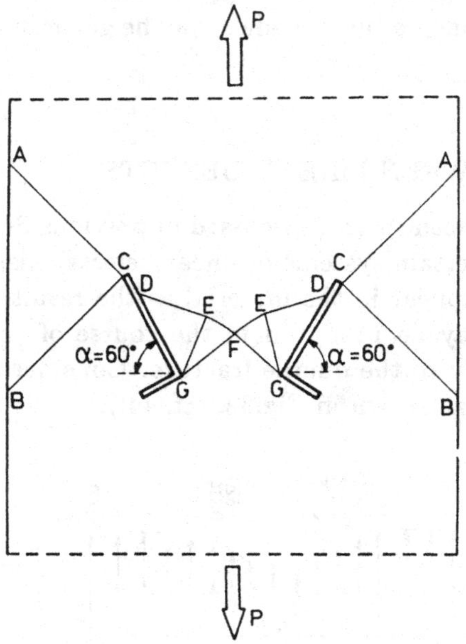

Fig.33, Slip-line mechanism of deformation for a model with a pair of inclined cracks.

stress field presented in Fig.32 is statically admissible. It is also kinematically admissible. The deforming region in the central part is bounded by the lines of velocity discontinuity G-C and H-C. In the two outer strips plastic deformation consists in rigid block sliding along the velocity discontinuity lines M-D and N-D for the left strip and along the lines K-A and L-A for the right-hand strip.

Thus the solution shown in Fig.32 is complete and gives the exact theoretical value of the load carrying capacity of the model in question. This load carrying capacity may be calculated either by integration of the external stresses shown in the figure, or simply by calculating the sum of resultant forces acting in the three narrowest cross-sections. Finally the limit force per unit thickness may be written as

$$P^* = [a + c + e(1 + \frac{\pi}{6})]\sigma_{pl}, \tag{7.1}$$

while the limit force for the strip without cracks is $P_0 = \sigma_{pl} d$. Thus the damage parameter is

$$\omega^* = 1 - \frac{P^*}{P_0} = \frac{1}{d}(m + n - e\frac{\pi}{6}). \tag{7.2}$$

So defined damage parameter is evidently smaller than that calculated with the use of the definition (3.1), according to which in the case in question we obtain

$$\omega_0 = \frac{1}{d}(m+n). \tag{7.3}$$

The slip-line solution presented in Fig.32 is valid also for numerous other configurations of cracks, one of which is shown in Fig.33.

Now the deforming region in the central part is bounded by a pair of lines of velocity discontinuity D-E-F-G , which coincide with appropriate slip-lines of the stress field shown in Fig.32.

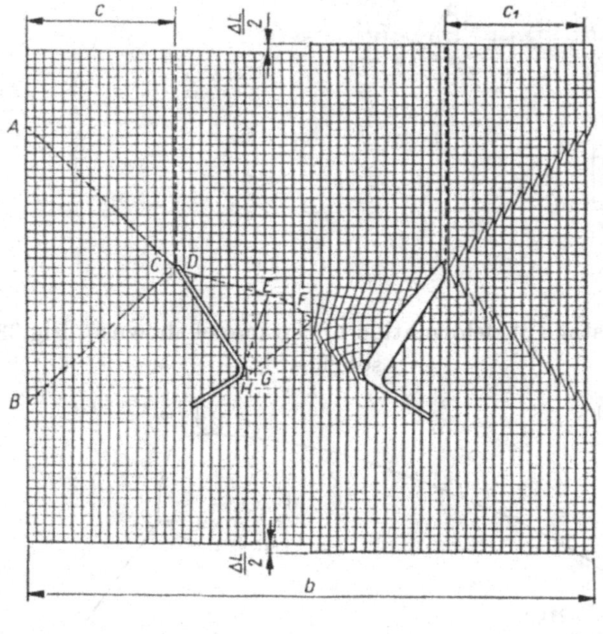

Fig.34, Initial stage of deformation of the model shown in Fig.33 - theoretical solution.

The right-hand side of Fig.34 shows deformation of a square grid for small total elongation $\Delta L = 0.032b$, where b is the width of the model shown in the figure. At this stage of deformation the conventional lateral strain in the narrowest cross-section reaches the value 0.22, while in the outer strips such a conventional lateral strain, defined as $(c - c_1)/c$, is more than four times smaller reaching the value 0.05. Strain concentration in the ligament between the two slits increases with the advancement of the process of deformation. For example, for more advanced total elongation $\Delta L = 0.081b$ (Fig.35) conventional lateral strain in the

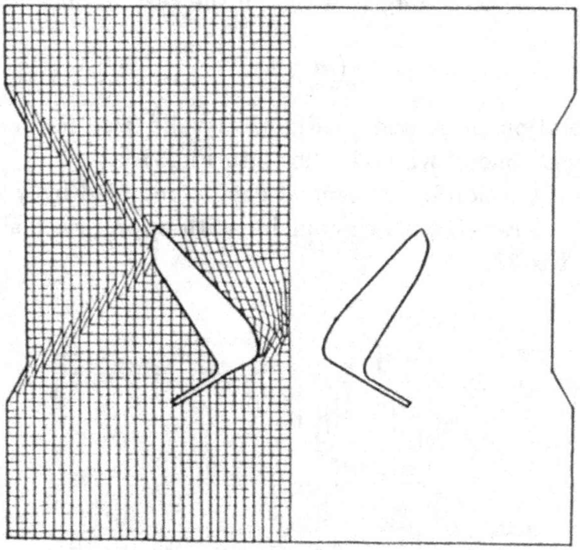

Fig.35, Advanced stage of deformation of the model shown in Fig.33 - theoretical solution.

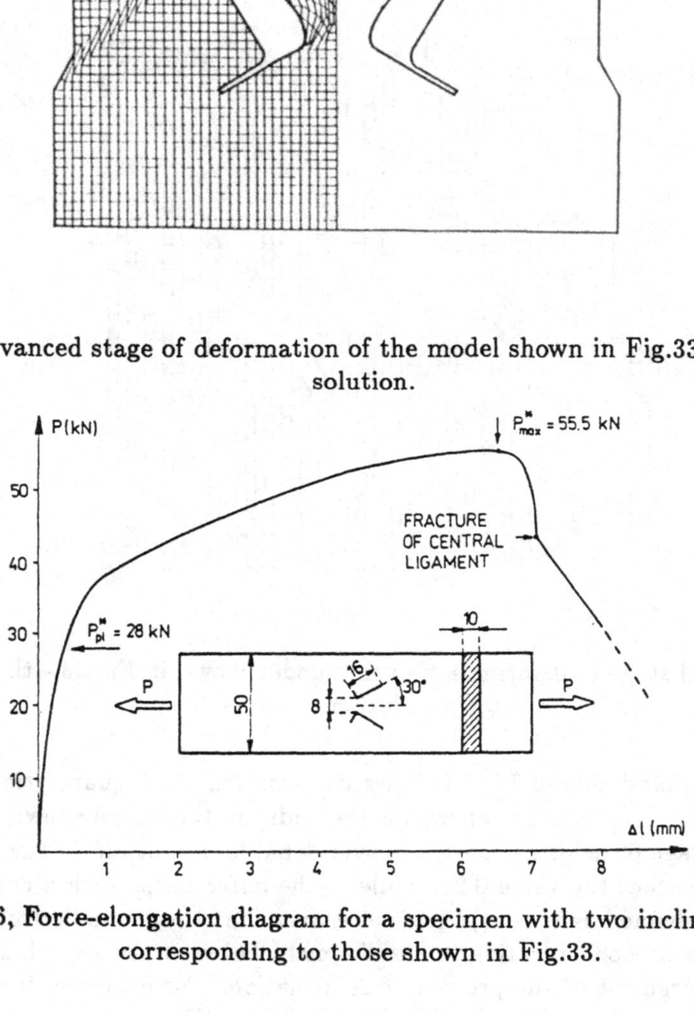

Fig.36, Force-elongation diagram for a specimen with two inclined slits corresponding to those shown in Fig.33.

central ligament is over five times larger than that in the outer strips. The strain concentration factor, defined as the ratio of the conventional lateral strain in the central part to that in the outer strips, tends to infinity when the width of the ligament between the two cavities reduces to a point. Thus present analysis leads to the local internal separation of the material.

a b

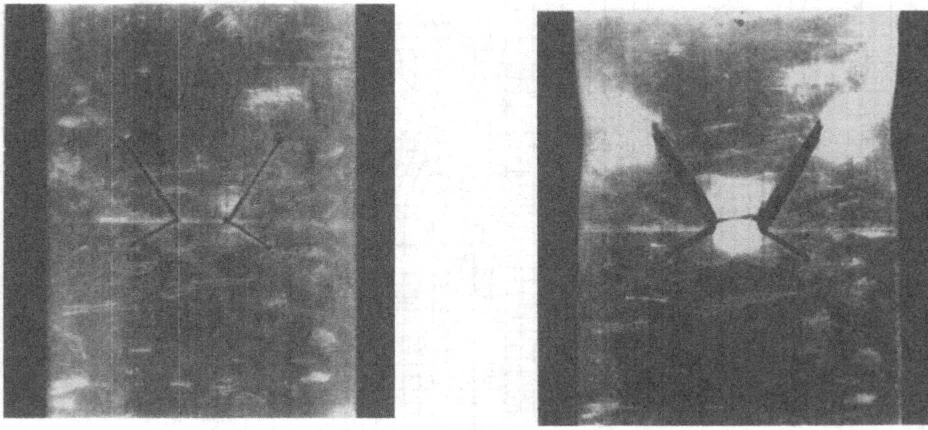

Fig.37, Specimen shown in the inset in Fig.36 before and after the tension test.

The particular configuration of cracks shown in Fig.33 was used in a simple experimental test performed in order to demonstrate the practical significance of the previous theoretical considerations. A specimen made of a rolled bar of a ductile Al-2% Mg aluminium alloy was tested. A pair of narrow slits simulating cracks were machined in the specimen as shown in the inset in Fig.36 and in Fig.37a.

The specimen was pulled in tension and the force-elongation diagram shown in the figure was recorded. The conventional yield force $P_{pl}^* = 28kN$ corresponds to that point of the diagram at which the tangent modulus equals 0.3 of the initial elastic modulus.

For an analogous specimen which, however, was not weakened by any slits the conventional yield force was found to be equal $P_{pl}^0 = 37kN$. Thus the damage parameter related to the conventional yield forces obtained experimentally is

$$\omega_1^* = 1 - \frac{P_{pl}^*}{P_{pl}^0} = 0,243.$$

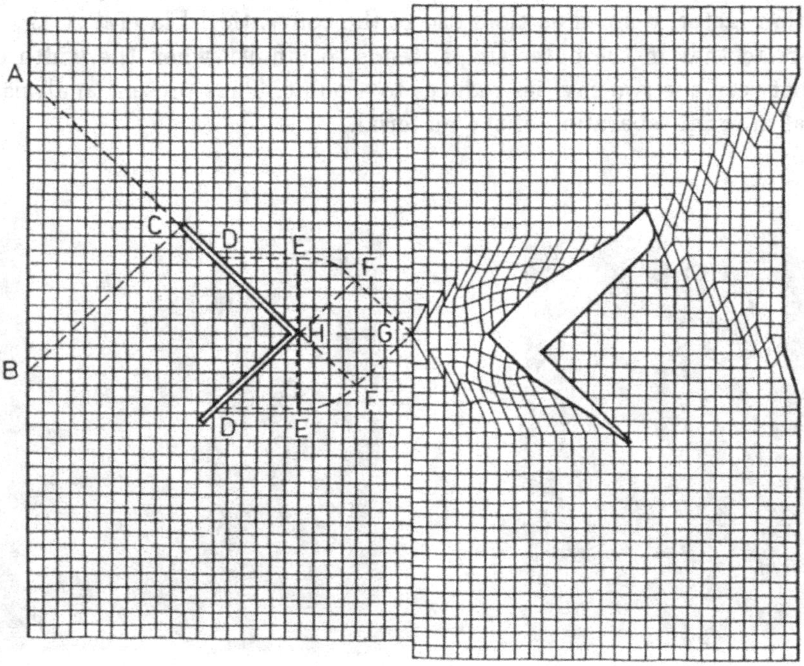

Fig.38, Theoretical deformation mode for a model with a pair of inclined cracks.

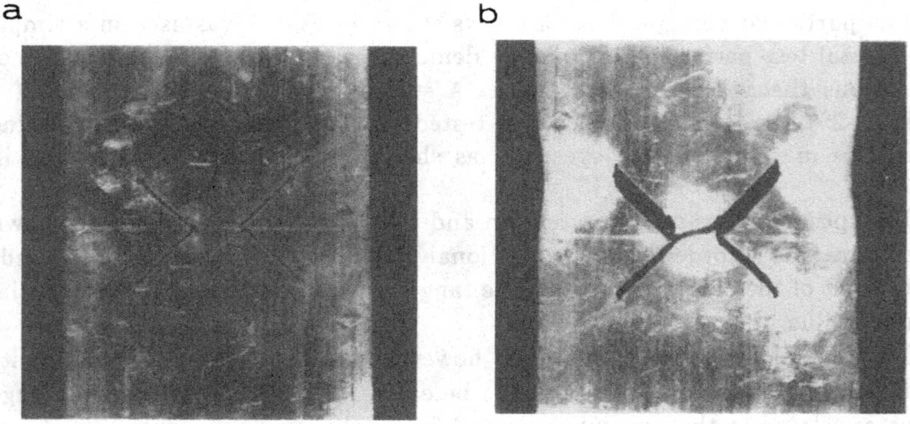

Fig.39, Specimen with two slits prepared according to the configuration shown in Fig.38 before and after tension test.

Similarly we can calculate the value of the damage parameter related to the maximum forces

$$\omega_2^* = 1 - \frac{P_{max}^*}{P_{max}^0} = 0,306.$$

Both experimentally obtained values of the damage parameter are smaller than theoretical value calculated from the formula (7.3). In the case in question we have $m = n = 8mm$ and $d = 50mm$. Substituting these values to (7.3) we obtain $\omega_0 = 0,320$.

Formula (7.2) based on the theory of the load carrying capacity gives the value

$$\omega^* = \frac{1}{d}(m + n - e\frac{\pi}{6}) = 0,236,$$

very close to the experimentally obtained value $\omega_1^* = 0,243$. Note, however, that the latter value depends on the convention used for determining the yield forces P_{pl}^* and P_{pl}^0. Nevertheless this simple experiment indicates that the damage parameter not always corresponds simply to the relative reduction of the working area.

In Fig.37a is shown the central part of the specimen before the test. The distance between the slits in the narrowest cross-section was equal to 8 mm while the specimen was 10 mm thick. Under such conditions the plane strain solution discussed above constitutes fairly good approximation of the real process of deformation (cf. [38], [39]).

Fig.37b presents the specimen after deformation causing fracture of the ligament between the slits. The deformation pattern is in general similar to the theoretical one shown in Figs.34 and 35. Owing to the limited ductility of the alloy used in the test, fracture occured along the band of strain concentration.

The slip-line solution for another configuration of slits is presented in Fig.38. The left-hand side of the figure shows the initial pattern of slip-lines. The central zone of plastic flow is bounded by lines of velocity discontinuity DEFG. The right-hand side of the figure shows the deformation pattern after relatively small total elongation (cf. [1]). The conventional lateral strain for this stage of deformation equals 0.31 in the central ligament, while in the outer strips it is equal to 0.136. Thus the strain concentration factor equals approximatelly 2.3.

Fig.39a presents the specimen with a pair of slits corresponding to the initial configuration shown in Fig.38. Material of the specimen was the same as in the previous tests. In Fig.39b is presented the specimen after deformation. The deformation pattern is in general similar to the plastic deformation mode predicted by the slip-line solution shown in Fig.38. The fracture in the central part, where we have strong strain concentration, is connected with the limited plastic deformability of the alloy of which the specimen was made.

8. FINAL REMARKS

Some hypothetical mechanisms of formation of certain systems of cracks are discussed. Simple slip-line analysis indicates that in the presence of such systems of cracks there appear strong strain concentrations inside the material leading to local microfractures. Simple experimental models demonstrate that mechanisms of ductile fracture discussed here should be considered as possible factors in the complex processes of ductile fracture.

APPENDIX
ON MODELLING INTERGRANULAR SLIDING AND FRACTURE.

Consider a simple structural model composed of n rigid regular hexagons in horizontal rows and of m hexagons in vertical rows as shown in Fig.40 (cf.[20], [21]). The hexagons simulating grains in a polycrystalline structure are assumed to be

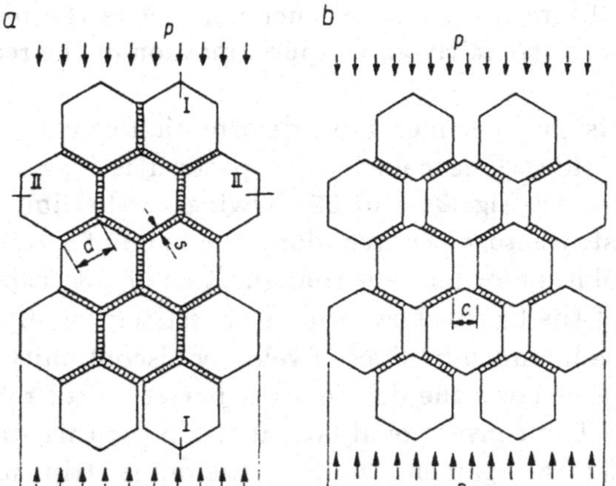

Fig.40, Idealized model of a polycrystal before and after hypothetical deformation; gaps are formed along vertical boundaries.

joined together by thin layers of the thickness s simulating the grain boundaries. The model may be used for simulation of the behaviour of metals with weak grain boundaries. We shall discuss here the behaviour of the model under uniaxial compression as shown in the figure. A more general case of biaxial loading is discussed in [20].

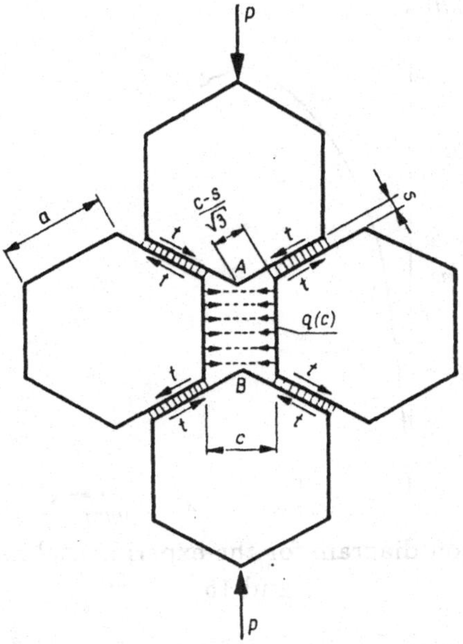

Fig.41, Stresses acting during deformation of the model.

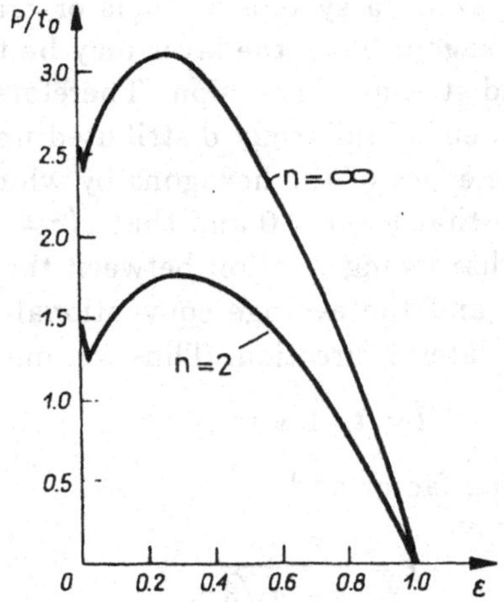

Fig.42, Force-deformation diagram for the theoretical model shown in Fig.40.

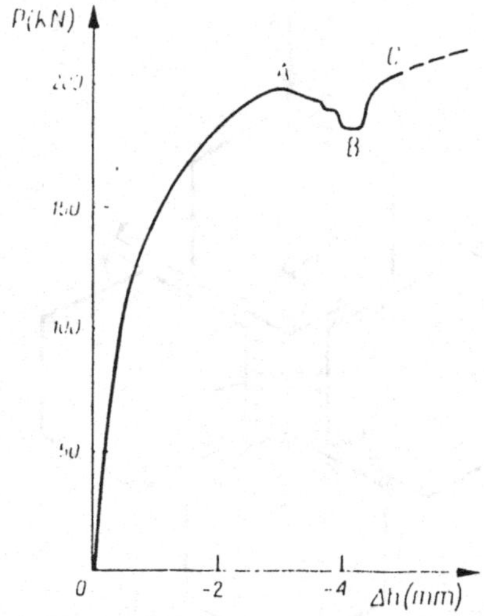

Fig.43, Force-deformation diagram for the experimental model shown in Figs.14 and 15.

The ratio s/a is assumed to be small. The material of the layers between hexagons is taken as rigid-plastic with linear strain hardening. Fig.41 presents the configuration of four hexagons after a certain amount of deformation defined by the distance c between the vertical edges ($c > s$). Assume that in each layer between the hexagons there exists a system of voids or cracks perpendicular to the respective edge of the hexagon. Thus the layer may be treated as a system of individual elements of limited strength in tension. Therefore the stress $q(c)$ along the boundary AB may be taken as uniformly distributed up to the limit value c_0 of the distance c between the edges of the hexagons by which the cohesion forces decrease to zero. This means that $q(c_0) = 0$ and that $q(c = s) = q_0$.

Assume a linear strain hardening relation between the shear stress t on the inclined surface of hexagons and the average conventional strain ϵ of the entire aggregate of hexagons in the lateral direction. Thus we may write

$$t = t_0(1 + r\epsilon), \qquad (a)$$

where r is the strain hardening factor and

$$\epsilon = \frac{c - s}{s + a\sqrt{3}}.$$

Writing the equation of virtual work for small deformations ($s < c < c_0$), we arrive at the relation between the compressive unit load p and the lateral strain ϵ

$$pn(m-1) = [m(n-2) + (m-1)(n-1)]q(c) + \frac{4}{\sqrt{3}}(m-1)(n-1)(1+r\epsilon)(1-\epsilon)t_0. \qquad (b)$$

The cohesion force per unit length $q(c)$ along AB changes its value from q_0 at the beginning of the deformation process ($c = s$) to $q = 0$ for $c = c_0$. Thus the first member of the right-hand side of the yield condition (b) changes its value from

$$[m(n-2) + (m-1)(n-1)]q_0$$

for $c = s$ to zero for $c = c_0$.

The model displays, therefore, the phenomenon of the upper and lower yield point in the initial stage of the deformation process. This is shown in Fig.42 in which the diagram compressive stress versus lateral strain of the aggregate is presented.

For the advanced stage of deformation the first member of the right-hand side of the condition (b) vanishes and we obtain the following relation between the compressive stress p and the lateral conventional strain:

$$p = \frac{4}{\sqrt{3}} \frac{n-1}{n+(n-1)\epsilon}(1+r\epsilon)(1-\epsilon)t_0. \qquad (c)$$

This relation is represented in Fig.42 for the particular value of the strain hardening factor $r = 5$ and for two values of n, namely for $n = 2$ and $n = \infty$, respectively. These stress-strain diagrams display two kinds of instability. At the beginning there appears a sudden decrease in stresses connected with the decohesion along the vertical boundaries between hexagons.

The second type of instability appearing in the advanced stage of deformation results from the reduction of the contact area along the inclined boundaries where sliding takes place. The physical significance of this second type of instability seems to be doubtful. It should be rather treated as the property of our idealized theoretical model without connection with the phenomena taking place in real polycrystalline metals.

The condition (c) and Fig.42 show how strongly the yield stress in our theoretical model depends on the number n of hexagons. It is seen that the larger the number of hexagons, the higher the yield point of the aggregate. So if we compare two aggregates of the same external dimensions, the one composed of a large number n of hexagons with small length of the side a will be stronger than the other composed of a smaller number of larger hexagons. This corresponds to the well-known property of technical metals. Structures of metals with fine grains behave better than those with coarse grains.

In Fig.43 is presented the force-displacement diagram for the experimental model shown previously in Figs.14 and 15. There appears instability between the points A and B. This is connected, similarly as in the theoretical model, with the process of decohesion along the vertical rows of holes simulating the grain boundaries. After total separation along these vertical rows the compressive force increases. This should be attributed to the hardening effect along the inclined rows of holes. This effect is also similar to that obtained for the theoretical model.

More experiments of this kind are described in the previous paper [21].

REFERENCES:

1. Lee, E.H.: Plastic flow in a V-notched bar pulled in tension, J.Appl.Mech., 11 (1952), 331-336.
2. Hellan, K.: Introduction to Fracture Mechanics, McGraw-Hill, 1984.
3. Hill, R.: On discontinuous states, with special references to localized necking in thin sheets, J.Mech.Phys.Solids, 1 (1952), 19-20.
4. Hill, R.: The Mathematical Theory of Plasticity, Oxford at the Clarendon Press, Oxford 1956.
5. Marciniak, Z. and K. Kuczyński: Limit strains in the process of stretch-forming sheet metal, Int.J.Mech.Sci., 9 (1967), 609-620.
6. Marciniak, Z. and K. Kuczyński: The forming limit curve for bending processes, Int.J.Mech.Sci., 21 (1979), 609-621.
7. Rice, J.R. and D.M. Tracey: On the ductile enlargment of voids in triaxial stress fields, J.Mech.Phys.Solids, 17 (1969), 201-217.
8. McClintock, F.A.: A criterion for ductile fracture by the growth of holes, J.Appl. Mech., 35 (1968), 363-371.
9. Thomason, P.F.: A theory for ductile fracture by internal necking of cavities, J.Inst. Metals, 96 (1968), 360-365.
10. Puttick, K.E.: Ductile fracture in metals, Phil.Mag., Series B, 4 (1959), 964-969.
11. Nagpal, V., McClintock, F.A., Berg, C.A. and M. Subudhi: Traction - displacement boundary conditions for plastic fracture by hole growth, Proc. Symp. Foundations of Plasticity, Warsaw 1972, Noordhoff Intern., Publ., Leyden 1973.
12. Szczepiński W.: On the role of strain concentrations in the mechanics of ductile fracture of metals, Arch.Mech., 40 (1988), 149-161.
13. Garr, L., Lee, E.H. and A.J. Wang: The pattern of plastic deformation in a deeply notched bar with semicircular roots, J.Appl.Mech., 23 (1956), 56-58.
14. Wang, A.J.: Plastic flow in a deeply notched bar with semicircular roots, Q.Appl. Math., 11 (1954), 427-438.
15. Garofalo, F.: Ductility in Creep, in: Ductility, papers presented at the seminar of the American Society for Metals, 1967, ASM, Metals Park, Ohio 1968.
16. Sklenicka, V., Saxl, I., Popule, J. and J. Cadek: Strain components in high temperature creep of a Cu-30% Zn alloy, Material Science and Engineering, 18 (1975), 271-278.

17. Dyson, B., Loveday, M.S. and M.I. Rodgers: Grain boundary cavitation under various states of stress, Proc.Roy.Soc.London, A.349 (1976), 245-259.
18. Marcinkowski, M.J. and L. Larsen: The effect of atomic order on fracture surface morphology, Metall.Trans., 1 (1970), 1034-1036.
19. Rellick, J.R. and C.J. McMahon Jr.: The elimination of oxygen-induced intergranular brittleness in iron by addition of scavengers, Metall.Trans., 1 (1970), 929-937.
20. Szczepiński, W.: On experimental two-dimensional models of intercrystalline sliding and fracture in polycrystalline metals, Arch.Mech., 34 (1982), 502-514.
21. Szczepiński, W.: Experimental simulation of intercrystalline sliding and fracture in metals, Arch.Mech., 37 (1985), 691-704.
22. McClintock, F.A. and A.S. Argon: Mechanical Behavior of Materials, Addison-Wesley, 1966.
23. Hirth, J.P. and J. Lothe: Theory of Dislocations, McGraw-Hill, 1968.
24. Rogers, C.: The effect of materials variables on ductility, in: Ductility, Proc. Seminar ASM 1967, ASM Metals Park, Ohio 1968.
25. Backofen, W.A.: Deformation processing, Addison-Wesley, 1972.
26. Hult, J. and L. Travnicek: Carrying capacity of fibre bundles with varying strength and stiffness, J.Méc.Théor. et Appl., 2 (1983), 643-657.
27. Martin, J.W.: Micromechanisms in Particle-Hardened Alloys, Cambridge University Press, 1980.
28. Brown, L.M. and J.D. Embury: The initiation and growth of voids at second phase particles, Proc. 3rd Conf. on Strength of Metals and Alloys "The Microstructure and Design of Alloys", paper 33 (1973), 164-169.
29. Trefilov, W.I. (Editor): Hardening and Fracture of Polycrystalline Metals (in Russian), Naukova Dumka, Kiev 1987.
30. Szczepiński, W.: On the Mechanisms of Ductile Fracture of Metals, in: Defects and Fracture, Proc. First International Symposium on Defects and Fracture, held at Tuczno, Poland, October 13-17, 1980, G.C. Sih and H. Zorski - editors, Martinus Nijhoff Publ.,1982, 155-163
31. Szczepiński, W.: On the mechanism of local internal necking as a factor of the process of ductile fracture of metals, Journal de Mécanique Théorique et Appliquée, Numero Special, 1982, 161-174.
32. Szczepiński, W.: Internal micronecking as a factor of the process of ductile fracture of metals, Arch.Mech., 35 (1983), 533-540.
33. Mróz, Z.: On generalized kinematic hardening rule with memory of maximal prestress, J. Mécanique Appliquée, 5 (1981), 241-260.
34. Szczepiński, W.: Plasticity approach to the mechanics of softening and ductile fracture of metals, Proc. of the Symposium Plasticity Today, held in Udine, June 27-30, 1983.

35. Szczepiński, W.: On the mechanisms of ductile microfacture in metals; experimental modelling, Arch.Mech., 36 (1984), 569-586.
36. Lee, E.H.: Plastic flow in a rectangularly notched bar subjected to tension, J.Appl. Mech., 21 (1954), 140-146.
37. Lee, E.H. and A.J. Wang: Plastic flow in deeply notched bars with sharp internal angles, Proc. 2nd U.S.Nat.Congr.Appl.Mech., 1954, 489-497.
38. Drucker, D.C.: On obtaining plane strain or plane stress conditions in plasticity, Proc. 2nd U.S.Nat.Congr.Appl.Mech., 1954, 485-488.
39. Szczepiński, W. and J. Miastkowski: Plastic straining of notched bars with intermediate thickness and small shoulder ratio, Int.J.Non-Linear Mech., 3 (1968), 83-97.
40. Szczepinski, W., Dietrich, L., Drescher, E. and J. Miastkowski: Plastic flow of axially-symmetric notched bars pulled in tension, Int.J.Solids Structures, 2 (1966), 543-554.
41. Dietrich, L. and W. Szczepiński: Plastic yielding of axially-symmetric bars with non-symmetric V-notch, Acta Mechanica, 4 (1967), 230-240.
42. Lippmann, H.: Ductility caused by progressive formation of shear cracks, in: Three Dimensional Constitute Relations and Ductile Fracture, S. Nemat-Nasser (Editor), North-Holland, 1981, 389-404.
43. Dietrich, L.: Theoretical and experimental analysis of load-carrying capacity in tension of bars weakened by non-symmetric notches, Bull.Acad.Polon.Sci., Série Sci.Tech., 14 (1966), 363-372.
44. Szczepiński, W.: On modelling interaction between linear defects in the ideally ductile fracture mechanics, in: Inelastic Solids and Structures, M. Kleiber and J.A. Kõnig, eds, Pineridge Press 1989, 33-46.
45. Bishop, J.F.W.:On the complete solution of deformation of a plastic-rigid material, J.Mech.Phys.Solids, 2 (1953), 43-53.

MATHEMATICAL MODELLING OF NONLINEAR PHENOMENA IN FRACTURE MECHANICS

M.P. Wnuk

University of Wisconsin-Milwaukee, Milwaukee, Wisconsin, USA

ABSTRACT

In damage-tolerant and structural-integrity analyses, the residual strength of a flawed component must be evaluated. For brittle materials, linear-elastic fracture mechanics (LEFM) concepts, such as plane-strain fracture toughness, K_{Ic}, are used. For materials that exhibit large amounts of plasticity at the crack tip prior to failure, LEFM techniques must be extended to incorporate new notions such as J-integral, CTOD, CTOA, R-curve and C_t-integral. Introduction of these quantities and the demonstration of their applications in materials and designs pertinent to the engineering fields will be the primary theme of these lectures.

CONTENTS

INTRODUCTION. NONLINEAR EFFECTS IN FRACTURE MECHANICS

When failure occurs in a material which is not perfectly brittle and capable of dissipating a certain amount of energy prior to decohesion, the classical Griffith theory of fracture is inadequate in explaining various phenomena directly associated with nonlinear material behavior. For example, elastic-plastic materials, with or without strain-hardening, exhibit an ability to sustain a slowly growing stable crack which precedes the transition into a catastrophic failure. This continuing sub-critical crack extension is described in the realm of nonlinear theory by a certain differential equation relating the rate of load increase $\dot{\sigma}$, the load σ, the rate of growing crack $\dot{c}$ and the current crack length c, say $f(\dot{c},\dot{\sigma},c,\sigma) = 0$. Two versions of such equation will be presented, one corresponding to the model suggested by Cherepanov and Wnuk, and another resulting from the Wnuk-Rice-Sorensen equation of an R-curve valid within the contained yielding range. Both models contain the Griffith result as a limiting case obtained when the size of the Neuber particle or the so-called process zone, Δ, shrinks to a point.

For a dissipative medium the apparent fracture toughness K_R becomes dependent on the crack size, and it may be modeled by the Barenblatt's formula:

$$K_R(c) = \sqrt{\frac{2}{\pi}} \int_0^{R(c)} \frac{S(x_1,c)dx_1}{\sqrt{R(c) - x_1}}$$

The variable x_1 denotes the distance from the crack tip to a material point located on the perspective fracture path. The symbol $S(x_1,c)$ denotes a distribution of the so-called restraining or cohesive stresses which simulate the nonlinear material response over the distance R from the crack front. Note that for a quasi-static crack both K_R and the size of the nonlinear zone R vary with time, thus they become certain a priori unknown functions of the current length c. Siroe R and c vary at ditrerent rates, the ratio dR/dc is introduced and related to the properties of the system, say $dR/dc = F(\Delta/R)$. In this way the material resistance curve is defined and subsequently used in the analysis of stability in tearing fracture.

Ample experimental evidence exists to demonstrate that the critical level of the material resistance to cracking, K_f, at which a transition from stable to unstable propagation occurs, is influenced not only by the size and geometrical configuration of the specimen, but also by the manner in which the external loads are applied. Such size and geometry dependence cannot be predicted by the linear theory. New near-crack-tip asymptotic fields have been proposed for 2D elasto-plastic fracture; the Hutchinson-Rice-Rosengren (HRR) field for a stationary crack, and the Gao-Hwuang-Dai (GHD) field for a quasi-static crack. The thrust of the recent research in nonlinear fracture mechanics is directed at experimental verification of these fields and at extending the results into the 3D case.

Subcritical crack growth phenomena are well documented for viscoelastic materials, both linear and nonlinear, in which the deformation process requires energy dissipation and, of course, it is time-dependent. The Knauss-Wnuk-Schapery equation

$$D_{creep}(\Delta/\dot{c})/D_{creep}(0) = (K/K_{glassy})^2$$

derived for glassy polymers in which existence of small process zone (Δ) was postulated, describes motion of the crack at a constant external load. Rate of an extending crack was shown to be proportional to a certain power of the applied K factor, $\dot{c} \propto K^q$. The exponent q is directly related to the material constant which equals the slope of the creep compliance function D_{creep} plotted vs. time in double logarithmic scale. Similar relations were discovered in the studies of metal creep fracture at elevated temperatures. Here again, the rate $\dot{c}$ is proportional either to a certain power of the K factor, or J integral, or C^* integral. For the case of strain-controlled creep crack growth in ductile metals numerous investigators have found

$$\dot{c} \propto (C^*)^{n/n+1}/\epsilon_f$$

Here ϵ_f denotes the strain at rupture, C^* is time-dependent equivalent of J integral and n denotes the work hardening exponent appearing in the Norton's law $\dot{\epsilon} = B\sigma^n$, or suggested by the Ramberg-Osgood constitutive relation

$$\epsilon/\epsilon_o = \sigma/\sigma_o + \alpha(\sigma/\sigma_o)^n$$

Finally, failure in damaging materials provides yet another example of subcritical crack growth, during which the progress of fracture is inhibited by accumulation of micro-defects which tend to localize in the narrow band ahead of the crack front, thus providing an energy screening mechanism. To predict fracture evolution process an integro-differential equation was derived by Wnuk and Kriz, based on a modified version of the Kachanov's hypothesis and using the basic concepts of the continuous damage mechanics (CDM). Purpose of the modification was to allow some degree of interaction between the dominant macroscopic crack and the randomly distributed micro-defects. Such an assumption describes a synergistic effect of both the macro and micro forms of damage. Interaction between these two forms of damage and the associated microstructure degradation account for the time-dependent material response needed to explain continuing growth of the dominant crack. An R-curve for damaging materials is shown to obey a logarithmic expression strongly influenced by the size of the macroscopic crack and the material properties which appear in the damage evolution law.

Nonlinear Effects in Fracture Mechanics. Introduction

The principle of energy conservation

$$\delta\left[\int_S T_i\ u_i\ dS\right] = \delta\left[\frac{1}{2}\int_V \sigma_{ij}\ \epsilon_{ij}\ dV\right] + \delta SE \tag{i}$$

applied by Griffith to the problem of brittle fracture occurring in an elastic continuum ($\underline{T}$ and $\underline{u}$ are traction and displacement vectors, while $\underset{\sim}{\sigma}$ and $\underset{\sim}{\epsilon}$ are stress and strain tensors) included the surface energy term "SE". This is perhaps the most essential feature of the Griffith theory: an assumption that an additional energy is required to produce fracture. In a more compact form Eq. (i) reads

$$\delta W = \delta U + \delta SE \tag{ii}$$

Here W denotes the work done by external forces and U denotes the stored elastic energy. Variations "δ" are associated with a virtual extension of crack from an initial value of its half-length c to the new length $c+\delta c$. Alternatively, Eq. (ii) can be written as follows

$$\delta(W-U) = \delta SE \tag{iii}$$

or

$$\partial(W-U)/\partial c = \partial SE/\partial c \tag{iiia}$$

The left hand side of this equation represents the increment (or rate) of the energy available for a virtual extension of fracture, $c \rightarrow c + \delta c$, while on the right hand side we have the increment (or rate) of energy required for the decohesion, or separation, process to occur. Up-dated version of equation (iiia) has indeed a very brief form

$$G = G_f \tag{iv}$$

For linearly elastic solids the expression for energy release rate

$$G = \frac{1}{2t}\frac{\partial(W-U)}{\partial c} = -\frac{1}{t}\frac{\partial U}{\partial \ell} \tag{v}$$

(t is the thickness of specimen, c is used to denote the half-length of the crack, while $\ell = 2c$ and π denotes the potential energy of the system considered) reduces to

$$G = \begin{cases} \dfrac{1}{2t}\dfrac{\partial U}{\partial c}, & \text{fixed load condition, (a)} \\[2ex] -\dfrac{1}{2t}\dfrac{\partial U}{\partial c}, & \text{fixed grips condition, (b)} \end{cases} \tag{vi}$$

Obviously in a load-controlled experiment, i.e. case (a), the work $W = 2U$, while for the case of fixed displacement, case (b), we deal with a closed system, and then $W = 0$. For the Griffith crack the energy release <u>per unit thickness</u>

$$G = \frac{1}{2}\frac{\partial}{\partial c}\left(\frac{\pi\sigma^2c^2}{E}\right) = (\pi c)\frac{\sigma^2}{E} = \frac{K^2}{E} \tag{vii}$$

(K denotes the stress intensity factor) while the material resistance to cracking G_f is identified with the specific fracture energy, which in the atomistic model of fracture equals 2γ, with γ denoting the surface tension.

To illustrate evaluation process aimed at the quantity G, let us mention here one related to the classic experiments of splitting mica crystals reported by Obreimov (1930). In this case the strain energy of a double cantilever beam, see Figure I, was evaluated as

$$U = \frac{1}{2} \text{ force} \times \text{displacement}$$

Fig. I

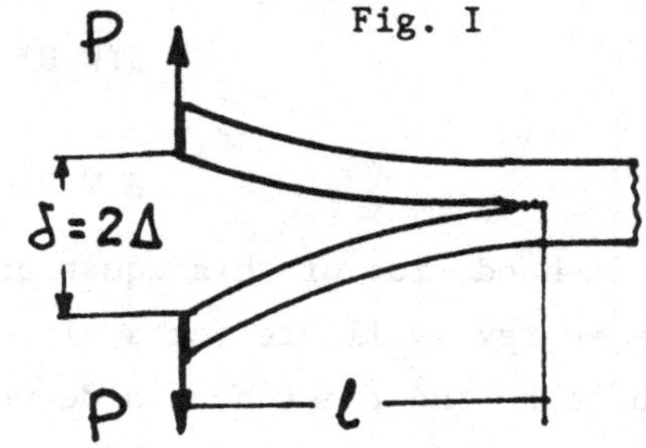

or

$$U = \left(\frac{P\Delta}{2}\right)2 = \left(\frac{P}{2}\right)\left(\frac{P\ell^3}{3EI}\right)2 = \frac{P^2\ell^3}{3EI} \qquad \text{(viii)}$$

Factor 2 written behind the parentheses accounts for two, upper and lower, beams. Thus, the energy release rate of Obreimov's specimen is derived

$$G = \frac{1}{t}\frac{\partial U}{\partial \ell} = \frac{P^2\ell^2}{EIt} \qquad \text{(ix)}$$

Note that the same result can be obtained from Irwin's compliance formula (elastic compliance C_e is defined as the ratio of load line displacement to load)

$$G = \frac{P^2}{2t}\frac{\partial C_e(\ell)}{\partial \ell} = \frac{P^2}{2t}\frac{\partial}{\partial \ell}\left(\frac{2\ell^3}{3EI}\right) = \frac{P^2\ell^2}{EIt} \qquad \text{(x)}$$

If the material used in the experiment was perfectly elastic, then the onset of fracture in the specimen shown in Fig. I is predicted by the equation $G = 2\gamma$, or

$$P_{cr} = \sqrt{2EI\gamma t}/\ell \qquad \text{(xi)}$$

What happens, however, when the material is not perfectly linear?

One important term is conspicuously missing in the energy balance equation postulated by Griffith. And that is the dissipation of energy, say D. The energy dissipation does not exist in the realm of the linear elastic fracture mechanics (LEFM), but it is a source of a very signifieant contribution in any real material capable to sustain irreversible deformation process. Plasticity in metals, crazing in polymers, microcracking within large process zones associated with fracture in cementitious composites, matrix/fiber debonding and bridging effects encountered in fiber reinforced composite materials are the dominant dissipative mechanisms, all of which contribute to the term D, omitted by Griffith. When equation (iii) is re-written so that the non-linear and irreversible material response are incorporated, one obtains

$$\delta(W-U) = \delta SE + \delta D \tag{xii}$$

Here the term δD represents all non-elastic contributions to the energy balance associated with a virtual crack extension. This term can no longer be derived from linear theories, and once its existence is acknowledged, we enter into the realm of non-linear fracture mechanics. Frequently, Eq. (xii) is written in this form (dot is used to denote time derivative):

$$\dot{W} = \dot{U} + S\dot{E} + \dot{D} \tag{xiii}$$

Another essential modification of the Griffith concept of fracture when applied to inelastic materials is an acknowledgement that the infinite stress cannot indeed be sustained by the material in the immediate vicinity of the crack tip. The concept of crack tip itself undergoes a major modification; it is no longer a point as Griffith has envisioned, but it becomes "diffused" into a finite domain of intensive irreversible straining, hereafter referred to as the process zone.

Let us briefly discuss some examples of nonlinear phenomena which do occur in real materials. The rate of energy dissipation $\dot{D}$ is non-zero for elastic-perfectly plastic materials whenever plastic flow occurs in the vicinities of the crack tips which -- within the realm of nonlinear fracture mechanics -- may of course be moving. If the simplest model of Dugdale is applied to a quasi-static crack, then one may show that

$$U + D = \frac{\sigma^2 A}{2E} + \frac{8\sigma_o^2}{\pi E} c^2 \ \ell n \sec(\pi\sigma/2\sigma_o) \tag{xiv}$$

and that (A denotes the finite area of the sheet containing crack)

$$\dot{W} = \frac{1}{E}\left\{\sigma\dot{\sigma}A + \frac{16}{\pi}\sigma_o^2 c^2\left[\frac{\dot{c}}{c} Q \tan Q + \frac{1}{2}\dot{Q}Q \sec^2 Q\right]\right\} \tag{xv}$$

On substituting the relations (xiv) and (xv) into Eq. (xiii) it can be shown that the crack is driven by the external load, measured by a nondimensional parameter, $Q = \pi\sigma/2\sigma_o$, in which σ_o denotes the yield point or the so-called flow stress, defined as mean value of the yield and ultimate stresses. The rate with which such a stable crack extends, dc/dQ, is determined by the Cherepanov-Wnuk's equation

$$\left\{4G_f - \frac{16\sigma_o^2}{\pi E} c(Q \tan Q - \ln \sec Q)\right\} \frac{dc}{dQ} = \frac{8\sigma_o^2}{\pi E} c^2(Q \sec^2 Q - \tan Q) \quad \text{(xvi)}$$

or

$$\frac{d\zeta}{dQ} = \frac{\zeta(Q \sec^2 Q - \tan Q)}{1-2\zeta(Q \tan Q - \ln \sec Q)} \quad \text{(xvii)}$$

where nondimensional crack length ζ and the loading parameter Q are defined as follows

$$\zeta = \frac{4\sigma_o^2 c}{\pi E G_f}, \qquad Q = \frac{\pi\sigma}{2\sigma_o} \quad \text{(xviii)}$$

The symbol G_f is used to denote the specific fracture energy measured at the onset of fracture growth. This quantity may exceed 2γ by several orders of magnitude. The range of the stable phase of the preliminary crack extension ($c_o \le c \le c_f$, $Q_{ini} \le Q \le Q_f$) can be established by imposing the requirement of a non-negative rate dQ/dc. This leads to an R-curve analysis. Details of analyses of such a nature are described in the following sections.

Finally, we note that the Griffith locus $\sigma_{critical}$ vs. $c_{critical}$ is now replaced by the domain shown by the cross-hatched area in Fig. II, contained between the propagation threshold (lower curve) and the terminal criticality locus (upper curve). Within this domain a stable crack extension will occur. There is an ample experimental evidence for such a

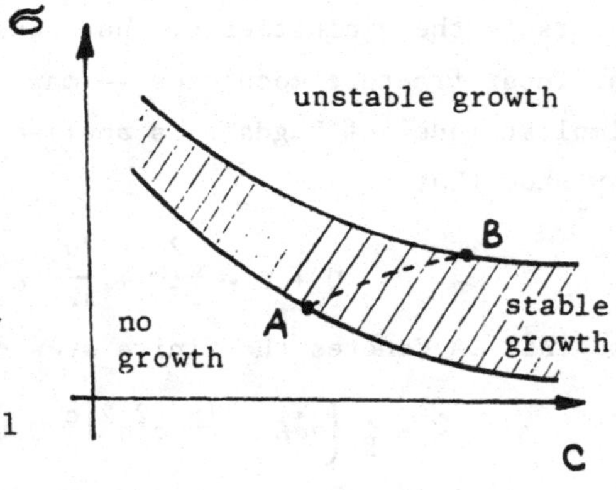

Fig. II

phenomenon to exist. Onset of crack extension can be predicted using the Griffith result, $\sigma_{ini}^2(\pi c_o) = EG_f$. However description of crack growth process lies entirely outside the linear Theory of Fracture. Only in the limit of infinitely large yield stress, i.e. for $Q \to 0$, one recovers the Griffith result, as then the rate $dQ/d\zeta$ reduces to

$$\left(\frac{dQ}{d\zeta}\right)_{Q\to 0} = \frac{1 - \zeta(Q^2 + \frac{1}{2}Q^4 + \ldots)}{\frac{2}{3}\zeta Q^3 + \ldots} \tag{xix}$$

Setting it to zero gives $1 = \zeta Q^2$ and this is identical with the classic formula relating σ and c at the onset of unstable fracture, i.e.,

$$\sqrt{\pi c}\,\sigma = \sqrt{EG_f} \tag{xx}$$

Similar effect is predicted by Wnuk's equation

$$\frac{dQ}{d\zeta} = \frac{\ln[n/\zeta Q^2] - Q^2}{2\zeta Q} \tag{xxi}$$

which resulted from the "final stretch" criterion postulated for the continuing crack growth, Wnuk (1972, 1974). The coefficient n defines the ratio of the steady-state toughness, i.e., the upper plateau of the R-curve, R_{ss}, to its initiation level, R_{ini}, namely

$$n = \frac{R_{ss}}{R_{ini}} = \frac{1}{4\rho_i}\exp(2M-1) \tag{xxii}$$

Here

$$\rho_i = R_{ini}/\Delta \simeq \epsilon_p^f/\epsilon_o, \qquad \epsilon_o = \sigma_o/E \tag{xxiia}$$

and M denotes the tearing modulus. The domain of stable crack extension process is shown in Fig. III. It may be shown that stable cracking may occur only if the modulus M is greater than a certain threshold value

$$M_{min} = \frac{1}{2}\log(4e\rho_i) \sim \log(\epsilon_p^f/\epsilon_o) \tag{xxiii}$$

Since the ductility parameter ρ_i is closely approximated by the ratio of the plastic component of strain at fracture ϵ_p^f to the yield strain ϵ_o, it is evident that the minimum value of the tearing modulus depends logarithmically on the ductility measure ϵ_p^f/ϵ_o, as indicated by Eq. (xxii). The toughness at which the transition from stable to unstable

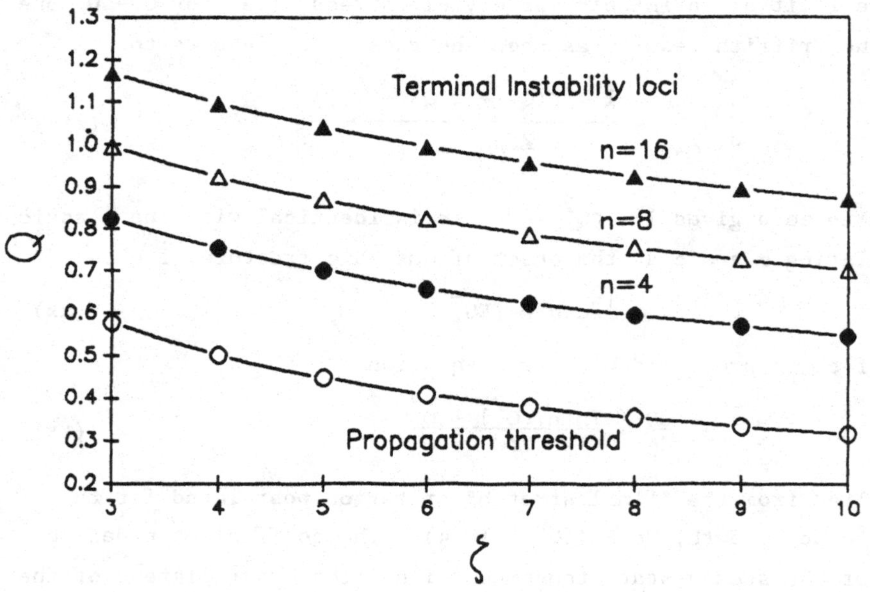

Fig. III Lower and upper bounds on load levels associated with quasi-static fracture occurring under small scale yielding conditions. The domain contained between both bounds represents the set of load/crack length values for which stable cracking is possible. The lower bound results from the equation $Q = \zeta^{-1/2}$, while upper bounds satisfy the equation $\ln(n/Q^2\zeta)-Q^2 = 0$. Here, $n = R_{ss}/R_{ini}$, Q is the loading parameter ($= \pi\sigma/2\sigma_0$), and ζ denotes the nondimensional crack length, $\zeta = c/2R_{ini}$. Equations used here resulted from Wnuk's final stretch model of ssy quasi-static fracture.

crack propagation will take place, say R_f, is contained within the interval

$$R_{ini} < R_f < R_{ss} \tag{xxiv}$$

After R-curve analysis is performed, it will become evident that the value of R_f is not merely a material property, dependent on tearing modulus and ductility, but it is also influenced by the geometrical configuration, size of the specimen and the manner in which external loads are applied. None of such effects were described by the standard linear theory, the LEFM.

Again, if we investigate the brittle fracture limit, which is contained in Eq. (xxi), then $R_{ss} \to R_{ini}$, $m \to 1$, $Q \to 0$ (because $\sigma_o \to \infty$), and the catastrophic fracture is predicted when $dQ/d\zeta \to 0$, and we recover the condition $\zeta Q^2 = 1$. This leads to the result identical with the classic expression (xx).

Other nonlinear effects, not predicted by the Griffith theory of fracture, abound in the studies of quasi-static crack extension in polymers at room temperature and in metals at elevated temperature (creep fracture). Such time dependent effects will be discussed in a little more detail in the sections following those devoted to the R-curve analysis.

1. OUTLINE OF GENERAL PROCEDURES

Most statistical theories of fracture under tensile or bending stresses of brittle materials are based on the weakest link concept originally proposed by Weibull (1939, 1951). This view is consistent with the essential assumptions underlying the classic theory of fracture of brittle solids proposed by Griffith (1921). According to Griffith a brittle material contains a large number of inherent flaws (or microcracks), and thus fracture occurs at the flaw whose stress intensity factor first reaches the fracture toughness of the material, $K_{APPL} = K_{IC}$. Indeed, in a more recent study Trustrum and Jayatilaka (1983) have shown that the Weibull distribution of strength is the limiting case for structures containing a large number of flaws if their size distribution decays in the form of an inverse power law. In this case both classic and statistical theories of fracture predict a sudden onset of catastrophic propagation of macroscopic fracture once the K-factor of the applied stress field reaches the critical level as dictated by the largest pre-existing flaw, i.e., the weakest link in the structure.

Such a description of fracture is adequate for brittle homogeneous materials like glass. However, in other less homogeneous materials like ceramics, mortar and concrete, the first crack formed does not necessarily lead to immediate fracture of structure. Experience with those materials suggest an existence of "early warning" period which manifests itself through the phenomenon of slow stable cracking occurring prior to the terminal fracture and accompanied by a large number of secondary micro-cracks. The classic theory of Griffith, or even its modifications by Irwin and Orowan, do not predict anything resembling this time-dependent and energy-absorbing process.

At the other extreme with respect to the weakest link theory is the bundle model proposed originally by Daniels (1945). In this theory, which best applies to cables or continuous fiber reinforced materials, the material is assumed to consist of a bundle of fibers whose strengths have a particular distribution. When the weakest fiber fails, the load it carried is distributed among those that survive. Complete failure of the structure does not occur until the load on the remainder of the fibers exceeds their total load-carrying capacity. Although the bundle theory has been applied to the fracture of inhomogeneous brittle materials, Hasofer (1968), it is not believed to be entirely appropriate. Obviously, fracture in a perfectly homogeneous brittle material is dependent on its weakest link. Therefore, any theory of fracture in an inhomogeneous material should incorporate the weakest link theory as its limiting case.

In the theory introduced in this paper we assume that the small volume of the material immediately adjacent to the front of any pre-existing crack obeys the rules of the bundle model, see Fig. 1, and thus the fracture on the macroscopic scale appears to be a gradual displacement- or load-controlled process rather than a single event of

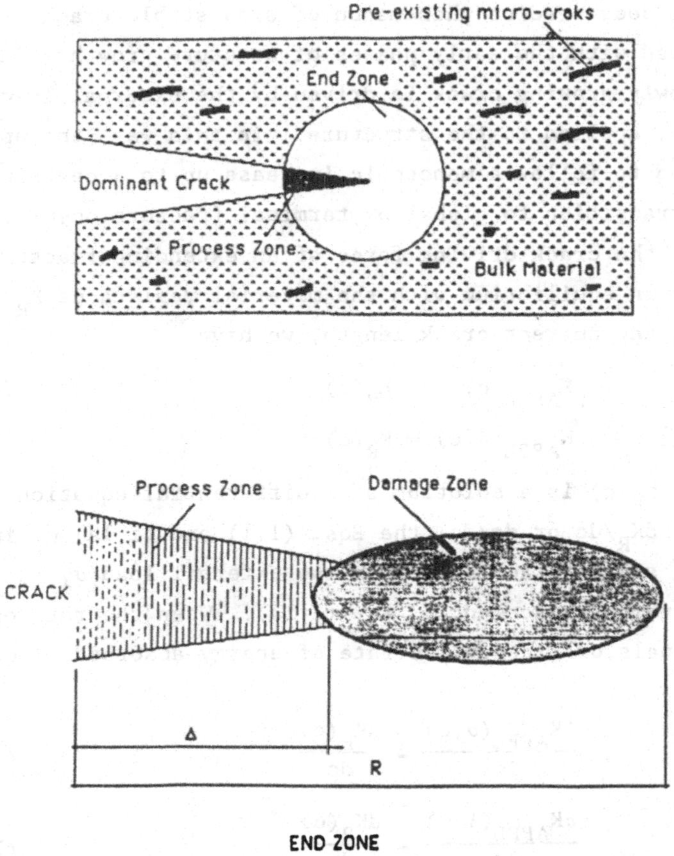

Fig. 1 Structured nature of the end-zone associated with a dominant crack in an inelastic solid. Note that the highly localized deformation is confined to the process zone, while the plastic action and/or secondary cracking (in damaging materials) are present within the damage zone. According to the "bundle model", which is used to describe the deformation preceding fracture, the energy absorption processes are represented as the work done by a certain cohesive force $S(x_1)$ active within the end-zone.

sudden fracture initiation followed by an immediate and unstable propagation. Contrary to what the classic theories of brittle fracture suggest, our model does predict phenomenon of slow stable crack extension associated with the early phases of fracture growth. During this process a slowly growing crack is driven by the external load, σ, or displacement, δ, applied to the structure. In this way, the applied K-factor is allowed to follow a monotonic increase up to a certain point at which a transition to global or terminal fracture occurs. Since at all time, the crack driving force of an extending fracture, say K_{APPL}, remains in equilibrium with the material resistance $K_R(c)$, in which c denotes the current crack length, we have

$$K_{APPL}(\sigma,c) = K_R(c) \tag{1.1}$$

or

$$K_{APPL}(\delta,c) = K_R(c) \tag{1.1a}$$

Since the function $K_R(c)$ is a solution of a differential equation involving the rate dK_R/dc or $d\sigma/dc$, the Eqs. (1.1) and (1.1a) do indeed describe motion of stable crack driven by the external load σ, Eq. (1.1) or the external displacement δ, Eq. (1.1a). When the rate of energy supplied equals or exceeds the rate of energy absorbed, i.e., when

$$\frac{\partial K_{APPL}(\sigma,c)}{\partial c} = \frac{dK_R(c)}{dc} \tag{1.2}$$

or

$$\frac{\partial K_{APPL}(\delta,c)}{\partial c} = \frac{dK_R(c)}{dc} \tag{1.2a}$$

the initially stable crack reaches the point of transition into a terminally unstable fracture. The level of K-factor corresponding to such a transition is no longer a material property (such as was K_{IC} for brittle materials), but it depends on the geometry of the cracked specimen, type of loading conditions, i.e. stress- or displacement-controlled external loading, and finally, on the shape of the K_R-curve, or J_R-curve, which replaces now the single toughness parameter (K_{IC}), and serves as a mathematical vehicle for representation of "apparent" or "global" material fracture toughness, the so-called R-curve, e.g. Wnuk (1974,1983) and more recently Wnuk and Read (1986) and Narasimhan et al. (1987).

In this paper we focus attention on the influence of the constitutive equations of material on the shape of the R-curve. Quantitative description of quasi-static crack extension leads to determination of a specific R-curve for any given material. The correlation between the stable crack growth phenomenon and the R-curve viewed as the final outcome of the theory proposed here requires that certain basic assumptions are made. These assumptions concern:

(a) Stress-strain relation obtained from a tensile test. Specifically, two possible forms of σ vs. ϵ relationships will be considered, one that corresponds to strain-hardening and another which implies strain-softening behavior. Input information derived from the experimental data dictates different distributions of strength among the "bundles" which restrain extension of fracture within the small volume of material adjacent to crack front, the so-called "process zone," see Fig. 1. These distributions, in turn, determine the form of the differential equation governing an apparent increase of resistance to fracture associated with the early phase of fracture development.

(b) Local criterion for crack opening which must suggest a certain quantity invariant to the crack growth. Existence of such an invariant material characteristic implies that local material resistance to fracture is determined solely by microstructure, and it remains a material property contrary to the global or apparent fracture toughness which is given by the entire R-curve. (Propagation threshold and the initial slope of this curve are perhaps the two most important parameters which enter the theory.)

If the size of the zone of secondary cracking is small versus crack length, the resulting R-curve is "universal", or in other words, it is believed to constitute characteristic "signature" of material. By analogy with fracture in metals, these conditions are referred to as those of "small scale yielding". The scope of our model discussed in this section is restricted to the ssy range, and the emphasis is placed on fracture behavior of non-hardening, hardening and strain-softening materials.

2. MATHEMATICAL MODELING OF QUASI-STATIC CRACKS AND R-CURVE IN INELASTIC SOLIDS

When fracture initiates in a solid capable to sustain a stable crack growing over a certain range of the applied load, the apparent material resistance to cracking continues to rise steeply with crack advance. Such variations of fracture toughness cannot be explained solely on the basis of the linear elastic fracture mechanics, and certain other mechanisms such as ductile void growth and coalescence in some models, or creation of secondary microcracks some distance ahead of the front of a dominant crack (as in strain-softening solids) have to be admitted in order to explain the early stages of fracture development. It should be noted that even if the initiation parameters, such as K_{IC} or J_{IC}, are known from the observations made at the onset of crack growth, they alone are not sufficient means for predicting the instability which eventually terminates the process of slow stable cracking. Therefore, it is necessary to devise a technique which would aupply more complete information regarding material response to fracture within the range of mechanical behavior in which the laws of linear elastic fracture mechanics break down.

The model proposed here refers to cracking in nonhomogeneous brittle solids, and it combines the weakest link concept of classical theory of fracture with the bundle structure used to describe the final stage of irreversible deformation which occurs within the end zone, see Fig. 1, just prior to fracture advance. The bulk of the solid is visualized as an imperfect solid containing a large number of preexisting microcracks, one of which generates the highest K-factor and develops into a dominant macrocrack. This is the "weakest link" concept consistent with classical LEFM. From that point on, however, our model incorporates a certain structured end zone or damage zone, associated with the dominant crack. This is where the intensive irreversible deformation takes place prior to the terminal decohesion, and tkis is where a localized band of secondary microcracks builds up. The physical law within the damage zone resembles that corresponding to the "bundle model" closely related to the basic concepts of the continuous damage mechanics (CDM). Energy dissipated within the camage zone is "diverted" from the crack tip, and thus existence of the damage band leads to energy screening effect which in turn has a beneficial effect on the observable or "apparent" fracture toughness. The toughness parameters measured globally, such as K_R or J_R, associated with an instantaneous equilibrium system load-dominant crack-damage zone, show a distinct increase as crack progresses. Note, however, that the crack tip -- a point in the realm of classical LEFM -- has been now replaced by a small domain of finite dimensions, characterized by the length Δ. In what follows we postulate that the energy absorbed within this zone determines the true, or <u>local</u> material resistance to cracking, measured by the decohesion energy release <u>rate</u>, G^{Δ} or J^{Δ}. If $S(x_1)$ denotes the restraining stress developed within the end zone ($0 \leq x_1 \leq R$), and u_y is the work-conjugate opening displacement for the same range of x_1, then the energy absorbed within the process zone can be represented by the integral

$$J^{\Delta} = -2 \int_0^{\Delta} S(x_1)\left(\frac{du_y}{dx_1}\right)dx_1 = -2 \int_{\delta_t/2}^{u_{\Delta}} S(x_1)\delta u_y \tag{2.1}$$

Here, δ_t denotes the crack tip opening displacement (CTOD), while $2u_{\Delta}$ is the opening displacement a fixed distance Δ ahead of the crack front. Note also, that the crack considered here is quasi-static, i.e., it is continually moving, driven by the applied K-factor. In lieu of time, the variable x_1 is used. This variable is defined as the current distance of any stationary point within the end zone from the approaching crack front. The variation δu_y is associated with an incremental crack extension, and therefore it is defined as follows

$$\delta u_y = u_y[c+\delta c,\ x_1-\delta c,\ R(c+\delta c,\ x_1-\delta c)] - u_y[c,\ x_1,\ R(c,x_1)] \tag{2.2}$$

The quantities x_1 and "c" are not independent, as they are related at all times, for any given control point P,

$$x_P = \text{const.} = c + x_1, \quad -\delta c = \delta x_1 \tag{2.3}$$

Thus the increment δu_y can be expressed by either of two expressions

$$\delta u_y = \begin{cases} u_y[c+\delta c,\ R(c+\delta c)] - u_y[c,\ R(c)] \\ \\ u_y[x_1-\delta c,\ R(x_1-\delta c)] - u_u[x_1,\ R(x_1)] \end{cases} \tag{2.4}$$

The Sneddon integral

$$u_y = \frac{4a}{\pi E_1}\int_x^1 \frac{t\,dt}{\sqrt{t^2-x^2}}\int_0^1 \frac{p(u)du}{\sqrt{t^2-u^2}} \tag{2.5}$$

is used to evaluate (see Appendix A) the opening displacement u_y within the end zone, $c < \bar{x} < a$, or $0 \le x_1 \le R$, for a specific form of the restraining stress $S(x_1)$. Here E_1 equals E for plane stress while it is $E(1-\nu^2)^{-1}$ for plane strain. The restraining stress S enters equation (2.5) through the surface traction or the pressure p(u) defined over the end zone associated with the dominant crack.

In what follows we consider three possible distributions of the cohesive stress S, associated with a quasi-static crack, namely

(a) non-hardening case for which S = constant ($=\sigma_o$) over the entire end zone as in Dugdale model,

(b) hardening case for which

$$S(x_1) = \begin{cases} \sigma_o\left(\frac{R}{\Delta}\right)^{\beta} = \sigma_* & 0 \le x_1 \le \Delta \\ \\ \sigma_o\left(\frac{R}{x_1}\right)^{\beta} & \Delta \le x_1 \le R \end{cases} \tag{2.6}$$

(Here β is the hardening exponent, $0 \le \beta \le 1/2$, while σ_o denotes the initial yield point), and

(c) softening case for which S is modeled by the step-like function

$$S(x_1) = \begin{cases} \sigma_{coh} = \alpha\sigma_o & 0 \le x_1 \le \Delta \\ \\ \sigma_o & \Delta \le x_1 \le R \end{cases} \tag{2.7}$$

Here σ_{coh} denotes the level of the cohesive stress within the process zone while α is either the strain-softening parameter ($\alpha < 1$) or the strain-hardening parameter ($\alpha > 1$), see Fig. 2. To combine these two notions we shall sometimes refer to the quantity α as "cohesive stress distribution parameter".

Evaluation of the integral (2.5) for these three hypothetical

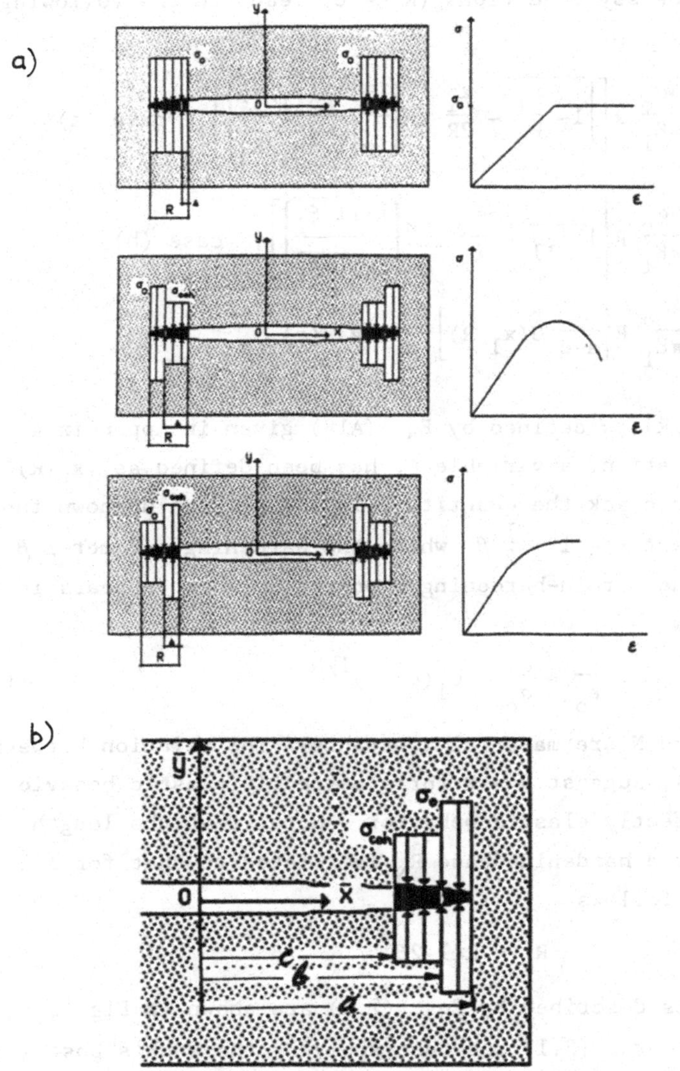

Fig. 2 (a) Bi-level cohesive force distributions simulating three basic modes of material behavior. From the top: (1) non-hardening, (2) strain-softening, and (3) strain-hardening.

(b) Details of the $S(x_1)$ distribution over the end-zone.

cohesive stress distributions (see Appendix A) performed under the restrictions of the ssy conditions ($R \ll c$) leads to the following results

$$u_y(x_1,R) = \begin{cases} \dfrac{4\sigma_o}{\pi E_1} R\left[\sqrt{1-\dfrac{x_1}{R}} - \dfrac{x_1}{2R}\ln\left(\dfrac{1+\sqrt{1-(x_1/R)}}{1-\sqrt{1-(x_1/R)}}\right)\right], & \text{case (a)} \\ \dfrac{4\sigma_o}{\pi E_1} R\left[\sqrt{1-\xi_1} - \dfrac{\xi_1}{2}\ln^{\kappa}\left(\dfrac{1+\sqrt{1-\xi_1}}{1-\sqrt{1-\xi_1}}\right)\right], & \text{case (b)} \\ \dfrac{4\sigma_o}{\pi E_1} R\left[\dfrac{a}{a-c}\, G(x_1/R)\right], & \text{case (c)} \end{cases} \tag{2.8}$$

The function $G(x_1/R)$ is defined by Eq. (A18) given in Appendix A. To abbreviate the notation, a variable ξ_1 has been defined as $(x_1/R)^{1-\beta}$. For a quasi-static crack the quantity R is, as yet, an unknown function of x_1. The constant $\kappa = 1 - \frac{1}{2}\beta$, while the hardening parameter β is in turn related to the strain-hardening exponent N, which appears in the Ramberg-Osgood law

$$\frac{\epsilon}{\epsilon_o} = \frac{\sigma}{\sigma_o} + C_1(\sigma/\sigma_o)^{1/N} \tag{2.9}$$

Here C_1, ϵ_o, σ_o and N are material constants. The relation between β and N, $\beta = N/(N+1)$, suggest $\beta \to 0$ for a perfectly plastic behavior, and $\beta \to 0.5$ for a perfectly elastic behavior. Note, that the length of end zone evaluated for a hardening case R_N is related to that for a non-hardening case as follows

$$R_N = R(1-2\beta) \tag{2.10}$$

Displacement curves described by Eqs. (2.8) are shown in Fig. 3.

Using the integral (2.1) and then applying the Wnuk's postulate of J^{Δ} remaining invariant to the slow stable crack growth, or the so-called "final stretch" growth criterion suggested by Wnuk (1972, 1974), one arrives at the following equations ($\Delta \ll R$ and $R \ll c$) relating the local toughness J^{Δ} to other pertinent material parameters

$$J^{\Delta} = \frac{8\sigma_o^2}{\pi E_1} \begin{cases} \Delta \dfrac{dR}{dc} + \dfrac{\Delta}{2} + \dfrac{\Delta}{2} \ln(\dfrac{4R}{\Delta}), & \text{case (a)} \\ \Delta \dfrac{dR}{dc} + \dfrac{\Delta}{2} (\dfrac{R}{\Delta})^{\beta} + \dfrac{\Delta}{2} (\dfrac{R}{\Delta})^{\beta} \ln^{\kappa}[4(\dfrac{R}{\Delta})^{1-\beta}], & \text{case (b)} \\ \Delta \dfrac{dR}{dc} + R - RF(\Delta/R) \ , & \text{case (c)} \end{cases} \quad (2.11)$$

Definition of the function $F(\Delta/R)$ is given by Eq. (A24) in Appendix A. Note that in all these expressions the length of the damage zone R is as yet an unknown function of the time-like variable x_1 or the current crack length "c". The function $R = R(c)$ will be evaluated through the application of the growth criterion for a quasi-static crack. Once the function R(c) is determined, we can substitute it back to the Eqs. (2.8) and find out the profiles of the quasi-static cracks corresponding to the constitutive responses of the materials considered here, see Fig. 3. At the same time, as we proceed to demonstrate, the material resistance to slow stable cracking can be established quantitatively, since the function $R = R(c)$ will be used here as a representation of the global fracture toughness, the so-called R-curve.

Equation (2.11a) was first proposed by Wnuk (1972, 1974), then by Rice and Sorensen in 1978. Equation (2.11b) was derived by Wnuk and Hunsacharoonroj (1987), while the details of derivation of the last form given above, Eq. (2.11c), are given in Appendix A. Now, setting the quantity J^{Δ} equal to the true energy of fracture, i.e.

$$J^{\Delta} = J_f = (\text{stress over the } \Delta\text{-zone}) \times (\text{final stretch } \hat{\delta}) \quad (2.12)$$

we arrive at the following nonlinear differential equations which define the R-curves for each of the constitutive responses considered

$$\frac{dR}{dc} = \begin{cases} M - \dfrac{1}{2} \ln[4e(R/\Delta)], & \text{case (a)} \\ M - \dfrac{1}{2} (\dfrac{R}{\Delta})^{\beta} - \dfrac{1}{2} (\dfrac{R}{\Delta})^{\beta} \ln^{\kappa}[4(\dfrac{R}{\Delta})^{1-\beta}], & \text{case (b)} \\ M - \dfrac{R}{\Delta} + \dfrac{R}{\Delta} F(\Delta/R), & \text{case (c)} \end{cases} \quad (2.13)$$

These equations represent the "bottom line" of the present

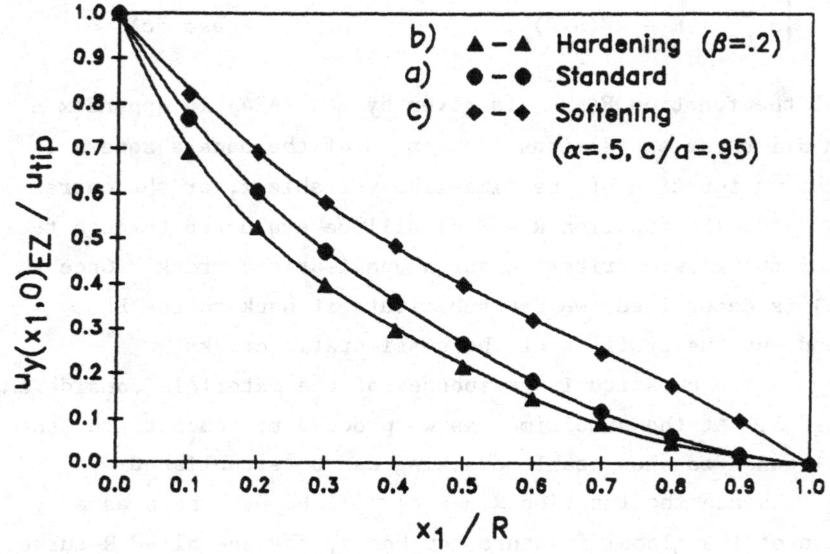

Fig. 3(a) Opening displacement, or so-called crack profile, within the end-zone. Curves (a), (b) and (c) correspond to the three cases described by Eqs. (2.8).

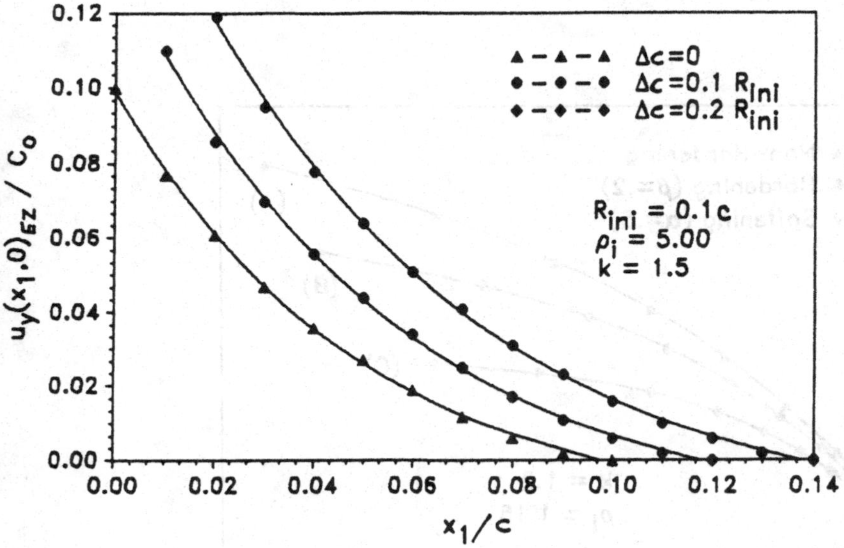

Fig. 3(b) Crack profiles corresponding to a growing crack. Three successive locations of the crack tip are shown: c, c + .01c, c + .02c. Note that the end of the damage zone moves at a different rate than the crack tip. The ratio dR/dc, therefore, is not invariant to the crack extension. Normalizing constant is $C_o = 4\sigma_o R_{ini}/\pi E_1$.

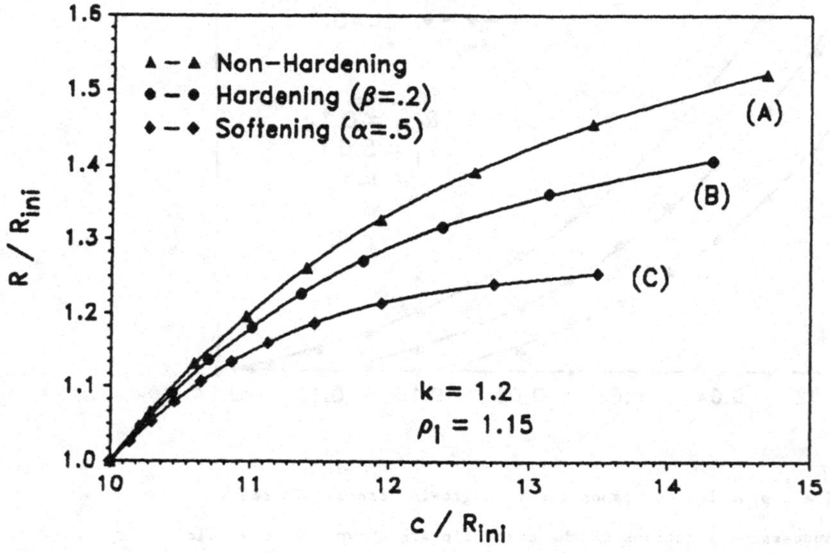

Fig. 4(a) Material resistance curves resulting from Eqs. (2.13) at a fixed ductility parameter $\rho_i = R_{ini}/\Delta = \epsilon_p^f/\epsilon_o$.

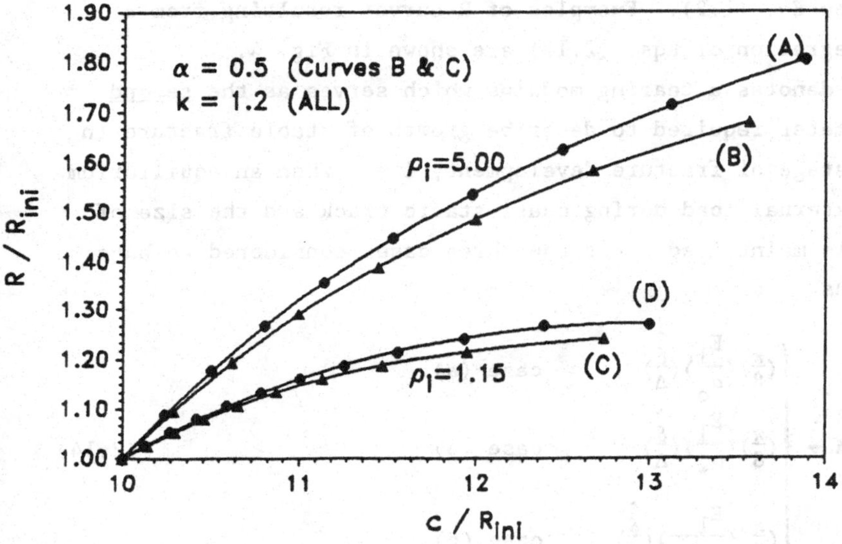

Fig. 5 R-curves according to the "standard", i.e., non-hardening model. Top curve (A) is obtained at high ductility level while the bottom curve (D) corresponds to brittle fracture occurring at $R_{ini} = \Delta$. Note that curves (A) and (D), obtained from the asymptotic expressions (3.3a) and (3.3b) closely match the results obtained from the complete differential equation of the R-curve for a strain-softening solid with $\alpha = .5$, Eq. (2.13).

research. They provide a tool for a detailed study of the effect of the material mechanical characteristics on the shape of the R-curve. This in turn allows one to predict the onset of fracture initiation, the amount of slow stable crack growth, range of external loads for which such a phenomenon exists, and finally, occurrence of the terminal instability under either load- or displacement-controlled end conditions, see Eq. (1.2). Examples of R-curves resulting from a numerical integration of Eqs. (2.13) are shown in Fig. 4.

Symbol M denotes a tearing modulus which serves as the <u>second</u> material parameter required to describe growth of stable fracture in the "latent" stage of fracture development, i.e., when an equilibrium between the external load during quasi-static crack and the size of such a crack is maintained. For the three cases considered we have these relations

$$M = \begin{cases} (\frac{\pi}{8})(\frac{E_1}{\sigma_o})(\frac{\hat{\delta}}{\Delta}), & \text{case (a)} \\ (\frac{\pi}{8})(\frac{E_1}{\sigma_*})(\frac{\hat{\delta}}{\Delta}), & \text{case (b)} \\ (\frac{\pi}{8})(\frac{E_1}{\sigma_{coh}})(\frac{\hat{\delta}}{\Delta}), & \text{case (c)} \end{cases} \qquad (2.14)$$

3. LIMITING CASE OF BRITTLE RESPONSE, $\Delta \simeq R$

Expressions (2.11) and (2.13) were obtained under two restrictive assumptions:

a) the ssy conditions prevail during the phase of quasi-static crack propagation ($R \ll c$), and

b) the process zone (Δ) is entirely embedded within the end zone (R), i.e., $\Delta \ll R$.

Here, we shall not attempt to relax the assumption (a). However, in this particular section we will alter the second assumption.

Assumption (b) implies "ductile" response of the material during the early stages of fracture, cf. Wnuk and Mura (1983). Such assumption reflects appropriately the conditions existing in metal fracture, or any materials which exhibit the strain-hardening effect. For the strain-softening solids, however, such as concrete, cementitious composites, rock, mortar, certain ceramics, fracture appears as a brittle phenomenon despite the fact that the final instability is preceded by a certain amount of stable cracking. This stable cracking process is made possible by the energy screening effect due to the formation of the damage zone containing a large number of the secondary cracks which tend to increase the true area of fracture by several orders of magnitude. Thus, even in the absence of plastic deformation a mechanism exists by which the material can demonstrate a substantial increase in the global (or apparent) fracture toughness after the initiation of the stable crack. Of course, the shape of the resistance curve is very sensitive to the particular mechanism of energy screening activated within the end zone associated with a quasi-static crack. Two distinctly different mechanisms are possible:

(i) Plastic deformation realized by the growth and coalescence of voids enclosed by the end zone. Here the inequality $\Delta \ll R$ is implied, and the case is referred to as "ductile fracture".

(ii) Formation of secondary cracks within the volume of the damage band preceding the dominant crack. Here, fracture has an appearance of a brittle phenomenon, even though it is accompanied by some measurable quantity of stable cracking. Experimental data suggest that for such "brittle fracture" the process zone nearly equals the end-zone, i.e., $\Delta \approx R$. Since now both parameters, Δ and R, are of the same order of magnitude, we shall have to alter assumption (b) listed above.

To illustrate this point further let us discuss in some detail the results obtained for the non-hardening case $S(x_1) = \text{const.} = \sigma_o$. For this material the local energy release rate associated with a quasi-static crack can be expressed, cf. Wnuk and Mura (1983), Wnuk, Bazant and Law (1984)

$$J^{\Delta} = \frac{8\sigma_o^2}{\pi E_1}\left\{\Delta \frac{dR}{da} + R - \sqrt{R(R-\Delta)} - \frac{\Delta}{2}\ln\left(\frac{\sqrt{R} - \sqrt{R-\Delta}}{\sqrt{R} + \sqrt{R-\Delta}}\right)\right\} \qquad (3.1)$$

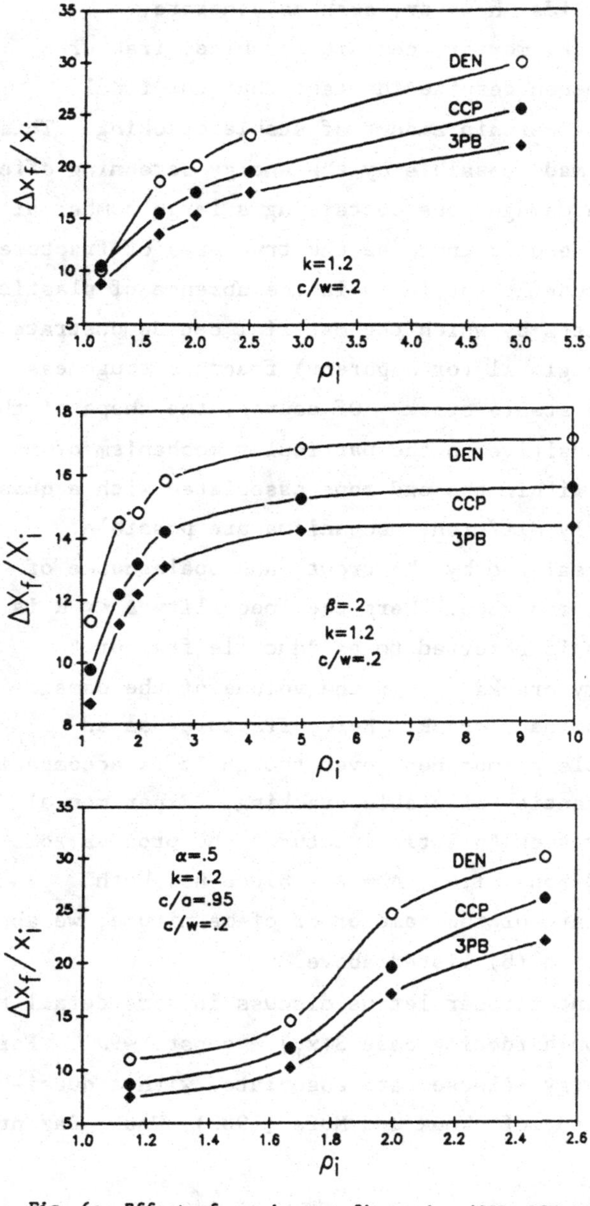

Fig. 6a Effect of specimen configuration (CCP, DEN & 3PB) and material ductility on amount of stable crack extension, $\Delta X_f/X_{ini}$ = $(c_f - c_{ini})/c_{ini}$.

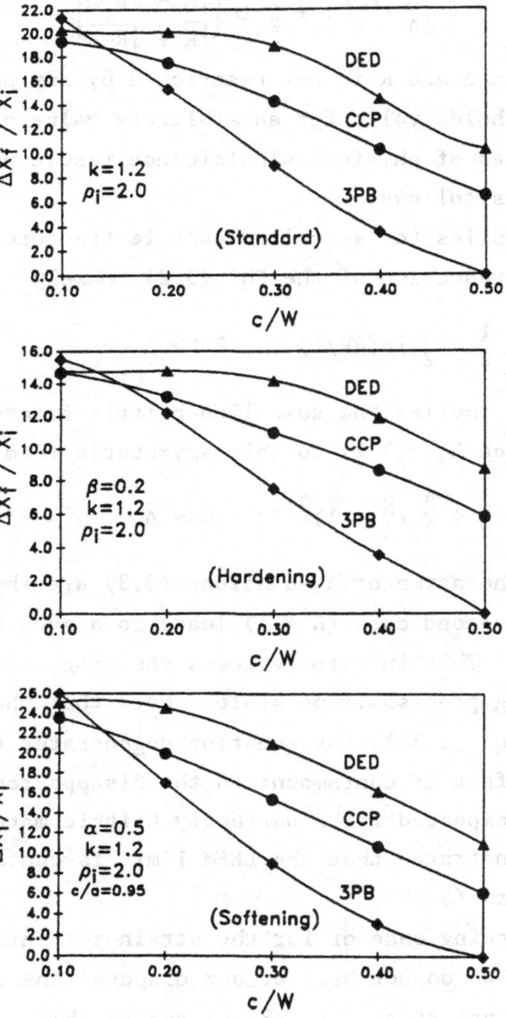

Fig. 7a Effect of specimen configuration (CCP, DEN and 3PB) and the initial crack size $(c/W)_i$ on amount of stable crack extension, $\Delta X_f/X_{ini} = (c_f - c_{ini})/c_{ini}$.

This leads to the following differential equation defining an R-curve

$$\frac{dR}{dc} = M - \frac{R}{\Delta} + \sqrt{\frac{R}{\Delta}\left(\frac{R-\Delta}{\Delta}\right)} + \frac{1}{2}\ln\left(\frac{\sqrt{R} - \sqrt{R-\Delta}}{\sqrt{R} + \sqrt{R-\Delta}}\right) \tag{3.2}$$

Here the relation between Δ and R is not restricted by any particular requirement. Eq. (3.2) holds valid for an arbitrary value of the ratio Δ/R. Two asymptotic cases of physical significance result when we restrict the ratio Δ/R as follows.

Case 1, $\Delta << R$, which implies the so-called ductile fracture, and which leads to a considerable reduction of the Eq. (3.2), namely

$$\frac{dR}{dc} = M - \frac{1}{2} - \frac{1}{2}\ln(4R/\Delta)\ , \quad R >> \Delta \tag{3.3a}$$

and Case 2, $\Delta \simeq R$, which implies the so-called brittle fracture, and which allows one to reduce Eq. (3.2) to this asymptotic form

$$\frac{dR}{dc} = M - \frac{R}{\Delta} + \frac{2}{3}\left(\frac{R}{\Delta} - 1\right)^{3/2}\ , \quad R \simeq \Delta \tag{3.3b}$$

R-curves determined by the asymptotic equations (3.3) are shown in Fig. 5. As can be seen, the second case ($R \simeq \Delta$) leads to a much flatter, less developed, R-curve. This in turn reflects the range of loads for which the stable cracking process is possible. Note that when one substitutes $R = \Delta$ into Eq. (3.3b), the equation degenerates to dR/dc = const.(= M-1), which in fact is tantamount to the disappearance of the R-curve, as it would be expected for a perfectly brittle material. In this manner we have demonstrated that the LEFM limit is contained within the asymptotic form (3.3b).

For the strain-hardening case or for the strain-softening material, unfortunately, we do not have at our disposal the closed-form asymptotic equations equivalent to Eqs. (3.3a) and (3.3b). Let us now focus the attention on strain-softening solids for which we do have the general expression (i.e. valid for an arbitrary Δ/R ratio) for the local energy release rate, see Eq. (2.11c) and the resulting R-curve equation, of Eq. (2.13). Therefore, it is possible to integrate Eq. (2.13c) numerically for the Δ/R ratio fixed at near unity level. This corresponds to the choice of parameter $\rho_i = R_{ini}/\Delta$ being close to one.

The outcome of such a procedure is shown in Fig. 5, where the results for the non-hardening case are superposed with those obtained for a strain-softening material. In both cases the trend is the same; the R-curve tends to a "flat" limit which would be expected for brittle fracture. Still, a certain amount of stable cracking does exist and the load attained at the point of terminal instability exceeds the load level at the onset of crack growth. An obvious similarity with the terminology used in the Theory of Plasticity should be pointed out. The "maximum elastic" value of the external load corresponds to the K-factor at the onset of crack extension, while the limit load predicted through analysis of the load-bearing capacity corresponds to the K-factor level attained at the point of terminal instability, K_f.

It is noteworthy that the quantity K_f is not a material property, just like the terminal magnitude of the external load parameter Q_f and the amount of stable cracking Δc_f. All quantities denoted with a subscript "f" refer to the terminal instability point and they strongly depend on the specimen/crack and loading configuration. The values of K_f, Q_f and Δc_f are predicted from the equations (1.2). Interplay between the material curve K_R vs. Δc (or J_R vs. Δc) and the applied K (or J) factor result in a distinct geometry dependence of all quantities pertinent to the point of transition into catastrophic fracture. To illuminate this point, some limited numerical data are gathered in Table 1 and the pertinent graphs are shown in Figs. 6 and 7. Note that in order to generate the numbers collected in Table 1 and to construct the diagrams shown in Figs. 6 and 7 the following pieces of information had to be provided.

(a) material response to fracture as described by the R-curve, say "ductile" vs. "brittle" case;

(b) specimen-crack geometry and configuration of the external load;

(c) the manner in which loads are applied reflected by "load-controlled" or "displacement-controlled" conditions chosen for the particular test, or imposed on the actual structure.

TABLE 1. Effect of specimen configuration (CCP, DEN & 3PB) and the initial crack size $(a/W)_i$ at fixed material ductility (ρ_i and k) on (a) amount of stable crack extension, $\Delta X_f/X_{ini} = (c_f - c_{ini})/c_{ini}$; (b) increase of load above the initiation threshold level, $\Delta Q_f/Q_{ini} = (\sigma_f - \sigma_{ini})/\sigma_{ini}$; (c) increase in the apparent toughness due to stable cracking process, $\Delta Y_f/Y_{ini} = (R_f - R_{ini})/R_{ini}$. Numbers represent percentages, and W is the width of the specimen.

1. Non-Hardening Case (Standard Model)

$(c/W)_i$	CCP			DEN			3PB		
	$\Delta X_f/X_i$	$\Delta Q_f/Q_i$	$\Delta Y_f/Y_i$	$\Delta X_f/X_i$	$\Delta Q_f/Q_i$	$\Delta Y_f/Y_i$	$\Delta X_f/X_i$	$\Delta Q_f/Q_i$	$\Delta Y_f/Y_i$
0.1	19.3	6.5	35.99	20.3	6.8	37.09	21.3	7.31	38.14
0.2	17.5	5.5	33.29	20.1	6.79	36.89	15.3	4.97	30.55
0.3	14.3	4.24	28.85	18.9	6.06	34.92	9	2.53	20.67
0.4	10.3	2.9	22.79	14.5	5.15	30	3.5	0.69	9.45
0.5	6.5	1.33	15.45	10.3	2.99	22.61	0	0	0

$\beta = 0$ $\quad k = 1.2 \quad \rho_i = 2$

2. Hardening Case (Ramberg-Osgood Power Hardening)

$(c/W)_i$	CCP			DEN			3PB		
	$\Delta X_f/X_i$	$\Delta Q_f/Q_i$	$\Delta Y_f/Y_i$	$\Delta X_f/X_i$	$\Delta Q_f/Q_i$	$\Delta Y_f/Y_i$	$\Delta X_f/X_i$	$\Delta Q_f/Q_i$	$\Delta Y_f/Y_i$
0.1	14.7	5.5	28.14	14.8	5.92	28.72	15.5	6.38	29.55
0.2	13.2	4.9	26.33	14.8	5.89	28.66	12.2	4.37	24.81
0.3	11	4.06	23.55	14.2	5.39	27.58	7.5	2.57	18.01
0.4	8.6	2.77	19.54	11.9	4.42	24.56	3.5	0.66	9.38
0.5	5.8	1.37	14.09	8.7	2.89	19.6	0	0	0

$\beta = 0.2 \quad k = 1.2 \quad \rho_i = 2$

3. Softening Case (Bi-level Cohesive Stress Model)

$(c/W)_i$	CCP			DEN			3PB		
	$\Delta X_f/X_i$	$\Delta Q_f/Q_i$	$\Delta Y_f/Y_i$	$\Delta X_f/X_i$	$\Delta Q_f/Q_i$	$\Delta Y_f/Y_i$	$\Delta X_f/X_i$	$\Delta Q_f/Q_i$	$\Delta Y_f/Y_i$
0.1	23.5	6.32	40.49	25	6.82	42.47	26	7.45	43.66
0.2	20	5.45	36.4	24.5	6.82	42.03	17	4.78	32.5
0.3	15.4	4.16	30.26	21	6.4	38.38	9	2.33	20.19
0.4	10.7	2.57	22.7	16.2	4.83	31.64	3.1	0.51	8.1
0.5	6.1	1.25	14.42	11	2.37	22.35	0	0	0

$\alpha = 0.5 \quad m_c = 0.95 \quad k = 1.2 \quad \rho_i = 2$

4. DUCTILE FRACTURE IN METALS

Although we shall continue to use the parameter R, i.e., the extent of the end-zone, as a measure of material resistance to stable cracking, other alternatives exist. These are: the CTOD or δ_t and Rice's J integral. In fact, within the range of small scale yielding applications, δ_t, J and R are all equivalent, and they differ only by a multiplicative constant. Let us now briefly review some results derived in terms of the J integral.

Assuming the process zone size to be small vs. the length of the plastic zone, Wnuk (1979a) derived the following expression for the separation rate

$$J_I^\Delta = \left(\frac{8\sigma_o^2}{\pi E_1}\right)\left\{\Delta \frac{dR}{dc} + \frac{\Delta}{2}\log\left(\frac{4eR}{\Delta}\right)\right\} \tag{4.0}$$

valid within the small scale yielding range ($\Delta \ll R \ll c$). In the same paper this formula was given for the energy separation rate

$$J_I^\Delta = \left(\frac{8\sigma_o^2}{\pi E_1}\right)\left\{\frac{c}{c+R}\frac{dR}{dc} + \frac{\Delta}{2}\log\left[\frac{4eRc}{\Delta(2R+c)}\right]\right\} \tag{4.1}$$

valid for an arbitrary R/c ratio. This of course includes the limiting case of the fully yielded specimen ($\Delta \ll R \geq c$), as then eq. (4.1) can be readily reduced to the form

$$J_I^\Delta = \left(\frac{8\sigma_o^2}{\pi E_1}\right)\left\{\frac{c}{R}\,\Delta\,\frac{dR}{dc} + \frac{\Delta}{2}\log\left(\frac{2ec}{\Delta}\right)\right\} \tag{4.2}$$

In the derivation of the last three expressions given above only the simplest geometry of a center-cracked plate of infinite width was considered. It would be desirable to extend such analyses so that other common geometrical configurations, such as center-cracked finite width panel, double edge tension specimen, three point bend, compact tension, and a single edge notched tension would be incorporated. In order to analyze an arbitrary geometrical configuration Wnuk (1980) has suggested a somewhat more specific form of defining the material resistance curve, J_R vs. Δc. It reads as follows:

$$\frac{dJ_R}{dc} = \left(\frac{\sigma_o^2}{E_1}\right)\left\{T_\delta - \frac{4}{\pi}\log\left[\frac{\Lambda(\Delta,c)}{\Delta}\right]\right\} \tag{4.3}$$

Here, the symbol T_δ denotes the tearing modulus

$$T_\delta = \begin{cases} (E_1/\sigma_o)(\hat{\delta}/\Delta) \\ \text{or} \\ (\text{CTOA})(E_1/\sigma_o) \end{cases} \tag{4.4}$$

The acronym "CTOA" is derived from the term "crack tip opening angle", while the geometrical factor Λ depends on the given crack configuration. For a center-cracked infinite width plate Wnuk (1979b) gave the factor Λ in terms of the extent of the plastic zone R and the current crack length c (with "e" denoting the base of natural logarithm) as follows:

$$\Lambda(\Delta,c) = 2ec\,\frac{R(R+2c)}{(c+R)^2} \tag{4.5}$$

Extending these considerations to a finite boundary case Wnuk and Sedmak (1980) calculated the factor Λ for a center-cracked plate of width 2h, i.e.

$$\Lambda(\Delta,c) = 2ec\,\frac{\sin^2\left[\frac{\pi(c+R)}{2h}\right] - \sin^2\left[\frac{\pi c}{2h}\right]\tan\left(\frac{\pi c}{2h}\right)}{\sin^2\left[\frac{\pi(c+R)}{2h}\right]\left(\frac{\pi c}{2h}\right)} \tag{4.6}$$

This latter expression reduces to a simple form

$$\Lambda(\Delta,c) = \frac{2eh}{\pi}\sin\left(\frac{\pi c}{h}\right) \tag{4.6a}$$

when it is assumed that all of the unbroken ligament has yielded, i.e., when $c + R \cong h$.

Application of the crack growth law (2.12), in which the quantity J^Δ is replaced by the forms (4.0) and (4.1), for small and large scale yielding, respectively, leads to these differential equations describing extension of a stable quasi-static crack

$$\frac{dR}{dc} = \frac{\pi}{8}T_\delta - \frac{1}{2}\log(4eR/\Delta) \tag{4.7}$$

or

$$\frac{dJ_R}{dc} = (4\sigma_o^2/\pi E_1)\log(J_{ss}/J_R) \tag{4.7a}$$

for a contained yield situation, and

$$\frac{dR}{dc} = \frac{R}{c}\left\{\frac{\pi}{8}\,T_\delta - \frac{1}{2}\log\left(\frac{2R^2}{ec\Delta}\right)\right\} \tag{4.8}$$

or

$$\frac{dJ_R}{dc} = \left(\frac{\sigma_o^2}{E_1}\right)\left\{T_\delta - \frac{4}{\pi}\log\left(\frac{2ec}{\Delta}\right)\right\} \tag{4.8a}$$

for a fully yielded tensile specimen of infinite width ($\Delta \ll R \gg c$). Note that the length of the plastic zone R is used here as an alternative measure of the material resistance to cracking developed during the early stages of ductile fracture. The constants T_δ, σ_o, E_1 and J_{ss} have been already explained with the exception of the last one, J_{ss}. This quantity represents the steady-state limit of the apparent fracture toughness, attained when $dJ_R/dc \to 0$. In other words, the J_{ss} level defines the upper plateau of the J_R-curve. It can be related to other material constants by either one of the following expressions:

$$J_{ss} = \begin{cases} \left(\dfrac{2\sigma_o^2\Delta}{\pi e E_1}\right) \exp\left\{\dfrac{\pi}{4}\,T_\delta\right\} \\[2ex] \left(\dfrac{2\sigma_o^2\Delta}{\pi e E_1}\right) \exp\left\{\left(\dfrac{\pi E_1}{4\sigma_o^2}\right)(\hat{J}/\Delta)\right\} \end{cases} \tag{4.9}$$

Expressions quoted here enable one to study the instabilities occurring in ductile fracture process. A non-dimensional parameter, names stability index

$$\lambda = (\pi E_1/8\sigma_o^2)(dJ_R/dc - \partial J_A/\partial c) \tag{4.10}$$

provides a useful measure of the "degree of stability" of any given stress state associated with a growing crack. As is seen from the formula (4.10) the quantity λ is positive for a stable crack, as only then the demand for the energy flow into the process zone, dJ_R/dc, exceeds the actual rate of energy supply, $\partial J_A/\partial c$ (J-integral labeled with an index "A" denotes the intensity of the external field).

Before we conclude this section let us investigate two forms of the differential equations defining material resistance curve for a large scale yielding occurring in a center-cracked panel (CCP). Setting J_I^{Δ}, as given by Eq. (4.1), to the constant $\hat{J} = \hat{\delta}\sigma_o$, yields

$$\frac{dR}{dc} = \frac{R}{c}\left[\frac{\hat{R}}{\Delta} - \frac{1}{2}\,\ell n\,(\frac{2R^2}{ec\Delta})\right], \qquad \hat{R} = \left(\frac{\pi E_1}{8\sigma_o}\right)\hat{\delta} \tag{4.11}$$

Using the transformation valid within the large scale yielding range

$$J_R(c) = \frac{8\sigma_o^2 c}{\pi E_1}\,\ell n\left[1 + \frac{R(c)}{c}\right] \tag{4.12}$$

One can readily reduce Eq. (4.11) to this form

$$\frac{dJ_R}{dc} = \frac{\hat{J}}{\Delta} - \frac{4\sigma_o^2}{\pi E_1}\,\ell n\,(\frac{2ec}{\Delta}), \qquad \hat{J} = \hat{\delta}\sigma_o \tag{4.13}$$

Somewhat unexpectedly both equations (4.11) and (4.13) can be integrated in a closed form. The results are

$$R(c) = R_{ini}(\frac{c}{c_o})^{1/2} \exp\left\{\frac{1}{2}\,(1+2\lambda_i)(1 - \frac{c_o}{c})\right\} \tag{4.14}$$

and

$$J_R(c) = J_{ini}\left\{\frac{c}{c_o} + (qc/J_{ini})\left[-\frac{1}{2}\,\ell n\,(\frac{c}{c_o}) + \frac{1+2\lambda_i}{2}\,(1 - \frac{c_o}{c})\right]\right\} \tag{4.15}$$

Here, the constant $q = 8\sigma_o^2/\pi E_1$, and the ductility index λ_i is related to the initial crack size c_o/Δ, the tearing modulus M and the ductility parameter ρ_i as follows, cf. Wnuk (1979b)

$$\lambda_i = M - \frac{1}{2}\,\ell n(2\Delta/ec_o) - \ell n(\rho_i) \tag{4.16}$$

Fig. 8 illustrates a distinct difference between an R-curve obtained from Eq. (4.14) and that valid for the small scale yielding situation, as defined by Eq. (2.13a). It is noteworthy that even small differences in the appearance of the R-curve usually lead to significant differences in the Q-curves, i.e., the load vs. crack

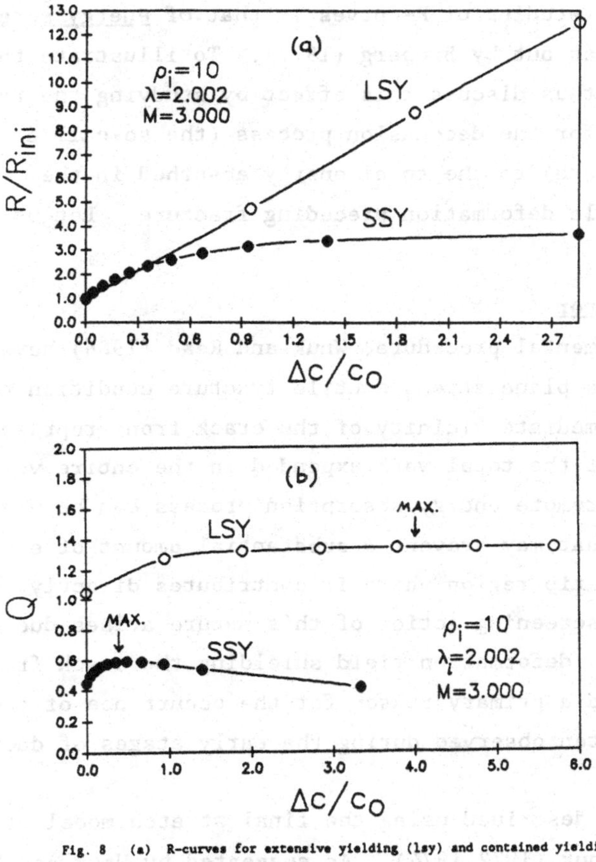

Fig. 8 (a) R-curves for extensive yielding (lsy) and contained yielding (ssy) situations. Parameters used:

$$\alpha = a_o/\Delta = \begin{cases} 10 \text{ for lsy} \\ 100 \text{ for ssy} \end{cases}$$

$$\rho_i = R_{ini}/\Delta = 10 \text{ for both cases}$$

(b) Q-curves for the lsy and ssy limits shown in (a). Data pertaining to the critical state

$$\Delta c_f/c_o = \begin{cases} 400\% \text{ for lsy} \\ 49\% \text{ for ssy} \end{cases}$$

$$\Delta\sigma_f/\sigma_{ini} = \begin{cases} 27.6 \text{ for lsy} \\ 32.4 \text{ for ssy} \end{cases}$$

All data shown are derived from Wnuk's final stretch model of a quasi-static crack.

length relationships. For details we refer the reader to earlier work of Wnuk (1979b) and to the Figures 8a and 8b which dramatically illustrate this point.

Yet another plausible interpretation of the "toughening" effect that is revealed by the studies of R-curves is that of energy screening phenomenon, first pointed out by Broberg (1977). To illustrate this point quantitatively let us discuss this effect by studying the ratio of the energy consumed for the decohesion process (the so-called essential work of fracture) to the total energy absorbed in the processes of irreversible deformation preceding fracture. Let us agree to name this ratio

Energy Transmission Factor

In a simple experimental procedure, Wnuk and Read (1986) have established that under a plane stress ductile fracture condition the work absorbed in the immediate vicinity of the crack front represents only a small fraction of the total work expended in the entire volume of the specimen. Such remote energy absorption process can be viewed as a screening action that may prevent a substantial amount of energy from reaching the crack tip region where it contributes directly to fracture process. The screening action of this nature arises due to existence of the plastic deformation field shielding the crack front, and it may be considered a primary reason for the occurrence of the slow stable cracking often observed during the early stages of ductile fracture.

This effect can be described using the final stretch model of a slowly growing crack, Wnuk (1972,1974). As suggested by Wnuk and Read (1986), the so-called "energy transmission factor" (ETF for short)

$$\text{ETF} = w_e/w_f = \text{essential work of fracture/total work of fracture}$$

is to be used as a suitable quantitative measure of the energy screening phenomenon associated with the ductile fracture process. It can be shown that the quotient ETF is always less than or equal to one, and that it decreases from its initial value recorded at the onset of crack propagation as the stable crack growth progresses (a decrease in

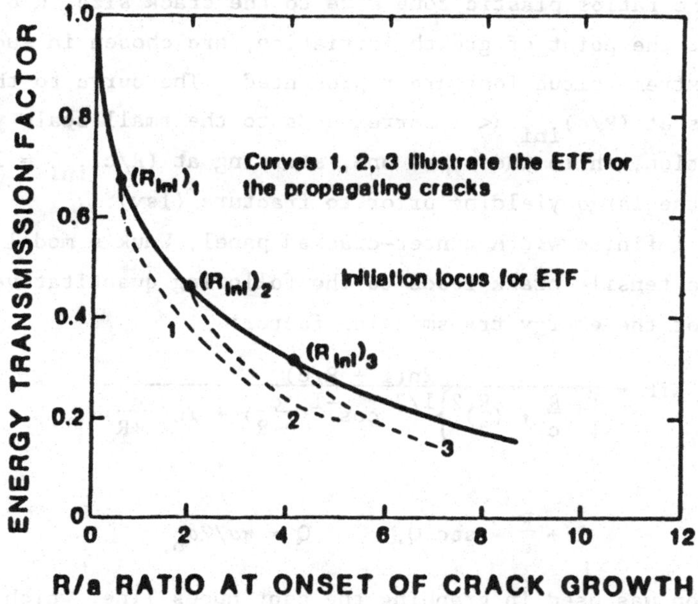

Fig. 9 Energy transmission factor for a stationary (solid line) and for quasi-static cracks (broken lines) contained in elastic-perfectly plastic solid. Data shown are derived from Wnuk's model of "final stretch."

ETF signifies a more intensive screening action). This situation is depicted in Fig. 9 in which the solid line denotes the energy transmission factor for a <u>static</u> crack, while the broken lines show the variations of the ETF <u>after</u> the onset of stable crack extension. The values of the ratios plastic zone size to the crack size, R/c, prevalent at the point of growth initiation, are chosen in such a way that both extreme situations are represented. The curve to the left, which begins at $(R/c)_{ini} \ll 1$ corresponds to the small scale yield (ssy) situation, while the other one, starting at $(R/c)_{ini} \simeq 1$ represents the large yielding prior to fracture (lsy).

For an infinite width center-cracked panel, Wnuk's model of a quasi-static tensile crack leads to the following quantitative prediction of the energy transmission factor:

$$\mathrm{ETF} = \frac{\ell n(1 + R/c)}{\left\{2\frac{R}{c} + (\frac{R}{c})^2\right\}^{1/2} \cos^{-1}(\frac{c}{c+R}) + \ell n(\frac{c}{c+R})} \tag{4.17}$$

in which

$$1 + \frac{R}{c} = \sec Q, \qquad Q = \pi\sigma/2\sigma_o \tag{4.17a}$$

This equation was used in graphing the continuous line, which represents ETF for a static crack as shown in Fig. 9. However, when the crack begins to move, the extent of the plastic deformation field, R, is no longer a quantity depending only on the magnitude of the external load, Q, but it is <u>also</u> a certain function of time. As it is usual in all quasi-static problems, time may be replaced by a time-like variable, i.e. the current crack length "c". The functional relationship R = R(c) is then defined by an R-curve equation which for the geometry discussed here results from a numerical integration of the R-curve equation resulting from Eq. (4.1) and valid for an arbitrary R/c ratio

$$\frac{dR}{dc} = \frac{c+R}{c}\left\{M - \frac{1}{2}\,\ell n\left[\frac{2eR(2c+R)}{c\Delta}\right]\right\} + \frac{R}{c} \tag{4.18}$$

To simplify the task of integration we may reduce the Eq. (4.18) to the following limiting forms

$$\frac{dR}{dc} = M - \frac{1}{2}\ \ell n(4eR/\Delta), \qquad R \ll c \ (ssy) \tag{4.19}$$

and

$$\frac{dR}{dc} = \frac{R}{c}\left[M - \frac{1}{2}\ \ell n(\frac{2R^2}{e\Delta c})\right], \qquad R \gg c \ (lsy) \tag{4.20}$$

Of course, the general equation for the ETF, Eq. (4.17), valid for an arbitrary ratio R/c, should then be reduced correspondingly. Expansions for R << c and R >> c lead to these forms

$$(ETF)_{ssy} = 1 - \frac{2}{3}\ (\frac{R}{c}) + \frac{23}{45}\ (\frac{R}{c})^2 + \ldots \tag{4.21}$$

and

$$(ETF)_{lsy} \simeq \frac{\ell n(\frac{R}{c})}{(\frac{R}{c})\cos^{-1}(\frac{c}{R}) + \ell n(\frac{R}{c})} \tag{4.22}$$

The quantity "R" in the equations (4.21) and (4.22) for a quasi-static crack should be evaluated from the differential equations (4.19) and (4.20), correspondingly.

It is easily seen that while for $R/c \to 0$, the energy transmission factor approaches one, i.e. the screening effect disappears for the case of brittle fracture. The opposite situation is encountered when $R/c \geq 1$, i.e., when the ETF given by the expression (4.22) approaches zero indicating a maximum screening for the case of ductile fracture.

Tests performed on 0.317 cm. thick aluminum sheets (Al 5052) containing two colinear edge cracks has provided a qualitative support for the trend in the variation of ETF with the amount of plastic straining as predicted by Eqs. (4.17), (4.21) and (4.22), compare Wnuk and Read (1986).

5. CREEP FAILURE IN POLYMERS

Ability of viscoelastic solids to fracture when exposed for a long period of time to loads a few orders of magnitude smaller than the level of the critical load required to cause instantaneous failure has puzzled researchers for a good number of years.

While studying response of <u>linear</u> viscoelastic solids to the existing "dormant" cracks Wnuk and Knauss (1970) have found that the process of failure development can be divided into two stages: 1) latent stage ($\dot{c} = 0$), and 2) crack propagation stage for which the growth rate $\dot{c}$ is controlled by the external load (or some other field parameter such as K-factor or C^*-integral) and by the rheological properties of the material. These findings were later confirmed by Nuismer (1974), Schapery (1975) and Swanson (1976).

In more recent studies done by numerous researchers and reviewed by Riedel (1986,1989), similar laws of material response have been established pertaining to fracture in non-linear viscoelastic solids.

In this section we shall briefly review the essential results of the theories based on the continuum mechanics. Behavior of many glassy polymers may be adequately described by the constitutive equations of linear visco-elasticity, namely

$$\begin{aligned} s_{ij}(\underline{x},t) &= \int_{-\infty}^{t} G_1(t-\tau)\,\frac{\partial e_{ij}(\underline{x},t)}{\partial\tau}\,d\tau \\ s(\underline{x},t) &= \int_{-\infty}^{t} G_2(t-\tau)\,\frac{\partial e(\underline{x},t)}{\partial\tau}\,d\tau \end{aligned} \tag{5.1}$$

Here, $G_1(t)$ and $G_2(t)$ denote the relaxation moduli* for shear and dilatation, respectively, t denotes times, s_{ij} and e_{ij} are the deviatoric parts of the stress and strain tensors, while $s\delta_{ij}$ and $e\delta_{ij}$ are their hydrostatic parts (with δ_{ij} denoting the Kronecker delta). With these constitutive equations, Wnuk and Knauss (1970) were able to predict the onset of crack propagation at time t_* after the instant of load application. The prediction involves rheological material properties incorporated in the creep compliance**, D(t), the

*When the Poisson ratio does not appreciably change with time, one may show that $G_1(t) = (1-2\nu/1+\nu)G_2(t)$. Usually, though, both moduli G_1 and G_2 must be determined experimentally.

**Creep compliance D(t) is defined as the strain per unit stress induced by a step-load of magnitude σ_o, i.e. $D(t) = \epsilon(t)/\sigma_o$. The function D(t) may be obtained from the moduli G_1 and G_2 (if known), or determined experimentally.

instantaneous critical K-factor K_{Ig}, and the applied K-factor. Time t_*, which is tantamount to the termination of the dormant stage of crack development, can be thus obtained as a solution to the following equation, proposed by Wnuk and Knauss

$$\frac{D(t_*)}{D(0)} = \left(\frac{K_{Ig}}{K}\right)^2 \tag{5.2}$$

For a step-load of magnitude σ_o ($\sigma_o < \sigma_{Griffith}$) the Eq. (5.2) reduces to

$$\psi(t_*) = \left(\frac{\sigma_G}{\sigma_o}\right)^2 \tag{5.3}$$

To simplify the notation, a normalized creep compliance function $\psi(t) = D(t)/D(o)$ is used, while σ_G denotes the Griffith critical stress evaluated for the polymer in its "glassy" state, i.e. when the viscoelastic moduli such as G_1, G_2 or D are obtained at t = 0, implying the so-called "instantaneous" material response. Eq. (5.3) suggests that delayed fracture may occur only within a certain range of loads for which a solution of Eq. (5.3) may be found, mainly

$$\sigma_{min} \le \sigma_o \le \sigma_G \tag{5.3a}$$

The lower and upper bounds of this range are determined by the rheological properties evaluated at the "rubbery" state ($t \to \infty$) and the "glassy" state ($t \to 0$) of the material. Since

$$\frac{D(t)}{D(0)} \simeq \frac{E_{rel}(0)}{E_{rel}(t)} \tag{5.4}$$

in which $E_{rel}(t)$ denotes the time-dependent relaxation modulus, we have as the lower bound of delayed failure, or so-called long-time tensile strength

$$\sigma_{min} = \sigma_G\left(\frac{E_e}{E_g}\right)^{1/2} \tag{5.5}$$

Here the relaxation modulus at very long time (rubbery state) has been denoted by E_e, while E_g stands for the glassy value of the relaxation modulus. It is noteworthy that the ratio E_e/E_g, as measured for instance for the solid rocket propellents, may be very small (on the order of 10^{-6}), implying a substantial reduction of strength when loads are allowed to act over prolonged periods of time.

Next phase of time-dependent failure begins with an onset of crack propagation. Several authors attempted to derive the equation of motion of such a sub-critical crack penetrating a viscoelastic medium. In independent studies, Knauss (1970), Wnuk (1971) and Schapery (1975), using a framework of continuum mechanics, but applying somewhat different approaches, arrived at the common result

$$\frac{D(\Delta/\dot{c})}{D(0)} = \left(\frac{K_{Ig}}{K}\right)^2 \tag{5.6a}$$

or

$$\frac{E_{rel}(0)}{E_{rel}(\Delta/\dot{c})} = \left(\frac{K_{Ig}}{K}\right)^2 \tag{5.6b}$$

We note that the ratio of the process zone size Δ to the creep crack growth rate $\dot{c}$ represents time needed by the tip of a moving crack to traverse the adjacent process zone.

In order to derive the equations (2.6) Knauss (1970) employed the Irwin-type local criterion of failure involving energy release rate. Wnuk (1971) used the COD concept extended to encompass a viscoelastic medium, while Schapery (1975) followed the classic energy balance approach, based on the first law of thermodynamics. Eq. (5.6) was subjected to an experimental verification, cf. Knauss (1970), Mohanty and Wnuk (1972), and with a certain number-matching procedure, the results showed an excellent agreement with the theory, as illustrated by Fig. 10, extracted from the original work of Knauss (1970). Knauss' estimate of the process zone size Δ for Solithane 113 (50/50) was 3.56 $\times 10^{-9}$ cm, which unfortunately is outside the limits encompassed by the mechanics of continua.

Under certain restrictions and with additional assumptions regarding the functional form of the modulus of relaxation, Schapery (1975) and then Swanson (1976) have shown that both Eqs. (5.6) may be reduced to an extremely simple and plausible result

$$\dot{c} = AK^m \tag{5.7}$$

Coefficients A and m are related to the constants D(o), $\dot{D}$(o) and the exponent

$$m = \frac{d \log D(t)}{d \text{ lot } t} \tag{5.8}$$

respectively. For example, the creep compliance measured by Knauss for Solithane 50/50 (a cross-linked, amorphous polyurethane rubber) is of the form

$$D(t) = \begin{cases} C_o + C_2 t^m \\ 10^{-4.8} + 10^{-1.7} \quad [1/\text{psi}] \end{cases} \tag{5.9}$$

This relation was shown to be valid for the time interval (t in minutes) stretching over several decades

$$-5 \leq \log_{10} t \leq -2.5 \tag{5.10}$$

Within the same time interval the power law (5.7) is valid. Thus with the material constants

$$C_o = D(0) = 10^{-4.8}, \quad mC_2 = \dot{D}(0) = \frac{1}{2}10^{-1.7} \quad \text{and} \quad m = \frac{1}{2} \tag{5.11}$$

the power law provides a very close approximation of the general equation of motion (5.6a). Shapery (1975) was able to relate the constant A to the rheological material characteristics C_o, C_2 and m, namely

$$A = \left[\frac{c_2 \lambda_m}{G_f(1-K^2/K_g^2)}\right]^{1/m} \left(\frac{\pi}{8\sigma_m^2 I_1}\right) \tag{5.12}$$

Here G_f is the specific fracture energy of the polymer in its glassy state, σ_m is the maximum stress within the process zone, while I_1 and λ_m are numerical factors which can be found by carrying out certain integrals given by Shapery (1975). If the level of the applied constant load σ_o is low compared to the Griffith stress σ_G, treated as an instantaneous material characteristic, then for the CCP specimen

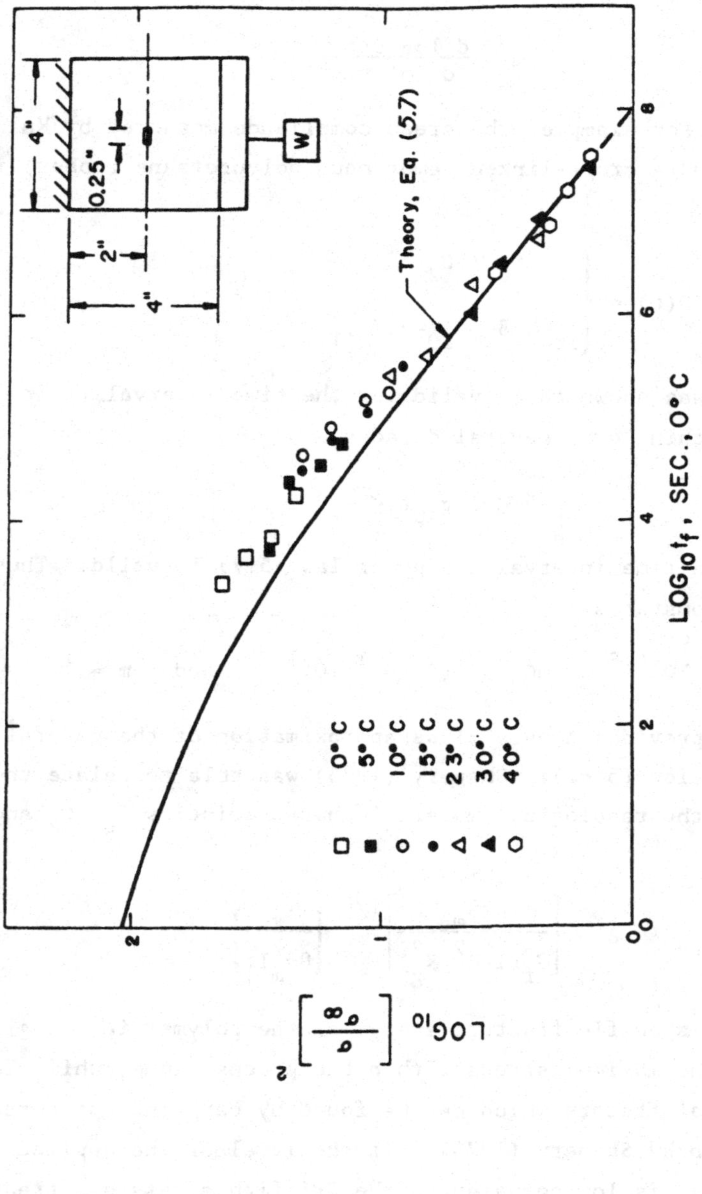

Fig. 10 Predicted, Eq. (3.7), and measured failure times for a CCP Solithane 50/50 specimen (courtesy of W. G. Knauss).

configuration the failure time may be given in a closed form

$$t_f = \frac{2m\sigma_m^2 I_1^2}{\pi(2+1/m)} \left[\frac{4G_f}{\lambda_m C_2 c_o}\right]^{1/m} \sigma^{-2(1+1/m)} \tag{5.13}$$

For the case shown in Fig. 10, the pertinent constants were evaluated as follows: $\sigma_m I_1 = 172$ ksi, $I_1 \simeq 2$, $\lambda_m \simeq .11$, $\sigma_m = 86$ ksi, $E_\infty \simeq D^{-1}(\infty) = 430$ psi, while with the initial crack size $a_o = .125$ in. the propagation threshold,

$$\sigma_{min} = \sqrt{4G_f E_\infty / 3\pi a_o} \tag{5.14}$$

was found to be 8.40 psi, and the specific fracture energy was estimated as $G_f = 4.82 \times 10^{-2}$ lb/in. Equation (5.13) brings out the effect of the applied stress σ and the initial crack size c_o on the lifetime, i.e.,

$$t_f \propto \left(\frac{1}{c_o^m}\right)\left[1/\sigma^{2(1+1/m)}\right] \tag{5.15}$$

It is seen that agreement between the predicted and measured failure times is very good, particularly in low stress range. Schapery (1975) estimates the largest value of the time increment $\Delta/\dot{c}$ to occur at the instant of crack initiation

$$\left(\frac{\Delta}{\dot{c}}\right)_{max} = 14 \times 10^{-2} \text{ min} \tag{5.16}$$

which is still within the range of the assumed linear relation between the log of the creep compliance and log of time, and yet sufficiently small to justify the basic assumptions underlying the theory (i.e. that the change in the applied stress intensity factor is small over each time increment $\Delta/\dot{c}$ for the entire experimental range of behavior). One third of $(\Delta/\dot{c})_{max}$ gives the initiation time. As expected, the initation time t_* turned out to be several orders of magnitude smaller than the failure time and therefore may be neglected in the evaluation of the life span. The length of the nonlinear zone was estimated to vary from about 10^{-8} cm at low stress level at the early stage of crack propagation to about 32×10^{-8} cm at a higher stress level in the advanced stage of propagation, say at $t = 0.9\ t_f$. This provides some evidence supporting a concept of an R-curve associated with fracture in a linear viscoelastic medium.

The basic equations governing slow extension of fracture in a viscoelastic medium imply that the propagation rate $\dot{c}$ depends on the instantaneous stress intensity factor and not on either its history of the stress history. Therefore, it is possible to study crack growth under cyclic loadings of any complex waveform in a way similar to that used for the case of constant or monotonic loads.

6. CREEP FAILURE IN METALS AT HIGH TEMPERATURES

The state of understanding in the area of time-dependent fracture mechanics (TDFM) is now at a level approaching current state-of-the-art of elasto-plastic fracture mechanics (EPFM). Critical components of structures such as gas and steam turbines, power-plant boilers and nuclear reactors operate at elevated temperatures over prolonged periods of time. They tend to develop cracks during their service life, and some components are known to have crack-like defects even at the time they go into service. These defects can grow and cause failure. Some examples of major failures in the power industry involving elevated temperature components, for which creep was a significant contributing factor, include turbine rotors and steam pipes. Thus the topic presented in this section, concerning extension of cracks under creep conditions at higher temperature service environment, has an important industrial relevance, cf. Wells (1986).

Webster (1989) describes briefly the life prediction techniques at high temperatures. He considers basically two separate types of fracture: (1) the net section rupture occurring in a defectless component, or in a situation where the non-growing cracks are negligibly small compared to the ligament, and (2) fracture caused by growth of a crack which is strongly influenced by the deformation field developed near the crack tip under creep conditions. Net section rupture is most likely to occur in structures that are initially defect free and in which damage can develop relatively uniformly through the thickness. Failure by crack growth, on the other hand, is most likely to be observed in the components which contain an initial defect (or, which are capable to initiate such a defect under nominal service conditions), and which are located at thicker sections where the prevailing state of stress favors crack propagation. In certain circumstances crack initiation and subsequent growth may be terminated and followed by net section rupture after some amount of creep crack extension has taken place.

In this section we shall focus attention on the description of the creep crack growth phenomenon. By analogy with ductile fracture, we proceed now to show that the rate of crack growth in nonlinear viscoelastic solids, such as metals at elevated temperatures, may no longer be simply correlated with the K-factor, as it was implied by Eq. (5.7), but rather it should be tied up with C^*-integral which represents a natural extension of J-integral into the domain of visco-plastic solids. Let us begin with the constitutive equations believed to describe material behavior under nonlinear creep conditions. Using the standard notation we have

$$\dot{e}_{ij} = \frac{1}{2G}\dot{s}_{ij} + \frac{3}{2}Bs_{ij}\,\sigma_e^{n-1} \tag{6.1}$$

Here e_{ij} and s_{ij} are the deviatoric strain and stress rate tensors, σ_e is the equivalent tensile stress defined in a usual way, e.g., $(\frac{3}{2} s_{ij}s_{ij})^{1/2}$. The creep exponent n and the constant B must be obtained from the empirical Norton's law which relates the rate of strain $\dot{\epsilon}$ and the applied constant stress σ under uniaxial creep test conditions, i.e. $\dot{\epsilon} = B\sigma^n$. Obviously, the second term in Eq. (6.1) is a generalization of 1D Norton's law onto a 3D state of stress.

Let us consider two limiting cases, that of a crack progressing faster than the growing creep zone r_{cr}, i.e., when $\Delta c > \Delta r_{cr}$, or $\dot{c} > \dot{r}_{cr}$, and the second case when the crack is stationary or it moves very slowly compared to the time-dependent expansion of the creep zone, $\Delta c \ll \Delta r_{cr}$ or $\dot{c} \ll \dot{r}_{cr}$. The asymptotic near-crack-tip fields corresponding to these two limiting cases have been derived based on the constitutive equation (6.1). For a moving crack Hui and Riedel (1981) and Hui (1983) have given these equations (superscript "v" denotes viscous part of strain rates)

$$\sigma_{ij} = \alpha_n \left[\frac{\dot{a}(t)}{BEr}\right]^{1/n-1} \tilde{\sigma}_{ij}(\theta,n)$$
$$\dot{\epsilon}^v_{ij} = \frac{\alpha_n}{E}\left[\frac{\dot{a}(t)}{BEr}\right]^{n/n-1} \tilde{\dot{\epsilon}}_{ij}(\theta,n) \tag{6.2}$$

while for a stationary crack, or an almost stationary crack, the HRR field applies with one essential difference that the amplitude of the singular terms valid in an elasto-plastic solid, and given by the J integral, is now replaced by the C^* integral, namely

$$\sigma_{ij} = \sigma_o \left[\frac{C^*}{BI_n r}\right]^{1/n+1} \Sigma_{ij}(\theta,n)$$
$$\dot{\epsilon}^v_{ij} = \epsilon_o \left[\frac{C^*}{BI_n r}\right]^{1/n+1} \dot{E}_{ij}(\theta,n) \tag{6.3}$$

Here B and n are material properties extracted from the Norton law, α_n and I_n are certain numerical factors, cf. Shih (1983), the symbols indicating dependence on the polar coordinate θ and the exponent n denote the angular distributions of the respective fields. Superscript "v" has been added to denote the viscous part of the strain rate. As it is seen from Eqs. (6.3) the characterization of the singular behavior of the asymptotic fields within the creep zone adjacent to the crack tip ($r \to 0$), is now determined by the path-independent contour integral C^*, defined in close analogy with the J-integral, namely

$$C^* = \int_\Gamma W^* \, dy - T_i \frac{\partial \dot{u}_i}{\partial x} ds$$
$$W^* = W^*(\dot{\epsilon}_{ij}) = \int_0^{\dot{\epsilon}_{ij}} \sigma_{ij} \, d\dot{\epsilon}_{ij} \tag{6.4}$$

The coordinates (x,y) are attached at the crack tip, contour Γ encompasses the crack tip and integration proceeds in a counterclockwise sense. Dimensions of C^* correspond to energy per unit area and per unit time; thus a complete analog of the J-integral has been defined.

Just like J integral, the C^* integral can be determined either analytically or experimentally in a manner following the pioneering work of Landes and Begley (1976), and the more recent studies of Saxena, Shih and Ernst (1984).

For transient creep processes, which in certain cases are dominant during the life of a component exposed to high temperature deformation, the C^* integral has to be replaced by the C(t) integral

$$C(t) = \left(1 + \frac{t_1}{t}\right)C^* \tag{6.5}$$

where t_1 denotes the transition time, and it can be computed from the Riedel's expression

$$t_1 = \frac{K_I^2}{EC^*(n+1)} \tag{6.6}$$

Obviously, for $t >> t_1$, C(t) reduces to the steady-state value, C^*. Yet another parameter, C_t, has been suggested by Atluri (1981), Bassani, Hawk and Saxena (1986), Bassani et al. (1986), and shown to be perhaps the most successful parameter which correlates a wide range of experimental creep data. This effect of "collapsing" the data is best illustrated by Figs. 11(a) and (b), reprinted here by courtesy of Saxena (1989). The physical interpretation of the parameter C_t is very similar to the classic concept of energy release rate, namely

$$C_t = -\frac{1}{B}\frac{\partial W_t^*}{\partial c} \tag{6.7}$$

in which W_t^* represents the instantaneous stress-power dissipation rate, and B is the specimen thickness. For small scale creep (ssc) regime, Saxena (1986) provides the following formula which relates C_t to the rate of deflection due to creep deformation, $\dot{v}_{cr}$, measured at the load-line:

$$(C_t)_{ssc} = \frac{P\dot{v}_{cr}}{BW} F'/F \tag{6.8}$$

Here P denotes applied load, W is the specimen width while F and F' are calculated as follows (K is the stress intensity factor):

$$F = (K/P)BW^{1/2}, \qquad F' = dF/d(c/W) \tag{6.9}$$

An essential property of the C_t parameter is that it can relate the load and deflection rate measurements made at the load-point (remote from the crack tip) to the creep zone expansion rate, $\dot{r}_{cr}$. Thus if $\dot{v}_{cr}$ can be measured, the C_t for small scale creep range is determined by Eq. (6.8). In components where $\dot{v}_{cr}$ cannot be measured, it can be estimated on the basis of finite element studies of Bassani et al. (1986), namely

$$\dot{v}_{cr} = \frac{2B}{E}\frac{K^2}{P}\cdot\beta\,\dot{r}_{cr}(1-\nu^2) \tag{6.10}$$

Here β_n is a scaling factor which approximately equals 1/3. Finally, according to Saxena (1989), the quantities C(t) and C_t are related under the small scale creep condition

$$C_t = \left[\beta_n(1-\nu^2)\,\frac{n+1}{n-1}\,(F'/F)\,\frac{r_{cr}}{W}\right]C(t) \tag{6.11}$$

Interestingly, neither C_t nor C(t) are uniquely related to the crack tip displacement rate, $\dot{\delta}_t$, under the small scale creep regime. Saxena (1989) notes, however, that such unique relations do exist in the extensive creep region.

To conclude this section let us quote the recent result of Webster (1989) who related the rate of creep crack growth $\dot{c}$ to the steady-state value of the C(t) integral, i.e., the C^*. In essence Webster's formula reduces to

$$\dot{c} = (3/\epsilon_f^*)(C^*)^{0.85} \tag{6.12}$$

Here $\dot{c}$ is in mm/h, C^* is in MJ/m^2h and ϵ_f^* denotes the so-called effective creep ductility. Eq. (6.12) suggests a power law relating $\dot{c}$ and C^*. In more general terms, this law reads

$$\dot{c} = D_o(C^*)^{\phi} \tag{6.13}$$

in which the proportionality coefficient was given by Webster (1989) as

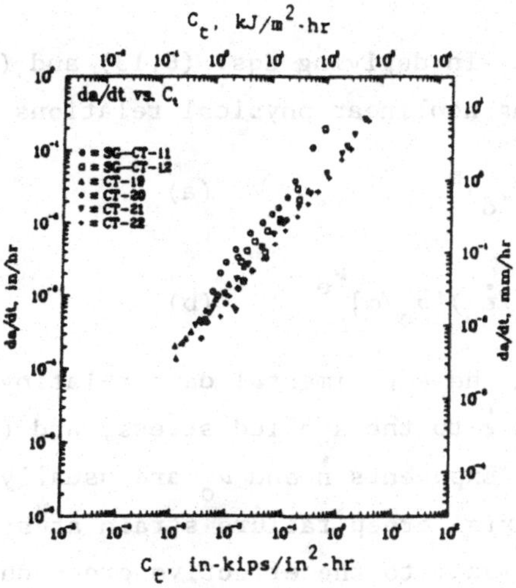

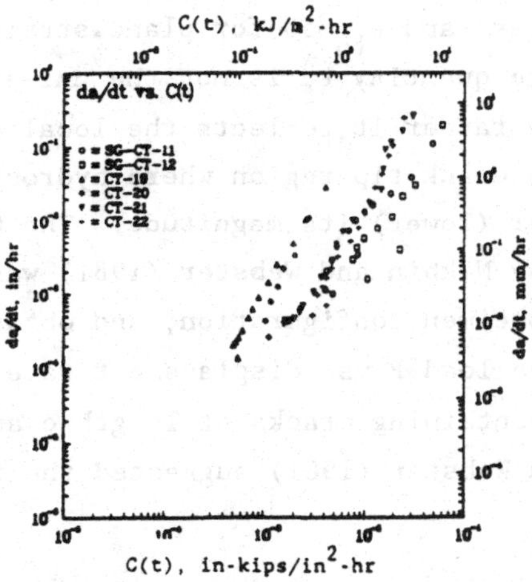

Fig. 11 (a) Creek crack growth rate as a function of C_t in Cr-Mo-V steel at 594°C.

(b) Creep crack growth rate as a function of the C(t) in Cr-Mo-V steel at 594°.

Figures are reproduced from Saxena (1989) with author's permission.

$$D_o = (n+1) \frac{\dot{\epsilon}_o}{\epsilon_f^*} \left[\frac{1}{I_n \sigma_o \dot{\epsilon}_o}\right]^{n/(n+1)} r_{cr}^{1/(n+1)} \tag{6.14}$$

and the exponent $\phi = n/(n+1)$. In deriving Eqs. (6.13) and (6.14) Webster has used the following nonlinear physical relations

$$\dot{\epsilon} = \dot{\epsilon}_o(\sigma/\sigma_o)^n \qquad \text{(a)}$$

$$t_R = (\epsilon_{fo}/\dot{\epsilon}_o)[\sigma_o/\alpha]^{\nu_o} \qquad \text{(b)} \tag{6.15}$$

Both expressions are based on the experimental data relating (a) rate of creep components of strain $\dot{\epsilon}$ to the applied stress, and (b) the rupture time t_R to stress σ. Exponents n and ν_o are usually the same numbers, ϵ_{fo} denotes the uniaxial creep failure strain at stress σ_o. Since D_o is inversely proportional to the effective creep ductility ϵ_f^*, estimated as ϵ_{fo} for plane stress and $\epsilon_{fo}/50$ for plane strain, a note of caution should be added; the quantity ϵ_f^* is not the uni-axial creep ductility of the material, but rather it reflects the local value of rupture strain existing at the crack tip region where hydrostatic tension may significantly alter (lower) its magnitude. The C^* integral in the experiments conducted by Nikbin and Webster (1981) was measured in a double-cantilever-beam specimen configuration, and obtained as a difference in areas between the load P vs. displacement rate $\dot{v}_{cr}$ diagrams for two test-pieces containing cracks of length c and c+dc, respectively. Thus Nikbin and Webster (1981) suggested the following simple formula

$$C^* = [P/(n+1)B]d\dot{v}_{cr}/dc \tag{6.16}$$

where B is the thickness of the grooved DCB-specimen through which a crack propagates. They have demonstrated that the creep crack growth rate could indeed be correlated with C^*. The measurements were performed within the temperature range 100°C to 200°C in DCB-specimens cut out of aluminum alloy RR58 which was of interest to the aircraft industry.

As we can see, the creep fracture mechanics has grown and developed into a substantial body of scientific knowledge. However, a lot remains to be done. Quantitative description of the transition from slow creep crack growth to fast fracture needs further analysis. Accurate methods are still needed for estimating C_t for nonlinear constitutive laws which properly account for effects due to primary creep, cyclic loading and crack growth. The existing solutions for evaluation of C^* are limited to only a few geometries and almost exclusively to 2D crack configurations. Research aimed at extension of these results onto complex configurations encountered in industrial applications will require more attention in the near future.

7. TIME DEPENDENT FRACTURE IN DAMAGING MATERIALS

The dominant mechanism of energy dissipation associated with the deformation process preceding and accompanying fracture in damaging materials is the evolution of microdefects randomly distributed within the bulk of the specimen. Should a macroscopic crack initiate at a certain site, it will interact with the existing field of microdefects and thus the secondary cracks will tend to localize in a narrow region located directly ahead of the crack. High stress levels σ_Σ and the stress gradients $\partial\sigma_\Sigma/\partial a$ which exist in the neighborhood of the dominant crack front, see Fig. 12, accelerate the process of damage accumulation. As shown by Wnuk and Kriz (1985), this interaction process has a synergistic effect on fracture extension, and thus even at a constant level of the applied load, the fracture will propagate at a monotonically increasing rate. Phenomenologically, this time dependent effect caused by gradual degradation of microstructural toughness and the ensuing crack extension closely resembles the creep crack extension described in the previous two sections.

A modified version of the Kachanov damage accumulation law is employed to study the damage kinetics in composite materials. The purpose of this study is to quantify the two basic phases of the failure process in composites, namely (1) localization of microdefects and (2) spread of the dominant crack. Phase one leads to an apparent

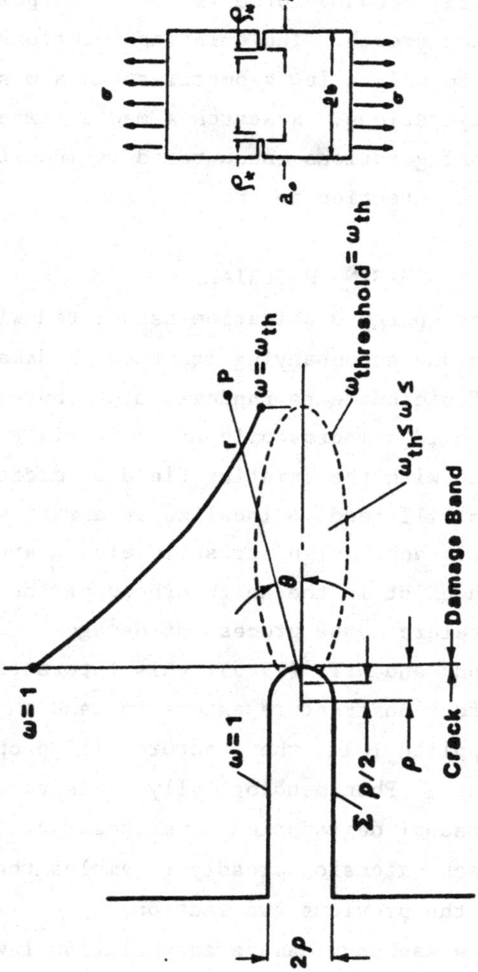

Fig. 12 Specimen geometry (DEN) and details of the damage zone ($\omega_{th} \leq \omega \leq 1$) adjacent to the crack front. It is assumed that when the crack propagates, the root radius ρ assumes a constant value ρ_* which is regarded as a microstructural characteristic.

increase in the material fracture toughness observed in the early stages of crack extension and reflected by the so-called "resistance" curve, or an R-curve. The phenomenon of R-curve suggests that there is no unique value of fracture toughness in materials capable of microcracking or "plastic-fracturing" such as various types of mortars, concretes, and other cementitious composites. If, however, a continuous damage mechanics (CDM) approach is employed, then the mathematical model implies certain new measures of material resistance to cracking which are invariant to the specimen geometry and size (saturation level of the microcrack density is one such measure). Consideration of the interaction of the dominant crack with the field of the micropores generated ahead of its front leads to the following nonlinear integro-differential equation defining the rate, $\dot{a}$, of the dominant crack growth, cf. Wnuk and Kriz (1985),

$$\dot{a} = \frac{-C\sigma_{\Sigma}^{\nu}(a)}{C\int_{a_o}^{a} \frac{\partial}{\partial a}\left[\sigma_{\Sigma}^{\nu}(a,a')\right]\frac{da'}{\dot{a}(a')} + d\Omega_1/da} \tag{7.0}$$

Symbol "a" will be used to designate crack size throughout this section. Equation (7.0) involves the stress σ_{Σ} and the stress gradient $\partial\sigma_{\Sigma}/\partial a$ which exist ahead of a dominant crack (basically it is a LEFM solution modified by the presence of a cluster of the microdefects adjacent to the crack front), while a scalar parameter $\Omega_1(a)$ serves as a measure of the damage generated during the incubation phase of the failure process involving formation and localization of the micropores around an initial macrodefect of length a_o. A numerical approach to an effective integration of Eq. (7.0) is necessary. An iterative procedure has been worked out based on a FORTRAN program. The program has been tested and shown to give satisfactory convergence. Upon completion of the numerical part of this study, it has been shown that the Kachanov parameters, C and ν, can be chosen in a certain range to obtain a fair correlation with a large amount of the available experimental data, cf. Wnuk (1984) and Wnuk and Kriz (1985).

Derivation of the Governing Equation

According to the continuous damage mechanics (CDM) view of fracture, the presence of the dominant crack is always accompanied by a region of microdefects in which damage is being continually built up. Integration of the Kachanov law of damage accumulation

$$\frac{d\omega}{dt} = C\left(\frac{\sigma_\Sigma}{1-\omega}\right)^\nu , \qquad 0 \leq \omega \leq 1 \tag{7.1}$$

in which $\omega = \omega(t)$ is a scalar function describing the intensity of damage, σ_Σ denotes the stress ahead of the dominant crack while C and ν are material parameters, leads to the following criterion oa fracture

$$C\int^t \sigma_\Sigma^\nu[r(t'),t']dt' = \Omega_c \tag{7.2}$$

$$\sigma_\Sigma[r(t'),t'] = \sigma_\Sigma(a,a') = \frac{K_I(a')}{\sqrt{2\pi\rho_*}} Y(\rho_*+a-a')$$

This criterion may be verbalized as follows: "... for a collapse of a material element located at the distance r from a dominant crack tip of half-length 'a' it is necessary that the time integral of ν-th power of the stress at that point attains the critical value $\Omega_c = (1+\nu)^{-1}$."

Decomposing the integral in Eq. (7.2) into two parts:

$$C\int_0^{t_1} [\sigma_\Sigma^\nu]_{\dot{a}=0} dt' + C\int_{t_1}^t [\sigma_\Sigma^\nu]_{\dot{a}\neq 0} \tag{7.3}$$

and identifying the first part, say Ω_1, with the damage generated during the latent phase of the failure process associated with a buildup of microdefects while the dominant crack remains stationary, we obtain

$$C\int_{t_1}^t [\sigma_\Sigma^\nu]_{\dot{a}\neq 0} dt' = \Omega_c - \Omega_1 \tag{7.4}$$

The first critical time t_1 is defined as the time at which the material element adjacent to the tip of an initial crack collapses. This occurs when the integral

$$C\int_0^{t_1} \sigma_\Sigma^\nu(a_o)dt', \qquad \sigma_\Sigma(a_o) \equiv \sigma_\Sigma(a_o,a_o) \tag{7.5}$$

attains the critical value, Ω_c. Since the stress $\sigma_\Sigma(a_o) = K_I(a_o)Y_*/\sqrt{2\pi\rho^*}$ does not depend on time, the integral (5) equals C $\sigma_\Sigma^\nu(a_o)t_1$, and therefore

$$t_1 = (\Omega_c/C)\,\frac{(K_I(a_o)Y_*)^\nu}{(2\pi\rho_*)^{\nu/2}}\,, \qquad Y_* = Y(\rho_*) \tag{7.6}$$

Applying the fracture criterion ((7.2) to the second phase of the failure process, we can evaluate the amount of damage generated within the time interval $t_1 \le t' \le t$. Here, t coincides with the instant at which the material element located at point P collapses. Now, the stress distribution σ_Σ varies not only with the distance from the current crack tip but, also, it is time dependent, since at any fixed point in space, say P, the stress becomes elevated when the crack front propagates toward this point. The integral

$$\Omega_2(t) = C\int_{t_1}^{t} [\sigma_\Sigma^\nu(P,P')]_{\dot{a}\neq 0}\, dt' \tag{7.7}$$

may be replaced by an integral with respect to the current crack length

$$\Omega_2(a) = C\int_{a_o}^{a} \sigma_\Sigma^\nu(a,a')\,\frac{da'}{\dot{a}(a')} \tag{7.8}$$

When the nondimensional crack length $x = a/\rho_*$ and the nondimensional time $\theta = t/t_1$ are introduced, the expression (7.7) assumes the form

$$\Omega_2(x) = \Omega_c\int_{x_o}^{x} \Phi^\nu(x')[Y(x,x')/Y_*]^\nu\,\frac{dx'}{\dot{x}(x')}\,, \qquad \dot{x} = dx/d\theta \tag{7.9}$$

in which the geometry dependent function Φ is defined as the ratio of the stress intensity factor for a current crack and that for the initial crack, i.e.,

$$\Phi(x) = K_I(x)/K_I(x_o) \tag{7.10}$$

The function Y(x,x') is the familiar nondimensional stress ahead of a moving crack, $Y(\xi)$, in which the distance ξ is expressed as $r/\rho_* = 1+x-x'$, see Fig. 13. We note that the rate of crack propagation $\overset{\circ}{x}$ appearing in the expression (7.9) has to satisfy the governing equation (7.0), or (7.15).

Next, we eliminate time by the time-like variable, a(t). Replacing dt' by $da'/\dot{a}(a')$, we arrive at the following equation of a moving crack

$$C\int_{a_o}^{a} \sigma_\Sigma^\nu(a,a')\,\frac{da'}{\dot{a}(a')} = \Omega_c - \Omega_1(a) \tag{7.11}$$

When this equation is differentiated with respect to time, remembering that $d[\ ]/dt = \dot{a}d[\ ]/da$, one arrives at

$$\dot{a}C\int_{a_o}^{a} \sigma_\Sigma^\nu(a,a')[da'/\dot{a}(a')] + C\sigma_\Sigma^\nu(a,a) = -\dot{a}d\Omega_1/da \tag{7.12}$$

Solving for the rate of crack growth gives

$$\dot{a} = \frac{-C\sigma_\Sigma^\nu(a)}{C\displaystyle\int_{a_o}^{a} \frac{\partial}{\partial a}\left[\sigma_\Sigma^\nu(a,a')\right]\frac{da'}{\dot{a}(a')} + d\Omega_1/da} \tag{7.13}$$

which is identical with equation (7.0). We note that both terms appearing in the denominator of (7.13) involve not only the stresses $\sigma_\Sigma(a,a')$ and $\sigma_\Sigma(a,a_o)$ but also their gradients.

These gradients are always negative; therefore, the entire expression on the right-hand side of Eq. (7.13) is positive.

Let us introduce the nondimensional variables

$$x = a/\rho_*\ , \qquad \theta = t/t_1$$
$$x' = a'/\rho_*\ , \qquad x_o = a_o/\rho_* \tag{7.14}$$

in which ρ_* denotes the characteristic structural length (such as aggregate size), while the critical time t_1 is defined by Eq. (7.6). Substitution of (7.14) into Eq. (7.13) yields the nondimensional version of the governing equation, useful in further numerical studies, namely

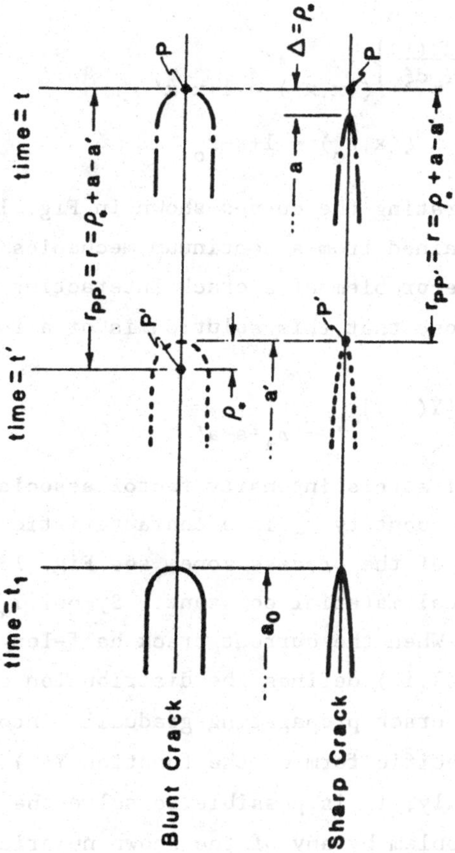

Fig. 13 Location of crack front at

t_1 - onset of crack propagation,

t' - intermediate time,

t - current instant.

Note that the damage accumulation process at the material point P and time t is influenced by the preceding stages of crack growth.

$$\frac{dx}{d\theta} = \frac{\nu^{-1}\Phi^{\nu}(x)}{\int_{x_o}^{x} \Phi^{\nu}(x')F_1(x,x')[dx'/\overset{\circ}{x}(x')] + F_2(x)} \tag{7.15}$$

Here, the function $\Phi(x)$ is geometry dependent as indicated by Eq. (7.10), while the other two auxiliary functions F_1 and F_2 are defined as follows:

$$F_1(x,x') = -\left\{\left(\frac{Y(\xi)}{Y_*}\right)^{\nu-1} \frac{dY(\xi)}{Y_* d\xi}\right\}_{\xi(x,x') = 1+x-x'}$$

$$F_2(x) = F_1(x,x_o), \quad \text{or} \quad \xi(x,x_o) = 1+x-x_o \tag{7.16}$$

These forms were used in generating the curves shown in Fig. 14.

The function Y(r) is obtained from a continuum mechanics or a finite element solution to the problem of a crack interacting with its own damage zone. Let us suppose that this solution is of a form

$$\sigma_{\Sigma}(a,a') = \frac{K_I(a')}{\sqrt{2\pi\rho_*}} [Y(r')]_{r' = \rho_*+a-a'} \tag{7.17}$$

in which K_I denotes the Mode I stress intensity factor associated with crack of half-length a'. The quantity ρ_* is a characteristic length parameter, such as the length of the process zone, see Fig. 13, and it is treated here as an additional material constant. Symbol Y_* denotes the value of Y(r) at $r = \rho_*$. When the current crack half-length a' varies between a_o and a, Eq. (7.17) defines the distribution of the near-tip stress σ_{Σ} ahead of a crack propagating gradually into its own damage zone. Although the specific form of the function Y(r) may be difficult to obtain analytically, it is possible to solve the appropriate boundary value problem by any of the known numerical methods. For the simplest assumption of a valid LEFM field, the function Y(r) is given by a familiar $\sqrt{\rho_*/r}$ expression. Curves shown in Fig. 14 were generated by assuming a double-edge notch crack configuration (see Fig. 12) and the $\sqrt{\rho_*/r}$ form for the Y-function. The K-factor for such crack configuration is given by the expression, cf.

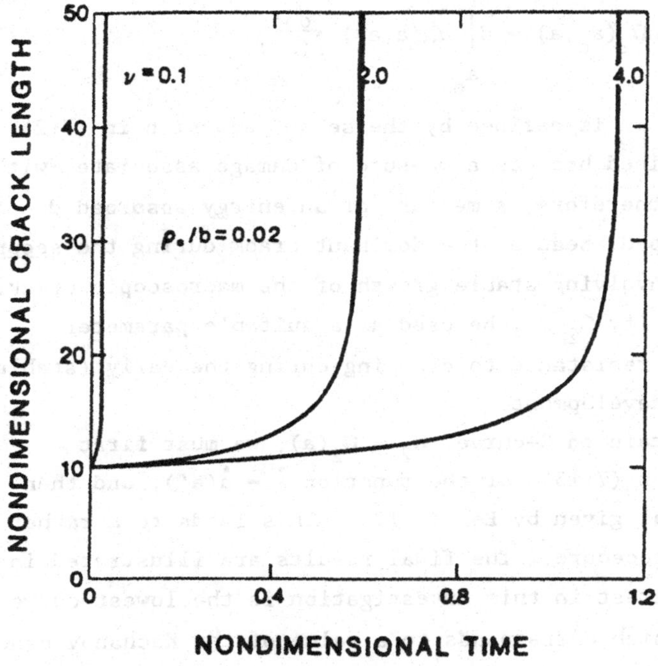

Fig. 14 **Three crack growth histories obtained by numerical integration of Eq. (7.0) for the double-edge-notch specimen configuration when Kachanov's exponent ν equals 0.1, 2.0 and 4.0.**

the K-factors catalog by Tada, Paris and Irwin (1973):

$$K_I(a') = \sigma(t') \sqrt{\pi a'} \left\{ (1 - \frac{a'}{b})^{-1/2} [1.122 - 0.561 (\frac{a'}{b}) - 0.205 (\frac{a'}{b})^2 + 0.471 (\frac{a'}{b})^3 - 0.190 (\frac{a'}{b})^4] \right\} \quad (7.18)$$

Symbol 2b denotes the width of a component, see Fig. 12, while $\sigma(t)$ is the stress applied remotely from the crack site.

<u>Resistance Curve in Damaging Materials</u>

If we denote the left-hand side of Eq. (7.11) by Ω_2, i.e.,

$$\Omega_2(a_o, a) = C \int_{a_o}^{a} \sigma_\Sigma^\nu(a, a') \frac{da'}{\dot{a}(a')} \quad (7.19)$$

in which the stress σ_Σ is defined by the second equation in (7.2), then the quantity so defined becomes a measure of damage associated with crack growth, and, therefore, a measure of an energy absorbed due to microdefects developed ahead of the dominant crack during the <u>second</u> phase of fracture involving stable growth of the macroscopic crack. Therefore, the quantity Ω_2 may be used as a suitable parameter describing material resistance to cracking during the early (stable) stages of fracture development.

In order to obtain an R-curve, $\Omega_2 = \Omega_2(a)$, we must first substitute expression (7.13) for the function $\dot{a} = \dot{a}(a')$, and then evaluate the integral given by Eq. (7.19). This leads to a rather lengthy numerical procedure. The final results are illustrated in Fig. 15. Of special interest in this investigation is the lowest curve shown in Fig. 15, which corresponds to a value of the Kachanov exponent ν approaching zero. This is a case of fracture occurring in a brittle material. Despite the apparent complicated algebra, an equation describing the R-curve when $\nu \to 0$ can be derived in a closed form. Referring to this case as a "limit case," we proceed as follows.

First, we omit* the integral appearing in the denominator of Eq. (7.15), which, with $\xi'(x', x_o) = 1+x'-x_o$, gives

*The validity of such simplification was proven numerically.

$$\overset{\circ}{x}(x') \simeq \frac{\frac{1}{\nu}\,\Phi^{\nu}(x')}{y^{\nu-1}(1+x'-x_o)y'(1+x'-x_o)} \tag{7.20}$$

Here, a compact notation is introduced, namely

$$y(\xi) = Y(\xi)/Y_{*}\ , \qquad \overset{\bullet}{x} = dx/d\theta$$
$$y' = dy/d\xi \quad , \qquad \xi = 1+x-x' \tag{7.21}$$

Next, using expression (7.9) we rewrite Eq. (7.19) in a nondimensional form

$$\Omega_2(x_1,x_o) = -\nu\Omega_c\int_{x_o}^{x} \Phi^{\nu}(x')y^{\nu}(1+x-x')y^{\nu-1}(1+x'-x_o)$$
$$y'(1+x'-x_o)\,\frac{dx'}{\Phi^{\nu}(x')} \tag{7.22}$$

It is seen now that the geometry dependent function Φ cancels out. For ν approaching zero, expression (7.22) can be further simplified, namely

$$[\Omega_2]_{\nu\to 0} \simeq -\nu\Omega_c \int_{x_o}^{x} \frac{y'(1+x'-x_o)}{y(1+x'-x_o)}\,dx' \tag{7.23}$$

We will attempt to integrate the latter expression in a closed form. Particularly attractive appears the case of a LEFM field for which

$$y(\xi) = 1/\sqrt{\xi}$$
$$y'(\xi) = -1/(2\xi^{3/2}) \tag{7.24}$$

Thus, the integral (7.23) reduces to

$$[\Omega_2]_{\nu\to 0} = \frac{\nu\Omega_c}{2}\int_{x_o}^{x} \frac{dx'}{1+x'-x_o} \tag{7.25}$$

which is elementary. Now, our R-curve is defined by a logarithmic function

$$[\Omega]_{\nu\to 0} = \Omega_1 + [\Omega_2]_{\nu\to 0} = \frac{\nu\Omega_c}{2}\,\ell n[1-x_o+x] + \Omega_1 \tag{7.26}$$

This, indeed, is a "universal" geometry independent R-curve, as usually is the case for an R-curve obtained for quasi-brittle solids. Since $\rho_*(x-x_o)$ represents the increment in crack length, Δa, we may rewrite Eq. (7.26) in this way

$$[\Omega]_{\nu\to 0} = \Omega_1 + \frac{\nu\Omega_c}{2}\,\ell n[1 + \frac{\Delta a}{\rho_*}] \tag{7.27}$$

The slope of this curve $d\Omega/da$, is given by

$$[\frac{d\Omega}{da}]_{\nu\to 0} = \frac{\nu\Omega_c}{2}\,\frac{1}{1+\frac{\Delta a}{\rho_*}} \tag{7.28}$$

Equations (7.27) and (7.28) were used to construct the curves shown in Fig. 15. The R-curves experimentally obtained for various cementitious composites resemble very closely the R-curve shown in Fig. 15, cf. Wnuk (1984). An obvious drawback of all the models developed so far, including the present one, is their deterministic nature. Further research aimed at incorporating stochastic features into the mathematical model of damage accumulation process is underway.

Conclusions

The Kachanov damage accumulation law is modified so that the effective stress which enters in the damage evolution equation reflects the elevation of stress due to the presence of a dominant macro-crack. This, in turn, triggers an interaction process between the damage zone, which precedes the crack front, and the crack itself. Coupling between micro- and macro-defects provides the time dependent mechanism responsible not only for an accelerated damage accumulation but also for the propagation of the macro-crack. Total damage Ω has been partitioned into the damage Ω_1 accumulated at a given material point when the macro-crack is not propagating (during the so-called latent stage of fracture), and Ω_2 which represents the damage build-up due to an increase in stresses and stress gradients observed at a stationary control point while the crack front is approaching. The quantity Ω has

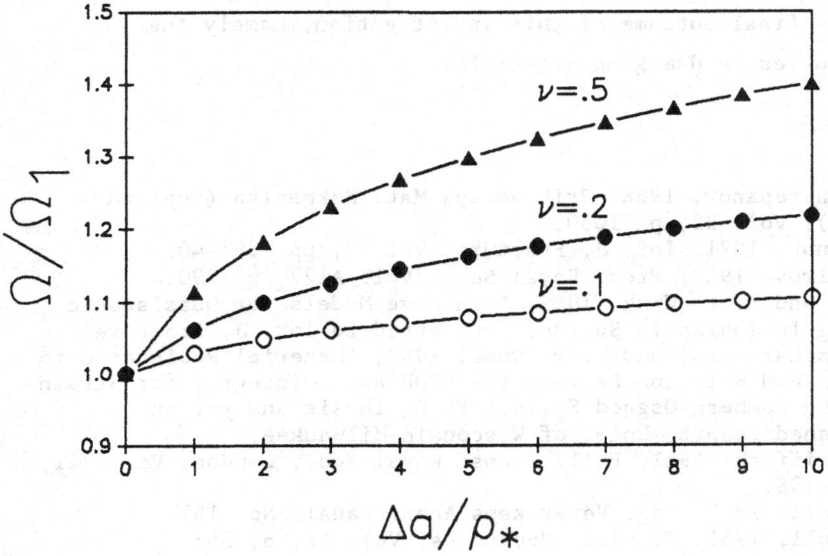

Fig. 15 R-curves in damaging material as predicted by Eq. (7.27). Total damage Ω at the material point is represented as a sum of damage accumulated in the latent period of fracture development, Ω_1, and the damage Ω_2 generated while the crack front approaches the given control point P. Note a pronounced sensitivity of the R-curve shape to the magnitude of Kachanov's exponent ν.

been evaluated in a closed form for the case $\Omega_1 >> \Omega_2$, which is the so-called "graceful fracture" case. The opposite situation, when $\Omega_1 << \Omega_2$, corresponding to the "sudden death" type of fracture, requires further numerical studies. The diagrams Ω vs. Δa, shown in Fig. 15, represent the final outcome of this investigation, namely the resistance curves in damaging materials.

REFERENCES

1. G. P. Cherepanov, 1968, Prikladnaya Mat. Mekhanika (English version), Vol. 32, p. 1050.
2. M. P. Wnuk, 1971, Int. J. Fracture, Vol. 7, pp. 383-407.
3. W. Obreimov, 1930, Proc. Royal Soc., Vol. A127, p. 290.
4. H. Feng and M. P. Wnuk, 1989, "Cohesive Models for Quasistatic Cracking in Inelastic Solids," submitted to Int. J. Fracture.
5. I. Hunsacharoonroj and M. P. Wnuk, 1987, "Material Resistance to Cracking and Relation Between the CTOD and J-integral for Strain-hardening Ramberg-Osgood Solid," Ph.D. Thesis and yet an unpublished report, Univ. of Wisconsin-Milwaukee.
6. A. A. Griffith, 1921, Phil. Trans. Royal Soc., London, Vol. 221, pp. 163-198.
7. W. Weibull, 1939, Ing. Veterskaps Akad. Hanal, No. 151.
8. W. Weibull, 1951, J. Appl. Mechanics, Vol. 18, p. 293.
9. A. de S. Jayatilaka and K. Trustrum, 1983, J. Mater. Sci., Vol. 18.
10. H. E. Daniels, 1945, Proc. Royal Soc. London, Vol. A183, p. 405.
11. A. M Hasofer, 1968, Int. J. Fracture, Vol. 4.
12. M. P. Wnuk, 1974, J. Appl. Mechanics, Vol. 41, pp. 234-242.
13. M. P. Wnuk, 1983, in "Modelling Problems in Crack Tip Mechanics," Editor J. T. Pindera, Martinus Nijhoff Publishers, pp. 91-109.
14. M. P. Wnuk and D. T. Read, 1986, Int. J. Fracture, Vol. 31, pp. 161-171.
15. R. Narasimhan, A. J. Rosakis and J. F. Hall, 1987, J. Appl. Mechanics, Vol. 109, pp. 838-845 (Part I), and pp. 846-853 (Part II).
16. M. P. Wnuk, 1972, in Proceedings of Int. Conf. on Dynamic Crack Propagation, Editor G. C. Sih, published by Noordhoff, The Netherlands.
17. J. R. Rice and E. P. Sorensen, 1978, J. Mech. Phys. Solids, Vol. 25, pp. 163-186.
18. M. P. Wnuk and T. Mura, 1983, Mechanics of Materials, Vol. 2, pp. 33-46.
19. M. P. Wnuk, Z. P. Bazant and E. Law, 1984, J. Theor. Appl. Fracture Mechanics, Vol. 2, pp. 259-286.
20. W. W. Gerberich, 1977, Int. J. Fracture, Vol. 13, pp. 535-538.
21. I. N. Sneddon and M. Lowengrub, 1969, "Crack Problems in the Classical Theory of Elasticity," SIA Series in Appl. Math., publ. by John Wiley and Sons.

22. M. P. Wnuk, 1979a, in Proceedings of ICM3, Vol. 3, Cambridge, Pergamon Press, pp. 549-561.
23. M. P. Wnuk, 1979b, Int. J. Fracture, Vol. 15, pp. 553-581.
24. M. P. Wnuk, 1980, "Stability of Tearing Fracture," lecture given at Int. Symposium on Plasticity and Nonlinear Mechanics, Dourdan, France, eds. D. Zarka and S. Nemat-Nasser.
25. M. P. Wnuk and S. Sedmak, 1980, ASTM STP 743, pp. 500-508.
26. K. B. Broberg, 1977, in Proceedings of Int. Conference on "Fracture Mechanics and Technology" held in Hong Kong, Vol. 2, pp. 837-862, eds. G. C. Sih and C. L. Chow, publisher Sijthoff and Noordhoff.
27. M. P. Wnuk and W. G. Knauss, 1970, Int. J. Solids and Structures, Vol. 6, pp. 995-1009.
28. R. J. Nuismer, 1974, J. Appl. Mechanics, Vol. 41, pp. 631-634.
29. R. A. Schapery, 1975, Int. J. Fracture, Vol. 11, pp. 141-159 (Part I) and pp. 549-562 (Part II).
30. S. R. Swanson, 1976, J. Spacecraft, Vol. 13, No. 9.
31. H. Riedel, 1987, "Fracture at High Temperatures," Springer-Verlag, Berlin.
32. H. Riedel, 1989, "Recent Advances in Modelling Creep Crack Growth," in Proceedings of ICF7, Vol. 2, pp. 1495-1523, eds. K. Salama, K. Ravi-Chandar, D.M.R. Taplin, P. Rama Rao, publisher Pergamon Press.
33. A. Saxena, 1989, ibid., pp. 1675-1688.
34. G. A. Webster, 1989, ibid., pp. 1689-1697.
35. W. G. Knauss, 1970, Int. J. Fracture, Vol. 6, pp. 7-20.
36. r. Mohanty and M. P. Wnuk, 1972, "Experimental Verification of Equation for Creep Crack Motion," unpublished Progress Report for NSF, SDSU, Grant No. GH-43605.
37. C. A. Wells, 1986, "On Life Analysis of Longitudinal Seam Welds in Hot Reheat Piping," RTI Report, Palo Alto, CA.
38. C. Y. Hui and H. Riedel, 1981, Int. J. Fracture, Vol. 17, pp. 409-425.
39. C. Y. Hui, 1983, in ASTM STP 803, pp. 1573-1593.
40. J. D. Landes and J. A. Begley, 1979, in ASTM STP 590, pp. 128-148.
41. A. Saxena, T. T. Shih and H. A. Ernst, 1984, in ASTM STP 833, pp. 516-531.
42. J. L. Bassani, D. E. Hawk and A. Saxena, 1986, to appear in ASTM STP 995 on "Nonlinear Fracture Mechanics."
43. A. Saxena, 1986, in ASTM STP 905, pp. 185-201.
44. K. M. Nikbin and G. A. Webster, 1981, in Proceedings of Symposium on Micro and Macro-Mechanics of Crack Growth, Metallurgical Society of AIME, pp. 107-117.
45. M. P. Wnuk, 1984, in Proceedings of Int. Conference on Computer-Aided Analysis and Design of Concrete Structures, Part I, held in Split, Yugoslavia, eds. F. Damjanić, E. Hinton, D.R.J. Owen, N. Bićanić and V. Simović, pp. 163-177.
46. H. Tada, P. C. Paris and G. R. Irwin, 1973, "The Stress Analysis of Cracks Handbook, Del Research Corp., Hellertown, PA.
47. M. P. Wnuk and R. D. Kriz, 1985, Int. J. Fracture, Vol. 28, pp. 121-138.

48. C. F. Shih, 1983, "Tables of the Hutchinson-Rice-Rosengren Singular Field Quantities," Brown University Report MRL E-147, Providence, RI.
49. S. N. Atluri, 1981, "Path Independent Integrals in Finite Elasticity and Inelasticity with Body Forces, Inertial and Arbitrary Crack Face Conditions," ONR Progress Report, Georgia Institute of Technology, Atlanta, GA; also in Eng. Fract. Mech., 1982, Vol. 16, No. 3, pp. 341-364.
50. M. P. Wnuk, 1981, J. Appl. Mechanics, Vol. 48, pp. 500-508.

APPENDIX A

An essential feature of the cohesive fracture model is the requirement of finite stresses at the end point of the extended crack, $\bar{x} = a$, see Fig. 2. To assure fulfillment of this condition, the singular terms in the LEFM expressions for the stress components, i.e. the terms containing the K-factor, are required to vanish. Using the Green function representation for K-factor, and substituting both the applied stress σ and the restraining stresses σ_{coh} and σ_o into the pertinent integrals, one obtains the following relation

$$2\sqrt{\frac{\bar{a}}{\pi}}\left\{\sigma\int_0^a \frac{dx}{\sqrt{a^2-x^2}} - \sigma_{coh}\int_c^b \frac{dx}{\sqrt{a^2-x^2}} - \sigma_o\int_b^a \frac{dx}{\sqrt{a^2-x^2}}\right\} = 0 \qquad \text{(A1)}$$

The integrals involved are elementary and it follows

$$\frac{\pi}{2}\sigma - (\sigma_o-\sigma_{coh})\cos^{-1}(b/a) - \sigma_{coh}\cos^{-1}(c/a) = 0 \qquad \text{(A2)}$$

If the nondimensional variables

$$\sigma_{coh}/\sigma_o = \alpha$$

$$c/a = m_c, \qquad b/a = m_b, \qquad Q = \pi\sigma/2\sigma_o \qquad \text{(A3)}$$

are introduced, then Eq. (A2) reads

$$Q = \alpha\cos^{-1}(m_c) + (1-\alpha)\cos^{-1}(m_b) \qquad \text{(A4)}$$

The same result was given by Gerberich (1977).

Next, we shall address the problem of the shape of the profile of the crack in its extended part, which corresponds to the structured end-zone. This can be done best by applying the integral representation of Sneddon

$$[u_y(x)]_{y=0} = \frac{4a}{\pi E_1}\int_x^1 \frac{f(t)dt}{\sqrt{t^2-x^2}} \qquad E_1 = E \text{ or } E(1-\nu^2)^{-1}, \quad c/a \le x \le 1 \qquad \text{(A5)}$$

in which $x = \bar{x}/a$, and the auxiliary function f(t) accounts for the assumed distribution of the restraining stress $S(x_1)$ which for the strain-softening material is given by Eq. (2.7). Using (2.7) and the Sneddon definition of the function f(t), one obtains

$$
f(t) = \begin{cases} f_1(t) = \displaystyle\int_0^t \frac{\sigma du}{\sqrt{t^2-u^2}} = \sigma\frac{\pi}{2}, & 0 < t < m_c \\ f_2(t) = \displaystyle\int_0^t \frac{\sigma du}{\sqrt{t^2-u^2}} - \int_{m_c}^t \frac{\sigma_{coh} du}{\sqrt{t^2-u^2}}, & m_c < t < m_b \\ f_3(t) = \displaystyle\int_0^t \frac{\sigma du}{\sqrt{t^2-u^2}} - \int_{m_c}^{m_b} \frac{\sigma_{coh} du}{\sqrt{t^2-u^2}} - \int_{m_b}^t \frac{\sigma_o du}{\sqrt{t^2-u^2}}, & m_b < t < 1 \end{cases} \tag{A6}
$$

Since we are considering the opening displacement within end-zone only, the functions $f_2(t)$ and $f_3(t)$ are all one needs to complete the calculations, thus

$$
f(t) = \begin{cases} f_2(t) = \sigma\frac{\pi}{2} - \sigma_{coh}\cos^{-1}(m_c/t) \\ f_3(t) = \sigma\frac{\pi}{2} - \sigma_{coh}\cos^{-1}\left(\frac{m_c}{t}\right) + (\sigma_{coh}-\sigma_o)\cos^{-1}\left(\frac{m_b}{t}\right) \end{cases} \tag{A.7}
$$

Substituting these expressions into (A5) we arrive at

$$
[u_y(x)]_{EZ} = \frac{4a}{\pi E_1}\begin{cases} \displaystyle\int_x^{m_b} \frac{f_2(t)t\,dt}{\sqrt{t^2-x^2}} + \int_{m_b}^1 \frac{f_3(t)t\,dt}{\sqrt{t^2-x^2}}, & m_c \le x \le m_b \\ \displaystyle\int_x^1 \frac{f_3(t)t\,dt}{\sqrt{t^2-x^2}}, & m_b \le x \le 1 \end{cases}
$$

$$
= \frac{4a}{\pi E_1}
\begin{cases}
\left(\frac{\pi\sigma}{2}\right)\int_x^1 \frac{t\,dt}{\sqrt{t^2-x^2}} - \sigma_{coh}\int_x^1 \frac{\cos^{-1}\left(\frac{m_c}{t}\right)t\,dt}{\sqrt{t^2-x^2}} \\
\qquad + (\sigma_{coh}-\sigma_o)\int_{m_b}^1 \frac{\cos^{-1}\left(\frac{m_b}{t}\right)t\,dt}{\sqrt{t^2-x^2}}, \quad m_c \le x \le m_b \\
\left(\frac{\pi\sigma}{2}\right)\int_x^1 \frac{t\,dt}{\sqrt{t^2-x^2}} - \sigma_{coh}\int_x^1 \frac{\cos^{-1}\left(\frac{m_c}{t}\right)t\,dt}{\sqrt{t^2-x^2}} \\
\qquad + (\sigma_{coh}-\sigma_o)\int_x^1 \frac{\cos^{-1}\left(\frac{m_b}{t}\right)t\,dt}{\sqrt{t^2-x^2}}, \quad m_b \le x \le 1
\end{cases}
\tag{A8}
$$

The first integral equals $\sqrt{1-x^2}$, while the other three are as follows

$$
I_2 = \int_x^1 \frac{\cos^{-1}\left(\frac{m_c}{t}\right)t\,dt}{\sqrt{t^2-x^2}}
$$

$$
I_3' = \int_{m_b}^1 \frac{\cos^{-1}\left(\frac{m_b}{t}\right)t\,dt}{\sqrt{t^2-x^2}}
\tag{A9}
$$

$$
I_3'' = \int_x^1 \frac{\cos^{-1}\left(\frac{m_b}{t}\right)t\,dt}{\sqrt{t^2-x^2}}
$$

All three integrals can be evaluated in closed-form. We have

$$
I_2 = \int_x^1 \frac{\cos^{-1}\left(\frac{m_c}{t}\right)t\,dt}{\sqrt{t^2-x^2}} =
$$

$$
- \left[\sqrt{t^2-x^2}\cos^{-1}\left(\frac{m_c}{t}\right)\right]_x^1 - \int_x^1 \frac{\sqrt{t^2-x^2}\left(\frac{m_c}{t^2}\right)dt}{\sqrt{1-\left(\frac{m_c}{t}\right)^2}}
\tag{A10}
$$

With

$$z = t^2, \qquad 2dt/t = dz/z$$

the integral in (A10) becomes

$$-\frac{m_c}{2}\left[\int_{x^2}^{1} \frac{dz}{\sqrt{(z-x^2)(z-m_c^2)}} - x^2 \int_{x^2}^{1} \frac{dz/z}{\sqrt{(z-x^2)(z-m_c^2)}}\right] \tag{A11}$$

$$= -\frac{m_c}{2}\left[2\ln\left|\frac{\sqrt{1-x^2}+\sqrt{1-m_c^2}}{\sqrt{x^2-m_c^2}}\right| - \frac{x}{m_c}\ln\left|\frac{x + m_c\sqrt{\frac{1-x^2}{1-m_c^2}}}{x - m_c\sqrt{\frac{1-x^2}{1-m_c^2}}}\right|\right]$$

Thus (A10) can be re-written as follows

$$I_2 = \sqrt{1-x^2}\cos^{-1}(m_c) - m_c\ln\left|\frac{\sqrt{1-x^2}+\sqrt{1-m_c^2}}{\sqrt{x^2-m_c^2}}\right| + \frac{x}{2}\ln\left|\frac{x + m_c\sqrt{\frac{1-x^2}{1-m_c^2}}}{x - m_c\sqrt{\frac{1-x^2}{1-m_c^2}}}\right| \tag{A12}$$

In an analogous way the other two integrals can be evaluated. We obtain

$$I_3' = \sqrt{1-x^2}\cos^{-1}(m_b) - m_b\ln\left|\frac{\sqrt{1-x^2}+\sqrt{1-m_b^2}}{\sqrt{m_b^2-x^2}}\right| + $$

$$+ \frac{x}{2}\ln\left|\frac{2x^2m_b^2-m_b^2-x^2-2xm_b\sqrt{(1-x^2)(1-m_b^2)}}{x^2-m_b^2}\right| \tag{A13}$$

for $m_c \le x < m_b$, and

$$I_3'' = \sqrt{1-x^2}\cos^{-1}(m_b) - m_b \ell n \left| \frac{\sqrt{1-x^2} + \sqrt{1-m_b^2}}{\sqrt{x^2-m_b^2}} \right| -$$

$$- \frac{x}{2} \ln \left| \frac{2x^2m_b^2 - x^2 - m_b^2 - 2xm_b\sqrt{(1-x^2)(1-m_b^2)}}{m_b^2 - x^2} \right| \tag{A.14}$$

for $m_b < x \leq 1$. If the absolute values for the difference $x^2 - m_b^2$ is used, then both the integrals assume a common form, namely

$$I_3 = \sqrt{1-x^2}\cos^{-1}(m_b) - m_b \ell n \left| \frac{\sqrt{1-x^2} + \sqrt{1-m_b^2}}{\sqrt{|x^2-m_b^2|}} \right| -$$

$$+ \frac{x}{2} \ell n \left| \frac{2x^2m_b^2 - x^2 - m_b^2 - 2xm_b\sqrt{(1-x^2)(1-m_b^2)}}{|x^2 - m_b^2|} \right| \tag{A.15}$$

With the integrals I_2 and I_3 defined by Eqs. (A12) and (A15), respectively, the opening displacement within the end-zone assumes the form

$$[u_y]_{EZ} = \frac{4a}{\pi E_1}\left\{\sigma \frac{\pi}{2}\sqrt{1-x^2} - \sigma_{coh} I_2 + (\sigma_{coh} - \sigma_o) I_3\right\} \tag{A16}$$

Finally, when the finiteness condition (A4) is incorporated, the profile of the crack within the end-zone can be described as follows

$$[u_y]_{EZ} = \frac{4\sigma_o R}{\pi E_1} \frac{1}{1-m_c} G(x_1/R) \tag{A17}$$

Here R stands for the total length of the end-zone (R = a-c), and the coordinate x is replaced by x_1, i.e. the distance measured directly from the crack tip. Note that when x varies from m_c to 1, the coordinate x_1 covers the range (0,R). The function $G(x_1/R)$ is defined by comparing (A16) with (A17), and with (A4) taken into account. It reads

$$G(x_1/R) = \left[Q\sqrt{1-x^2} - \alpha I_2(x_1) + (\alpha-1)I_3(x_1)\right]_{x_1=ax-c} \tag{A18}$$

When x_1 approaches zero (or, when $x \to m_c$), the expressions for I_2 and I_3 simplify

$$I_2\Big|_{x=m_c} = I_2^{tip} = \sqrt{1-m_c^2}\cos^{-1}(m_c) - m_c\,\ell n\left(\frac{1}{m_c}\right)$$

$$I_3\Big|_{x=m_c} = I_3^{tip} = \sqrt{1-m_c^2}\cos^{-1}(m_b) - m_b\,\ell n\left[\frac{\sqrt{1-m_c^2}+\sqrt{1-m_b^2}}{\sqrt{m_b^2-m_c^2}}\right] \tag{A19}$$

$$- \frac{m_c}{2}\,\ell n\left[\frac{2(m_c m_b)^2-(m_c^2+m_b^2)-2m_c m_b\sqrt{(1-m_c^2)(1-m_b^2)}}{m_b^2 - m_c^2}\right]$$

Thus, one obtains the expression for the opening displacement at the crack tip

$$u_{tip} = \frac{4\sigma_o R}{\pi E_1}\frac{1}{1-m_c}\left\{Q\sqrt{1-m_c^2} - \alpha\, I_2^{tip} + (\alpha-1)I_3^{tip}\right\} \tag{A20}$$

Clearly, when $\alpha = 1$, and for the case of small scale yielding ($m_c \to 1$), the third term in the curly bracket in (A20) drops out, while $I_2^{tip} \to m_c \ell n(1/m_c)$, and thus one recovers the well known expression derived for a Dugdale crack model, i.e.,

$$[u_{tip}]_{\substack{ssy\\ \alpha=1}} = \frac{4\sigma_o R}{\pi E_1}\frac{m_c}{1-m_c}\,\ell n\left(\frac{1}{m_c}\right) \tag{A21}$$

As $m_c \to 1$, this expression becomes $4\sigma_o R/\pi E_1$ as expected for the small scale yielding range.

When the opening displacement $[u_y]_{EZ}$ is renormalized, say

$$\frac{\pi E_1}{4\sigma_o R}[u_y]_{EZ} = F(x_1/R) \tag{A22}$$

and when these auxiliary quantities are introduced

$$A = \frac{1}{2} - \frac{x_1}{R}\left[\frac{x_1}{R} C + \alpha D\right] \ln(\alpha\, m_c)$$

$$g = 1 - \frac{x_1}{R} \ln(\alpha) \tag{A23}$$

$$B = 70.1482 - 277.4247 m_c + 359.2092 m_c^2 - 152.7439 m_c^3$$

$$C = -.5618 + 7.2474\left(\frac{x_1}{R}\right) - 16.3782\left(\frac{x_1}{R}\right)^2 + 9.5807\left(\frac{x_1}{R}\right)^3$$

$$D = .3285 + 3.0163\,\alpha - 11.1455\,\alpha^2 + 7.6632\,\alpha^3$$

The "crack profile" function $F(x_1/R)$ can be written in a more compact form (certain curve-fitting techniques have been used), namely

$$F\left(\frac{x_1}{R}\right) = \sqrt{1-\left(\frac{x_1}{R}\right)^g} - A\left(\frac{x_1}{R}\right)^g \ln\left(\frac{1+\sqrt{1-\left(\frac{x_1}{R}\right)^g}}{1-\sqrt{1-\left(\frac{x_1}{R}\right)^g}}\right) - B\,\frac{x_1}{R}\left[1-\left(\frac{x_1}{R}\right)^g\right]^{2/3} \ln(\alpha m_c) \tag{A24}$$

Using Wnuk's (1974) criterion of final stretch, which amounts to evaluations of the difference in the opening displacement at a fixed microstructural distance $x_1 = \Delta$ from the crack tip, while the crack traverses its own process zone, we obtain the following equation of motion for a quasi-static crack

$$[u_y(c+\delta c,\ R+\delta R,\ x_1)]_{x_1=0} - [u_y(c,R,x_1)]_{x_1=\Delta} = \hat{\delta}/2 \tag{A25}$$

The constant $\hat{\delta}$ is the final stretch; a material property on the order of magnitude of CTOD at the crack growth initiation. The growth step δc is identified with Δ, and thus the Taylor expansion for the quantity $R(c+\delta c)$ gives

$$R(c+\delta c) = R + (dR/dc)\Delta + \ldots\ , \quad R = R(c) \tag{A26}$$

and the criterion of continuing crack growth, Eq. (A25) reduces to the differential equation

$$\frac{dR}{dc} = M - \frac{R}{\Delta} + \frac{R}{\Delta} F(\Delta/R) \tag{A27}$$

This equation defines the material R-curve, $R = R(c)$, for either strain-hardening ($\alpha > 1$) or strain-softening ($\alpha < 1$) solids. Function $F(\Delta/R)$ is defined by Eq. (A24) with x_1 replaced by Δ, while the material constant $M = (\pi/8)(E_1/\sigma_{coh})(\hat{\delta}/\Delta)$ assumes the role of the tearing modulus. The limits of applicability of Eq. (A27) are determined as follows

$$\Delta/R \text{ - arbitrary}, \quad 0.10 \leq \alpha \leq 1.25$$

Since small scale yielding is implied, both the process zone Δ and the damage zone R should be small compared to the crack size and the specimen dimensions.

APPENDIX B

In order to predict the terminal instabilities occurring at the end of the stable cracking process in ductile solids, the following information is required:

a) equation defining material resistance curve, $R = R(c)$,
b) relation between the applied field parameter, such as R or the J-integral, crack length, load and geometry of the specimen,
c) conditions at which the loading process is executed, e.g., load controlled or displacement controlled test, if we are to mention only the extreme possibilities.

To illustrate the procedure leading to prediction of

- critical amount of stable crack extension, $\Delta c_f = c_f - c_o$,
- critical increase of the external load, $\Delta\sigma_f = \sigma_f - \sigma_{ini}$,
- increase in the apparent material toughness associated with the phase of stable cracking, $\Delta R_f = R_f - R_{ini}$,

we shall focus on the CCP geometry for which $c/W \to 0$, i.e., we will consider the classic Griffith crack but spreading in a ductile solid. We shall also limit the analysis to a load controlled fracture test, implying that the terminal instability occurs at $d\sigma/dc \to 0$. As far as ductile response of the material itself is concerned, we shall consider two limiting cases, namely

a) small scale yielding (ssy),

b) large scale yielding (lsy).

For quasi-brittle materials, such as most metals, the case (a) is relevant. Wnuk (1972, 1974) derived the differential equation defining an R-curve, namely

$$\frac{dR}{dc} = M - \frac{1}{1+\lambda} - \frac{1}{2}\,\ell n\left(\frac{1+\lambda}{1-\lambda}\right) \tag{Bi}$$

$$\lambda^2 = 1 - \Delta/R$$

Two limits of material behavior, ductile and brittle, are contained in the Wnuk equation quoted above. The ductile limit results when the RHS of Eq. (Bi) is expanded near the point $\Delta/R = 0$, and the brittle limit is obtained as an expansion into a Taylor series at the point $\Delta/R = 1$. From Eq. (Bi) it follows

$$\frac{dR}{dc} = \begin{cases} M - \frac{1}{2}\,\ell n(4eR/\Delta) + \ldots & \text{for } \Delta/R \ll 1, \text{ ductile limit} \\ M - \frac{R}{\Delta} + \frac{2}{3}\left(\frac{R}{\Delta} - 1\right)^{3/2} + \ldots, & \text{for } \Delta \simeq R, \text{ brittle limit} \end{cases} \tag{Bii}$$

For an infinite width plate, the <u>initiation locus</u> is given by

$$X_{ini} = 2/Q_{ini}^2, \qquad X = c/R_{ini} \tag{Biii}$$

and the <u>terminal instability locus</u> is defined by

$$X_f = \frac{2Y_{ss}}{Q_f^2}\exp(-Q_f^2), \qquad Q = \pi\sigma/2\sigma_o \tag{Biv}$$

In usual applications the amount of slow stable cracking $\Delta X_f = X_f - X_o$ is a small fraction of the initial crack length. Recall that for a valid K_{Ic} test the ASTM standard mandates that $\Delta X_f/X_o$ should not exceed 2%. For $\Delta X_f/X_o$ less than 30% of the following approximation holds, Wnuk (1990)

$$\Delta X_f = \frac{1}{B}\,\ell n[(1-\alpha)(1+BX_o)] \tag{Bv}$$

in which the constants are

$$\alpha = \frac{B}{A+B}, \quad A = \frac{1}{2}\ \ell n(Y_{ss}), \quad B = A/(Y_{ss}-1)$$

$$Y_{ss} = \frac{1}{4\rho_i}\exp(2M-1), \quad \rho_i = R_{ini}/\Delta \qquad \text{(Bvi)}$$

$$M = k\ M_{min} = k[\frac{1}{2}\ \ell n(4e\rho_i)]$$

To illustrate the process of numerical search for the set of critical parameters encountered at the point of transition from stable cracking to terminal structural instability

X_f - critical crack length

Q_f - critical load

Y_f - critical toughness parameter

we shall solve a simple numerical example. Consider the following input data

- ductility index, $\rho_i = 5$
- initial crack length, $X_o = 10$
- tearing modulus, $M = k\ M_{min}$, in which we assume $k = 1.2$.

First, let us evaluate the constants involved in Eqs. (Bv) and (Bvi), namely

$$M = (1.2)[.5 + .5\ \ell n(4 \times 5)] = 2.3974$$

$$Y_{ss} = \frac{1}{(4)(5)}\exp[(2)(2.3974) - 1] = 2.2236$$

$$A = \frac{1}{2}\ \ell n(2.2236) = .39943 \text{ and } B = \frac{.39943}{2.2236 - 1} = .3266 \qquad \text{(Bvii)}$$

$$\alpha = (.3266)/(.39943 + .3266) = .4498$$

Now using Eq. (Bv) we may compute the increase in the crack length caused by the processes of stable crack growth, i.e.,

$$\Delta X_f = \frac{1}{.3266}\ \ell n[(1 - .4498)(1 + .3266 \times 10)] = 2.6121 \qquad \text{(Bviii)}$$

$$X_f = X_o + \Delta X_f = 12.6121$$

Now to find the critical load Q_f one needs to search for the root of a transcendental equation which results from the requirement $(dR/dc)_f = (R/c)_f$ at the point of terminal instability

$$X_f\ \ell n(Y_{ss}/Y_f) - 2Y_f = 0 \qquad \text{(Bix)}$$

With $X_f = 12.6121$ and $Y_{ss} = 2.2236$ we find $Y_f = 1.6985$, and we use this value to determine Q_f, namely

$$Q_f = \sqrt{2Y_f/X_f} = [(2)(1.6985)/(12.6121)]^{\frac{1}{2}} = 0.51903 \tag{Bx}$$

This concludes the evaluation process. We note that the obtained numbers imply 26.1% increase in the crack length and 16.1% increase in the load level prior to beginning at the initiation point and ending at the terminal structural instability.

To facilitate the process of root finding in Eq. (Bix) the following estimate is suggested as a "first guess"

$$Y_f^* = (A+B)/[B + (1/X_f)] \tag{Bxi}$$

With the input data used here, Y_f^* turns out to be 1.7887, and when this value is used as a first estimate, the computer quickly converges to the correct value of $Y_f = 1.6985$. The errors involved in using the equations (Bv) and (Bvi) do not exceed a fraction of 1%, as long as $\Delta X_f/X_o$ is less than 30%. Note that by using these equations we are able to extract all the pertinent data describing the early stages of the ductile fracture process without a need to execute the integration involved in the R-curve and Q-curve analyses.

Fracture Under Contained Yielding Conditions

For the first case we shall employ the Wnuk (1972, 1974) equation of the R-curve, based on the final stretch criterion for quasi-static fracture, namely

$$\frac{dR}{dc} = \begin{cases} M - \frac{1}{2}\,\ell n(4eR/\Delta), & \text{or} \\ \frac{1}{2}\,\ell n(R_{ss}/R), & R = R(c) \end{cases} \tag{Bl}$$

Here, R_{ss} denotes the steady-state value of the material toughness R, which is theoretically attained when $\Delta c \to \infty$, and which can be related to other material characteristics, such as tearing modulus M, toughness level at the onset of crack growth, R_{ini}, and the size of the process zone. Comparing the right-hand sides of the equations (Bl) at the onset of crack growth one readily obtains

$$R_{ss} = \begin{cases} \frac{\Delta}{4} \exp(2M-1), & \text{or} \\ \frac{R_{ini}}{4\rho_i} \exp(2M-1), & \rho_i = R_{ini}/\Delta \end{cases} \tag{B2}$$

Next, nondimensionalizing the quantities R and c

$$Y = R/R_{ini}, \quad X = c/R_{ini} \tag{B3}$$

we obtain

$$\frac{dY}{dX} = \begin{cases} M - \frac{1}{2} \ell n(4e\rho_i Y), & \text{or} \\ \frac{1}{2} \ell n(n/Y), & n = R_{ss}/R_{ini} \end{cases} \tag{B4}$$

We note that the form of this equation is identical to the plane strain result given by Rice and Sorensen (1978) and the plane stress result of Narasimhan, Rosakis and Hall (1987). Unfortunately, Eq. (B3) cannot be integrated in a closed form expressible in terms of elementary functions. However, a very close approximation was suggested by Feng and Wnuk (1989), namely

$$Y = 1 + A(n) \ \ell n[1 + B(n)(X-X_o)^3] \tag{B5}$$

in which

$$A(n) = -.083398 + .078098n - .000621n^2$$

$$B(n) = 98.8461 - 24.6653n + 2.0596n^2 - .05667n^3 \tag{B6}$$

Examining these results, see Fig. B1, we conclude that the R-curve, represented by the function Y defined above, depends on two variables: (1) amount of crack growth, $X-X_o = \Delta X$, and (2) the material characteristic n. The quantity n, in turn, is affected by the tearing modulus M and the ductility parameter $\rho_i = R_{ini}/\Delta$, which is directly proportional to the normalized plastic strain at fracture, i.e., $\rho_i \approx \epsilon_p^f/\epsilon_o$. The relation binding n, M and ρ_i results from Eq. (B2), namely

$$n = \frac{1}{4\rho_i} \exp(2M-1) \tag{B7}$$

If the modulus M is replaced by the product kM_{min}, where k is a number greater than one, and M_{min} (= $.5\ell n[4e\rho_i]$) is that value of M below

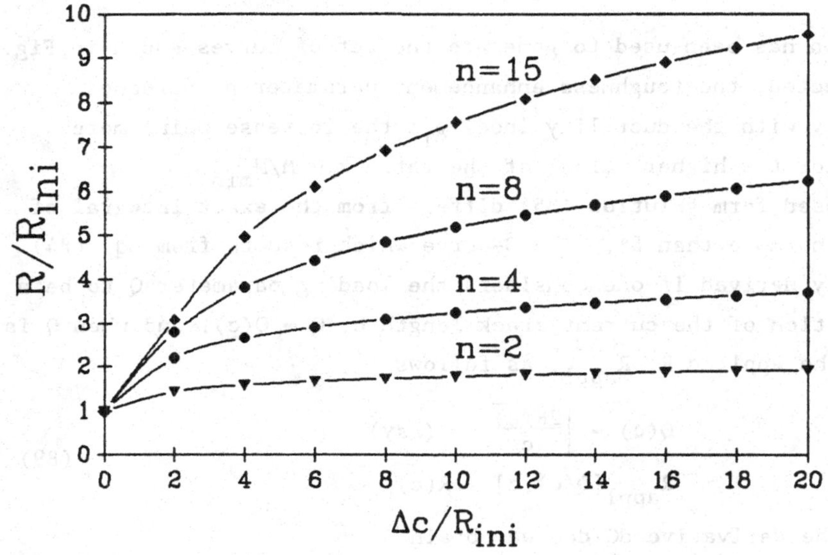

Fig. B1. Resistance curves for quasi-static fracture occurring under small scale yielding condition. Closed form solution (B5) was used to generate the curves shown.

which stable crack growth cannot occur, then Eq. (B7) reads

$$n = \begin{cases} \frac{1}{4\rho_i} \exp[k\ln(4e\rho_i) - 1], & \text{or} \\ (4e\rho_i)^{k-1} & \end{cases} \tag{B8}$$

This relation has been used to generate the set of curves shown in Fig. B2. As expected, the toughness enhancement parameter n increases monotonically with the ductility index ρ_i; the increase being more pronounced for the higher values of the ratio $k = M/M_{min}$.

The closed form solution (B5) differs from the exact integral of Eq. (B4) by no more than 6%. The Q-curve which results from Eq. (B4) can be easily derived if one considers the loading parameter Q to be a certain function of the current crack length c, $Q = Q(c)$, and when Q is related to the applied R, R_{appl}, as follows

$$\begin{aligned} Q(c) &= \sqrt{\frac{2R(c)}{c}} \quad \text{(ssy)} \\ R_{appl}[Q(c),c] &= R(c) \end{aligned} \tag{B9}$$

Evaluating the derivative dQ/dc, we obtain

$$cQdQ/dc = dR/dc - R/c \tag{B10}$$

and thus Eq. (B1) is transformed into a new equation defining progress of a stable crack associated with an increasing load

$$\frac{dQ}{dc} = \frac{\ell n(\frac{2R_{ss}}{cQ^2}) - Q^2}{2cQ^2} \tag{B11}$$

If c is replaced by the nondimensional quantity, $\varsigma = c/2R_{ss}$, then Eq. (B11) reads

$$\frac{dQ}{d\varsigma} = \frac{\ell n(1/\varsigma Q\ )^2 - Q^2}{2\varsigma Q^2} \tag{B12}$$

which is identical to Eq. (xxi). If, however, $X = c/c_o$ is used as a non-dimensional crack length, then Eq. (B11) assumes this form

$$\frac{dQ}{dX} = \frac{\ell n\left[\frac{2n}{XQ^2}\right] - Q^2}{2XQ^2}, \quad \text{or}$$

$$\frac{dQ}{dx} = \frac{\ell n\left(\frac{2n\rho_i}{\gamma_o xQ^2}\right) - Q^2}{2xQ^2} \tag{B13}$$

Here, $x = c/c_o$ and the auxiliary parameters γ_o and ρ_i are defined as follows

- $\rho_i = R_{ini}/\Delta$ ($\approx \epsilon_p^f/\epsilon_o$); provides a suitable ductility measure,
- $\gamma_o = c_o/\Delta$; serves as a nondimensional measure of the initial crack size.

Fracture initiates at $R_{appl} = R_{ini}$, which for the Griffith crack configuration and under the restrictions of small scale yield, gives

$$\frac{c_o}{2} Q_{ini}^2 = R_{ini} \tag{B14}$$

or

$$Q_{ini} = \sqrt{\frac{2R_{ini}}{c_o}} = \sqrt{\frac{2\rho_i}{\gamma_o}} \tag{B15}$$

This establishes the threshold of the loading parameter at which the stable cracking begins. To find the transition from stable to unstable crack propagation, one needs to solve two equations

$$R_{appl} = R$$
$$\frac{\partial R_{appl}(\sigma,c)}{\partial c} = \frac{dR}{dc} \tag{B16}$$

We note that for the specimen configuration considered here $\partial R_{appl}/\partial c$ may be replaced by R_{appl}/c. Using nondimensional symbols, Eqs. (B16) assume the form

$$\left(\frac{1}{2} XQ^2\right)_f = Y_f$$
$$\left(\frac{Y}{X}\right)_f = \left(\frac{dY}{dX}\right)_f \tag{B17}$$

Subscript "f" has been added to emphasize that the equations (B17) holds at the onset of <u>unstable</u> fracture. To solve equations (B17) we proceed as follows. First, using Eq. (B5) the derivative dY/dX is evaluated

$$\frac{dY}{dX} = \frac{3AB(\Delta X)^2}{1 + B(\Delta X)^3}, \qquad \Delta X = X - X_o \tag{B18}$$

Then the equality Y/X = dY/dX is used, yielding a transcendental equation for the critical amount of crack extension ΔX_f, namely

$$\frac{1}{2}\,\ell n\left\{\frac{n}{1 + A\ell n[1 + B(\Delta X_f)^3]}\right\} - \frac{1 + A\ell n[1 + B(\Delta X_f)^3]}{\Delta X_f + X_o} = 0 \tag{B19}$$

Obviously the root of this equation, ΔX_f, is affected by the material characteristic n, contained in the parameters A and B, and the initial crack size X_o. The results of the computations aimed at finding the root ΔX_f are shown in Fig. B3. To facilitate the numerical process of finding the root of Eq. (B19), Feng and Wnuk (1989) have suggested the following algorithm (error of the approximations involved does not exceed 3%):

$$\Delta X_f = A_1\ \ell n|B_1 + C_1 X_o|$$

$$\begin{aligned} A_1(n) &= 3.9447287 - .80400395n + .18507561n^2 + .00028735877n^3 \\ B_1(n) &= -274.868222 + 81.23533n - 7.0114402n^2 + .15974438n^3 \\ C_1(n) &= 19.429665 - 5.9168634n + .54226595n^2 - .014038872n^3 \end{aligned} \tag{B20}$$

This formula was tested and found valid within a range of variables n and X_o defined as follows

$$\begin{aligned} 1.5 \le n \le 10 \\ 30 \le X_o \le 1000 \end{aligned} \tag{B21}$$

Once we know ΔX_f (or $X_f = \Delta X_f + X_o$), we can evaluate the remaining quantities pertinent in the description of the terminal instability point attained in ductile fracture. The load increase due to the process of stable cracking is obtained from Eq. (B13). Setting the rate dQ/dX equal to zero, we obtain

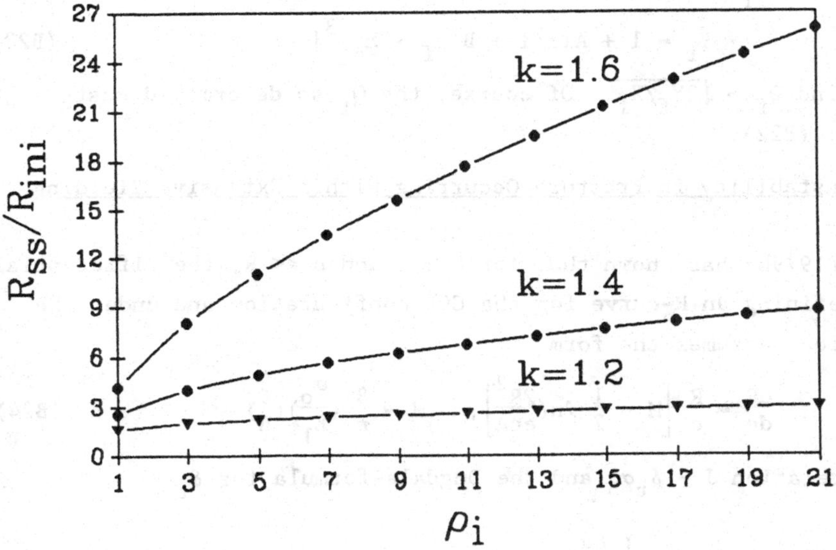

Fig. B2. Toughness enhancement parameter, $n = R_{ss}/R_{ini}$ shown as a function of ductility index $\rho_i = \epsilon_p^f/\epsilon_o$ and the coefficient of "capacity for stable cracking", given by the ratio $k = M/M_{min}$

$$\left(\frac{\exp Q^2}{Q^2}\right)_f = \frac{2n}{X_f} \tag{B22}$$

This can be readily solved for Q_f, and once we have Q_f, the effective toughness at the point of transition to unstable crack propagation, $Y_f = R_f/R_{ini}$, can be evaluated $Y_f = \frac{1}{2} X_f Q_f^2$. An alternative way is to first evaluate Y_f from the equation

$$Y_f = 1 + A\ell n[1 + B(X_f - X_o)^3] \tag{B23}$$

and then find $Q_f = \sqrt{2Y_f/X_f}$. Of course, the Q_f so determined must satisfy Eq. (B22).

Terminal Instability in Fracture Occurring Within Extensive Yielding Range

Wnuk (1979b) has shown that for $R \geq c$ and $\Delta \ll R$, the differential equation defining an R-curve for the CCP configuration and under the lsy conditions assumes the form

$$\frac{dR}{dc} = \frac{R}{c}\left[M - \frac{1}{2}\,\ell n(\frac{2R^2}{ec\Delta})\right], \quad M = \frac{8}{\pi}\,(\frac{\sigma_o}{E_1})(\frac{\hat{\delta}}{\Delta}) \tag{B24}$$

Using the relation $J = \delta_t \sigma_o$ and the Dugdale formula for δ_t

$$\delta_t = \begin{cases} \dfrac{8\sigma_o}{\pi E_1}\, c\ \ell n[\sec(Q)], & \text{or} \\[2ex] \dfrac{8\sigma_o}{\pi E_1}\, c\ \ell n(\dfrac{c+R}{c}) & \end{cases} \tag{B25}$$

one can transform Eq. (B24) into this form

$$\frac{dJ_R}{dc} = M_J - \frac{4\sigma_o^2}{\pi E_1}\,\ell n(\frac{2ec}{\Delta}), \quad M_J = (\frac{\pi}{8})E_1 M \tag{B26}$$

Somewhat unexpectedly, both equations, (B24) and (B26), can be integrated in a closed form, yielding the solutions for R-curves in terms of the toughness parameters, $R(c)$ and $J_R(c)$, respectively. These solutions are

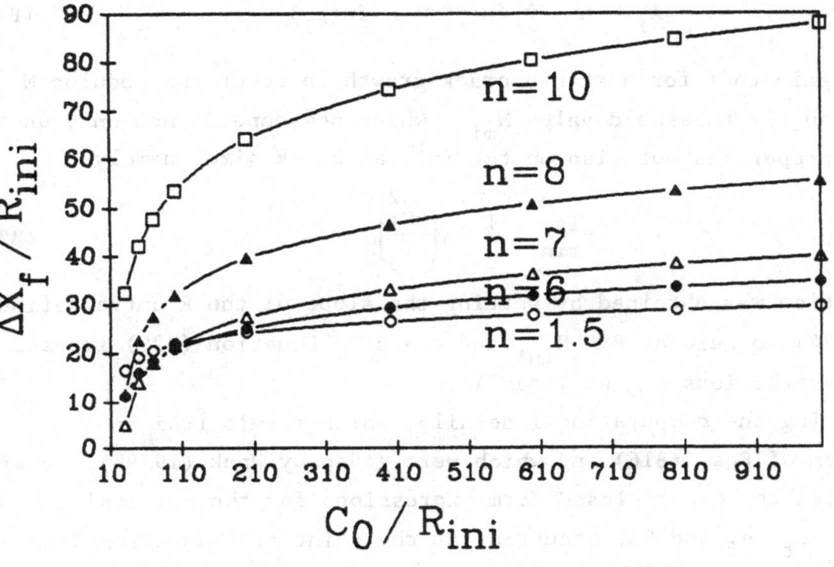

Fig. B3. Critical amount of stable cracking, $\Delta X_f = \Delta c_f/c_o$, shown as a function of the initial crack size, $X_o = c_o/R_{ini}$ and the material parameter, $n = R_{ss}/R_{ini}$. Graphs were generated on the basis of Eq. (B20).

$$R(c) = R_{ini}\left(\frac{c}{c_o}\right)^{1/2} \exp\left[\frac{1+2\lambda_i}{2}\left(1 - \frac{c_o}{c}\right)\right]$$

$$J_R(c) = J_{ini}\left\{\left(\frac{c}{c_o}\right) + \frac{qc}{J_{ini}}\left[\frac{1}{2}\ \ell n\left(\frac{c_o}{c}\right) + \frac{1+2\lambda_i}{2}\left(1 - \frac{c_o}{c}\right)\right]\right\} \tag{B27}$$

Here, symbol q is used to denote a constant, i.e., $q = 8\sigma_o^2/\pi E_1$, while the parameter λ_i depends on the tearing modulus M, the initial crack length $\gamma_o = c_o/\Delta$, and the ductility index $\rho_i = R_{ini}/\Delta$. The following relation holds

$$\lambda_i = M - \frac{1}{2}\ \ell n\left(\frac{2}{e\gamma_o}\right) - \ell n(\rho_i) \tag{B28}$$

We note again that for a stable crack growth to occur the modulus M must exceed the threshold value M_{min}, which now depends not only on the material properties but also on the initial crack size, namely

$$M_{min}^{lsy} = \frac{1}{2}\ \ell n\left(\frac{2\rho_i^2}{e\gamma_o}\right) \tag{B29}$$

This relation was obtained by setting the slope of the R-curve defined by Eq. (B24) to zero at $R = R_{ini}$ and $c = c_o$. Equation (B29) imposes certain restrictions on the index λ_i.

Omitting the computational details, which result from an application of Eqs. (B16) and which were given by Wnuk (1979b), we will mention here the exact closed form expressions for the critical quantities c_f, σ_f and R_f, occurring at the point of transition from stable to unstable crack propagation. The point of terminal instability for the load controlled fracture test, when extensive yielding of the ligament prevails, is defined as follows

$$\begin{aligned} c_f &= c_o(1 + 2\lambda_i) \\ \sigma_f &= \left(\frac{2}{\pi}\sigma_o\right)\cos^{-1}\left[1 + (\rho_i/\gamma_o)(1 + 2\lambda_i)^{-1/2}\exp(\lambda_i)\right]^{-1} \\ R_f &= R_{ini}\sqrt{1 + 2\lambda_i}\ \exp(\lambda_i) \end{aligned} \tag{B30}$$

Load level at the onset of crack growth, say σ_{ini}, results from Eqs. (B25). At $Q = Q_{ini}$, $c = c_o$ and $R = R_{ini}$, it follows

$$\sigma_{ini} = (\frac{2}{\pi}\sigma_o)\cos^{-1}[1 + (\rho_i/\gamma_o)]^{-1} \quad \text{(B31)}$$

Of course, during the course of stable cracking process these inequalities are satisfied

$$c_o \le c \le c_f$$

$$\sigma_{ini} \le \sigma \le \sigma_f \quad \text{(B32)}$$

$$R_{ini} \le R \le R_f$$

Since the logarithmic term appearing on the right-hand-side of Eq. (B26) is rather insignificant compared to the large values of the tearing modulus M_J encountered in high ductility materials, the J_R vs. Δc curve very closely resembles a straight line* slope of which equals approximately M_J. This observation can be used to relate the moduli M and M_J to the Paris tearing modulus T_J $(= [dJ_R/dc][E_1/\sigma_o^2])$, namely

$$M_J^{lsy} = (\sigma_o^2/E_1)T_J$$

$$M^{lsy} = (\frac{8}{\pi})(\frac{\sigma_o}{E_1})^2 T_J \quad \text{(B33)}$$

The significant differences between quasi-static cracking processes occurring in solids of limited ductility (ssy) as opposed to those capable of extensive yielding prior to fracture (lsy) are illustrated in Figs. 8a and 8b.

<u>Numberical Example</u>

(a) For the case of ssy let us assume the initial crack to be 10 times larger than the length parameter R_{ini}, which in turn is 10 times greater than the process zone size Δ. With the tearing modulus M = 3 we have the following input data

---

*Similar observation was also made for other geometrical configurations, cf. Wnuk and Sedmak (1980) and Wnuk (1981).

$$\gamma_o = c_o/\Delta = 100, \quad \rho_i = R_{ini}/\Delta = 10$$

$$Q_{ini} = \sqrt{2\rho_i/\gamma_o} = \sqrt{.2} = .4472, \ M = 3 \tag{B34}$$

$$M_{min} = .5 \ \ln(4e\rho_i) = 2.3444, \ k = M/M_{min} = 1.2796$$

The critical increment of the crack length generated during the stable cracking phase of fracture development, $\Delta X_f = \Delta c_f/R_{ini}$ is obtained either by seeking the root of Eq. (B19) [or by using Eq. (B20) if $X_o \geq 30$]. With $n = (1/4\rho_i)\exp(2M-1) = 3.710329$, and $X_o = c_o/R_{ini} = 10$ we find

$$\Delta X_f = 4.6341 \quad \text{and} \quad X_f = X_o + \Delta X_f = 14.6341 \tag{B35}$$

The terminal value of the loading parameter, Q_f, results as the root of Eq. (B22), namely

$$Q_f = .5961 \tag{B36}$$

Finally, the terminal value of the toughness parameter, $Y_f = R_f/R_{ini}$, is calculated

$$Y_f = \frac{1}{2} X_f Q_f^2 = \frac{1}{2} (14.6341)(.5961)^2 = 2.6005 \tag{B37}$$

Alternatively, the value of X_f given by (B35) is substituted into Eq. (B23) yielding $Y_f = 2.6005$, and then the critical load Q_f is found as $\sqrt{2Y_f/X_f}$.

(b) For the case of lsy let us assume the initial crack equal in size to the nonlinear zone, $c_o = R_{ini}$, and both are assumed 10 times greater than the microstructural parameter, Δ. Let tearing modulus M be equal to 3, like in the previous example. Thus our initial input data becomes

$$\gamma_o = c_o/\Delta = 10, \ \rho_i = R_{ini}/\Delta = 10, \ M = 3 \tag{B38}$$

Next, we need to evaluate the parameter λ_i, defined by Eq. (B28). We obtain

$$\lambda_i = 3 - .5 \ \ln(\frac{2}{10e}) - \ln(10) = 2.0021 \tag{B39}$$

The load level at the onset of crack growth is evaluated from Eq. (B31)

$$Q_{ini} = \cos^{-1}\left[\frac{1}{1+1}\right] = 1.0472 \tag{B40}$$

The load parameter Q increases during the stable crack growth to attain the value Q_f at the point of terminal instability as given by Eq. (B30)

$$Q_f = \cos^{-1}[1 + (1 + (2)(2.0021))^{-1/2} \exp(2.0021)]^{-1}$$
$$= \cos^{-1}(0.2320) = 1.3366 \tag{B41}$$

The critical crack length is also given by Eq. (B30)

$$X_f = \frac{c_f}{R_{ini}} = \frac{c_o}{R_{ini}}(1+2\lambda_i) = (1)[1+2(2.0021)] = 5.0042 \approx 5 \tag{B42}$$

The toughness parameter at the point of transition to unstable crack propagation, $Y_f = R_f/R_{ini}$, is predicted by Eq. (B30) as follows

$$Y_f = \sqrt{1+2\lambda_i}\, \exp(\lambda_i) = (5.0042)^{1/2} \exp(2.0021) = 16.5641 \tag{B43}$$

The same numbers were derived on the basis of the graphs shown in Figs. 8a and 8b, which represent the R-curves and Q-curves for the ssy and lsy cases discussed here.

Printed in the United States
By Bookmasters